Nutritional Neurosciences

Series Editor

Mohamed Essa, Sultan Qaboos University
Qaboos, Oman

This book series aims to publish volumes focusing on both basic and clinical research in the field of nutritional neuroscience with a focus on delineating the effect of nutrition on brain function and behavior. The books will examine the role of different nutrients, food agents and supplements (both macro and micro) on brain health, neurodevelopment, neurochemistry, and behaviour. The books will examine the influence of diet, including phytochemicals, antioxidants, dietary supplements, food additives, and other nutrients on the physiology and metabolism of neurons, neurotransmitters and their receptors, cognition, behavior, and hormonal regulations.

The books will also cover the influence of nutrients and dietary supplements on the management of neurological disorders. It details the mechanism of action of phytonutrients on signaling pathways linked with protein folding, aggregation, and neuroinflammation. The books published in the series will be useful for neuroscientists, nutritionists, neurologists, psychiatrists, and those interested in preventive medicine.

Wael Mohamed • Aynur Özge
Editors

Feeding the Mind: The Connection Between Diet, Drugs, and Mental Health Volume 1

Foundations and Clinical Connections in Diet and Mental Health

Springer

Editors
Wael Mohamed
Basic Medical Science Department
Kulliyyah of Medicine
International Islamic University
Malaysia (IIUM)
Kuantan, Pahang, Malaysia

Aynur Özge
Mersin University School of Medicine
Mersin, Türkiye

NOROM Neuroscience
and Excellence Center
Ankara, Türkiye

ISSN 2730-6712 ISSN 2730-6720 (electronic)
Nutritional Neurosciences
ISBN 978-981-95-1413-7 ISBN 978-981-95-1414-4 (eBook)
https://doi.org/10.1007/978-981-95-1414-4

This Springer imprint is published by the registered company Springer Nature Singapore Pte Ltd.
The registered company address is: 152 Beach Road, #21-01/04 Gateway East, Singapore 189721, Singapore

To my beloved chaos crew, Eyad, Ahmed, and Aser—my tiny tornadoes of joy, mischief, and questions that outnumber the stars. As the Egyptian proverb goes, “A house with a child is a marketplace”*... and ours? Let’s just say it’s a very lively souk (complete with snack raids and sock-stealing ninjas). Your laughter fuels my soul, your curiosity keeps Google in business, and your endless energy makes me question my caffeine intake. Yet, in your chaos, I find magic: your love is my compass, your giggles my favorite soundtrack, and your sticky-fingered hugs my daily reset button.*

This work is a testament to the fact that even pyramids weren’t built in a day—and neither was my patience. *But like the Nile, you three flow through my life, carving purpose into every moment. May this remind you that dedication, perseverance, and a solid Wi-Fi connection can turn dreams into reality... or at least keep the LEGO battles raging. Shukran for being my “albi” (heart) and my greatest adventure.*

Wael Mohamed
Pahang, Malaysia

To all those who have unjustly lost the vast repository of their lifelong knowledge to the relentless grip of neurodegenerative diseases, especially to my beloved elders in my family, whom I lost to Alzheimer's and dementia. May it not only illuminate their fading memories but also serve as a guiding light for the devoted relatives who endure the weight of this affliction with unwavering resilience. Collaborating with a group of exceptionally talented young authors from diverse corners of the world has been an enlightening and profoundly enriching experience. As Atatürk so eloquently stated, the youth shall forever remain our beacon of hope and the steadfast guardian of our future.

Aynur Özge
Mersin, Türkiye

Foreword

Throughout millennia, diet has played a fundamental role in human evolution, adapting to the human species' ever-changing physical and mental demands. Modern research has unveiled intricate interactions between diet and mental health, grounded in clinical observations of indigenous populations known for their exceptional longevity and low incidence of cognitive and cerebrovascular disorders. At a time when brain health has become a central pillar in managing many chronic neurodegenerative diseases, the value of this volume lies in highlighting the importance of understanding the fundamental mechanisms that support proper dietary habits and broader lifestyle behaviors. These behaviors contribute to overall well-being, making it more stable and widely accessible, in alignment with the United Nations' *Sustainable Development Goal No. 3*.

The concept of diet as a determinant of well-being is shifting away from a homogenized, globalized approach to nutrition, instead favoring a model rooted in local dietary traditions. This paradigm shift prioritizes a qualitative perspective on nutrition, challenging the dominant global nutritional dogma. It is crucial to recognize that an inadequate diet can disrupt the delicate balance of the Brain-Body Axis, potentially triggering systemic diseases that extend beyond *neurological disorders*, leading to metabolic and vascular dysfunctions.

I am confident that the detailed insights presented by the well-renowned editors of this book will provide valuable guidance for independent critical analysis, particularly in questioning dietary models imposed by globalization—models that often lack alignment with effective public health strategies and fail to safeguard brain health. A rigorous evaluation of nutritional epidemiology and multi-omics evidence will be instrumental in refining research priorities in brain health and addressing the critical clinical questions that this remarkable volume presents to its readers.

Unitelma Sapienza University
Rome, Italy

Paolo Martelletti

Preface

> Exploring the intricacies of illness without books is like embarking on an uncertain voyage, while delving into books without the insights of patients is akin to remaining anchored to the shore.
>
> —Sir William Withey Gull (1816-1890)

In recent years, the impact of nutrition and food intake on brain function and development has been extensively investigated, as evidence increasingly suggests that the diet we consume plays a pivotal role in shaping our mental well-being. Beyond its well-documented effects on cardiovascular diseases and cancer, nutrition has been shown to significantly contribute to the preservation of cognitive function and behavioral health. Recent studies indicate that the onset or exacerbation of certain mental disorders is closely linked to nutritional imbalances, as reflected in the emergence of their symptoms and clinical manifestations.

These insights have been driven by the revolutionary application of various "omics" disciplines in studying the central nervous system (CNS), which has broadened our understanding of fundamental neurobiological processes. This approach has enabled the identification of proteins and pathways involved in the complex molecular mechanisms underlying various CNS disorders. Furthermore, metabolomics and microbiome research advancements have shed light on the gut-brain axis, particularly the role of serotonin secretion and its modulation by beneficial gut bacteria, which influence sleep cycles, mood, and pain regulation.

Upon surveying the literature, we identified a significant knowledge gap regarding the interplay between brain health and nutrition. While we do not suggest that dietary modifications should replace established medical interventions for brain health, we seek to underscore the profound influence of a balanced diet and proper nutrition in supporting neurological function and potentially alleviating certain conditions. Drawing from our extensive research backgrounds in brain health, the editors (*Drs. Mohamed and Aynur*) have collaborated with esteemed colleagues specializing in brain disorders and nutrition to address these gaps.

This book presents an updated and innovative perspective on the intricate relationship between mental disorders and food intake. We believe this compilation will be of great interest to researchers and clinical scientists working in neurology, nutrition, and biochemistry.

Finally, we extend our heartfelt gratitude to all the contributing authors for their invaluable efforts in crafting these excellent chapters. We also sincerely appreciate their patience and dedication throughout the compilation and final editing process of this book.

Pahang, Malaysia
Mersin, Türkiye

Wael Mohamed
Aynur Özge

Preface for the First Volume

In recent decades, scientific advancements have dramatically reshaped our understanding of the brain-gut axis, illuminating the intricate pathways through which diet influences neural development, emotional regulation, and mental health trajectories. Volume 1 of this two-volume compendium lays the foundational framework for nutritional psychiatry by integrating biological principles, lifespan perspectives, and clinical relevance. It bridges the neurochemical mechanisms of nutrients with their behavioral and cognitive correlates, offering a comprehensive exploration of how specific dietary components—from omega-3 fatty acids to micronutrients and probiotics—modulate mental wellness.

This volume is distinguished by its deliberate attention to vulnerable populations and nuanced clinical scenarios. From childhood through senescence and across conditions ranging from food allergies to hydration deficits, the included chapters emphasize not only the diversity of dietary impacts but also their intersectionality with comorbidities such as substance use disorders and chronic pain syndromes.

By converging insights from neuroscience, clinical nutrition, and public mental health, we aim to establish a common vocabulary that enables clinicians, researchers, and policy-makers to collaborate in promoting brain health through dietary intervention. The volume culminates in integrative chapters addressing nutritional strategies for individuals with alcohol and opioid use disorders, thereby exemplifying the translational potential of this field.

We believe that this foundational volume will not only inform but also inspire the development of evidence-based dietary models for the prevention and management of mental and neurological conditions.

Pahang, Malaysia Wael Mohamed
Mersin, Türkiye Aynur Özge

Acknowledgments

First and foremost, we extend our most profound appreciation to all the authors who contributed to this timely and significant project. Their unwavering dedication and collaboration with the editors have made writing this book an intellectually enriching and rewarding journey.

We are especially grateful to our esteemed contributors from medical psychiatry and neuropsychiatric research, whose extensive expertise and years of dedicated work in various domains of psychiatric disorders have culminated in this insightful and elegantly crafted volume. The topics explored in this book hold immense value for nutrition, psychiatry, neurological disorders, and neurodegeneration.

Lastly, we express our heartfelt gratitude to our friends and colleagues, whose unwavering support, encouragement, and inspiration have been invaluable throughout this endeavor. Your belief in this project has been a source of great motivation, and we sincerely thank you.

Wael Mohamed
Aynur Özge

Contents

Editors and Contributors

About the Editors

Wael Mohamed a Physician Neuroscientist, earned his PhD from PSU, USA, and presently holds the position of Professor Madya at IIUM Medical School, Malaysia. With over 200 speaking engagements both domestically and internationally, Dr. Mohamed is a seasoned lecturer. He has authored more than 170 peer-reviewed papers in the field of Neuroscience/Psychiatry, boasting an h-index of 25. Additionally, he serves as an editor for several international journals, having contributed to numerous journal special issues focusing on brain disorders. Furthermore, he is involved in editing several neuroscience books with several publishers. Dr. Mohamed has received research grants from esteemed national and international organizations including IBRO, ISN, MJF, STDF, FRGS, and INDO-ASEAN, totaling a million US dollars in research funding. He is also the founder of the AfrAbia-PD-Genomic Consortium (AA-PD-GC).

Aynur Özge is an internationally renowned neurologist specializing in neurology, algology, and clinical neurophysiology. She is a founding faculty member of the Department of Neurology at Mersin University Faculty of Medicine and has played a pivotal role in advancing headache and neurodegenerative disease research globally. Prof. Özge is a Board Member of the International Headache Society (IHS) (2019-2025) and the founding president of the Global Migraine and Pain Society (GMPS) and the MENAA Headache Society, demonstrating her leadership in international headache and pain research. She has been actively involved in patient advocacy as the founding president of the Mersin Branch of the Alzheimer Association of Turkey and the Mersin Alzheimer Association, furthering awareness and care for neurodegenerative disorders. With an extensive academic career, Prof. Özge has published widely in neurology, pain management, and neurodegeneration, contributing to numerous peer-reviewed journals, book chapters, and international research collaborations. She has served as the editor of multiple high-impact books in the fields of headache disorders, neurological scales, and pandemic-related neurological conditions, published under reputable publishers such as Springer. Her expertise spans from migraines and primary headaches to Alzheimer's disease and cognitive dysfunction, and she has played a crucial role in advancing research on the gut-brain axis, neuroinflammation, and nutrition's impact on neurological health. Prof. Özge's work has significantly contributed to understanding neurogenesis, cognitive resilience, and neuroprotective interventions, making her a key figure in the evolving landscape of nutritional neuroscience. As a researcher, clinician, and leader in neurology and pain management, Prof. Özge continues to shape the future of brain health research with her interdisciplinary approach, integrating nutrition, neurobiology, and clinical practice to improve patient outcomes worldwide.

Contributors

Zakirah Zainal Abidin School of Graduate Studies, Postgraduate Centre, Management & Science University, Shah Alam, Selangor, Malaysia

Semiat Adedimeji Department of Biochemistry, Faculty of Pure and Applied Sciences, Kwara State University, Malete, Kwara State, Nigeria

Ceren Alis Department of Neurology, Sisli Hamidiye Etfal Training and Research Hospital, Istanbul, Turkey

Rawan Ashraf Allam Biotechnology Program, Faculty of Agriculture, Ain Shams University, Cairo, Egypt

Salma Allam Faculty of Medicine, Galala University, Suez, Egypt

Buse Cagla Ari Sarioglu Istanbul Medipol University, Neurology, Istanbul, Turkey

Ummul Asfeen Georgia Cancer Center, Augusta University, Augusta, GA, USA

Abraham Olufemi Asuku Bioresourses Development Centre, National Biotechnology Research and Development Agency, Ogbomoso, Nigeria

Nada E. Atia Faculty of Pharmacy, King Salman International University (KSIU), Ras Sudr, South Sinai, Egypt

Nour E. Bayali Faculty of Medicine, Galala University, Suez, Egypt

Gülce Coşku Yılmaz Çakan İzmir University of Economics, Faculty of Medicine, Department of Neurology, Balçova – İzmir, Türkiye

Esraa Ayman Elemary Biotechnology Program, Faculty of Agriculture, Ain Shams University, Cairo, Egypt

Nancy Ashraf Abd El-Fattah Biotechnology Program, Faculty of Agriculture, Ain Shams University, Cairo, Egypt

Abdullrahman Elgarawany Faculty of Medicine, Galala University, Suez, Egypt

Jana S. Elhelw Faculty of Pharmacy, New Giza University, Giza, Egypt

Nada Fath School of Basic Medical Sciences, Fujian Medical University, Fuzhou, Fujian Province, China

Parisa Gazerani Department of Life Sciences and Health, Faculty of Health Sciences, Oslo Metropolitan University, Oslo, Norway

Department of Health Science and Technology, Faculty of Medicine, Aalborg University, Gistrup, Denmark

Hamit Genc Gaziantep City Hospital, Health Application and Research Center, Department of Neurology, Gaziantep, Turkey

Amanina Nurjannah Binti Atan Hamdan School of Graduate Studies, Postgraduate Centre, Management & Science University, Shah Alam, Selangor, Malaysia

Mohab Ali Hassan Faculty of Pharmacy, King Salman International University (KSIU), Ras Sudr, South Sinai, Egypt

Faculty of Pharmacy, Zagazig University, Zagazig, Egypt

Hanife Karakaya Usak Research and Training Hospital, Usak, Turkey

Doaa S. R. Khafaga Department of Basic Medical Sciences, Health Sector, Galala University, New Galala City, Suez, Egypt

Özge Yılmaz Küsbeci İzmir University of Economics, Faculty of Medicine, Department of Neurology, Balçova – İzmir, Türkiye

Ismail Louragli Higher Institute of Nursing and Health Techniques, Ministry of Health and Social Protection, Dakhla, Morocco

Rana Mansour Faculty of Medicine, Galala University, Suez, Egypt

Wael Mohamed Basic Medical Science Department, Kulliyyah of Medicine, International Islamic University Malaysia (IIUM), Kuantan, Pahang, Malaysia

Clinical Pharmacology Department, Menoufia Medical School, Menoufia University, Menoufia, Egypt

Ibrahim Adeola Moronfolu Department of Chemical Sciences, Biochemistry Unit, Summit University, Offa, Kwara State, Nigeria

Salah Abdelfatah Mostafa Biotechnology Program, Faculty of Agriculture, Ain Shams University, Cairo, Egypt

Leyla Mustafayeva Department of Physical Therapy and Rehabilitation, Istanbul University-Cerrahpasa, Cerrahpasa Medical Faculty, Fatih, Turkey

Nihat Mustafayev Ozel Capa Private Hospital, Istanbul, Türkiye

Asiat Na'Allah Department of Biochemistry, Faculty of Pure and Applied Sciences, Kwara State University, Malete, Kwara State, Nigeria

Adeyemi Fatai Odetayo Department of Physiology, Faculty of Basic Medical Science, Federal University of Health Sciences, Ila Orangun, Nigeria

Mosunmola Busayo Oyeleke Department of Physiology, Faculty of Basic Medical Sciences, Afe Babalola University, Ado-Ekiti, Nigeria

Aynur Özge Mersin University School of Medicine, Mersin, Türkiye

NOROM Neuroscience and Excellence Center, Ankara, Türkiye

Muhammad Danial Bin Che Ramli Department of Diagnostic and Allied Health Science, Faculty of Health and Life Sciences, Management and Science University, University Drive, Shah Alam, Selangor, Malaysia

Rahma Mahmoud Saad Faculty of Medicine Modern University for Technology and Information-MTI, Cairo, Egypt

Saheed Sabiu Department of Biotechnology, Faculty of Applied Sciences, Durban University of Technology, Durban, South Africa

Mohamed M. Sarhan Faculty of Pharmacy, King Salman International University (KSIU), Ras Sudr, South Sinai, Egypt

Jia Whei See Universitas Sriwijaya, Palembang, Indonesia

Ahmed H. Shaaban Faculty of Science, Galala University, Suez Governorate, Egypt

Nermin Tanik Yozgat Bozok University Medical School, Department of Neurology, Yozgat, Turkey

Hasna Tirichen Institutes of Biomedical Sciences, Shanxi University, Taiyuan, China

Genevieve Simboh Anak Umbar Department of Diagnostic and Allied Health Science, Faculty of Health and Life Sciences, Management and Science University, University Drive, Shah Alam, Selangor, Malaysia

Al-Hassan Soliman Wadan Oral Biology Department, Faculty of Dentistry, Galala University (15888), Galala Plateau, Attaka, Suez Governorate, Egypt

Hasnaa Yaigoub Institutes of Biomedical Sciences, Shanxi University, Taiyuan, China

Sophie Merve Onerli Yener Migraine Foundation Australia, Keilor East, VIC, Australia

Department of Neurology, Australian Institute of Migraine, Pascoe Vale South, VIC, Australia

Department of Neurology, Monash Health, Clayton, VIC, Australia

Department of Neurology, Ankara Bilkent City Hospital, Cankaya, Ankara, Turkey

Pelin Yenilmez Yeşildaş Gaziantep City Hospital, Neurology Clinic, Gaziantep, Turkey

Gülcan Göçmez Yılmaz Mersin City Training and Research Hospital, Mersin, Turkey

Babalola Ola Yusuf Department of Biochemistry, Faculty of Science, Federal University of Health Sciences, Ila-Oragun, Osun, Nigeria

Department of Biotechnology, Faculty of Applied Sciences, Durban University of Technology, Durban, South Africa

Introduction: The Mind-Body Connection

1

Hamit Genc and Nermin Tanik

Abstract

The mind-body connection has been a topic of debate for centuries. This topic has been among the subjects of many disciplines, such as philosophy, quantum physics, and medicine. In philosophy, mind-body dualism assumes that mind functions are not physical or that the mind and body are separable. In recent years, developments in neuroscience have transformed this situation from a philosophical problem to a scientific issue. The brain and nervous system are at the center of the mind-body connection. Neuro-endocrine, neuro-chemical, and neuroimaging studies have revealed the continuity in the mind-body connection. Many mind-body practices change mental and emotional states, such as meditation, yoga, tai chi, qigong, Pilates, and their combination with exercise. Multidisciplinary research is needed to understand the neurobiological mechanisms of these practices.

Keywords

Dualism · Mind-body interventions · Mind-body exercises · Nutrition · Plasticity · Yoga · Tai Chi · Qigong · Dance/movement therapy · Aerobic/resistance exercises

H. Genc (✉)
Gaziantep City Hospital, Health Application and Research Center, Department of Neurology, Gaziantep, Turkey

N. Tanik
Yozgat Bozok University Medical School, Department of Neurology, Yozgat, Turkey

W. Mohamed, A. Özge (eds.), *Feeding the Mind: The Connection Between Diet, Drugs, and Mental Health Volume 1*, Nutritional Neurosciences,
https://doi.org/10.1007/978-981-95-1414-4_1

Abbreviations

4E cognition	Embodied, Embedded, Enactive, and Extended cognition
ACC	Anterior Cingulate Cortex
ADM	Affective Directed Methods
aINS	Anterior Insula
aMCC	Anterior Midcingulate Cortex
BERN	Behavior, Exercise, Relaxation, and Nutrition
CBT	Cognitive-Behavioral Therapy
CDM	Cognitive Directed Methods
CLBP	Chronic Low Back Pain
dACC	Dorsal Anterior Cingulate Cortex
dlPFC	Dorsolateral Prefrontal Cortex
dmPFC	Dorsomedial Prefrontal Cortex
DMT	Dance/Movement Therapy
EEG	Electroencephalography
FA Meditation	Focused Attention Meditation
GABA	Gamma-Aminobutyric Acid
IGF-1	Insulin-like Growth Factor-1
MBE	Mind-Body Exercises
MBI	Mind-Body Interventions
MBSR	Mindfulness-Based Stress Reduction
MMBIs	Movement-Based Mind-Body Interventions
mPFC	Medial Prefrontal Cortex
NDM	Null Directed Methods
OM Meditation	Open Monitoring Meditation
pINS	Posterior Insula
rACC	Rostral Anterior Cingulate Cortex
S1	Primary Somatosensory Cortex
S2	Secondary Somatosensory Cortex
sgACC	Subgenual Anterior Cingulate Cortex
TM Meditation	Transcendental Meditation
vlPFC	Ventrolateral Prefrontal Cortex

1.1 The Key to Holistic Health: The Mind-Body Connection

The mind and body are inseparable elements of human life in constant communication. The connection between these two areas represents a complex relationship that shapes individuals' physical, mental, and emotional health. Scientific research shows that the mind-body connection is a theoretical concept and a measurable and manageable reality (Littrell 2008; Bringmann et al. 2021).

Mind-body harmony emerges as a powerful tool for optimizing human health, reducing stress, and improving quality of life. Thanks to the knowledge provided by modern psychology, neuroscience, biology, and technology, the dynamics of this connection are better understood, and practical applications are being developed (Garland et al. 2020; Frith et al. 2020; St. Marie and Talebkhah 2018; Borghesi et al. 2024; Voss et al. 2023). In addition to traditional mind-body interventions *(MBI)* such as meditation, yoga, tai chi, and qigong, modern approaches such as virtual reality and mobile-based applications diversify the solutions in this area (Voss et al. 2023; Dong et al. 2024; Park et al. 2020).

In this context, efforts to ensure the harmony of mind and body are key to a journey supporting individual health and social well-being. This text invites the reader to a more holistic perspective by examining the scientific basis of the connection between mind and body and the essential applications in this field.

1.2 Historical Perspective: The Mind-Body Dilemma

> Cogito, ergo sum. (Descartes)

> Mens sana in corpore sano. (Juvenal)

Throughout the history of philosophy, many philosophers have discussed whether there is a connection between the mind and the body. Many questions have been sought to be answered regarding this situation.

- Are the mind and the body two completely separate entities?
- Is there an interaction between the mind and the body?
- How can this be explained if there is an interaction?
- Can mental processes affect the body?
- Can bodily processes affect the mind? (Chambliss 2018).

There are classical theories such as materialism, behaviorism, and dualism that try to explain this connection. Materialism assumes that the mind is a physical entity. Behaviorism states that behavior is an indicator of mental states. Dualism, on the other hand, positions the mind and the body as separate. The dualism of Descartes, one of the representatives of this field, is known as Cartesian dualism (Tripathi 2024). Descartes defined the mind as "self-aware and self-thinking" with the sentence "I think, therefore I am." He stated that thinking and existing are the same thing (Duncan 2000). Gualtiero Piccinini defines the mind as a functional and complex system. Contemporary theories of the mind include functionalism and computational functionalism. Functionalism compares the mind to a computer program. According to this theory, sensory inputs and behavioral outputs determine the mind, and the result is important. The computational functionalist approach treats mental activity as a static process of symbol manipulation. It ignores the idea that the environment shapes mental activity. According to Maurice Merleau-Ponty, an existential phenomenologist, the mental state is embodied. The embodied mind theory suggests that mental states are tied to the physical body (Tripathi 2024). 4E *(embodied, embedded, enactive, and extended cognition)* is a new approach that sheds a different light on mind-body dynamics. 4E cognition synthesizes old and contemporary ideas by considering both philosophical and scientific developments. Functionality and purpose are essential in this approach (Stanciu 2023). Many disciplines, such as philosophy, quantum physics, neuroscience, and psychiatry, have dealt with this issue. It took a long time for this relationship to be established in medicine. Because it was difficult to relate the functioning of the mind to the functioning of the body. For many years, the biomedical model, which focused solely on the body, was at the forefront of healthcare services. This model corresponds to the mind-body dualism in philosophy. *George Engel* proposed a new medical model, known as the biopsychosocial model, which contrasts with the biomedical model. *Engel* believed that the person's experience of the disease, whether the person or others perceive the situation as a disease, and the impact of environmental conditions on the onset and course of the disease were significant for some diseases (Engel 1977; Switankowsky 2000). *Ludolf Krehl*, an internal medicine specialist, suggested that patients should be considered whole. His slogan was, *"We treat patients, not diseases."* (Tesarz 2024). Mental processes and the mind-body connection are the subject of many studies conducted in cognitive science and neuroscience today. The fundamental discussions in the philosophy of mind, which have been ongoing for centuries, continue to evolve in light of neuroscience. The complex structure of the nervous system makes this process difficult. In addition, investigating of the autonomic nervous system and immune, endocrine, and gut-brain relationships and the biomarkers associated with these processes is essential in evaluating these applications and determining their potential role in treatment (Kim et al. 2013).

1.3 Neuroscientific and Physiological Foundations

The brain and nervous system are at the center of the mind-body connection. The brain's complex structure functions as a control center, managing physiological processes, emotions, thoughts, and behaviors. Scientific studies have attempted to discover changes that may impact the mind-body connection by studying neurotransmitters, the hormonal system, the vagus nerve, homeostasis, and neuroplasticity.

1.3.1 The Role of Neurotransmitters and Hormones

An integrative psychophysiological model is suggested in which bidirectional autonomic and neuroendocrine pathways transmit signals from the central nervous system and peripheral tissues (Taylor et al. 2010), expressing hormonal, autonomic, emotional, and immune responses (Morgan et al. 2014). Neurotransmitters, hormones, and the nervous system play critical roles in this process. For example, dopamine, serotonin, and norepinephrine neurotransmitters regulate mood and directly affect stress management and happiness. Changes in their levels are thought to play a role primarily in the behavioral manifestations of psychic states (Blows 2000). The nervous system, especially the autonomic one, is a concrete indicator of the mind-body connection. The autonomic nervous system integrates bodily sensation with emotion by integrating structures in the upper brainstem, hypothalamus, amygdala, insula, and anterior cingulate cortex *(ACC)*. It enables the production of homeostatic autonomic responses (Cortelli et al. 2013). The sympathetic nervous system activates stress responses, while the parasympathetic system promotes relaxation and healing. This balance affects both the mental and physical health of individuals through several effects, including pain anticipation, pain response, and relief of symptoms of depression (St. Marie and Talebkhah 2018; Cortelli et al. 2013; Stefano et al. 2019a, b; Braden et al. 2016).

Stress disrupts the release periods of certain hormones, primarily cortisol and melatonin, and the autonomic nervous system (Littrell 2008). Prolonged high levels of stress hormones such as cortisol can suppress the immune system and lead to chronic diseases (Sephton and Spiegel 2003; Rosmond and Björntorp 2000). However, MBIs support health by reducing cortisol levels (Antoni et al. 2000, 2001). Neurobiologically, strategies used to cope with stress increase the release of neurotransmitters such as dopamine, endocannabinoids, endorphins, and oxytocin. These processes also modulate stress hormones *(adrenaline/noradrenaline, and cortisol)* and have a positive effect on overall health. Most of these act by releasing nitric oxide and other messenger molecules (Esch 2002, 2022; Esch and Stefano 2004, 2010). Gallegos et al. (2013) found that participants in a mindfulness-based stress reduction (MBSR) study who practiced yoga and mindfulness more frequently had increased post-treatment insulin-like growth factor 1 (IGF-1), a biomarker associated with human lifespan.

1.3.2 Vagus Nerve and Homeostasis

The vagus nerve is a crucial mediator of the mind-body connection facilitating its adaptable and flexible interaction with the amygdala, prefrontal cortex, and peripheral organs (Reyes del Paso et al. 2013). Activation of the vagus nerve is known to enhance parasympathetic nervous system responses, support homeostasis, improve mood and energy states, and lead to neuroendocrine, metabolic, cognitive, and immune responses (Tyagi and Cohen 2016).

1.3.3 Brain Volume Changes and Plasticity

Neuroimaging studies have demonstrated that MBI has a significant impact on brain volume changes and plasticity. The effects of MBI may be based on neuroplasticity. Structural and functional changes have been observed in brain regions involved in pain perception and attention processing. Mind-body exercises *(MBE)* can stimulate the regeneration of hippocampal neurons and alter the hypothalamic-pituitary-adrenal axis function (Wegner et al. 2014). Studies on the effects of MBE on pain management have observed significant changes in the prefrontal cortex and emotion-processing regions. During pain, increased activity in the posterior insula *(pINS)* and secondary somatosensory cortex *(S2)* has been associated with reduced pain discomfort during an alert state (Gard et al. 2012). In contrast, some studies have observed increased activity in the anterior insula *(aINS)* and anterior midcingulate cortex *(aMCC)* and decreased activity in these regions during pain anticipation (Lutz et al. 2013). It has been suggested that the aINS is necessary for subjective awareness of internal sensations (Critchley et al. 2004), while the pINS plays a more significant role in perceiving pain intensity (Apkarian et al. 2011). MBI may lead to changes in neural connectivity in response to chronic pain. For example, Tai Chi increases connectivity between the dorsolateral prefrontal cortex *(dlPFC)* and rostral anterior cingulate cortex/medial prefrontal cortex *(rACC/mPFC),* supporting self-regulatory processes for pain management (Kong et al. 2019). Furthermore, this practice reduces rACC/mPFC connectivity to the periaqueductal gray matter and hippocampus, contributing to the conceptualization of chronic pain as a form of long-term learning (Apkarian et al. 2011). Within mindfulness-based practices, pain management programs reduced dlPFC and S2 deactivation during pain anticipation, which correlated with improved mental health and pain controllability (Brown and Jones 2013). In another study, MBSR improved functional pain outcomes by increasing activity in the subgenual anterior cingulate cortex *(sgACC),* ventrolateral prefrontal cortex *(vlPFC),* and dorsomedial prefrontal cortex *(dmPFC)* during a sadness induction task. In an emotional Stroop task involving pain-related sensory and emotional words, MBSR reduced activity in the primary somatosensory cortex *(S1),* precuneus, and dlPFC (Braden et al. 2016; Hatchard et al. 2021). During a cognitive task, MBSR reduces activity in the cuneus and parietal operculum, a region that includes the pINS (Seminowicz et al. 2020).

Meditation was associated with more significant deactivation in prefrontal regions. Meditators with the lowest pain sensitivity *(requiring higher stimulus intensity)* had weak functional connectivity between the dlPFC and dorsal anterior cingulate cortex *(dACC)* (Grant et al. 2011). Increased cortical thickness in the dACC and S2 in long-term meditators has been suggested to reduce pain sensitivity (Grant et al. 2010). Additionally, increased activity in the orbitofrontal cortex facilitates pain reappraisal, allowing the sensory experience to be framed more positively. A study assessing experimental pain during meditation found that reduced pain unpleasantness and intensity were associated with increased activity in the ACC and aINS, areas involved in thalamic deactivation and cognitive regulation of nociception (Zeidan et al. 2011). The brain regions in which MBI caused changes are shown in Fig. 1.1.

Meditation, mindfulness, and other MBIs have also been observed to lead to significant changes in electroencephalography *(EEG)* activity. For example, focused attention meditation *(FA)* is associated with increased beta frequency power bands, while open monitoring meditation *(OM)* increases frontal theta bands. Similarly, transcendental meditation *(TM)* affects occipital alpha bands. All of these

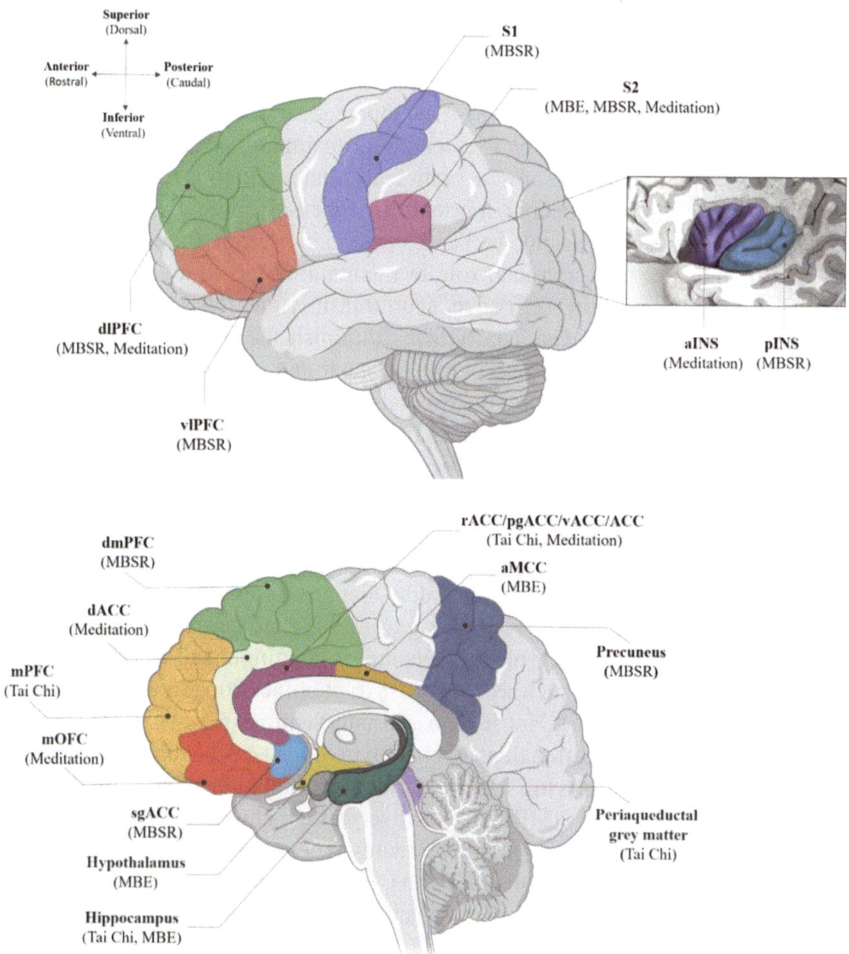

Fig. 1.1 The responsible regions of the brain that are affected by MBI in terms of brain volume and neuronal plasticity (see Sect. 1.3.3 for detailed explanations) (S1 primary somatosensory cortex, S2 secondary somatosensory cortex, aINS anterior insula, pINS posterior insula, sgACC subgenual anterior cingulate cortex, dACC dorsal anterior cingulate cortex, ACC anterior cingulate cortex, aMCC anterior midcingulate cortex, rACC rostral anterior cingulate cortex, dlPFC dorsolateral prefrontal cortex, mPFC medial prefrontal cortex, vlPF ventrolateral prefrontal cortex, dmPFC dorsomedial prefrontal cortex, MBSR mindfulness-based stress reduction, MBE mind-body exercises)

techniques increase parasympathetic activity in the autonomic nervous system, reducing sympathetic activity and enhancing stress-coping mechanisms (Borghesi et al. 2024).

1.3.4 Nutrition and the Mind-Body Connection

It is now widely accepted that nutrition can have a significant impact on cognitive functions, emotions, and even the development of mental health disorders. Adequate and balanced nutrition is essential for maintaining healthy mental functions, promoting the growth of new brain cells, and facilitating synaptic flexibility. Our brain requires specific nutrients to function optimally. For example, essential fatty acids such as omega-3 and omega-6, zinc, iron, folate, and vitamins A, C, B6, and B12 play a crucial role in maintaining the structural integrity of brain cells and supporting effective neurotransmission (NEMO Study Group 2007; Sandua 2023). Omega-3 fatty acids diminish the risk of depression and anxiety, while a Western diet rich in refined sugar and processed foods increases the risk of mental health problems. Minerals such as zinc, magnesium, and iron affect cognition and psychology. An example of their effects on neurotransmission is the conversion of tryptophan to serotonin. In this way, while regulating mood, a deficiency can negatively affect mental health. In addition, foods rich in omega-3 fatty acids, vegetables, and whole grains can positively impact diseases characterized by chronic inflammation, such as depression and schizophrenia, due to their anti-inflammatory effects. A healthy diet also supports the gut microbiota, which has neuroprotective effects. The gut microbiome communicates with the brain via the gut-brain axis, which affects mood, cognition, and behavior by producing neurotransmitters such as 5-hydroxytryptamine, dopamine, and gamma-aminobutyric acid *(GABA)*. Nutrition-based interventions may be more effective when combined with traditional mental health treatments. A healthy diet is critical in regulating mood, preserving cognitive functions, and improving overall mental well-being (Sandua 2023). MBI can positively affect chronic pain management, stress reduction, and general cognitive health. These practices are safe, widely accepted, and have a low risk of side effects. However, more research is needed on the biological and psychological mechanisms underlying the impact of these interventions.

1.4 Mind-Body Connection-Based Healthy Living and Applied Approaches

The mind-body connection is a complex and multidimensional network of interactions that affect human health at both physical and mental levels. This relationship is vital to understanding how biological, psychological, and social processes shape an individual's holistic health. Today, practices that strengthen the mind-body connection to promote healthy living are increasingly gaining attention.

MBI include a variety of types: *(a) meditation approaches* (e.g., *mindfulness, mindfulness-based stress reduction), (b) hypnosis or guided imagery, (c) Movement-Based Mind-Body Interventions (MMBIs)* (e.g., *yoga, tai chi, and qigong), and (d) body-based approaches* (e.g., *acupuncture, mobilization, spinal manipulation, and massage)* (Park et al. 2020).

1.4.1 Meditation

Meditation methods are "emotional and attentional regulatory training regimens developed for various purposes, including improving well-being and emotional balance" (Lutz et al. 2008).

1.4.1.1 Meditative State

The meditative state is defined as transitioning from the ordinary state of wakefulness to a more *"deep"* mental state, such as calmness, focus, insight, happiness, emptiness, etc. (Nash and Newberg 2023). There are various subtypes of meditation based on different cultures and belief systems. Some studies have been done on the classification of meditation; in one review, meditation methods are divided into two: simple and complex. Simple meditation methods are categorized into one of three areas according to their orientation within the Affect and Cognition paradigm: Affective-Directed Methods (ADM), Cognitive-Directed Methods (CDM), and Null-Directed Methods (NDM). During the meditation session, ADMs *[Anapanasati (concentration on the breath), walking mindfulness meditation, and vipassana]* aim to create an improved emotional state, CDMs *(karuna, metta, bodhisattva, and Tibetan)* aim to create an enhanced cognitive state, and NDMs *(TM meditation, Soham japa, rigpa, and Mahamudra)* aim to create an improved zero state (Nash and Newberg 2023). Despite the studies, there is still no clear consensus on this subject. To gain a better understanding of the topic, it is necessary to mention some common meditation types and their key features briefly.

Focused Attention Meditation

It involves focusing on a specific object, sound, breath, or thought (Travis and Shear 2010).

Open Monitoring Meditation

It aims to observe thoughts, feelings, or sensory experiences in the mind without judgment (Travis and Shear 2010).

Transcendental Meditation

A mantra is a word or phrase usually repeated during meditation or prayer. This type of meditation aims to calm the mind through mantra thought (Travis and Shear 2010).

Loving-Kindness Meditation *(Metta meditation)*

It aims to develop compassion and love for oneself and others (Villines et al. 2017).

Kundalini Meditation

A physically active form of meditation that combines movements with mantra and deep breathing (Stanciu 2023). Kirtan Kriya's meditation is an example of this. It is a meditation mentioned in Kundalini yoga in which the tip of the thumb is touched with firm and gentle pressure with all the other fingers, along with mantra, sound, and visualization (Inzunza 2023).

Zen Meditation

A form of meditation that encompasses specific steps and postures, finding a comfortable position, focusing on breathing, and observing one's thoughts without judgment (Villines et al. 2017).

Mindfulness Meditation

Awareness is a form of meditation that encourages practitioners to be aware of the moment and stay in the moment rather than focusing on the past or fearing the future (Villines et al. 2017).

Visualization Meditation

It involves mentally imagining a favorable situation or goal (Fontana 2012).

Breath Awareness Meditation

Breath awareness is a meditation that encourages conscious breathing. In it, practitioners breathe slowly and deeply, counting their breaths or otherwise focusing on their breathing (Villines et al. 2017).

Body-Scan Meditation

The body scan is a meditation practiced at various speeds and sensitivity levels. It describes being *"compassionate, open-hearted, and interested"* in the body (Kabat-Zinn 2005).

1.4.1.2 The Effect of Meditation and Mindfulness Practices

Meditation and mindfulness practices are some of the most effective methods that strengthen the mind-body connection by training mental abilities such as calmness, concentration, sensitivity to mental processes, and differentiation (Sedlmeier et al. 2012). Studies on the effects of meditation-based lifestyle changes on increasing individuals' mental health and life satisfaction have shown that meditation reduces negative emotional states such as stress, anxiety, and depression in individuals, as well as improving self-awareness and emotional regulation (Bringmann et al. 2021; Sedlmeier et al. 2012). A study evaluating experimental pain during meditation showed that meditation decreased pain unpleasantness by 57% and pain vigor by 40% (Zeidan et al. 2011). A meta-analysis explicitly conducted on chronic pain patients revealed that mindfulness-based approaches offer a practical alternative to managing pain by reducing the need for opioid use (Garland et al. 2020). Meditation and other non-pharmacological practices may activate natural endogenous analgesic processes, or these effects may be attributed to distraction or changing expectations. Indeed, in evaluating the impact of mind-body practices on interoceptive sensitivity, mindfulness-based practices were found to have a more significant effect than MMBIs (Voss et al. 2023). However, the exact mechanisms underlying the pain-reducing effects of meditation are not yet fully understood.

1.4.2 Movement-Based Approaches

MMBIs encompass various forms, including yoga, tai chi, qigong, Pilates, and dance, with the primary purpose of enhancing the physical and mental health of the individual (Zhang et al. 2019a; Zou et al. 2019). Park et al. reported that yoga, tai chi, and qigong alleviated physical discomfort and improved overall quality of life in individuals with back pain (Park et al. 2020). Furthermore, a comprehensive meta-analysis by Dong et al. in older individuals showed that these practices had positive effects on mental health, significantly reducing anxiety and depression levels (Dong et al. 2024). The main strength of these methods lies in their ability to improve physical endurance and flexibility while increasing the individual's internal awareness (Stahl et al. 2015).

1.4.2.1 Yoga

Yoga is an ancient Indian MBI consisting of different components such as yamas *(restrictions from actions, words, or thoughts that may cause harm),* niyamas *(virtuous behaviors and practices),* asanas *(physical postures),* pranayama (breathing techniques), pratyahara *(control of the senses),* dharana *(concentration),* dhyana *(meditation),* and samadhi *(absorption)* (Varambally and Gangadhar 2012). It aims to improve individuals' physical, mental, and spiritual health by helping to establish the balance between mind, body, and soul (Dong et al. 2024). On a physical level, yoga improves flexibility and muscle strength and corrects posture disorders (Nayak and Shankar 2004). In addition, it emerges as a positive pain management method by addressing the physical, cognitive, and emotional components of chronic pain through the interoceptive mechanism. Scientific studies support the effectiveness of yoga in improving general well-being and show that this practice offers significant benefits, especially for older adults and individuals with chronic diseases (Gnall et al. 2024). When its effect on headaches was examined, it was shown that yoga practice reduced the frequency of headaches in migraine patients (Wu et al. 2022). The *American Headache Consortium* has identified several non-pharmacological interventions, such as migraine prevention guidelines, relaxation training, electromyographic biofeedback, and cognitive-behavioral therapy *(CBT),* as Level A evidence for migraine treatment (Campbell et al. 2000). Yoga is practiced by 14.3% of the American population (Clarke et al. 2018); the percentage of these individuals who use this intervention for their headaches and the underlying pathophysiology remains to be investigated. Studies have also been conducted on the effects of yoga on sleep disturbances, depression/anxiety, and cognitive function. A meta-analysis found that MMBIs, especially those performed more than three times per week, effectively improved sleep quality (Wu et al. 2015). Yoga has positive effects on alleviating mental health problems such as depression and anxiety, possibly by reducing levels of the stress hormone cortisol (Smith et al. 2011; Lee et al. 2014; Taylor et al. 2020). It may also facilitate coping with anxiety and depressive symptoms in older adults by improving emotional management and self-awareness and increasing emotional balance (Taylor et al. 2020). In addition, positive effects have been shown on depression in family caregivers (Lavretsky et al. 2013). A randomized controlled trial of adults over 60 found that Hatha and Iyengar yoga *(light, non-aerobic)* moderately improved mental and physical health-related quality of life (Patel et al. 2012). Another study in older adults showed a positive effect on working memory and mental flexibility (Gothe et al. 2014). In a study evaluating the combined application of Kundalini yoga and Kirtan Kriya meditation on cognitive functions in older adults with MCI, both the yoga and memory training groups showed improved memory performance. Still, only the yoga group showed improvements in executive functions. These effects persisted for 3 months after treatment was discontinued (Eyre et al. 2017).

1.4.2.2 Tai Chi

Tai chi is an MBE rooted in the deep Chinese tradition and is characterized by a combination of slow, fluid movements, deep breathing, and meditation (Wang et al. 2016). Like other MBEs, tai chi acts by increasing serotonin and norepinephrine levels, activating the endocannabinoid system, altering the function of the hypothalamic-pituitary-adrenal axis, and reducing the response of glucocorticoid receptors (Wegner et al. 2014; Zheng et al. 2015; D'Ascola et al. 2018). One of the most notable benefits of tai chi is that it improves balance and reduces the risk of falls. This feature was observed in a study by Low et al. in elderly individuals (Low et al. 2009). In a randomized controlled trial, tai chi showed a therapeutic effect in patients with nonspecific chronic low back pain *(CLBP)* (Weifen et al. 2013). In another review, tai chi showed significant efficacy in improving pain intensity, functional disability, cervical mobility, and quality of life in patients with nonspecific neck pain (Gao et al. 2024). In addition, its effectiveness has been shown in individuals with fibromyalgia (Kong et al. 2019) and knee pain (Shen et al. 2022). Longer-term interventions seem to be needed for more effective results (Park et al. 2020).

Tai chi has multifaceted effects on mental health and cognitive functions. It provides mental stimulation through memorization, concentration, and meditation during different postures; it affects physical, cognitive, social, and spiritual functions (Fong et al. 2014; Wayne et al. 2014; Zhang et al. 2018). It has been reported that this exercise improves core executive functions such as working memory and cognitive flexibility (Diamond 2013). However, its effects in individuals with early-stage dementia remain variable (Lim et al. 2019; Yang et al. 2020). In addition, although traditional meta-analyses show high statistical heterogeneity, studies in elderly individuals have positive effects on depression and anxiety symptoms, probably through neuroendocrine changes, as mentioned above (Dong et al. 2024).

The effects of tai chi on sleep quality in elderly individuals have also been shown. Randomized controlled trials have demonstrated that tai chi significantly enhances overall sleep efficiency and mental quality of life through its effects on sleep latency, sleep duration, total sleep time, and daytime dysfunction (Lü et al. 2017; Chan et al. 2016). A 1-year follow-up study comparing tai chi with CBT concluded that CBT may be more effective than tai chi in treating insomnia in older age (Irwin et al. 2014).

1.4.2.3 Qigong

Qigong is a type of MBE originating from traditional Chinese medicine that aims to regulate energy flow (Hung et al. 2021). The most important advantages of qigong are that it requires minimal equipment and can be easily practiced by individuals of all ages (Liu et al. 2024). Qigong exhibits a mindfulness-based approach to breath and body awareness. These features make it an ideal tool for developing interoception and helping individuals to know themselves at a deeper level (Park et al. 2020). In addition, qigong types such as *Baduanjin work* by gently exercising the muscles and regulating blood circulation (Liu et al. 2023). Since qigong and tai chi share similar roots, many interventions incorporate aspects of both practices (Jahnke et al. 2010). Therefore, studies have shown similar effects to other MMBIs, such as yoga and tai chi, on pain (Park et al. 2020; Gao et al. 2024), depression (Dong et al. 2024), anxiety (Dong et al. 2024), sleep (Wu et al. 2015), and cognitive functions (Lim et al. 2019; Yang et al. 2020). In a study evaluating functional performance and health-related quality of life in patients with Parkinson's disease, it was shown to be the most effective method for improving walking speed compared to other MBEs (Mustafaoglu et al. 2022). Qigong was also effective in improving muscle strength, balance, and cardiovascular fitness (Lin et al. 2023) and reducing falls (Jones et al. 2022). Despite all this, there are some conflicting results in the literature, which may be due to the inconsistency of training or the lack of culturally adapted exercise components for the study population (Liu et al. 2024). As a result, a particular MBE may be more effective for some individuals but less suitable for others, thus explaining the differences in observed results. For example, some adults may find it easier to adopt the slow and gentle movements of tai chi, while others may be more receptive to the breathing regulation techniques of qigong. These individual differences and intervention preferences are important for creating personalized intervention plans. Because there are so few studies testing MBI *(especially qigong and tai chi)*, more rigorous studies of high methodological quality are needed to obtain consistent findings about the effects of the three interventions.

1.4.2.4 Dance and Movement Therapy

Dance/movement therapy *(DMT)* is a scientifically based therapy in which the mind and body are interdependent in function and functioning, working as an inseparable whole (Berrol 1992). DMT is based on a process that integrates the physical and emotional expressions of the individual, encompassing internal elements such as rhythm, space, weight, and flow (Fitt 1988). The neurological basis of DMT is related to the regulatory effect of rhythmic movements on the nervous system. Grillner et al. stated that the central neural network regulates movements through sensory feedback (Grillner 1985). The therapeutic power of dance lies in its ability to foster a sense of solidarity and harmony among individuals by synchronizing them through familiar rhythms and movements (Schmais 1985). In addition, it has similar effects on serotonin/norepinephrine levels, the endocannabinoid system, and the hippocampus-pituitary-adrenal axis, as other MBEs such as qigong, tai chi, and yoga (Wegner et al. 2014; Zheng et al. 2015; D'Ascola et al. 2018).

As mentioned above, MBEs have been shown to have significant effects on anxiety and depression in older adults. In a meta-analysis by Dong et al., the order of the impact of exercises on anxiety was *Tai Chi > Qigong > Yoga > Dance > Control group*. In contrast, the order of the effects on depression was as follows: *Tai Chi > Pilates > Yoga > Qigong > Dance > Control group* (Dong et al. 2024). In contrast, two meta-analyses have shown that dance interventions do not have a momentous effect on anxiety and depression (Wang et al. 2022; Zhang et al. 2019b). This may be due to many older adults' difficulties adapting to its specific rhythms and demands (Ventura et al. 2016). A study by Mustafaoğlu et al. on Parkinson's disease patients showed that dance is more effective than other MBEs in improving motor function (Mustafaoglu et al. 2022). Studies have emphasized the potential of physical movements to enhance higher-level cognition, such as creative thinking, by up to 90% (Frith et al. 2020). Oppezzo and Schwartz stated that simple movements can trigger creative processes and that structured activities strengthen this process (Oppezzo and Schwartz 2014). Numerous studies have confirmed that spontaneous and improvised movements, such as dance, increase divergent and convergent thinking and problem-solving skills (Abraham et al. 2019; Kuo and Yeh 2016; Friedman and Förster 2002; Kim 2015; Wang et al. 2018). DMT addresses the individual's physiological and psychosocial needs with a holistic approach. Supported by scientific research, this method takes place among modern therapeutic approaches within a model framework in which the mind and body work interactively.

1.4.2.5 Aerobic and Resistance Exercises

In addition to MBIs, classical aerobic and resistance exercises support brain plasticity (Li et al. 2019). Aerobic exercise includes various approaches, such as dynamic and isometric flexion, extension, or stretching (Van Middelkoop et al. 2011). It can also be used as a part of the combination of muscle exercise, coordinative exercise, social interaction, and meditation, which constitute Chinese MBEs (Chang et al. 2014). Aerobic exercises can positively impact cognitive functions by altering the volume of gray matter, particularly in the prefrontal, parietal, and temporal regions, as well as the white matter in the genu part of the corpus callosum (Colcombe et al. 2003). In the hippocampus, it regulates the expression of proteins related to synaptic plasticity (Li et al. 2019). In addition, an experimental study has observed that exercise increases the levels of brain-derived neurotrophic factor, a molecule critical for developing new neurons and learning and memory processes (Berchtold et al. 2001). It has been shown that 6 months of moderate aerobic exercise can reliably reverse age-related cognitive decline, that the benefits on cognitive function are more pronounced for executive control tasks, that it has a particular specificity for mental and brain function, and that it preserves the flexibility capacity and executive processes of older adults (Erickson and Kramer 2009). In addition, aerobic exercise has positive effects on depression and anxiety, although studies are reporting the opposite (Dergisi'nden 2015). In a meta-analysis, the effect size of aerobic exercise was found to be 0.34 for anxiety and 0.56 for depression (Wegner et al. 2014).

Another important aspect of aerobic and resistance exercise is that it reduces pain intensity and improves functional ability in patients with CLBP (Airaksinen et al. 2006). However, the literature remains uncertain regarding the most effective exercise dose *(intensity, frequency, and duration)* for CLBP patients. However, it has been observed that sessions performed five or more times per week provide greater effects compared to lower frequencies (Ren et al. 2021).

In conclusion, aerobic and resistance exercises have the potential to improve physical and mental health. With their multifaceted effects, including supporting neurogenesis in the hippocampus, reducing symptoms of depression and anxiety, and enhancing cognitive functions, these types of exercise make significant contributions to individuals' overall quality of life.

1.4.3 Combination of Meditation and Movement-Based Exercises

The combination of meditation and MMBIs plays a vital role in the foundation of mind-body health approaches. Meditation and MMBIs facilitate a two-way interaction between the body and mental awareness, promoting relaxation and a state of physical and spiritual well-being.

The results on the effects of movement-meditation combinations such as Kundalini yoga (Eyre et al. 2017), MBSR (Borghesi et al. 2024; Brown and Jones 2013), and Chinese MBE *(Tai Chi and qigong)* (Dong et al. 2024; Park et al. 2020; Kong et al. 2019; Wu et al. 2015; Low et al. 2009; Weifen et al. 2013; Gao et al. 2024; Shen et al. 2022; Fong et al. 2014; Wayne et al. 2014; Zhang et al. 2018; Diamond 2013; Lim et al. 2019; Yang et al. 2020; Lü et al. 2017; Chan et al. 2016; Mustafaoglu et al. 2022; Lin et al. 2023; Jones et al. 2022) on health have been mentioned above. Another movement-meditation combination is evaluated under the *"BERN"* framework. This framework consists of *(B) Behavior, (E) Exercise, (R) Relaxation, and (N) Nutrition components.* The BERN framework focuses on the effects of the mind on physical functioning and improving health while also aiming to develop individual self-knowledge and self-care capacity (Esch and Stefano 2022).

1.5 Future Perspectives on the Mind-Body Connection

Treatment methods focusing on the mind-body connection stand out as an approach with significant potential for individual and social health. Functional genomics, neuroimaging, and neurophysiological evaluations have been conducted to better understand the biological mechanisms of these methods (Muehsam et al. 2017). However, multidisciplinary research and applications supported by innovative technologies are necessary to determine the effects of these applications on different age groups and ethnicities, their long-term outcomes, and underlying neurobiological mechanisms. Another question that needs to be answered is which application is ideal for patients and healthy individuals, and how often and for how long it should be applied.

Another emerging field based on mind-body medicine is psychosocial genomics. Investigating the connections of MBIs with different disciplines, such as neuroscience, psychology, biotechnology, and medicine, can provide a better understanding of their mechanisms of action. New models are presented regarding the relationships between gene expression and neurogenesis, human behavior, and consciousness. It has been suggested that these approaches may stimulate epigenetic changes and gene expression associated with anti-inflammatory processes (Rossi 2002; Cozzolino and Celia 2021).

Another current field is the gut-mind connection. Psychogastroenterology investigates the connection between the gut, brain, and mind. The gut, which hosts a large community of microbiota, is closely associated with the immune system, the endocrine system, and the brain. This connection provides a two-way communication network (Boem et al. 2024).

Finally, new strategies need to be developed to increase the impact of these applications on public health and ensure their dissemination. Both their preventive and therapeutic properties can be used as practical tools in the fight against chronic diseases and psychological disorders.

1.6 Conclusion

The mind-body connection has a broad impact on both scientific and practical levels and is one of the cornerstones of healthy living. Methods (*see* Fig. 1.2) such as meditation, mindfulness, MMBIs, and healthy nutrition support not only the physical health of individuals but also their mental and spiritual well-being. Regularly adopting these practices provides a powerful tool for improving the quality of life of individuals and strengthening their ability to cope with difficult life circumstances. Future research can explore this connection and bring mind-body approaches to greater prominence in individual and community health strategies.

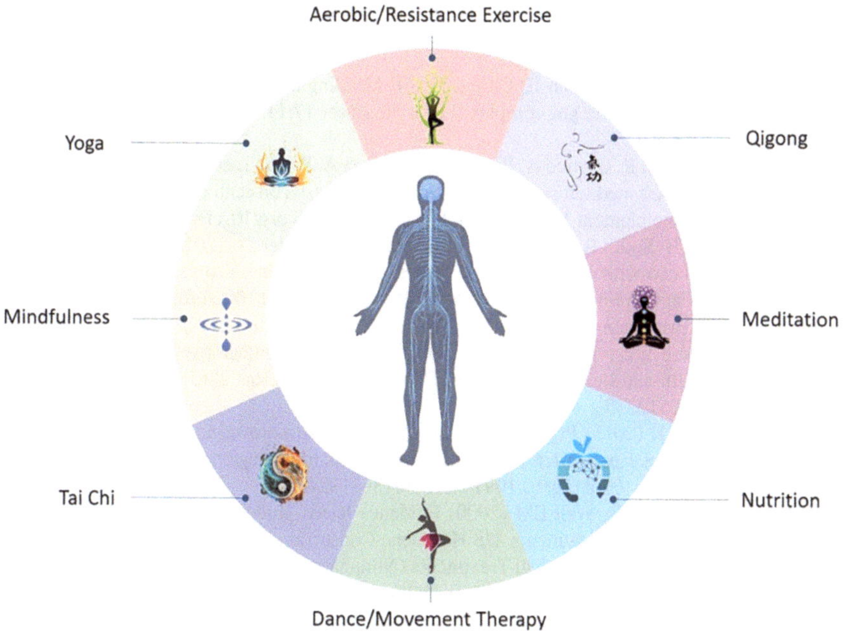

Fig. 1.2 The basic mind-body interventions

References

Abraham A, Asquith S, Ahmed H, Bourisly AK (2019) Comparing the efficacy of four brief inductions in boosting short-term creativity. J Cogn Enhanc 3:85–93. https://doi.org/10.1007/s41465-018-0085-0

Airaksinen O, Brox JI, Cedraschi C, Hildebrandt J, Klaber-Moffett J, Kovacs F et al (2006) European guidelines for the management of chronic nonspecific low back pain. Eur Spine J 15(Suppl 2):s192. https://doi.org/10.1007/s00586-006-1072-1

Antoni MH, Cruess S, Cruess DG, Kumar M, Lutgendorf S, Ironson G et al (2000) Cognitive-behavioral stress management reduces distress and 24-hour urinary free cortisol output among symptomatic HIV-infected gay men. Ann Behav Med 22(1):29–37. https://doi.org/10.1007/BF02895165

Antoni MH, Lehman JM, Kilbourn KM, Boyers AE, Culver JL, Alferi SM et al (2001) Cognitive-behavioral stress management intervention decreases the prevalence of depression and enhances benefit finding among women under treatment for early-stage breast cancer. Health Psychol 20(1):20. https://doi.org/10.1037/0278-6133.20.1.20

Apkarian AV, Hashmi JA, Baliki MN (2011) Pain and the brain: specificity and plasticity of the brain in clinical chronic pain. Pain 152(3):S49–S64. https://doi.org/10.1016/j.pain.2010.11.010

Berchtold NC, Kesslak JP, Pike CJ, Adlard PA, Cotman CW (2001) Estrogen and exercise interact to regulate brain-derived neurotrophic factor mRNA and protein expression in the hippocampus. Eur J Neurosci 14(12):1992–2002. https://doi.org/10.1046/j.0953-816x.2001.01825.x

Berrol CF (1992) The neurophysiologic basis of the mind-body connection in dance/movement therapy. Am J Dance Ther 14(1):19–29. https://doi.org/10.1007/BF00844132

Blows WT (2000) Neurotransmitters of the brain: serotonin noradrenaline (norepinephrine), and dopamine. J Neurosci Nurs 32(4):234–238. Retrieved December 17, 2024, from Journal of Neuroscience Nursing

Boem F, Greslehner GP, Konsman JP, Chiu L (2024) Minding the gut: extending embodied cognition and perception to the gut complex. Front Neurosci 17:1172783. https://doi.org/10.3389/fnins.2023.1172783

Borghesi F, Cremascoli R, Chirico A, Bianchi L, Di Moia A, Priano L et al (2024) Mind and body connection in expert meditators: a computational study based on central and peripheral nervous system. BMC Complement Med Ther 24(1):117. https://doi.org/10.1186/s12906-024-04413-5

Braden BB, Pipe TB, Smith R, Glaspy TK, Deatherage BR, Baxter LC (2016) Brain and behavior changes associated with an abbreviated 4-week mindfulness-based stress reduction course in back pain patients. Brain Behav 6(3):e00443. https://doi.org/10.1002/brb3.443

Bringmann HC, Bringmann N, Jeitler M, Brunnhuber S, Michalsen A, Sedlmeier P (2021) Meditation-based lifestyle modification: development of an integrative mind-body program for mental health and human flourishing. Complement Med Res 28(3):252–262. https://doi.org/10.1159/000512333

Brown CA, Jones AK (2013) Psychobiological correlates of improved mental health in patients with musculoskeletal pain after a mindfulness-based pain management program. Clin J Pain 29(3):233–244. https://doi.org/10.1097/AJP.0b013e31824c5d9f

Campbell JK, Penzien DB, Wall EM (2000) Evidence-based guidelines for migraine headache: behavioral and physical treatments. US Headache Consortium 1(1):1–29. Retrieved December 28, 2024, Behavioral and Physical Treatments (Nonpharmacological)

Chambliss B (2018) The mind–body problem. Wiley Interdiscip Rev Cogn Sci 9(4):e1463. https://doi.org/10.1002/wcs.1463

Chan AW, Yu DS, Choi KC, Lee DT, Sit JW, Chan HY (2016) Tai chi qigong as a means to improve night-time sleep quality among older adults with cognitive impairment: a pilot randomized controlled trial. Clin Interv Aging 11:1277–1286. https://doi.org/10.2147/CIA.S111927

Chang YK, Nien YH, Chen AG, Yan J (2014) Tai Ji Quan, the brain, and cognition in older adults. J Sport Health Sci 3(1):36–42. https://doi.org/10.1016/j.jshs.2013.09.003

Clarke TC, Barnes PM, Black LI, Stussman BJ, Nahin RL (2018) Use of yoga, meditation, and chiropractors among US adults aged 18 and over (pp. 1–8). US Department of Health and Human Services, Centers for Disease Control and Prevention, National Center for Health Statistics, Washington, DC

Colcombe SJ, Erickson KI, Raz N, Webb AG, Cohen NJ, McAuley E, Kramer AF (2003) Aerobic fitness reduces brain tissue loss in aging humans. J Gerontol Series A: Biol Sci Med Sci 58(2):M176–M180. https://doi.org/10.1093/gerona/58.2.M176

Cortelli P, Giannini G, Favoni V, Cevoli S, Pierangeli G (2013) Nociception and autonomic nervous system. Neurol Sci 34:41–46. https://doi.org/10.1007/s10072-013-1391-z

Cozzolino M, Celia G (2021) The psychosocial genomics paradigm of hypnosis and mind–body integrated psychotherapy: experimental evidence. Am J Clin Hypn 64(2):123–138. https://doi.org/10.1080/00029157.2021.1947767

Critchley HD, Wiens S, Rotshtein P, Öhman A, Dolan RJ (2004) Neural systems supporting interoceptive awareness. Nat Neurosci 7(2):189–195. https://doi.org/10.1038/nn1176

D'Ascola A, Bruschetta G, Zanghì G, Campo S, Medica P, Campana S et al (2018) Changes in plasma 5-HT levels and equine leukocyte SERT expression in response to treadmill exercise. Res Vet Sci 118:184–190. https://doi.org/10.1016/j.rvsc.2018.02.012

Dergis'inden AP (2015) Aerobic exercise in depression and anxiety disorders. Anatolian J Psychiatry 16(4):i. Retrieved December 28, 2024, from 0-Editorial.pdf

Diamond A (2013) Executive functions. Annu Rev Psychol 64(1):135–168. https://doi.org/10.1146/annurev-psych-113011-143750

Dong Y, Zhang X, Zhao R, Cao L, Kuang X, Yao J (2024) The effects of mind-body exercise on anxiety and depression in older adults: a systematic review and network meta-analysis. Front Psych 15:1305295. https://doi.org/10.3389/fpsyt.2024.1305295

Duncan G (2000) Mind-body dualism and the biopsychosocial model of pain: what did descartes really say? J Med Philos 25(4):485–513. https://doi.org/10.1076/0360-5310(200008)25:4;1-A;FT485

Engel GL (1977) The need for a new medical model: a challenge for biomedicine. Science 196(4286):129–136. https://doi.org/10.1126/science.847460

Erickson KI, Kramer AF (2009) Aerobic exercise effects on cognitive and neural plasticity in older adults. Br J Sports Med 43(1):22–24. https://doi.org/10.1136/bjsm.2008.052498

Esch T (2002) Health in stress: change in the stress concept and its significance for prevention, health and life style. Gesundheitswesen (Bundesverband der Arzte des Offentlichen Gesundheitsdienstes (Germany)) 64(2):73–81. https://doi.org/10.1055/s-2002-20275

Esch T (2022) The ABC model of happiness—neurobiological aspects of motivation and positive mood, and their dynamic changes through practice, the course of life. Biology 11(6):843. https://doi.org/10.3390/biology11060843

Esch T, Stefano GB (2004) The neurobiology of pleasure, reward processes, addiction and their health implications. Neuro Endocrinol Lett 25(4):235–251. Retrieved December 20, 2024, from The neurobiology of pleasure, reward processes, addiction and their health implications

Esch T, Stefano GB (2010) The neurobiology of stress management. Neuro Endocrinol Lett 31(1):19–39. Retrieved December 24, 2024, NEL310110A30.pdf

Esch T, Stefano GB (2022) The BERN framework of mind-body medicine: integrating self-care, health promotion, resilience, and applied neuroscience. Front Integr Neurosci 16:913573. https://doi.org/10.3389/fnint.2022.913573

Eyre HA, Siddarth P, Acevedo B, Van Dyk K, Paholpak P, Ercoli L et al (2017) A randomized controlled trial of Kundalini yoga in mild cognitive impairment. Int Psychogeriatr 29(4):557–567. https://doi.org/10.1017/S1041610216002155

Fitt SS (1988) Dance kinesiology. Schirmer Books, New York

Fong DY, Chi LK, Li F, Chang YK (2014) The benefits of endurance exercise and Tai Chi Chuan for the task-switching aspect of executive function in older adults: an ERP study. Front Aging Neurosci 6:295. https://doi.org/10.3389/fnagi.2014.00295

Fontana D (2012) Creative meditation & visualisation. Watkins Media Limited

Friedman RS, Förster J (2002) The influence of approach and avoidance motor actions on creative cognition. J Exp Soc Psychol 38(1):41–55. https://doi.org/10.1006/jesp.2001.1488

Frith E, Miller S, Loprinzi PD (2020) A review of experimental research on embodied creativity: revisiting the mind–body connection. J Creat Behav 54(4):767–798. https://doi.org/10.1002/jocb.406

Gallegos AM, Hoerger M, Talbot NL, Krasner MS, Knight JM, Moynihan JA, Duberstein PR (2013) Toward identifying the effects of the specific components of mindfulness-based stress reduction on biologic and emotional outcomes among older adults. J Alternat Complement Med 19(10):787–792. https://doi.org/10.1089/acm.2012.0028

Gao Q, Li X, Pan M, Wang J, Yang F, Guo P et al (2024) Comparative efficacy of mind–body exercise for treating chronic non-specific neck pain: a systematic review and network meta-analysis. Curr Pain Headache Rep:1–17. https://doi.org/10.1007/s11916-024-01218-6

Gard T, Hölzel BK, Sack AT, Hempel H, Lazar SW, Vaitl D, Ott U (2012) Pain attenuation through mindfulness is associated with decreased cognitive control and increased sensory processing in the brain. Cereb Cortex 22(11):2692–2702. https://doi.org/10.1093/cercor/bhr352

Garland EL, Brintz CE, Hanley AW, Roseen EJ, Atchley RM, Gaylord SA et al (2020) Mind-body therapies for opioid-treated pain: a systematic review and meta-analysis. JAMA Intern Med 180(1):91–105. https://doi.org/10.1001/jamainternmed.2019.4917

Gnall KE, Sinnott SM, Laumann LE, Park CL, David A, Emrich M (2024) Changes in interoception in mind-body therapies for chronic pain: a systematic review and meta-analysis. Int J Behav Med 31:833. https://doi.org/10.1007/s12529-023-10249-z

Gothe NP, Kramer AF, McAuley E (2014) The effects of an 8-week Hatha yoga intervention on executive function in older adults. J Gerontol Series A: Biomed Sci Med Sci 69(9):1109–1116. https://doi.org/10.1093/gerona/glu095

Grant JA, Courtemanche J, Duerden EG, Duncan GH, Rainville P (2010) Cortical thickness and pain sensitivity in zen meditators. Emotion 10(1):43. https://doi.org/10.1037/a0018334
Grant JA, Courtemanche J, Rainville P (2011) A non-elaborative mental stance and decoupling of executive and pain-related cortices predicts low pain sensitivity in zen meditators. Pain 152(1):150–156. https://doi.org/10.1016/j.pain.2010.10.006
Grillner S (1985) Neurobiological bases of rhythmic motor acts in vertebrates. Science 228(4696):143–149. https://doi.org/10.1126/science.3975635
Hatchard T, Mioduszewski O, Khoo EL, Romanow H, Shergill Y, Tennant E et al (2021) Reduced emotional reactivity in breast cancer survivors with chronic neuropathic pain following mindfulness-based stress reduction (MBSR): an fMRI pilot investigation. Mindfulness 12:751–762. https://doi.org/10.1007/s12671-020-01546-9
Hung SH, Hwang CY, Chang CY (2021) Is the Qi experience related to the flow experience? Practicing qigong in urban green spaces. PLoS One 16(1):e0240180. https://doi.org/10.1371/journal.pone.0240180
Inzunza K (2023) Kundalini yoga. Retrieved December 22, 2024, from Kristi-Inzunza-Final-Paper-0823-Globe-Inst-Kirtan-Kriya.pdf
Irwin MR, Olmstead R, Carrillo C, Sadeghi N, Breen EC, Witarama T et al (2014) Cognitive behavioral therapy vs. Tai Chi for late life insomnia and inflammatory risk: a randomized controlled comparative efficacy trial. Sleep 37(9):1543–1552. https://doi.org/10.5665/sleep.4008
Jahnke R, Larkey L, Rogers C, Etnier J, Lin F (2010) A comprehensive review of health benefits of qigong and Tai chi. Am J Health Promot 24(6):e1–e25. https://doi.org/10.4278/ajhp.081013-LIT-248
Jones C, Qi M, Xie Z, Moyle W, Weeks B, Li P (2022) Baduanjin exercise for adults aged 65 years and older: a systematic review and meta-analysis of randomized controlled studies. J Appl Gerontol 41(4):1244–1256. https://doi.org/10.1177/07334648211059324
Kabat-Zinn J (2005) Coming to our senses: healing ourselves and the world through mindfulness. Hachette. Retrieved December 26, 2024, Coming to our senses: healing ourselves and the world through mindfulness – Jon Kabat-Zinn – Google Kitaplar
Kim J (2015) Physical activity benefits creativity: squeezing a ball for enhancing creativity. Creat Res J 27(4):328–333. https://doi.org/10.1080/10400419.2015.1087258
Kim SH, Schneider SM, Kravitz L, Mermier C, Burge MR (2013) Mind-body practices for posttraumatic stress disorder. J Investig Med 61(5):827–834. https://doi.org/10.2310/JIM.0b013e3182906862
Kong J, Wolcott E, Wang Z, Jorgenson K, Harvey WF, Tao J et al (2019) Altered resting state functional connectivity of the cognitive control network in fibromyalgia and the modulation effect of mind-body intervention. Brain Imaging Behav 13:482–492. https://doi.org/10.1007/s11682-018-9875-3
Kuo CY, Yeh YY (2016) Sensorimotor-conceptual integration in free walking enhances divergent thinking for young and older adults. Front Psychol 7:1580. https://doi.org/10.3389/fpsyg.2016.01580
Lavretsky H, Epel ES, Siddarth P, Nazarian N, Cyr NS, Khalsa DS et al (2013) A pilot study of yogic meditation for family dementia caregivers with depressive symptoms: effects on mental health, cognition, and telomerase activity. Int J Geriatr Psychiatry 28(1):57–65. https://doi.org/10.1002/gps.3790
Lee M, Moon W, Kim J (2014) Effect of yoga on pain, brain-derived neurotrophic factor, and serotonin in premenopausal women with chronic low back pain. Evid Based Complement Alternat Med 2014(1):203173. https://doi.org/10.1155/2014/203173
Li J, Liu Y, Liu B, Li F, Hu J, Wang Q et al (2019) Mechanisms of aerobic exercise upregulating the expression of hippocampal synaptic plasticity-associated proteins in diabetic rats. Neural Plast 2019(1):7920540. https://doi.org/10.1155/2019/7920540
Lim KHL, Pysklywec A, Plante M, Demers L (2019) The effectiveness of Tai Chi for short-term cognitive function improvement in the early stages of dementia in the elderly: a systematic literature review. Clin Interv Aging:827–839. https://doi.org/10.2147/CIA.S202055

Lin H, Wan M, Ye Y, Zheng G (2023) Effects of Baduanjin exercise on the physical function of middle-aged and elderly people: a systematic review and meta-analysis of randomized controlled trials. BMC Complementary Med Ther 23(1):38. https://doi.org/10.1186/s12906-023-03866-4
Littrell J (2008) The mind-body connection: not just a theory anymore. Soc Work Health Care 46(4):17–37. https://doi.org/10.1300/J010v46n04_02
Liu J, Kowal I, Yang Y, Zhu Y, Chen S, Perez A et al (2023) Culturally tailored group qigong exercise in older Chinese immigrants: a feasibility study. Geriatr Nurs 51:245–252. https://doi.org/10.1016/j.gerinurse.2023.03.021
Liu J, Yang Y, Li C, Perez A, Raine A, Shi H, Zou L (2024) Effects of mind-body qigong exercise on overall health, fatigue/sleep, and cognition in older Chinese immigrants in the US: an intervention study with control. J Aging Res 2024(1):2481518. https://doi.org/10.1155/2024/2481518
Low S, Ang LW, Goh KS, Chew SK (2009) A systematic review of the effectiveness of Tai Chi on fall reduction among the elderly. Arch Gerontol Geriatr 48(3):325–331. https://doi.org/10.1016/j.archger.2008.02.018
Lü J, Huang L, Wu X, Fu W, Liu Y (2017) Effect of Tai Ji Quan training on self-reported sleep quality in elderly Chinese women with knee osteoarthritis: a randomized controlled trail. Sleep Med 33:70–75. https://doi.org/10.1016/j.sleep.2016.12.024
Lutz A, Brefczynski-Lewis J, Johnstone T, Davidson RJ (2008) Regulation of the neural circuitry of emotion by compassion meditation: effects of meditative expertise. PLoS One 3(3):e1897. https://doi.org/10.1371/journal.pone.0001897
Lutz A, McFarlin DR, Perlman DM, Salomons TV, Davidson RJ (2013) Altered anterior insula activation during anticipation and experience of painful stimuli in expert meditators. NeuroImage 64:538–546. https://doi.org/10.1016/j.neuroimage.2012.09.030
Morgan N, Irwin MR, Chung M, Wang C (2014) The effects of mind-body therapies on the immune system: meta-analysis. PLoS One 9(7):e100903. https://doi.org/10.1371/journal.pone.0100903
Muehsam D, Lutgendorf S, Mills PJ, Rickhi B, Chevalier G, Bat N et al (2017) The embodied mind: a review on functional genomic and neurological correlates of mind-body therapies. Neurosci Biobehav Rev 73:165–181. https://doi.org/10.1016/j.neubiorev.2016.12.027
Mustafaoglu R, Ahmed I, Pang MY (2022) Which type of mind–body exercise is most effective in improving functional performance and quality of life in patients with Parkinson's disease? A systematic review with network meta-analysis. Acta Neurol Belg 122(6):1433–1446. https://doi.org/10.1007/s13760-022-02070-4
Nash JD, Newberg AB (2023) An updated classification of meditation methods using principles of taxonomy and systematics. Front Psychol 13:1062535. https://doi.org/10.3389/fpsyg.2022.1062535
Nayak NN, Shankar K (2004) Yoga: a therapeutic approach. Phys Med Rehabil Clin N Am 15(4):783–798
NEMO Study Group (2007) Effect of a 12-mo micronutrient intervention on learning and memory in well-nourished and marginally nourished school-aged children: 2 parallel, randomized, placebo-controlled studies in Australia and Indonesia. Am J Clin Nutr 86(4):1082–1093
Oppezzo M, Schwartz DL (2014) Give your ideas some legs: the positive effect of walking on creative thinking. J Exp Psychol Learn Mem Cogn 40(4):1142. https://doi.org/10.1037/a0036577
Park J, Krause-Parello CA, Barnes CM (2020) A narrative review of movement-based mind-body interventions: effects of yoga, tai chi, and qigong for back pain patients. Holist Nurs Pract 34(1):3–23. https://doi.org/10.1097/HNP.0000000000000360
Patel NK, Newstead AH, Ferrer RL (2012) The effects of yoga on physical functioning and health related quality of life in older adults: a systematic review and meta-analysis. J Altern Complement Med 18(10):902–917. https://doi.org/10.1089/acm.2011.0473
Ren FF, Chen FT, Zhou WS, Cho YM, Ho TJ, Hung TM, Chang YK (2021) Effects of Chinese mind-body exercises on executive function in middle-aged and older adults: a systematic review and meta-analysis. Front Psychol 12:656141. https://doi.org/10.3389/fpsyg.2021.656141
Reyes del Paso GA, Langewitz W, Mulder LJ, Van Roon A, Duschek S (2013) The utility of low frequency heart rate variability as an index of sympathetic cardiac tone: a review with emphasis on a reanalysis of previous studies. Psychophysiology 50(5):477–487. https://doi.org/10.1111/psyp.12027

Rosmond RA, Björntorp P (2000) The hypothalamic–pituitary–adrenal axis activity as a predictor of cardiovascular disease, type 2 diabetes and stroke. J Intern Med 247(2):188–197. https://doi.org/10.1046/j.1365-2796.2000.00603.x

Rossi EL (2002) Psychosocial genomics: gene expression, neurogenesis, and human experience in mind-body medicine. Adv Mind-Body Med 18(2):22–30

Sandua D (2023) Mind-body connection: the impact of nutrition on mental health. https://books.google.com.tr/books?id=V7LdEAAAQBAJ

Schmais C (1985) Healing processes in group dance therapy. Am J Dance Ther 8:17–36. https://doi.org/10.1007/BF02251439

Sedlmeier P, Eberth J, Schwarz M, Zimmermann D, Haarig F, Jaeger S, Kunze S (2012) The psychological effects of meditation: a meta-analysis. Psychol Bull 138(6):1139. https://psycnet.apa.org/doi/10.1037/a0028168

Seminowicz DA, Burrowes SA, Kearson A, Zhang J, Krimmel SR, Samawi L et al (2020) Enhanced mindfulness-based stress reduction in episodic migraine: a randomized clinical trial with magnetic resonance imaging outcomes. Pain 161(8):1837–1846. https://doi.org/10.1097/j.pain.0000000000001860

Sephton S, Spiegel D (2003) Circadian disruption in cancer: a neuroendocrine-immune pathway from stress to disease? Brain Behav Immun 17(5):321–328. https://doi.org/10.1016/S0889-1591(03)00078-3

Shen CL, Watkins BA, Kahathuduwa C, Chyu MC, Zabet-Moghaddam M, Elmassry MM et al (2022) Tai chi improves brain functional connectivity and plasma lysophosphatidylcholines in postmenopausal women with knee osteoarthritis: an exploratory pilot study. Front Med 8:775344. https://doi.org/10.3389/fmed.2021.775344

Smith JA, Greer T, Sheets T, Watson S (2011) Is there more to yoga than exercise? Altern Ther Health Med 17(3):22

St. Marie R, Talebkhah KS (2018) Neurological evidence of a mind-body connection: mindfulness and pain control. Am J Psychiatry Resid J 13(4):2–5. https://doi.org/10.1176/appi.ajp-rj.2018.130401

Stahl JE, Dossett ML, LaJoie AS, Denninger JW, Mehta DH, Goldman R et al (2015) Relaxation response and resiliency training and its effect on healthcare resource utilization. PLoS One 10(10):e0140212. https://doi.org/10.1371/journal.pone.0140212

Stanciu D (2023) Consciousness, 4E cognition and aristotle: a few conceptual and historical aspects. Front Comput Neurosci 17:1204602. https://doi.org/10.3389/fncom.2023.1204602

Stefano GB, Esch T, Kream RM (2019a) Augmentation of whole-body metabolic status by mind-body training: synchronous integration of tissue-and organ-specific mitochondrial function. Med Sci Monit Basic Res 25:8. https://doi.org/10.12659/MSMBR.913264

Stefano GB, Esch T, Kream RM (2019b) Behaviorally-mediated entrainment of whole-body metabolic processes: conservation and evolutionary development of mitochondrial respiratory complexes. Med Sci Monit: Int Med J Exp Clin Res 25:9306. https://doi.org/10.12659/MSM.920174

Switankowsky I (2000) Dualism and its importance for medicine. Theor Med Bioeth 21:567–580. https://doi.org/10.1023/A:1026570907667

Taylor AG, Goehler LE, Galper DI, Innes KE, Bourguignon C (2010) Top-down and bottom-up mechanisms in mind-body medicine: development of an integrative framework for psychophysiological research. Explorer 6(1):29–41. https://doi.org/10.1016/j.explore.2009.10.004

Taylor J, McLean L, Korner A, Stratton E, Glozier N (2020) Mindfulness and yoga for psychological trauma: systematic review and meta-analysis. J Trauma Dissociation 21(5):536–573. https://doi.org/10.1080/15299732.2020.1760167

Tesarz J (2024) Körper, Psyche und Kultur. Der Schmerz:1–7. https://doi.org/10.1007/s00482-023-00787-6

Travis F, Shear J (2010) Focused attention, open monitoring and automatic self-transcending: categories to organize meditations from Vedic, Buddhist and Chinese traditions. Conscious Cogn 19(4):1110–1118. https://doi.org/10.1016/j.concog.2010.01.007

Tripathi RL (2024) The mind: from Cartesian dualism to computational functionalism. Philos Int J 7(3):8. Retrieved Semptember 27, 2024, from R. L. Tripathi, The mind: from Cartesian dualism to computational functionalism – PhilPapers

Tyagi A, Cohen M (2016) Yoga and heart rate variability: a comprehensive review of the literature. Int J Yoga 9(2):97–113. https://doi.org/10.4103/0973-6131.183712

Van Middelkoop M, Rubinstein SM, Kuijpers T, Verhagen AP, Ostelo R, Koes BW, van Tulder MW (2011) A systematic review on the effectiveness of physical and rehabilitation interventions for chronic non-specific low back pain. Eur Spine J 20:19–39. https://doi.org/10.1007/s00586-010-1518-3

Varambally S, Gangadhar BN (2012) Yoga: a spiritual practice with therapeutic value in psychiatry. Asian J Psychiatr 5(2):186–189. https://doi.org/10.1016/j.ajp.2012.05.003

Ventura MI, Barnes DE, Ross JM, Lanni KE, Sigvardt KA, Disbrow EA (2016) A pilot study to evaluate multi-dimensional effects of dance for people with Parkinson's disease. Contemp Clin Trials 51:50–55. https://doi.org/10.1016/j.cct.2016.10.001

Villines Z, Bubnis D, NASM-CPT NA (2017) What is the best type of meditation. Med News Today. Accessed on December, 22, 2017. Retrieved December 20, 2024, from Types of Meditation.pdf

Voss S, Boachie DA, Nieves N, Gothe NP (2023) Mind-body practices, interoception and pain: a scoping review of behavioral and neural correlates. Ann Med 55(2):2275661. https://doi.org/10.1080/07853890.2023.2275661

Wang H, Wei A, Lu Y, Yu B, Chen W, Lu Y et al (2016) Simplified tai chi program training versus traditional tai chi on the functional movement screening in older adults. Evid Based Complement Alternat Med 2016(1):5867810. https://doi.org/10.1155/2016/5867810

Wang X, Lu K, Runco MA, Hao N (2018) Break the "wall" and become creative: enacting embodied metaphors in virtual reality. Conscious Cogn 62:102–109. https://doi.org/10.1016/j.concog.2018.03.004

Wang LL, Sun CJ, Wang Y, Zhan TT, Yuan J, Niu CY et al (2022) Effects of dance therapy on non-motor symptoms in patients with Parkinson's disease: a systematic review and meta-analysis. Aging Clin Exp Res 34(6):1201–1208. https://doi.org/10.1007/s40520-021-02030-7

Wayne PM, Walsh JN, Taylor-Piliae RE, Wells RE, Papp KV, Donovan NJ, Yeh GY (2014) Effect of Tai Chi on cognitive performance in older adults: systematic review and meta-analysis. J Am Geriatr Soc 62(1):25–39. https://doi.org/10.1111/jgs.12611

Wegner M, Helmich I, Machado S, Nardi AE, Arias-Carrion O, Budde H (2014) Effects of exercise on anxiety and depression disorders: review of meta-analyses and neurobiological mechanisms. CNS Neurol Disord Drug Targets (Formerly Current Drug Targets-CNS & Neurological Disorders) 13(6):1002–1014. Retrieved December 20, 2024, from Effects of Exercise on Anxiety and Depression Disorders: Review o…: Ingenta Connect

Weifen W, Muheremu A, Chaohui C, MD LW, Lei S (2013) Effectiveness of tai chi practice for non-specific chronic low back pain on retired athletes: a randomized controlled study. J Musculoskelet Pain 21(1):37–45. https://doi.org/10.3109/10582452.2013.763394

Wu WW, Kwong E, Lan XY, Jiang XY (2015) The effect of a meditative movement intervention on quality of sleep in the elderly: a systematic review and meta-analysis. J Altern Complement Med 21(9):509–519. https://doi.org/10.1089/acm.2014.0251

Wu Q, Liu P, Liao C, Tan L (2022) Effectiveness of yoga therapy for migraine: a meta-analysis of randomized controlled studies. J Clin Neurosci 99:147–151. https://doi.org/10.1016/j.jocn.2022.01.018

Yang YJ, Deng Y, Liao LL, Peng J, Peng QH, Qin YH (2020) Tripterygium glycosides combined with leflunomide for rheumatoid arthritis: a systematic review and meta-analysis. Evid Based Complement Altern Med 2020(1):1230320. https://doi.org/10.1155/2020/1230320

Zeidan F, Martucci KT, Kraft RA, Gordon NS, McHaffie JG, Coghill RC (2011) Brain mechanisms supporting the modulation of pain by mindfulness meditation. J Neurosci 31(14):5540–5548. https://doi.org/10.1523/JNEUROSCI.5791-10.2011

Zhang Y, Li C, Zou L, Liu X, Song W (2018) The effects of mind-body exercise on cognitive performance in elderly: a systematic review and meta-analysis. Int J Environ Res Public Health 15(12):2791. https://doi.org/10.3390/ijerph15122791

Zhang Y, Loprinzi PD, Yang L, Liu J, Liu S, Zou L (2019a) The beneficial effects of traditional Chinese exercises for adults with low back pain: a meta-analysis of randomized controlled trials. Medicina 55(5):118. https://doi.org/10.3390/medicina55050118

Zhang Q, Hu J, Wei L, Jia Y, Jin Y (2019b) Effects of dance therapy on cognitive and mood symptoms in people with Parkinson's disease: a systematic review and meta-analysis. Complement Ther Clin Pract 36:12–17. https://doi.org/10.1016/j.ctcp.2019.04.005

Zheng G, Li S, Huang M, Liu F, Tao J, Chen L (2015) The effect of Tai Chi training on cardiorespiratory fitness in healthy adults: a systematic review and meta-analysis. PLoS One 10(2):e0117360. https://doi.org/10.1371/journal.pone.0117360

Zou L, Zhang Y, Yang L, Loprinzi PD, Yeung AS, Kong J et al (2019) Are mindful exercises safe and beneficial for treating chronic lower back pain? A systematic review and meta-analysis of randomized controlled trials. J Clin Med 8(5):628. https://doi.org/10.3390/jcm8050628

Understanding Nutritional Neuroscience

2

Muhammad Danial Bin Che Ramli, Zakirah Zainal Abidin, Amanina Nurjannah Binti Atan Hamdan, and Genevieve Simboh Anak Umbar

Abstract

Nutritional neuroscience is a newly developed multidisciplinary field of research exploring the intricate relationships between dietary intake, brain function, and mental health. The chapter below outlines how macronutrients (carbohydrates, proteins, and lipids) and micronutrients (vitamins, minerals, and antioxidants) regulate neurotransmission, neuroplasticity, and cognitive resilience. The gut-brain axis plays a critical role in the modulation of neuroinflammation and mood regulation, where diet affects the microbiota composition and, consequently, mental health outcomes. Dietary patterns such as the Mediterranean, Mediterranean–dietary approaches to stop hypertension intervention for neurodegenerative delay, and ketogenic diets have been linked with enhanced cognitive function and reduced risk of neurodegenerative disorder. In addition, the field of nutrigenomics recognizes the gene-derived foundation of food response in justification of personalized nutrition for mental well-being. The impact of processed food and excessive sugar intake on cognitive impairment, mood disorder, and neuroinflammation is also considered.

New evidence sheds light on the therapeutic role of diet in prevention of neurological disorders and emphasizes the use of functional foods, fasting diets, and personalized supplementation. The chapter endeavors to bridge the knowledge

M. D. B. Ramli (✉) · G. S. A. Umbar
Department of Diagnostic and Allied Health Science, Faculty of Health and Life Sciences, Management and Science University, University Drive, Shah Alam, Selangor, Malaysia
e-mail: muhddanial_cheramli@msu.edu.my

Z. Z. Abidin · A. N. B. A. Hamdan
School of Graduate Studies, Postgraduate Centre, Management & Science University, Shah Alam, Selangor, Malaysia

W. Mohamed, A. Özge (eds.), *Feeding the Mind: The Connection Between Diet, Drugs, and Mental Health Volume 1*, Nutritional Neurosciences,
https://doi.org/10.1007/978-981-95-1414-4_2

gap between nutrition and mental health by encouraging precision-based dietary care to support brain health, emotional well-being, and neuroprotection.

Keywords

Nutritional neuroscience · Gut-brain axis · Neuroinflammation · Dietary · Patterns · Mental health

2.1 Introduction to Nutritional Neuroscience

Nutritional neuroscience is an interdisciplinary field that studies the complex interactions among nutrition, brain health, and mental health. Over the past 10 years, this field has garnered a significant amount of interest, as our understanding of how nutrients impact the nervous system continues to grow. Modern lifestyles, which are often characterized by poor-quality diet and increased prevalence of mental illness, highlight the necessity for food to be considered a modifiable factor in the pursuit of cognitive and emotional well-being (Jacka et al. 2020). The human brain is a metabolically expensive organ that utilizes about 20% of the body's energy but only comprises roughly 2% of body weight. The nutritional demands of the brain further emphasize the importance of nutrition for optimal brain health (Gomez-Pinilla 2021). While glucose is the brain's primary energy substrate, its demands extend beyond that. Proteins and lipids, which are macronutrients, and a collection of micronutrients are necessary for the synthesis of neurotransmitters and myelination, synaptic plasticity, and neuroprotection.

2.1.1 Macronutrients and Brain Function

2.1.1.1 Carbohydrates

Carbohydrates, specifically complex carbohydrates, are essential to ensure stable glucose levels and thereby enable sustained cognitive functioning. Diets consisting of high simple sugars, on the other hand, are associated with a sharp rise and fall in glucose levels, hampering cognitive function and putting individuals at risk of mood disorders (Weston-Green et al. 2022).

2.1.1.2 Proteins

Proteins provide amino acids required for the synthesis of neurotransmitters such as serotonin, dopamine, and norepinephrine. For instance, tryptophan is a precursor to serotonin, a mood control neurotransmitter, while tyrosine is a precursor to dopamine, which regulates motivation and reward pathways (Fernstrom and Wurtman 2020).

2.1.1.3 Fats

Dietary lipids, particularly omega-3 fatty acids such as docosahexaenoic acid (DHA), are part of neuronal membrane fluidity and signaling transduction. Omega-3 deficiency has been associated with cognitive dysfunction, depression, and increased risk of neurodegenerative diseases (Morris et al. 2022).

2.1.1.4 Vitamins

Vitamins B6, B12, and folate play a critical role in homocysteine metabolism. Elevated homocysteine levels have been implicated with cognitive deficiency and neurodegenerative disease, emphasizing the need for adequate vitamin intake (Smith and Refsum 2021). Antioxidant vitamins C and E also protect neural tissue from oxidative damage, a pathogenic factor in aging and the majority of neurodegenerative diseases.

2.1.1.5 Minerals

Minerals like magnesium, zinc, and iron are essential for neurotransmitter synthesis and enzymatic processes. Iron deficiency, for example, disrupts dopamine metabolism, resulting in compromised cognitive function and attentional impairment (Beard et al. 2021).

2.1.1.6 The Gut-Brain Axis

The gut-brain axis refers to the two-way communication between the central nervous system (CNS) and the gastrointestinal tract. The gut microbiota influence this interaction by affecting neurotransmitter synthesis, immune modulation, and the production of neuroactive molecules such as short-chain fatty acids (SCFAs). The gut microbiome is significantly modulated by diet, and axis dysregulation has been involved in mood disorders and neurodegenerative diseases (Cryan et al. 2020).

2.1.2 Dietary Patterns and Mental Health

2.1.2.1 Western Diet

The Western dietary pattern, characterized by high consumption of processed food, refined sugar, and saturated fat, has been associated with high inflammation, oxidative stress, and a high risk of depression and anxiety (Jacka et al. 2020).

2.1.2.2 Mediterranean Diet

On the other hand, the Mediterranean diet, rich in fruits, vegetables, whole grains, lean protein, and healthy fats, is associated with reduced inflammation and improved cognitive function. Numerous studies suggest its protective effect against depression and cognitive decline (Psaltopoulou et al. 2022).

2.1.2.3 Plant-Based Diets

Plant-based diets, which emphasize nutrient-dense foods such as legumes, nuts, and vegetables, provide high levels of antioxidants and fiber, which equate to better mental health outcomes (Kahleova et al. 2021).

2.1.3 Emerging Evidence in Neurological Disorders

2.1.3.1 Ketogenic Diet

Originally designed for the management of epilepsy, the ketogenic diet (KD) has been used in neurodegenerative diseases such as Alzheimer's and Parkinson's. By inducing ketosis, the diet provides an alternative source of energy to glucose, exerting neuroprotective actions (Taylor et al. 2021).

2.1.3.2 Anti-inflammatory Diets

Anti-inflammatory dietary approaches, such as the Dietary Approaches to Stop Hypertension (DASH) diet, are becoming ever more noted for their role in reducing neuroinflammation, a common feature in many psychiatric and neurological conditions (Sarris et al. 2022).

The burgeoning field of nutritional neuroscience underscores the enormous impact of diet on brain health. Nutritional interventions not only support cognitive function and emotional well-being but also offer encouraging therapeutic avenues for neurological illness. As the science advances, tailored nutrition according to one's neurochemical profile may revolutionize dietary recommendations, placing nutrition at the cornerstone of mental and neurological health (Table 2.1).

2.2 The Gut-Brain Axis

2.2.1 How Nutrition Shapes Cognitive Function

The role of nutrition in cognitive functioning is important throughout the lifespan. The brain required a regular supply of nutrients to sustain its architecture and facilitate vital functions such as neurotransmission, synaptic plasticity, and neurogenesis. Nutrient deficiencies or imbalances that impede these processes can lead to cognitive decline and higher risk for neurodegenerative disorders. Datasets equivalent to the brain, such as nutrient datastores, give way to alternative as well food for an example. It has shown to impact brain functionality in studies from years past. Macronutrients include carbohydrates, fats, and proteins, providing the nutrients needed for optimal performance. Poor dietary consumption can contribute to both physical and cognitive functionally declines (Muth and Park 2021).

Carbohydrates are the brain's main fuel source, and they have a strikingly broad, and complex, effect on cognitive performance. Simple carbohydrates like sugars in processed foods and beverages are linked to cognitive declines. These rapidly absorbable sugars lead to rapid elevations of blood glucose, then drops that can lead

Table 2.1 Overview of nutrients and their neurological roles

Nutrient category	Key nutrient examples	Neurological mechanisms	Deficiency-linked conditions	Key references
Carbohydrates	Whole grains, legumes, oats, quinoa, fruits	Essential source of glucose; stabilize blood sugar; support hippocampal memory function	Cognitive fatigue, reduced attention span, mood instability	Muth and Park (2021), Jansen et al. (2020)
Proteins	Meat, fish, dairy, eggs, soy, legumes	Source of amino acids for neurotransmitters (serotonin, dopamine); supports synaptogenesis	Low mood, anxiety, impaired learning, and memory retention	Li et al. (2020), Gao et al. (2022)
Fats	DHA/ eicosapentaenoic acid (fish oil, flaxseed), olive oil, avocados	Contributes to membrane fluidity; supports neuroplasticity; anti-inflammatory	Memory loss, neurodegeneration risk	Morris et al. (2022), Gómez-Pinilla (2008)
Vitamins	B6, B12, folate, D, E, K, C	Cofactors in neurotransmitter synthesis; protects against antioxidant	Depression, cognitive decline, neuroinflammation	Smith and Refsum (2021), Suh et al. (2020)
Minerals	Iron, zinc, magnesium, copper, selenium	Contribute to neurotransmission, myelination, oxygen transport	ADHD, depression, attention deficits, developmental delays	Beard et al. (2021), Black (2003a)
Antioxidants	Vitamin C, E, flavonoids, polyphenols	Scavenge free radicals, preserve neuronal membranes	Neurodegeneration, accelerated aging, neuroinflammation	Pizzino et al. (2017), Franzoni et al. (2021)

to decreased attention and decreased overall brain power (Muth and Park 2021). Chronic intake of refined sugars has been linked to neurocognitive deficits in memory-related regions of the brain, such as hippocampus (Hawkins et al. 2018). On the other hand, complex carbs found in whole grains, legumes, fruit, and veggies provide a more stable source of energy. In turn, it has been reported that gradual increment in blood glucose levels might improve cognitive performance by enhancing attention and memory retention (Jansen et al. 2020). Additionally, high intake of complex carbohydrates has been associated with better cognitive profile and healthy brain ageing (Muth and Park 2021).

Proteins are essential for the synthesis of neurotransmitter such as serotonin and dopamine, both vital to neuronal communication, which in turn as an important role in mood and cognitive functions (Li et al. 2020). Higher protein consumption correlates positively with cognitive performance in older adults. It has been shown that dietary proteins from animal products, eggs, and legumes correlate with better transmission on cognitive tests to measure memory and fluency (Li et al. 2020). For instance, one study found that higher intakes of animal protein were associated with better cognitive assessment performances compared to lower intakes (Gao et al. 2022). While animal protein has been linked to better cognitive outcomes, higher intakes of plant proteins have shown an inverse relationship with cognitive performance, whereby increasing consumption of plant proteins shows a trend toward greater cognitive decline over time (Gao et al. 2022).

Studies in animals have found some evidence that the effects of stroke can result in a significant reduction in the ability to synthesize proteins in the brain (Adams et al. 2024). Research suggests that excessive consumption of saturated fats and trans fats, which are high in processed foods, appears to impact cognition negatively, diminishing memory and learning potential and reducing telomere size. In contrast, species of fats such as omega-3 polyunsaturated and monounsaturated are correlated with improved cognitive performance and less telomere shortening (Mostafa et al. 2023; Muth and Park 2021). High-diet saturated fat presents a serious threat to brain function, triggering cardiovascular and neurological disorders (Chianese et al. 2017).

However, a promising diet high in omega-3 can have a far-reaching positive influence, augmenting synaptic functions. These essential fatty acids are key to neuronal health and have been shown to enhance the expression of genes fundamental to synaptic plasticity (Gómez-Pinilla 2008). Studies in mice showed impairment of the learning and memory of those fed high-fat diets for several weeks. Such harmful effects are attributed to some other brain changes that happen during the particular time in general with close reference to the hippocampus, including decreased neurogenesis, heightened inflammation, altered blood-brain permeability, impaired neural plasticity, and reduced hippocampal insulin sensitivity (Gibson et al. 2013).

In addition to macronutrients, micronutrients play a crucial role in cognitive development. Vitamins and minerals exert their influences on cognitive functions throughout life, but their effects gain particular prominence during critical periods of brain development, such as early childhood, and in people above a certain age whose cognitive functions are at risk of further decline (Nieto 2024). The association between micronutrient deficiencies and early cognitive development draws considerable research attention due to specific physiological pathways linked to these micronutrients (Black 2003a). Vitamin A stands as an important factor for brain development and function. It is converted into retinoic acid, which then governs the expression of genes in synaptic plasticity, a fundamental process for learning and memory learning development (Wołoszynowska-Fraser et al. 2020). Vitamin D is a neuroprotective hormone thought to promote the growth of hippocampal volume and strengthens hippocampal-thalamus-prefrontal connectivity, therefore enhancing memory and brain function (Yang et al. 2020). Vitamin E acts particularly as an antioxidant protecting neuronal cells from oxidative stress. This is vital as oxidative damage is responsible for cognitive decline and neurodegenerative diseases (Suh et al. 2020). Vitamin E sufficiency additionally links to improved performance in the context of cognition, with evidence to support this found particularly among aging populations (Suh et al. 2020).

Vitamin K plays a role in the synthesis of sphingolipids, vital for myelin formation and optimal health of neurons. It offers protection to the neurons against apoptosis (programmed cell death) and might support cognitive performance by augmenting brain metabolism and mitigating inflammation (Suh et al. 2020). Iron is a vital element among minerals for transporting oxygen and the energy metabolism in the brain. Iron deficiency negatively affects cognitive functions, such as attention, reasoning, and memory during rapid periods of growth in children (Black 2003a). Zinc is another trace element that supports brain structure and function (Nieto 2024). The gut-brain axis, an intricate and complex bidirectional communication network, is essential for both physical and mental health. This system consists of the CNS, autonomic nervous system, and enteric nervous system that synthesize the connection between gut microbes and the way our brains function (Bhoomi 2024).

The gut microbiome, containing trillions of microorganisms, is extremely important for delivery digestive health. Recent studies have found these microbes may influence brain function through a number of mechanisms including the production of neurotransmitters, with gut bacteria allocated in making substances such as serotonin, which is essential for moods. Notably, about 90% of serotonin in the whole body is manufactured in the gut. Thus, it has a major link to emotional health (Appleton 2018). Other mechanisms include the membrane interaction and the interaction of the microbiome with the immune system, affecting how we set the inflammatory responses that will dictate the brain health.

Dysbiosis or imbalance in gut bacteria has been reported to heighten inflammation, which is linked with numerous neurodegenerative disorders (Suganya and Koo 2020). Besides, gut bacteria produce metabolite which bypass the blood-brain barrier and manipulate brain function (Loh et al. 2024). Recent studies have suggested a link between gut health and cognitive abilities such as memory and processing speed. About 1% article showed a moderate correlation between gut health indicators and cognitive performance metrics in adult subjects (Hameed et al. 2024).

2.3 Key Nutrients for Mental Well-Being

The brain cannot perform its function without sufficient intake of polyunsaturated fatty acids (PUFAs). These compounds are essential for normal membrane structure and functioning and are involved in brain development and neurotransmission processes. In other words, PUFAs in the human body are classified into two main classes: omega-6 and omega-3. The latter, especially eicosapentaenoic acid and DHA, have been largely linked with maintenance of mental health, and their deficiency has been associated with the development of mental disorders (Lange 2020). Amino acids account for around 50–60% of the weight of the brain, out of which omega-3 PUFAs represent 35%. Also, within neuronal tissue, DHA exceeds 40% of total omega-3 PUFAs, particularly in the grey matter (Dighriri et al. 2022).

The omega-3-family fatty acids take part in maintain membrane potentials that allow neuronal depolarization and repolarization to facilitate signal transmission throughout the cortex. Its deficiency possibly interferes with neurotransmitter synthesis and neuron firing, with consequent mental health issues, such as depression (Lawson and Watson 2014). The health benefits of omega-3 fatty acids originate in their anti-inflammatory, hypolipidemic, antithrombotic, and anti-arrhythmic characteristics. These PUFAs are critical to brain development from pregnancy and through infancy and childhood (Lange 2020). They enhance cognition and neuron preservation and protect against neurodegeneration (Dighriri et al. 2022). Those suffering from various depressive disorders generally present with low levels of omega-3 fatty acids. That is further supported by the generally acknowledged finding that deficits of omega-3 PUFAs lead to learning and memory impairments in rodents (Moriguchi et al. 2000).

B vitamins include a class of eight water-soluble vitamins that serve essential and highly interrelated roles in cellular function, acting as coenzymes in a great variety of catabolic and anabolic enzymatic reactions (Kennedy 2016). B vitamins, particularly B6, B9 (folate), and B12, are the essential nutrients for neurotransmitter biosynthesis, the chemical messengers that mediate signals within the brain. It is suggested that sufficient levels of these vitamins correlate with increased cognitive function and cognitive decline risk (Sawalha and Sarhan 2022). Vitamin B6 mediates amino acids metabolism, making it rate-limiting cofactor for synthesizing neurotransmitters (including, but not limited to, dopamine, serotonin, gama-aminobutyric acid [GABA], noradrenaline, and the hormone melatonin). The different synthesis of these neurotransmitters can be inhibited by vitamin B6. Even moderate deficiency alters activity such that GABA and serotonin reduce. This may allow GABA not to inhibit neural activity, leading to disordered sleep, behavior, and cardiovascular performance as well loss of hypothalamus-pituitary control over hormone excretion (Kennedy 2016). Sufficient levels of vitamin B6 are pivotal in the preservation of cognitive health (Nieto 2024).

Vitamin B12 is important in many cellular processes, including gene regulation, signaling, and replication. It aids in the breakdown of fatty acids, glucose, and amino acids (Hanna et al. 2022). This vitamin is also involved in neurotransmitter production and has been found to delay dementia symptoms; supplementation shows promise in enhancing cognitive function and brain health (Nieto 2024). Vitamin B12 is known to contribute to the mechanisms of depression through the involvement of several neurotransmitter systems: adrenergic, glutaminergic, serotonin, and dopamine (Ekinci and Sanlier 2022). A deficiency of B12 will also lead to a functional deficiency of folate as folate will be abandoned in its methylated form (Kennedy 2016). The proper functioning of the folate cycle is critical in producing and regenerating tetrahydrobiopterin, the key cofactor for the enzymes transforming amino acids into monoamine neurotransmitters and nitric oxide (Kennedy 2016). Folate is also important for nucleic acid synthesis, red blood cell production, and the conversion of homocysteine into methionine, which is critical for producing blood cells and preventing megaloblastic anemia (Hanna et al. 2022).

Table 2.2 Dietary patterns and neurological health outcomes

Dietary pattern	Core characteristics	Mechanisms supporting mental health	Linked mental health outcomes	Key references
Western diet	High processed food, refined sugars, trans fats, red meat	Increases systemic inflammation, increases oxidative stress, and alters gut microbiota	Depression, anxiety, cognitive decline	Jacka et al. (2020)
Mediterranean diet	Olive oil, nuts, fish, legumes, fruits, vegetables, whole grains	Has anti-inflammatory properties and antioxidant levels and promotes the integrity of gut-brain axis.	Reduced Alzheimer's risk, improved mood, slower cognitive aging	Psaltopoulou et al. (2022) Tsigalou et al. (2020)
Ketogenic diet	High fat, very low carbohydrate, moderate protein	Stimulates ketosis, reduces neuroinflammation, offer alternative brain fuel	Neuroprotection in epilepsy, Alzheimer's, Parkinson's	Taylor et al. (2021), Monda et al. (2024)
MIND diet	Leafy greens, berries, nuts, olive oil, fish, whole grains	Decrease amyloid plaque formation, antioxidant synergy	Decreased Alzheimer's risk, improved memory	Morris et al. (2015) Kheirouri and Alizadeh (2022)
DASH diet	Whole grains, low-fat dairy, lean meats, fruits, vegetables, nuts	Decrease hypertension and oxidative stress	Improved executive function, slower cognitive decline	Nestel and Mori (2024), Chu et al. (2022)
Flexitarian diet	Mostly plant-based, limited red meat, rich in fiber, nuts, legumes	Decrease inflammatory markers (CRP), improves gut microbiota	Reduced neuroinflammation, improved resilience	Wang et al. (2022), Kennedy et al. (2024)

These activities are fundamental for a healthy brain development, particularly during the perinatal period of life characterized by a lot of DNA synthesis and repair (Nieto 2024) (Table 2.2).

2.4 Nutritional Deficiencies and Cognitive Impairments

Nutritional deficiencies may drastically disrupt cognitive capabilities, which include memory, IQ scores, comprehension in both verbal and nonverbal modalities, attentiveness, and mental health. The absence of important vitamins and minerals necessary for brain health and development can lead to deficits of such a nature that conditions such as depression and anxiety will arise (Lam and Lawlis 2016). Brain imaging studies have shown continuing massive growth and development of the human brain during childhood and into the early adult years. During the gestational age, nutrients such as protein, vitamins B and D, iron copper, zinc, folate, and omega-3 fatty acids are functional for brain development (Puri et al. 2023), massively influencing brain and nervous system function (Zielińska et al. 2023). In this regard, insufficiency in the same group of nutrients may also account for mental health issues due to their vital role in brain functioning (Zielińska et al. 2023). Specifically, deficits in B vitamins, which including B6, B12, and folate, lead to cognitive impairments because of their role in neurotransmitter production and brain well-being (Zielińska et al. 2023).

Vitamin B deficiency can cause important psychological effects, including changes in mood, irritability, and disorientation. In some extreme cases, this may also develop into paranoia, dementia, depression, and anxiety (Sahu et al. 2022; Dommisse 1991). Vitamin D deficiency has been added to the list that includes commonly acknowledged contributors to dementia and cognitive decline influencing memory, attention, and mood. Neuroprotection and neurotransmitter regulation are important roles of Vitamin D in cognitive functioning. Several studies have demonstrated that decreased vitamin D level is correlated to lowered cognitive processes, including processing speed, mental flexibility, and executive function (Kang et al. 2021). Strong links have been drawn between vitamin D deficiency and cognitive impairment, especially in executive function and episodic memory. Cognitive assessment that included Mini-Mental State Examination (MMSE) and Montreal Cognitive Assessment (MoCA) revealed poor performance in affected individuals with very low vitamin D serum levels which is below 10 ng/mL (Lu et al. 2021). Moreover, it has also been noted that individuals with vitamin D deficiency reported more effects of depression (Kang et al. 2021).

Iron is fundamentally important for the functions of the brain, most notably in myelination, and it is therefore present at far greater concentrations than any of the other minerals. It is critical for mitochondrial electron transport, protein synthesis, and neurotransmitter formation and degradation (Alhazmi et al. 2024). Known deficits in iron have been linked to reduced attention spans, lowered intelligence levels, and diminished sensory perception (Table 2.3).

Studies show that children with iron deficiency anemia have longer response times and score lower on cognitive tests than their counterpart with normal iron levels (Jáuregui-Lobera 2014). Such cognitive deficit may continue into adulthood and further affect academic performance and functioning in daily life (Agrawal et al. 2019). Cognitive changes due to iron deficiency can also be expressed in increased anxiety, depression, and problems with socialization and attention. More recent research indicates strong association between iron deficiency and derangement of brain homeostasis, primarily affecting myelination and neurotransmission, particularly in monoamine metabolism (Kim and Wessling-Resnick 2014).

Copper is vital to normal brain function and indubitably plays and important role in synaptic plasticity, which is central to learning and memory. Memory retention and cognition may be severely impaired in patients with copper deficiency. People with low levels of copper may find it difficult to concentrate and learn, leading to poor academic performance and affecting daily activities (Choe et al. 2022). This

Table 2.3 Gene and nutrient interactions and mental health impact

Gene/pathway	Associated nutrients	Biological mechanism	Mental health impact
MTHFR	Folate, B12	Folate metabolism dysregulation influences methylation and serotonin/dopamine transmission	Elevated depression risk, poor stress resilience
COMT	Tyrosine, catechols	Impaired dopamine clearance in prefrontal cortex	Anxiety, poor cognitive flexibility, increased stress
APOE ε4	Omega-3 PUFAs	Impaired lipid metabolism and amyloid-β clearance	Elevated Alzheimer's disease risk
Epigenetic pathways	Choline, folate, polyphenols	DNA methylation and histone modifications regulating gene expression	Anxiety, depression, schizophrenia predisposition

has been related to the role of copper in the production of neurotransmitters, as it is necessary in synthesizing dopamine and norepinephrine, both of which are important in mood regulation. Copper deficiencies compromise such as neurotransmitter systems that may come to be involved in the onset of depressive symptoms (Li et al. 2023). Furthermore, recent studies provide strong evidence that copper dysregulation plays a very important role in the development of depression via the induction of oxidative stress and inflammatory responses in the body (Chen et al. 2023).

Various forms of cognitive development under attention, memory, and overall neuropsychological behavior can become impaired due to deficiency in zinc (Bhatnagar and Taneja 2001). Studies point out that children and adults with low zinc levels may results in reduced activity and increased emotional behaviors, compromising learning and memory capacities (Black 1998). Regarding long-term effects, chronic zinc deficiencies become crucial for emotional especially during the development stages (Black 2003b).

2.5 Neuroinflammation and Diet

Neuroinflammation is defined as the inflammatory response occurring through astrocytes, microglia, and endothelial cells inside CNS and disrupts the homeostasis (Xia et al. 2019). It is a major contributing factor to progressive neuronal damage (Amanullah et al. 2017). Many of these neurological diseases are manifested through neuroinflammation (Xia et al. 2019). In the past decades, a staggering rise has been witnessed in the number of mental disorders, such as depression, anxiety, cognitive disorders, and sleep disorders. Many pathologies of the brain recently commenced an investigation into the contribution of diet to the development of the gut microbiota in early life (Godos et al. 2020). This is a factor known to impose very high contributions to the global burden of disease due to dysbiosis through an unwholesome diet (Merlo et al. 2024).

Fortunately, there are modifiable risk factors, namely, diet, and its bioactive components, which are thought to play an important role in their pathogenesis

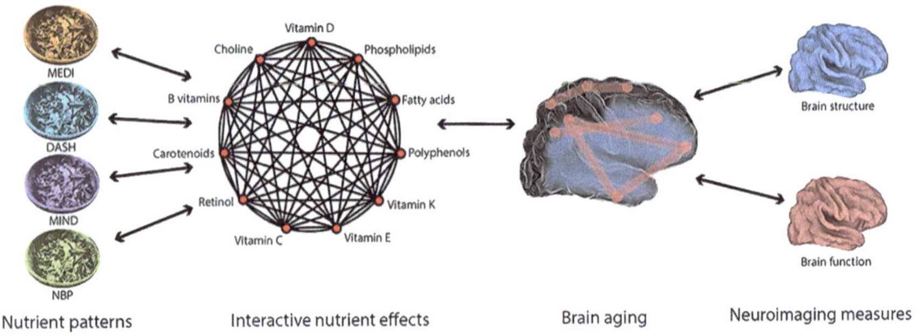

Fig. 2.1 The interactive effects of nutrients in the diet may be captured using nutrient patterns

(Godos et al. 2020). Foods in general, particularly rich in antioxidants, dietary fiber, and bioactive components such as berries, leafy greens, olive oil, whole grains, legumes, fatty fish rich in omega-3, nuts, especially walnuts, and fermented foods, are seen to help in modulating inflammation, support gut health, and promote neuroprotection (Van Zonneveld et al. 2024). Diets rich in refined carbohydrates are stated to induce brain inflammation and oxidative stress, contributing to cognitive decline and higher risk of neurodegenerative diseases (Zhao et al. 2022) (Fig. 2.1).

The Mediterranean–dietary approaches to stop hypertension intervention for neurodegenerative delay (MIND) diet study was a pioneering effort showing that more frequent consumption of berries, nuts, green leafy, vegetables, and whole-grain cereals, particularly rich in polyphenols, was associated with significant delay in cognitive decline (Godos et al. 2020). The MIND diet is a wholesome plan of foods, plant-based foods that have the potential to lessen the risk of the Alzheimer's disease (AD) and cognitive decline shortly (Morris et al. 2015). High adherence to the MIND diet has been linked to 53% reduction of AD, while even moderate adherence reduced it by 35%. The diet also associated with a lesser amount of Alzheimer's-related brain plaques and tangles (Van Zonneveld et al. 2024).

The Mediterranean diet is best described as the diet of people who reside along the coast of the Mediterranean Sea, mainly Greece, southern Italy, and southern Europe (Franco et al. 2023) characterized by consumption rates of fruits, vegetables, whole grains, and olive oil as the basic foundation of the diet, while meat was eaten in lesser amounts (Tsigalou et al. 2020). Studies show decreased blood inflammatory markers and decreased incidence of many chronic diseases in people consuming this diet (Van Zonneveld et al. 2024). Due to the nutrients with these anti-inflammatory properties, the Mediterranean diet has been suggested effective against neuroinflammation, possibly by protecting brain tissues against damages or chronic inflammation-related neurodegenerative diseases (Franco et al. 2023; Tsigalou et al. 2020). There is lower expression of proinflammatory pathways and higher expression of neuroprotective pathways in people consuming the Mediterranean diet. There is also preliminary evidence that the Mediterranean diet may protect against peripheral and central inflammation (Frye et al. 2024).

The KD also contributes considerably to the neuroinflammation and mental health aspect. This is a very high-fat, low-carbohydrate, and moderate-protein dietary intervention for a diagnosis. As a kind of fasting, it allows the switch of fat for carbohydrates as an energy source (Jiang et al. 2022a, 2022b). Recent accumulating studies in both humans and animal models have demonstrated that the KD could be beneficial for neurodegenerative diseases through modulation of central and peripheral metabolism, mitochondrial function, inflammation, oxidative stress, autophagy, and the gut microbiome (Koh et al. 2020). The ketogenic is the first diet to be used by Wilder as a nonpharmacological treatment for epilepsy in 1921 (Monda et al. 2024). Particularly, following their structural characteristics, ketone bodies can cross the blood-brain barrier to act as an alternative energy provider to the glucose-dependent brain. Therefore, ketone bodies do play a central role in enabling the organism to provide energy through metabolic processes during states of low carbohydrate intake such as the KD, enabling the organism to maintain homeostasis in energy level. This basic mechanism acting to effect changes in energy levels through which KD can influence certain neurodegenerative disease expression might form the generalized rationale for treating similarly oriented neurodegenerative diseases through additional supplementation with ketone bodies (Monda et al. 2024).

The flexitarian diet is based on a vegetarian diet with accessories of red meat consumption, and it is gaining popularity for its several health benefits, including those concerning neuroinflammation. When red meat consumption is reduced, mechanisms that reduce overall pro-inflammatory compounds may help to ameliorate neuroinflammation. High-temperature cooking of red met may yield compounds such as advanced glycation end products and N-glycolylneuraminic acid which promote inflammation (Samraj et al. 2014). Furthermore, there is evidence that it is thought that this reduces the C-Reactive Protein (CRP) levels, which is a marker of systemic inflammation. Some data have suggested some degree of positive correlation between rises in red meat intake with increases in CRP levels, therefore probably indicating greater inflammation. Higher total red meat intake has been associated with a rise in blood levels of CRP, according to a meta-analysis (Wang et al. 2022). Higher CRP levels have been associated with greater depressive symptoms in numerous studies (Wang et al. 2024), and they may add to anxiety disorders (De Azevedo Cardoso et al. 2023). Beyond that, research revealed that high red meat could harm the diversity of gut microbiota that is crucial in sustaining a healthy immune response. A gut microbiome that is diverse may synthesize SCFAs that exhibit anti-inflammatory properties. Dietary interventions that reduce red meat intake may allow for more favorable gut microbiome composition, thus supporting lessened plateaus of neuroinflammation (Kennedy et al. 2024).

2.6 The Role of Antioxidants in Brain Health

Antioxidants have a significant function in the defense against molecular oxidative damage and may have therapeutic significance in neurodegenerative disorders (Diniz et al. 2015) due to its stable molecule that may give an electron to a raging free radical and neutralize it, decreasing its ability to do harm (Lobo et al. 2010).

Flavonoids which are from polyphenol class are a diverse group of naturally occurring plant compounds. This flavonoids as vitamin C and vitamin E are considered as antioxidants which play a crucial role in protecting the brain from oxidative stress, which is one of the significant factors for neurodegenerative disease. Imbalance between free radicals such as reactive oxygen species (ROS) production and the body's ability to detoxify these harmful compounds are one of the reasons of how oxidative stress forming (Pizzino et al. 2017).

The brain is especially vulnerable to oxidative injury because of its high oxygen demand, lipid-rich composition, and limited antioxidant capability (Lee et al. 2020). To date, more than 10,000 distinct flavonoids have been described, showing considerable medicinal value (Minocha et al. 2022). The flavonoids occur in fruits and vegetables, nuts, plant-derived beverages, traditional eastern medicines, and dietary supplements based on herbs (Diniz et al. 2015). Flavonoids may be taken as able to influence brain function by crossing the blood-brain barrier (Calderaro et al. 2022). With an excellent potential to protect the brain health from oxidative stress injuries through antioxidant activity by the neutralization of harmful free radical species, enhancement of oxidative stress-confronting enzymes, or regulation of cellular signaling to counteract damage, flavonoids afford elaborate but multifaceted action in safeguarding the neuronal cells from oxidative stress-induced damage thereby preventing the possibility of neurodegenerative diseases (Minocha et al. 2022).

Vitamin C (ascorbic acid) is a water-soluble vitamin that consists of six-carbon compound as well as two acid-ionizing groups (Franzoni et al. 2021). It acts as an antioxidant and a necessary cofactor for collagen formation, for carnitine and catecholamine-specific metabolic pathways, and for dietary iron absorption (Abdullah et al. 2023). The total brain contains a specific concentration of 1–2 mM vitamin C, and the highest concentrations of vitamin C are associated with the cerebral cortex, hippocampus, and amygdala (Travica et al. 2017). An in-vitro study on the protective impact of vitamin C administration resulted in avoiding the age-induced oxidative damage in hippocampus neurons, while further analysis concluded that this frequency of vitamin C therapy safeguards hippocampal neurons from damage caused by free radicals (Franzoni et al. 2021). Equally and most importantly, vitamin C plays a crucial role in the synthesis of neurotransmitters like dopamine and norepinephrine. Supporting neurotransmitter synthesis helps in regulating neural signaling. This is of utmost importance as excessive neurotransmitter chaos can result in excitotoxicity, wherein neurons are overstimulated to cause oxidative stress (Zylinska et al. 2023).

Besides that, vitamin E is a fat-soluble compound derived from plants and is plentiful in foods predicated on the Mediterranean diet (Schirinzi et al. 2019). It comprises a family of eight molecules: α-, β-, γ-, and δ-tocopherols, as well as α-, β-, γ-, and δ-tocotrienols. These compounds provide potent antioxidant properties (Franzoni et al. 2021). Vitamin E is an effective ROS scavenger and has the ability to prevent oxidative damage which may accompany the occurrence of traumatic brain injury (Wu et al. 2009). Frequently, oxidative stress stimulates neuroinflammation enhancing the neuronal loss, and vitamin E probably impedes this process by inhibiting the activation of nuclear transcription factors like NF-κB, major players involved in the processes of inflammation that can create a more favorable environment for brain health (Behl 1999). Vitamin E interacts synergistically with other antioxidants, such as vitamin C, to boost their total efficiency against oxidative stress. An example of that would be vitamin C that can regenerate oxidized vitamin E to its active form, giving an extra boost to the cumulative antioxidant shield of the brain itself (Behl 1999).

2.7 Dietary Interventions for Neurodegenerative Diseases

With the increasing number of cases and the absence of cures, neurodegenerative diseases like AD and Parkinson's disease (PD) are a severe public health problem, but recent research shows that diet could play a role in both prevention and treatment. Diets such as ketogenic, DASH, and MIND are drawing much attention due to their perceived ability to protect the nervous system.

A high-fat, low-carbohydrate eating pattern, known as the KD, causes the body to use ketones as its primary fuel instead of glucose (Zhu et al. 2022). The neuroprotective potential of ketones is attributed to their ability to lessen excitotoxicity, inflammation, and oxidative stress, key factors in the development of neurodegenerative conditions (Jensen et al. 2020). The KD might improve cognitive function in AD by enhancing mitochondrial efficiency and reducing amyloid-beta plaques, a primary characteristic of the illness, according to recent research (Rusek et al. 2019). Randomized controlled trials included in Grammatikopoulou's 2020 found that 6 months of a KD led to notable improvements in memory and executive function in individuals with mild cognitive impairment, based on data from randomized controlled trials (Grammatikopoulou et al. 2020). There is evidence suggesting that the KD can be beneficial in alleviating motor symptoms and improving the quality of life for people with PD (Tidman et al. 2024). A 24-week ketogenic intervention led to enhancements in gait, mobility, self-care, and mood for PD patients, according to Tidman's 2024 research. These advantages were explained by the diet's positive impact on mitochondrial function and its ability to reduce neuroinflammation.

Focusing on fruits, vegetables, whole grains, lean proteins, and low-fat dairy and minimizing saturated fats and sodium, the DASH diet was created for blood pressure management (Nestel and Mori 2024). Because it is packed with nutrients, it delivers antioxidant and anti-inflammatory effects, vital for fighting brain cell decline (Makkar et al. 2020). Research on AD has shown that following the DASH diet is associated with a lower risk of cognitive decline (Van den Brink et al. 2019). The risk of AD was found to be lower in individuals with high adherence to the DASH diet, compared to those with low adherence, according to a study (Chu et al. 2022).

In PD, where oxidative stress is a major contributor, the DASH diet's emphasis on antioxidant might provide a protective effect (Bisaglia 2022; Bianchi et al. 2023). Experts in neurology recommend the DASH diet because it can slow down disease advancement and improve overall brain function (Neurology Solutions 2024). A diet called MIND, which takes principles from both the Mediterranean and DASH

Table 2.4 Dietary interventions for neurodegenerative disease management

Intervention	Primary mechanisms	Target neurological disorders	Clinical/functional outcomes	Key references
Ketogenic diet	Support ketone metabolism, decreases ROS, increases mitochondrial efficiency	Alzheimer's, Parkinson's, epilepsy	Improved cognition, motor function, reduced neuroinflammation	Rusek et al. (2019), Tidman et al. (2024)
DASH diet	Lessens oxidative stress, lower blood pressure; increases nitric oxide level.	Alzheimer's, MCI	Slower memory decline, improved executive function	Makkar et al. (2020), Chu et al. (2022)
MIND diet	Combines neuroprotective components of DASH and Mediterranean diets	Alzheimer's, parkinsonism	Reduced β-amyloid pathology, delayed cognitive decline	Kheirouri and Alizadeh (2022), Aderinto et al. (2023)
Intermittent fasting	Increases BDNF level, promotes neurogenesis, decrease systemic inflammation	Aging, Alzheimer's, Parkinson's	Enhanced hippocampal plasticity, improved memory, longevity	Jiang et al. (2022a, 2022b), Seidler and Barrow (2022)

diets, is thought to improve brain health by promoting the eating of leafy greens, berries, nuts, whole grains, fish, and olive oil and by discouraging the eating of red meat, butter, and sweets (Charisis et al. 2024). Following this diet has been associated with slower cognitive decline and a decreased chance of developing AD (Kheirouri and Alizadeh 2022). Even for those with a genetic vulnerability to AD, a study found that following the MIND diet resulted in considerably slower cognitive decline (Kheirouri and Alizadeh 2022). It has also been observed that people who follow the MIND diet have a lower chance of developing Parkinsonism, a disorder that mimics the movement symptoms of PD (Aderinto et al. 2023) (Table 2.4).

2.8 The Impact of Processed Foods and Sugar on Mental Health

The potential effects of sugary and processed diets on mental health, particularly depression and anxiety, are being scrutinized more closely. There is growing evidence that these eating patterns significantly raise the risk of mood disorders. Health problems such as obesity, type 2 diabetes, and cardiovascular disease are linked to excessive sugar intake. A connection between high sugar consumption and mental illness is increasingly supported by studies, such as the 2024 findings by Navratilova and her team, which revealed a 27% higher depression risk in those with a preference for sugary foods (Navratilova et al. 2024). It is suggested that sugar causes inflammation throughout the body, and this inflammation damages brain function.

Inflammation may hinder the activation of neurotransmitters and play a role in the onset of mood disorders (Xiong et al. 2024).

Mental health outcomes are negatively impacted by processed foods, with ultra-processed foods containing refined sugars, unhealthy fats, and additives being particularly problematic. Researchers conducting a meta-analysis discovered that adults who ate 10% more ultra-processed foods as part of their daily calories had an 11% higher chance of developing depression (Wiss and LaFata 2024). Disruptions to the gut microbiome and subsequent inflammation, along with effects on the brain, are possible consequences of consuming these foods, and these are key elements in the development of depression and anxiety (Jamar et al. 2021).

The relationship between food and mental health largely depends on the gut-brain axis (Berding et al. 2021). The balance of gut bacteria can be disrupted by diets rich in sugar and processed foods, which can cause dysbiosis and a more permeable intestinal barrier (Khoshbin and Camilleri 2020). Such imbalances can result in inflammation that affects the entire body and disrupt the formation of neurotransmitters that are vital for mood regulation (Yang et al. 2021). Also, large amounts of sugar in the diet can cause blood sugar to vary greatly, which can lead to mood swings and make anxiety worse (Sandua 2023).

The impact of dietary habits on mental health has gained attention, with discussions focusing on the association between ultra-processed food consumption and elevated risks of depression and anxiety. The British Medical Journal published a study showing that increased consumption of ultra-processed foods correlates with a higher risk of detrimental health effects, encompassing common mental health conditions (Lane et al. 2024). In 2024, studies indicated that consuming diets primarily composed of ultra-processed foods, which are high in sugar, salt, fat, and additives, can greatly contribute to the development of depression (Monda et al. 2024).

To achieve better mental health outcomes, diets focusing on whole, natural foods are beneficial, and the Mediterranean, DASH, and MIND diets, by emphasizing plant-based foods, lean meats, and low-fat dairy while limiting processed foods, have shown connections to reduced cognitive decline and improved mental well-being (Arora et al. 2023; Hu 2024). By increasing serotonin in the brain, complex carbohydrates, like whole grains, can aid in relaxation and the reduction of anxiety symptoms (Constantin and Fonseca 2020).

2.9 Nutrigenomics: Personalized Nutrition for Brain Health

As science has increasingly shown how genetic factors affect individual responses to nutrients, the connection between diet and mental health has become a topic of significant interest (Grajek et al. 2022). Understanding how genes and food influence each other, a field known as nutrigenomics, is key to developing personalized diets that can improve mental health (Lagoumintzis and Patrinos 2023). This field, through its exploration of genetic influences on food responses, epigenetics, and its broader applications, is fundamentally changing how we approach mental healthcare. Our genes play a major role in determining how we use nutrients, impacting the production of brain chemicals, hormone levels, and brain function, which ultimately affects our mental health (Godos et al. 2020). Variations within genes involved in folate processing, particularly the methylenetetrahydrofolate reductase (MTHFR) gene, can alter how effectively the body converts folate into its usable form (Menezo et al. 2022).

Reduced production of serotonin and dopamine, caused by this deficiency, can increase the risk of mood disorders, particularly depression (Liwinski and Lang 2023). The COMT gene, also known as catechol-O-methyltransferase gene, responsible for dopamine regulation, exhibits variations that can influence both stress response and cognitive function (Esmaiel et al. 2020). People with particular COMT gene variations might exhibit increased anxiety levels or be more prone to developing mood disorders (Tommasi et al. 2021). Through modifying how nutrients are digested and used, genetic variations, as seen in these examples, can influence mental health.

Epigenetics is crucial for revealing how nutrition affects mental health, as it explores modifications to gene expression that occur independently of DNA sequence changes (Bekdash 2021). Nutrients have the ability to epigenetically alter gene activity, which in turn affects brain function and mental wellness (Bekdash 2024). Certain foods, including those rich in folate, choline, and polyphenols, are involved in DNA methylation, a crucial epigenetic process that manages gene activity (Polverino et al. 2021). Issues with DNA methylation may contribute to the development of mental health conditions, including anxiety, depression, and schizophrenia (Alameda et al. 2022). The risk of mood disorders rises alongside poor DNA methylation when folate levels are deficient (Cao et al. 2022a, b; Lam et al. 2024). The epigenetic programming of a developing fetus can be influenced by maternal nutrition during pregnancy, which may have consequences for future mental health (Sebastiani et al. 2020).

This emphasizes how vital it is to have proper nutrition during those key developmental periods to secure lasting mental health. Integrating nutrigenomics into mental health treatment allows for a variety of personalized interventions (Singh 2023). Through genetic analysis, healthcare providers can develop dietary strategies that are specifically designed to support an individual's mental health (Singar et al. 2024). Supplementing with L-methyl folate could enhance mood and cognitive function in people whose genetics hinder their ability to metabolize folate effectively (Hoepner et al. 2021). Additionally, individuals with a genetic likelihood of developing mental health disorders can be identified through nutrigenomics, which then allows for targeted prevention and early intervention (Ahluwalia 2021). By taking this proactive approach, the development of those conditions was less likely. Furthermore, by adding personalized dietary recommendations, the impact of medications and therapies can be significantly increased (Strianese et al. 2020).

2.10 Dietary Patterns and Mental Disorders in Adolescents and Adults

Our food choices strongly influence our psychological health, spanning from childhood to our senior years. Eating a healthy, nutrient-rich diet is linked to improved mental health, while unhealthy eating can increase the risk of mental health problems, according to a new study. Scientific discoveries highlight the importance of studying diet's impact on mental health across different ages, which is supported by current scientific findings.

Proper nutrition during childhood and adolescence is fundamental for supporting both physical and mental development (Zavitsanou and Drigas 2021). This age group experiences poorer mental health outcomes when they engage in unhealthy eating, as evidenced by numerous studies. Children and teens eating diets high in processed foods, sugar, and unhealthy fats have a greater risk of behavioral problems, anxiety, and depression, according to a review of 12 studies (O'Neil et al. 2014). Unlike those with poor diets, individuals who prioritize fruits, vegetables, and whole grains usually demonstrate superior mental health. While the quality of food is important, the routines surrounding eating are equally essential for mental health (Cabiedes-Miragaya et al. 2021). Regular family meals and breakfast consumption were linked to a positive mental health outcomes in teenagers from five Nordic countries, as found in a cross-sectional analysis (Fismen et al. 2024). Improved mental health is achieved through these behaviors, as they promote habit, stability, and social connection. It was also revealed that making minor, healthy dietary shifts can lead to substantial mental health benefits.

Throughout adulthood, dietary choices remain crucial for mental well-being, with diets rich in fruits, health fats, vegetables, and whole grains significantly increasing the likelihood of positive mental health outcomes. The Mediterranean diet, with its focus on plant-based foods, fish, and healthy fats, was shown by Lassale's (2019) study to correlate with a decreased occurrence of depression (Lassale et al. 2019). The proposed benefits stem from the diet's ability to combat inflammation, and oxidative stress benefits brain health and minimizes inflammation, thereby influencing mood disorders. Jacka's (2017) study indicated that dietary adjustments can also serve as a valuable method for decreasing depression symptoms (Jacka et al. 2017). However, unhealthy eating patterns, such as those that include excessive processed foods and sugary beverages, correlate with a greater likelihood of anxiety and depression (de Melo Simplício et al. 2024).Older adults' mental health, especially cognitive decline and depression risk, relies heavily on diet quality (Dominguez et al. 2021).

Based on research involving older Iranians, improving dietary habits could be a useful strategy for reducing depression and enhancing quality of life within this demographic (Chegini et al. 2022). Older individuals who follow guidelines like the DASH, MIND, and Mediterranean diets tend to slower cognitive deterioration (Nishi et al. 2021). They center on a foundation of lean proteins, whole grains, fruits, vegetables, and low-fat dairy, while excluding processed foods, red meat, and excessive sodium (Lichtenstein et al. 2021). Following these dietary guidelines was linked to enhanced cognitive performance and a reduced rate of cognitive decline in a 6-year study of 6154 adults (Seago et al. 2024).

2.11 Supplements and Functional Foods in Mental Health Management

The human brain, with its intricate design, is the control center for mental health, emotional stability, and cognitive functions (Sultana et al. 2024). The potential brain-boosting effects of specific dietary supplements, including adaptogens, probiotics, and prebiotics, have become a subject of increased research. These supplements are thought to work by improving cognitive function, regulating stress responses, and influencing the communication pathway between the gut and brain.

Probiotics are live bacteria that, when consumed in adequate amounts, provide health benefits. Studies have demonstrated that probiotics can affect brain function via the gut-brain axis by regulating the gut microbiota. The 2024 comprehensive research by El Dib's team demonstrated the possibility of probiotics reducing anxiety and depression (El Dib et al. 2021). In this review, they indicated that strains of *Lactobacillus* and *Bifidobacterium* could significantly reduce psychological distress.

Probiotic administration was shown to enhance cognitive function in elderly individuals with mild cognitive impairment, according to Kim et al.' 2021 study (Kim et al. 2021). Following 12 weeks of probiotic use, significant improvements in executive functioning and memory cell were observed, suggesting probiotics may contribute to neuroprotection and cognitive enhancement, particularly in older individuals. These indigestible food substances, known as prebiotics, selectively foster the development of healthy gut bacteria (You et al. 2022).

Due to its ability to generate SCFAs and neurotransmitters, the gut microbiota significantly influence neurodevelopment and cognitive well-being (Qian et al. 2022). Inulin supplements were found to improve cognitive flexibility and reduce stress-related behaviors in animals, according to a 2024 study (González-Velázquez et al. 2024). Prebiotic supplementation in healthy adults led to improved mood and a significant reduction in cortisol levels, a stress indicator, according to a 2015 study by Schmidt and colleagues, supporting the previous animal findings (Schmidt et al. 2015). The hypothalamic-pituitary-adrenal axis, crucial for stress response regulation, may be positively affected by prebiotics, based on these results (Ansari et al. 2023).

Organic substances known as adaptogens contribute to physiological balance by strengthening the body's stress response (Panossian et al. 2021). Examples of well-known adaptogens include *Rhodiola rosea*, *Panax ginseng*, and *Withania somnifera* (Ashwagandha) (Esmaealzadeh et al. 2022). Studies have recently been exploring how these substances influence brain function, especially in terms of stress management and cognitive enhancement. The analysis of clinical trials using adaptogens revealed that ashwagandha led to a substantial decrease in anxiety and stress levels (Arumugam et al. 2024). It also led to enhanced attention and memory function in those experiencing mild cognitive decline. *Rhodiola rosea* was shown to enhance cognitive performance during stress and alleviate mental fatigue in a 2022 by Ivanova and Quintela (Ivanova Stojcheva and Quintela 2022). A 2024 review published by Shin and colleagues pointed to *Panax ginseng* as an adaptogen that significantly enhanced working memory and reaction speeds in healthy individuals (Shin et al. 2024). These results imply that adaptogens potentially improve mental clarity and enhance the body's capacity to withstand stress.

2.12 The Impact of Fasting and Caloric Restriction on Brain Health

Fasting is not eating or drinking for a period, whereas caloric restriction (CR) is the careful consumption of food with minimal activities. These strategies have drawn a lot of attention in the field of psychology due to their positive impacts as they increase neuroplasticity, lower inflammation, and change the metabolic processes. These changes are associated with greater longevity (Davaatseren et al. 2023) and enhanced cognitive functions (Perez-kast and Camacho-morales 2025). A point of note is the bioenergetic challenge of intermittent fasting (IF), which has been shown to promote longevity and health span through an adaptive stress response (Seidler and Barrow 2022).

Fasting, and more importantly IF, has also been well recognized for increasing adult neurogenesis and is considered one of the major factors for the improvement of cognitive function and the prevention of age-related neurogenic decline (Davaatseren et al. 2023). In other words, IF has been associated with improved survival and proliferation while decreasing cognitive decline for favorable ageing (Ye et al. 2024). Metabolic processes that change the gut-brain microbiota axis or undergo biosynthetic and ketone body degradation which lessen amyloid lesions are modulated by these effects. IF has been of value in the consideration for improving an energy deficit of the brain seen in AD with improved metabolism of ketone bodies.

Through the use of a variety of biological processes, including the regulation of free fatty acids, ketone bodies, neurotrophic factors, cytokines, leptin, reactive oxygen species, and autophagy, IF and CR provide notable neuroprotective advantages (Oliveira et al. 2023).

The capacity of these dietary therapies to improve metabolic pathways, decrease inflammation, and increase neuroplasticity, all of which lead to gains in cognition, is drawing more attention (Perez-kast and Camacho-morales 2025). In reality, the foundation of such advantages is neurotrophic substances like neurotrophin 3 and brain-derived neurotrophic factor (BDNF) (Ojha et al. 2023). Through alternate-day fasting (ADF) techniques, IF can therefore raise glial-cell-derived neurotrophic factor and BDNF expression, change gut microbiota within PD models (Jiang et al. 2022a, b), and restore BDNF levels, all of which in turn trigger the brain's neuroprotective pathways (Seidler and Barrow 2022). Additionally, fasting also depletes the cytosolic complexes known as inflammasomes, which are implicated in inflammation and cell death in acute and chronic neuropathological conditions. Antioxidant enzymes, protein chaperones, and neuroprotective trophic factors are also increased, protecting against age-related brain disorders (Poh et al. 2021).

Notably, the impact of further fasting and CR is amplified by impacting the nutrient-sensing mechanisms of the evolutionarily conserved hormones Insulin-like growth factor-1 (IGF-1), AMP protein kinase, and mTOR. Extended water fasting along with fasting mimicking diets allow for therapeutic dietary techniques with lowered oxidative stress and IGF-1 levels. These diets and treatments, which predominantly consist of low protein (30–50% of total daily caloric intake), have been effective in reducing age-related diseases and extending life expectancy (Fadhilah et al. 2024).

Hippocampus is known as one of the areas of the brain that facilitate memory and learning. It is highly responsive to IF and other cognitive functioning such as learning and memory. Fasting enhances neuronal survival, synaptic plasticity, and the stressed-activated response to neurons in the hippocampal CA1 and CA3 regions. Signs of increasing neural resilience against injury and illness are promoted by stimulating modest adaptive cellular stress responses. Mechanistically, IF promotes synaptic plasticity by upregulating neurotrophic and neurotransmitter signaling, especially that involving glutamate and BDNF, which aid in memory and learning through pathways such as cAMP response element-binding protein and nuclear factor kappa-B. Similarly, fasting enhances brain resilience and function by facilitating metabolic transition from glucose to ketone use, strengthening neuronal circuits (Solianik 2018).

Other than that, neurogenesis main location occurs in the hippocampus also known as the site of the creation of new neurons from neural progenitors or stem cells. Following are crucial for the emotional and cognitive functions (Melgar-locatelli et al. 2023). Lifelong learning is connected to synaptic plasticity, or the ability of synapses to become stronger or weaker over time (Wu et al. 2023). Time-restricted feeding further improves spatial memory by enhancing the electrophysiological properties of hippocampus synapses. Likewise, with ADF, there is supporting evidence for ADF facilitating synaptic plasticity during development of the nervous system due to the increased level of neural cell adhesion molecules (NCAM), especially in their polysialylated (PSA) form, PSA-NCAM, and also enhanced expression of synaptophysin in the hippocampus. Importantly ADF facilitates neurogenesis due to the increased production of BDNF, which aids in the differentiation of neural progenitor cells of neurons. It is plausible that this means that along with neural stem cell proliferation, ADF also enhances the survival of newly born neurons. Results may indicate that these eating patterns contributed to slowing down the aging of the brain, which is marked by loss of neuroplasticity and increased risk of neurodegenerative disease such as AD and stroke. As reported by the WHO (2017), each year 6 million people die from stroke while over 50 million people suffer from dementia and the estimated 60–70% are reported having AD.

Living a sedentary lifestyle and excessive calorie intake may accelerate the aging process; however, maintaining brain health lifelong is doable with intermittent bioenergetic challenges such as fasting, exercise, and cognitive stimulation (Cao et al. 2022a, 2022b). Greatly, IF improves cognitive function. It has been established that short-term fasting has lasting protective benefits against cognitive decline and cell death due to chronic insufficient blood flow to the brain. Not only that, IF inhibits the Chronic Cerebral Hypoperfusion (CCH)-induced increase in oxidative damage markers such as malondialdehyde by downregulating NADPH Oxidase 1 (NOX1) and increasing the expression of antioxidative enzymes like Superoxide Dismutase 2 (Mitochondrial) (SOD2) and Glutamate-Cysteine Ligase (GCL) subunits. These processes improve the neuronal microenvironment by decreasing structural and cognitive impairments after harsh oxidative and neuroinflammation damage at the hippocampus is lessened (Hu et al. 2017). As evidence of cognitive enhancement in humans keeps growing, many studies have, without questions, confirmed the efficacy of both CR and IF in extending health and lifespan in many species. These strategies also enhance the functioning of the brain and the ability to withstand neurodegenerative disease by encouraging metabolic switching and the use of ketone bodies, as well as reducing inflammation in the brain. In this way, IF fosters neurogenesis, strengthens neuronal networks, and builds cognitive resilience and general brain health through adaptive cellular stress response (Seidler and Barrow 2022).

2.13 Sleep, Nutrition, and Brain Health

The connection between sleep and nutrition plays vital role in maintaining the brain function. While sleeps support metabolic and cognitive processes, nutrition influences the quality of sleep and the overall brain function. Having better understanding of these interaction allows for better optimization of the brain health (Fjell et al. 2023).

2.13.1 Sleep and Mental Health

It has been shown that poor sleep has an impact on decision-making, emotional regulation, and memory consolidation. Those who sleep for less than 6 hours a day tend to perform worse on tests divided by general cognitive ability as per research, underscoring the importance of sufficient sleep toward optimal brain functioning (Fjell et al. 2023). Besides that, such state deprivation leads to an alteration in the function of the hypothalamus concerning hunger and appetite, thus being impaired by poor eating decisions that affect sleep cycles. Current research indicates that inadequate sleep affects the brain's ability to clear poisons and metabolites essential for memory formation. New formation is consolidated in the brain under deep sleep phases such as rapid eye movement, although learning and retention may be hindered by decreased clearance rates because of sleep deprivation (Miao et al. 2024). Importantly, sleep has effects on emotional regulation; poor metabolite clearance may result in mood disorders, anxiety, and greater emotional reactivity (Lokhandwala and Spencer 2022).

Moreover, changes in the brain's anatomy have been connected to sleep deprivation. Research shows that complaints of insomnia are associated with more surface area in the medial orbitofrontal cortex, which is important in decision-making and emotional control (Wang et al. 2023). However, those with a larger global cortical surface area have shorter sleep durations and are less prone to have insomnia, indicating that brain structure influences sleep quality (Mahalakshmi et al. 2020).

2.13.2 Nutrition and Brain Function

Among many, various nutrients have been found to be helpful in sustaining the cognitive health. One of it is the omega-3 fatty acid, which also promotes blood vessel health, cognitive functioning, and synaptic pliability. These can also enhance sleep via maximizing cerebrospinal fluid dynamics (Brodt et al. 2023). Other than that, omega-3 s also cause the reduction of neuroinflammation, associated with cognitive decline and sleep problems. Next are the antioxidants, which help fight against oxidative stress to the neurons and slow down cognitive decline while enhancing memory consolidation. Chronic low-grade inflammation due to poor diets usually is associated with anxiety, depression, and sleep disorders. Magnesium is such an essential substance that gives more profound sleep while letting loose neurotransmitters that are involved. It is found that leafy greens and almonds, both very high in magnesium, encouraged relaxation and quality sleep (Miao et al. 2024). Rather, low unaffordable diets in sugar and processed foods are destructive and lead to sadness, anxiety, and sleeplessness. It increases the blood sugar level due to high sugar intake because of the energy slumps and bad sleeping behavior. Processed food can worsen the mental state and cognitive ability through inflammation (Mason et al. 2022).

2.13.3 Combined Effects of Sleep and Nutrition

There are several ways in which sleep and diet interact in an interesting way as sleep phases are impacted by nutritional factors. For instance, foods high in magnesium promote relaxation, whereas excessive sugar and caffeine interfere with the structure of sleep (Lyons et al. 2024). High sugar consumption is connected with cognitive deficits and retarded sleep. Past studies indicated that eating a diet rich in whole grains, lean meats, and healthy fats improves the quality of sleep while reducing sleep disruptions due to stress (Fjell et al. 2023).

Nutrients of the Mediterranean diet tend to yield positive results concerning sleep and cognition through inflammation, oxidative stress, and neuroprotection (Miao et al. 2024). Did you know that metabolism does some of its work during sleep? It has become essential when the body removes the metabolic wastes in the brain through healthy eating.

Dietary measures such as calorie restriction and fasting can diminish neuroinflammatory processes that have been linked to poor quality of sleep (Lewis 2022). high-energy-dense (HED) diets have been demonstrated to have adverse effect on the brain activity. Sugars are one of the ingredients which are HED and considered to provoke neuroinflammation and cognitive impairment because of stimulating the innate immune system of the brain and producing several pro-inflammatory cytokines such as IL-1β and IL-6. Unhealthy diet furthers modified brain function as studies show that HED diets are associated with mood problems and cognitive deficits, such as sadness, anxiety, and addiction-like behaviors. In addition, it causes cognitive decline due neuroinflammation and metabolic instability, long-term HED diet intake is linked to memory and learning impairment. Furthermore, it causes impaired neurogenesis as these diets may lower BDNF levels, which might have a detrimental impact on cognitive function and the growth of new neurons which causes modifications in behavior. According to research on animals, eating HED diets may enhance anxiety and anhedonia, two symptoms that are frequently linked to mood disorders (Perez-kast and Camacho-morales 2025).

2.13.4 Stress and Brain Health

Neuroinflammation pathways, mood regulation processes, and cognitive performances are adversely affected by stress. Chronic stress impairs brain structure and functions in the hippocampus and prefrontal cortex (PFC). Additionally, stress inhibits learning and memory, injures neurons, and hampers neurogenesis when cortisol levels are raised (Brodt et al. 2023). The cognitive effects of stress particularly impair decision-making, executive function, and working memory. Other than that, it may also impair fine motor skills via psychomotor coordination. A dysregulation of neurotransmitters such as dopamine and serotonin potentiates anxiety and depressive symptoms followed by emotional regulation and mood disorders. Also, neuroinflammatory changes precipitated by stress may be linked to the neurodegenerative diseases of Alzheimer's and cognitive impairment (Solianik 2018). This opens the gates for different activities, such as physical activity, that can help build resilience and buffer the brain from deleterious effects of stress. Fasting may also alleviate stress. As evidence suggests, IF may provide resilience to stress-induced brain injury by promoting neuronal health via autophagy and neurogenesis (Solianik 2018).

2.13.5 Sleep and Cognitive Development

Sleep is critical for cognitive and brain development functions from infancy through adulthood. For infants and children, sleep cycles go from a polyphasic sleep structure to monophasic sleep structure, reflecting brain development (Lokhandwala and Spencer 2022). Electroencephalogram measurements of sleep help unravel the developmental mechanisms involved in some disorders, such as Autism Spectrum Disorder (ASD) and Attention-Deficit/Hyperactivity Disorder (ADHD). Slow-wave sleep has been linked to the synthesis of myelin and the strengthening of neuronal connections, making sleep indispensable for brain health. Sleep architecture is further modified by experiences-dependent plasticity, which in turn is affected by learning and environmental inputs.

2.14 Alcohol, Caffeine, and Brain Function

The two most commonly used psychoactive substances are alcohol and caffeine, which can have opposing influences in the brain. Caffeine is a stimulant that enhances alertness and attention, while alcohol is a depressant that impairs both cognitive and psychomotor functions. The respective mechanism, where interactions affect behaviors, development and neurobiology, with view on immediate as well as long-term effects on brain health is further explained (Kuhns 2022).

2.14.1 Alcohol

Initially, acute alcohol use disrupts cognitive functions, namely automatic processing and attention. This disturbance is made clear by event-related potential components such as the P3b Component of the Event-Related Potential (ERP) and Mismatch Negativity (MMN). Also, straying from the usual neural processing creates imbalances in mood, decision-making, and risk-taking. Crucially, the main cause of these effects is alcohol's interaction with GABA receptors, which at first cause relaxation but frequently result in aggressiveness after extended usage (Fairbairn et al. 2022). It is important to remember that alcohol has dose-dependent effects on cognition. As an example, very low doses are sufficient to alter cognitive performances, and this impairment becomes worse with increasing blood alcohol concentration. Behaviors like jitters or anxiety are more likely shown when a person is excessive dose-dependent on caffeine, while low levels increase attentiveness (Vore and Deak 2022).

Furthermore, alcohol causes mood swings, poor motor coordination, and cognitive impairments via increasing GABA activity and blocking excitatory neurotransmission (De Goede et al. 2021). Caffeine, on the other hand, inhibits adenosine receptors, which increases alertness and concentration.

Additionally, binge drinking and higher alcohol intake are associated with hereditary and anatomical predispositions, such as decreased global cortical thickness. A structure in the brain known as glutamatergic corticol neurons is affected causing structural differences by these behaviors (Karoly et al. 2024). The relationship of alcohol and brain use also supported by structure comes from genetic associations, such as those at the 17q21.31 locus (Mavromatis et al. 2022).

Short-term use of alcohol disrupts brain circuits that affect motor coordination, attentional control, and psychomotor speed. Due to the disruption of impulse control and attentional development caused by alcohol, adolescents are especially at risk. Furthermore, alcohol can disrupt neurodevelopment—when an adolescent consumes alcohol, it causes the brain to undergo major structural alterations, such as diminished white matter integrity and decreased grey matter volume. As a result, these changes bring about harmful effects on day-to-day life and educational performance, due to the long-term cognitive deficits associated with low processing speed and attentional control (Lees et al. 2021). Chronic alcohol intake has been associated with neurotoxicity and causes the reduction of brain volume particularly in the frontotemporal lobes. These structural modifications reduce brain volume varying from 3% to 9% and hinder cognitive and motor capacities. In the early stage, alcohol exposure causes an increase likelihood of globally degenerative diseases and persistent incapacitation. The demerit proportions usually exceed those advantages, even where modestly consumed alcoholic beverages might acquire some protective effects due to compounds like resveratrol (Peng et al. 2020).

Long-term alcohol use leads to increased risk of depression and dementia, that damages brain networks and promotes chronic inflammation, which also raises the risk of cognitive decline. These consequences draw attention to the wider impact of long-term alcohol consumption on mental health. The adolescent age is of vital importance for brain development (Boer et al. 2022). Indefinitely, alcohol consumption at this stage interferes with brain development which impact cognitive flexibility and decision-making. These deficits may be accentuated by co-use with other substances such as cannabis. Thus, the need for selective measures (Lees et al. 2021).

2.14.2 Caffeine

Caffeine reduces tiredness, increases wakefulness, and blocks adenosine receptors to improve alertness, concentration, and cognitive function. Its antioxidant qualities and effects on dopamine modulation may be the reason why moderate consumption has been associated with preventive benefits against neurodegenerative disorders like Parkinson's and Alzheimer's (Peng et al. 2020). However, it also shows that alcohol causes neuronal damage by elevating pro-inflammatory actions helping it to act in its neuroprotective capacity (Crews et al. 2024). The stimulating effects of caffeine can obscure the sedative effects of alcohol, thereby encouraging binge drinking and raising the odds of facing alcohol-related problems (Piekarski et al. 2022). Alcohol affects the GABA and glutamate neurotransmitter systems, and therefore it affects behavioral and cognitive mood. Instead, caffeine enhances alertness, improves cognitive function by increasing dopamine and norepinephrine activity (Walker and Pinares 2022). Moreover, issues with mental health are linked to excessive consumption of alcohol and coffee. While excessive coffee use can cause anxiety and dependence, chronic alcohol use adds to neuroinflammation and circuit deterioration. Notably, these results are influenced by personal genetic and behavioral variables (Fairbairn et al. 2022). Indeed, these two making opposite effects are also important considering that they have the above consequences on neurochemistry at this particular time, that is, during adolescence, which needs to be protected against long-term cognitive disturbances. Using caffeine should also be limited to prevent dependency and sleep disruption (Lees et al. 2020).

2.15 The Future of Nutritional Neuroscience: Artificial Intelligence and Precision Nutrition

2.15.1 Artificial Intelligence in Neuroscience

Artificial intelligence (AI) has progressed significantly within a very short span of time in various areas, such as economics, education, engineering, medical, and optics (Chen et al. 2022). In neuroscience, AI has upturned the definition of the whole science, for brain-functioning understanding certainly changes with AI. It is deep-learning models which mimic the brain cortex, lay a foundation toward understanding the vital processes such as memory, control of movement response, and visual processing (Macpherson et al. 2021). The collaborative interplay of AI with neuroscience offers an avenue for introducing new theories about how the brain operates to advance scientific research (Gupta et al. 2021). Furthermore, high-throughput analysis of massive datasets, incorporating whole-brain imaging in both humans and animals, has been made possible due to the merging of AI and neuroscience. Such invention speeds up the rate of research, particularly focusing on computational psychiatry, where machine learning and deep learning's approaches to normality are used by abnormal patters of neural activity in an individual's brain and stimulating the states of the brain using catechisms. Such patterns could be rigorous classifiers for brain diseases that hold the goading of a brighter future for diagnosis and therapy. AI technologies are also really revolutionizing visual neuroscience by bringing forth automatic capturing of behavior and brain activity. Today, brain-inspired neural networks are applied in real-world neuroscience applications, which display how cognitive science is indeed guiding the search for AI models that could speculate brain activity. All these developments are showing how revolutionary AI is, in that could cut across the boundaries of behavioral and neurological investigations.

2.15.2 AI and Precision Nutrition for Brain Health

Combining AI with precision nutrition, or personalized nutrition, opens a new frontier toward maximizing nutrition for brain health. Personalized dietary consideration may well pay further attention to cognitive function and resilience against neurodegeneration, now all taking accounts of genetics, microbiome, and lifestyle data. Nutrigenomics, as understood in this context, lies at the core of understanding how genetic variables influence the response of individual to specific dietary components. As an example, through an Apolipoprotein E (APOE) gene variant having the greatest impacts on omega-3 fatty acids. Such insight allows personalized dietary therapies, which can bring down protection against disease like Alzheimer's (Lyall et al. 2023).

Interestingly, wearable technology and smartphone apps are two examples of technological advancements that make it easier to track cognitive function in real time. These techniques make it possible to make quick dietary changes in response to demands that are identified. Nootropics and cognitive enhancement functional foods are also on the way to become available for cognitive wellness improvement. Times now have passed developments, but ethical issues are still in existence concerning the confidentiality of gene data and equitable distribution of technology for precision nutrition (Vatansever et al. 2021). Putting together AI and personalized nutrition has great potential to completely reshape public health through adaptation of diets to meet individual demands. This novel strategy manifests an exhilarating and rosy future by actually improving cognitive function, enhancing brain health, and turning the tide against cognitive decline (Winchester et al. 2024).

2.15.3 AI in Data Analysis and Healthcare Applications

Other than that, AI has brought revolutionary changes in the prediction of outcomes, not only in nutritional neuroscience but also in stroke research. Customized dietary recommendations would become more efficient due to AI's capability to examine large datasets and spot "complex patterns" connecting eating habits to brain health (Ho et al. 2021). These discoveries evince the fact that within the realm of nutrition, AI can very well revolutionize our entire concept of how nutrition affects neurological and cognitive function as far as the research goes. There should be extensive "individual-related data" to achieve the end of precise medicine, especially research on stroke and nutritional neuroscience. Data are necessary for nutritionist to devise individualized food plans that promote brain health and cognitive performance. Sufficient data quality is what is required of effective AI engineering. "Satisfactory data quality" is required to achieve progressed AI forecasts (Bonkhoff and Grefkes 2022).

Then, to ensure that customized nutrition does not exacerbate gaps, addressing bias in datasets, and ensuring fairness of AI models are essential for realizing equitable health outcomes among different populations (Richards et al. 2022). Moreover, the study of neurobiology of human-AI interaction provides an entirely new outlook into functioning of the brain. Social contact tends to activate the anterior insula, medial PFC, and middle temporal gyrus, which respectively represents aspects of social cognition and appraisal (Harris 2024). These findings emphasize the need to incorporate human-oriented AI approaches in nutrition and neuroscience, requiring the guidance of experts and input from users to verify recommendations proposed by AI.

2.15.4 The Broader Implications of AI Across Disciplines

AI's significant influence transcends academic boundaries, with the potential to change disciplines themselves (Surianarayanan et al. 2023). Rapid improvements in computational power now enable AI to tackled huge, complex datasets at speeds once considered impossible, leading primarily to breakthrough studies in neuroscience and meta-optics. AI's uses in precision nutrition also demonstrate its ability to customize and optimize methods for improving wellness. For instance, AI's accurate pattern recognition requires it to garner information from several sources, including real-time observations of food intake and cognitive performance registered by wearable technologies and smartphones apps.

Ongoing assessment of the technologies could ensure personalized recommendations, thus improving AI's value for precision nutrition (Richards et al. 2022). Yet, many ethical dilemmas face AI in the healthcare application setting, particularly issues related to the privacy of dietary and genetic data. Therefore, public trust and reduction of misuse rely on transparent processes, careful monitoring for errors, and stringent adherences to ethics (Monteith et al. 2023). With the capacity to stimulate the probabilistic model of human information-seeking behavior, AI increases its utility in decision-making tasks. The models predict humans' responses to dietary changes by assessing the importance and prioritizing information (Nakamura 2022). In highlighting the growth of an ever-present dynamic feedback loop geared toward promoting the brain and nervous systems' health, AI updates advice based on new information (Thompson 2025).

References

Abdullah M, Jamil RT, Attia FN (2023) Vitamin C (ascorbic acid). StatPearls – NCBI bookshelf. https://www.ncbi.nlm.nih.gov/books/NBK499877/

Adams MS, Mensink RP, Joris PJ (2024) Effects of dietary proteins on cognitive performance and brain vascular function in adults: a systematic review of randomised controlled trials. Nutr Res Rev 1–15. https://doi.org/10.1017/s0954422424000271

Aderinto N, Olatunji G, Abdulbasit M, Olajide TN, Kokori E (2023) Examining the efficacy of the Mediterranean-DASH diet intervention for neurodegenerative delay in mitigating cognitive decline. Egypt J Neurol Psychiatry Neurosurg 59(1):145

Agrawal S, Kumar S, Ingole V, Acharya S, Wanjari A, Bawankule S, Raisinghani N (2019) Does anemia affects cognitive functions in neurologically intact adult patients: two year cross sectional study at rural tertiary care hospital. J Fam Med Prim Care 8(9):3005. https://doi.org/10.4103/jfmpc.jfmpc_599_19

Ahluwalia MK (2021) Nutrigenetics and nutrigenomics—a personalized approach to nutrition. Adv Genet 108:277–340

Alameda L, Trotta G, Quigley H, Rodriguez V, Gadelrab R, Dwir D et al (2022) Can epigenetics shine a light on the biological pathways underlying major mental disorders? Psychol Med 52(9):1645–1665

Alhazmi S, Alshehri L, Alyahya M, AlMadani R, Aboalrasheeed R, Metwally R, Alotaishan S, AlSheikh R, Abdalla A, Abuzaid O, Sebastian T, Goja AM (2024) The association between iron deficiency anemia and cognitive abilities among female university students. Electron J Gen Medicine 21(2):em576. https://doi.org/10.29333/ejgm/14353

Amanullah S, Shivakumar K, Hassan S, Canfield A, Cole J (2017) Synthetic cannabinoids in dementia. Elsevier eBooks, pp 828–838. https://doi.org/10.1016/b978-0-12-800756-3.00100-9
Ansari F, Neshat M, Pourjafar H, Jafari SM, Samakkhah SA, Mirzakhani E (2023) The role of probiotics and prebiotics in modulating of the gut-brain axis. Front Nutr 10:1173660
Appleton J (2018) The gut-brain axis: influence of microbiota on mood and mental health. Integr Med (Encinitas, Calif) 17(4):28–32
Arora S, Santiago JA, Bernstein M, Potashkin JA (2023) Diet and lifestyle impact the development and progression of Alzheimer's dementia. Front Nutr 10:1213223
Arumugam V, Venugopal V, Balakrishnan A, Bhandari R, Boopalan D, Ponnurangam R et al (2024) Effects of ashwagandha (Withania Somnifera) on stress and anxiety: a systematic review and meta-analysis. Explore 20:103062
Beard JL, Wiesinger JA, Connor JR (2021) Iron and the brain. Nutrients 13(3):654
Behl N (1999) Vitamin E and other antioxidants in neuroprotection. Int J Vitam Nutr Res 69(3):213–219. https://doi.org/10.1024/0300-9831.69.3.213
Bekdash RA (2021) Early life nutrition and mental health: the role of DNA methylation. Nutrients 13(9):3111
Bekdash RA (2024) Epigenetics, nutrition, and the brain: improving mental health through diet. Int J Mol Sci 25(7):4036
Berding K, Vlckova K, Marx W, Schellekens H, Stanton C, Clarke G et al (2021) Diet and the microbiota–gut–brain axis: sowing the seeds of good mental health. Adv Nutr 12(4):1239–1285
Bhatnagar S, Taneja S (2001) Zinc and cognitive development. Br J Nutr 85(S2):S139–S145. https://doi.org/10.1079/bjn2000306
Bhoomi A (2024) The gut-brain axis: exploring the bidirectional communication between the gut microbiome and the brain. J Forensic Sci Res 8(1):047–057. https://doi.org/10.29328/journal.jfsr.1001064
Bianchi VE, Rizzi L, Somaa F (2023) The role of nutrition on Parkinson's disease: a systematic review. Nutr Neurosci 26(7):605–628
Bisaglia M (2022) Mediterranean diet and Parkinson's disease. Int J Mol Sci 24(1):42
Black MM (1998) Zinc deficiency and child development. Am J Clin Nutr 68(2):464S–469S. https://doi.org/10.1093/ajcn/68.2.464s
Black MM (2003a) Micronutrient deficiencies and cognitive functioning. J Nutr 133(11):3927S–3931S. https://doi.org/10.1093/jn/133.11.3927s
Black MM (2003b) The evidence linking zinc deficiency with children's cognitive and motor functioning. J Nutr 133(5):1473S–1476S. https://doi.org/10.1093/jn/133.5.1473s
Boer OD, El H, Franken IHA (2022) Brain morphology predictors of alcohol, tobacco, and cannabis use in adolescence: a systematic review. Brain Res 1795(July):148020. https://doi.org/10.1016/j.brainres.2022.148020
Bonkhoff AK, Grefkes C (2022) Precision medicine in stroke: towards personalized outcome predictions using arti fi cial intelligence. Brain 145:457
Brodt S, Inostroza M, Niethard N, Born J (2023) Review sleep — a brain-state serving systems memory consolidation. Neuron 111(7):1050–1075. https://doi.org/10.1016/j.neuron.2023.03.005
Cabiedes-Miragaya L, Diaz-Mendez C, Garcia-Espejo I (2021) Well-being and the lifestyle habits of the spanish population: the association between subjective Well-being and eating habits. Int J Environ Res Public Health 18(4):1553
Calderaro A, Patanè GT, Tellone E, Barreca D, Ficarra S, Misiti F, Laganà G (2022) The neuroprotective potentiality of flavonoids on Alzheimer's disease. Int J Mol Sci 23(23):14835. https://doi.org/10.3390/ijms232314835
Cao R, Xie J, Zhang L (2022a) Abnormal methylation caused by folic acid deficiency in neural tube defects. Open Life Sci 17(1):1679–1688
Cao S, Li M, Sun Y, Wu P, Yang W, Dai H et al (2022b) Intermittent fasting enhances hippocampal NPY expression to promote neurogenesis after traumatic brain injury. Nutrition 97:111621. https://doi.org/10.1016/j.nut.2022.111621

Charisis S, Yannakoulia M, Scarmeas N (2024) Diets to promote healthy brain ageing. Nat Rev Neurol:, 5–16 (2024)

Chegini M, Shirani P, Omidvar N, Eini-Zinab H, Pour-Ebrahim F, Rezazadeh A (2022) Relationship between diet quality and depression among Iranian older adults in Tehran. BMC Geriatr 22(1):708

Chen BY, Huang HS, Tsai KJ, Wu JL, Chang YT, Chang MC et al (2022) Protective effect of a water-soluble carotenoid-rich extract of Cordyceps militaris against light-evoked functional vision deterioration in mice. Nutrients 14(8). https://doi.org/10.3390/nu14081675

Chen J, Song W, Zhang W (2023) The emerging role of copper in depression. Front Neurosci 17. https://doi.org/10.3389/fnins.2023.1230404

Chianese R, Coccurello R, Viggiano A, Scafuro M, Fiore M, Coppola G, Operto FF, Fasano S, Laye S, Pierantoni R, Meccariello R (2017) Impact of dietary fats on brain functions. Curr Neuropharmacol 16(7):1059–1085. https://doi.org/10.2174/1570159x15666171017102547

Choe YM, Suh G, Lee BC, Choi I, Lee JH, Kim HS, Kim JW (2022) Association between copper and global cognition and the moderating effect of iron. Front Aging Neurosci 14. https://doi.org/10.3389/fnagi.2022.811117

Chu CQ, Yu LL, Qi GY, Mi YS, Wu WQ, Lee YK et al (2022) Can dietary patterns prevent cognitive impairment and reduce Alzheimer's disease risk: exploring the underlying mechanisms of effects. Neurosci Biobehav Rev 135:104556

Constantin ET, Fonseca S (2020) The effect of food on mental health. Revista Internacional de Educação, Saúde e Ambiente 3(2):1–17

Crews FT, Coleman LG, Macht VA, Vetreno RP (2024) Alcohol, HMGB1, and innate immune signaling in the brain. Alcohol Res 44(August):1–20

Cryan JF, O'Riordan KJ, Cowan CS et al (2020) The microbiota-gut-brain axis. Physiol Rev 99(4):1877–2013

Davaatseren A, Kleifeld J, Kep T (2023) Adult neural stem cells and neurogenesis are resilient to intermittent fasting:1–16. https://doi.org/10.15252/embr.202357268

De Azevedo Cardoso T, Silva RH, Fernandes JL, Arent CO, Amboni G, Borba LA, Padilha APZ, Botelho MEM, Maciel AL, Barichello T, Morales R, Soares SJB, Bagatini MD, Dallagnol C, Brighenti ME, Ignácio ZM, Quevedo J, Ceretta LB, Réus GZ (2023) Stress levels, psychological symptoms, and C-reactive protein levels in COVID-19: a cross-sectional study. J Affect Disord 330:216–226. https://doi.org/10.1016/j.jad.2023.03.019

De Goede J, Van Der Mark-reeuwijk KG, Braun KP, Cessie S, Durston S, Engels RCME et al (2021) Alcohol and brain development in adolescents and young adults: a systematic review of the literature and advisory report of the health council of the Netherlands. Adv Nutr 12(4):1379–1410. https://doi.org/10.1093/advances/nmaa170

Dighriri IM, Alsubaie AM, Hakami FM, Hamithi DM, Alshekh MM, Khobrani FA, Dalak FE, Hakami AA, Alsueaadi EH, Alsaawi LS, Alshammari SF, Alqahtani AS, Alawi IA, Aljuaid AA, Tawhari MQ (2022) Effects of omega-3 polyunsaturated fatty acids on brain functions: a systematic review. Cureus 14. https://doi.org/10.7759/cureus.30091

Diniz TC, Silva JC, De Lima-Saraiva SRG, De Almeida Ribeiro FPR, Pacheco AGM, De Freitas RM, Quintans-Júnior LJ, De Souza Siqueira Quintans J, Mendes RL, Da Silva Almeida JRG (2015) The role of flavonoids on oxidative stress in epilepsy. Oxid Med Cell Longev 2015:1–9. https://doi.org/10.1155/2015/171756

Dominguez LJ, Veronese N, Vernuccio L, Catanese G, Inzerillo F, Salemi G, Barbagallo M (2021) Nutrition, physical activity, and other lifestyle factors in the prevention of cognitive decline and dementia. Nutrients 13(11):4080

Dommisse J (1991) Subtle vitamin-B12 deficiency and psychiatry: a largely unnoticed but devastating relationship? Med Hypotheses 34(2):131–140. https://doi.org/10.1016/0306-9877(91)90181-w

Ekinci GN, Sanlier N (2022) The relationship between nutrition and depression in the life process: a mini-review. Exp Gerontol 172:112072. https://doi.org/10.1016/j.exger.2022.112072

El Dib R, Periyasamy AG, de Barros JL, França CG, Senefonte FL, Vesentini G et al (2021) Probiotics for the treatment of depression and anxiety: a systematic review and meta-analysis of randomized controlled trials. Clin Nutr ESPEN 45:75–90

Esmaealzadeh N, Iranpanah A, Sarris J, Rahimi R (2022) A literature review of the studies concerning selected plant-derived adaptogens and their general function in body with a focus on animal studies. Phytomedicine 105:154354

Esmaiel NN, Ashaat EA, Mosaad R, Fayez A, Ibrahim M, Abdallah ZY et al (2020) The potential impact of COMT gene variants on dopamine regulation and phenotypic traits of ASD patients. Behav Brain Res 378:112272

Fadhilah N, Aripin K, Zahid NI, Aizat M, Rahim A, Yaacob H et al (2024) Food science and human wellness a review of salivary composition changes induced by fasting and its impact on health. Food Sci Human Wellness 13(1):50–64. https://doi.org/10.26599/FSHW.2022.9250004

Fairbairn CE, Kang D, Federmeier KD (2022) HHS Public Access 89(10):990–1000. https://doi.org/10.1016/j.biopsych.2020.11.024.Alcohol

Fernstrom JD, Wurtman RJ (2020) Dietary precursors and brain neurotransmitter synthesis. Nutr Neurosci 23(8):621–630

Fismen A-S, Aarø LE, Thorsteinsson E, Ojala K, Samdal O, Helleve A, Eriksson C (2024) Associations between eating habits and mental health among adolescents in five Nordic countries: a cross-sectional survey. BMC Public Health 24:2640

Fjell AM, Sørensen Ø, Wang Y, Amlien IK, Baaré WFC, Drevon CA et al (2023) Is short sleep bad for the brain ?Brain structure and cognitive function in short sleepers. J Neurosci 43(32048):5241–5250

Franco GA, Interdonato L, Cordaro M, Cuzzocrea S, Di Paola R (2023) Bioactive compounds of the Mediterranean diet as nutritional support to fight neurodegenerative disease. Int J Mol Sci 24(8):7318. https://doi.org/10.3390/ijms24087318

Franzoni F, Scarfò G, Guidotti S, Fusi J, Asomov M, Pruneti C (2021) Oxidative stress and cognitive decline: the neuroprotective role of natural antioxidants. Front Neurosci 15. https://doi.org/10.3389/fnins.2021.729757

Frye BM, Negrey JD, Johnson CS, Kim J, Barcus RA, Lockhart SN, Whitlow CT, Chiou KL, Snyder-Mackler N, Montine TJ, Craft S, Shively CA, Register TC (2024) Mediterranean diet protects against a neuroinflammatory cortical transcriptome: associations with brain volumetrics, peripheral inflammation, social isolation, and anxiety in nonhuman primates (Macaca fascicularis). Brain Behav Immun 119:681–692. https://doi.org/10.1016/j.bbi.2024.04.016

Gao R, Yang Z, Yan W, Du W, Zhou Y, Zhu F (2022) Protein intake from different sources and cognitive decline over 9 years in community-dwelling older adults. Front Public Health 10. https://doi.org/10.3389/fpubh.2022.1016016

Gibson EL, Barr S, Jeanes YM (2013) Habitual fat intake predicts memory function in younger women. Front Human Neurosci 7. https://doi.org/10.3389/fnhum.2013.00838

Godos J, Currenti W, Angelino D, Mena P, Castellano S, Caraci F, Galvano F, Del Rio D, Ferri R, Grosso G (2020) Diet and mental health: review of the recent updates on molecular mechanisms. Antioxidants 9(4):346. https://doi.org/10.3390/antiox9040346

Gómez-Pinilla F (2008) Brain foods: the effects of nutrients on brain function. Nat Rev Neurosci 9(7):568–578. https://doi.org/10.1038/nrn2421

Gomez-Pinilla F (2021) Brain foods: the effects of nutrients on brain function. Nat Rev Neurosci 22(5):322–336

González-Velázquez G, Aguirre-Garrido JF, Oros-Pantoja R, Salinas-Velarde ID, Contreras I, Estrada JA, Soto-Piña AE (2024) Supplementation with inulin reverses cognitive flexibility alterations and modulates the gut microbiota in high-fat-fed mice. Front Behav Neurosci 18:1445154

Grajek M, Krupa-Kotara K, Białek-Dratwa A, Sobczyk K, Grot M, Kowalski O, Staśkiewicz W (2022) Nutrition and mental health: a review of current knowledge about the impact of diet on mental health. Front Nutr 9:943998

Grammatikopoulou MG, Goulis DG, Gkiouras K, Theodoridis X, Gkouskou KK, Evangeliou A et al (2020) To keto or not to keto? A systematic review of randomized controlled trials assessing the effects of ketogenic therapy on Alzheimer disease. Adv Nutr 11(6):1583–1602

Gupta R, Srivastava D, Sahu M, Tiwari S, Ambasta RK, Kumar P (2021) Artificial intelligence to deep learning: machine intelligence approach for drug discovery. Mol Divers 25. https://doi.org/10.1007/s11030-021-10217-3

Hameed M, Noor F, Hussain H, Khan RG, Rashid SKHU, Rashid SHU, Atiq A, Ali H, Rida SE, Abbasi MA (2024) Gut-brain axis to investigate the effects of gut health on cognitive functioning in adults. Cureus. https://doi.org/10.7759/cureus.64286

Hanna M, Jaqua E, Nguyen V, Clay J (2022) B vitamins: functions and uses in medicine. Perm J 26(2):89–97. https://doi.org/10.7812/tpp/21.204

Harris LT (2024) The neuroscience of human and artificial intelligence presence. Annu Rev Psychol 75:433

Hawkins MA, Keirns NG, Helms Z (2018) Carbohydrates and cognitive function. Curr Opin Clin Nutr Metab Care 21(4):302–307. https://doi.org/10.1097/mco.0000000000000471

Ho D, Quake SR, Mccabe ERB, Chng WJ, Edward K, Ding X et al (2021) HHS Public Access. 38(5):497–518. https://doi.org/10.1016/j.tibtech.2019.12.021.Enabling

Hoepner CT, McIntyre RS, Papakostas GI (2021) Impact of supplementation and nutritional interventions on pathogenic processes of mood disorders: a review of the evidence. Nutrients 13(3):767

Hu FB (2024) Diet strategies for promoting healthy aging and longevity: an epidemiological perspective. J Intern Med 295(4):508–531

Hu Y, Yang Y, Zhang M, Deng M, Zhang J (2017) Intermittent fasting pretreatment prevents cognitive impairment in a rat model of chronic cerebral hypoperfusion. J Nutr 147(7):1437–1445. https://doi.org/10.3945/jn.116.245613

Ivanova Stojcheva E, Quintela JC (2022) The effectiveness of Rhodiola rosea L. preparations in alleviating various aspects of life-stress symptoms and stress-induced conditions—encouraging clinical evidence. Molecules 27(12):3902

Jacka FN, O'Neil A, Opie R, Itsiopoulos C, Cotton S, Mohebbi M et al (2017) A randomised controlled trial of dietary improvement for adults with major depression (the 'SMILES' TRIAL). BMC Med 15:1–13

Jacka FN, O'Neil A, Opie R et al (2020) A review of the impact of dietary patterns on mental health. Mol Psychiatry 25(5):543–556

Jamar G, Ribeiro DA, Pisani LP (2021) High-fat or high-sugar diets as trigger inflammation in the microbiota-gut-brain axis. Crit Rev Food Sci Nutr 61(5):836–854

Jansen K, Tempes J, Drozdowska A, Gutmann M, Falkenstein M, Buyken AE, Libuda L, Rudolf H, Lücke T, Kersting M (2020) Short-term effects of carbohydrates differing in glycemic index (GI) consumed at lunch on children's cognitive function in a randomized crossover study. Eur J Clin Nutr 74(5):757–764. https://doi.org/10.1038/s41430-020-0600-0

Jáuregui-Lobera I (2014) Iron deficiency and cognitive functions. Neuropsychiatr Dis Treat 2087. https://doi.org/10.2147/ndt.s72491

Jensen NJ, Wodschow HZ, Nilsson M, Rungby J (2020) Effects of ketone bodies on brain metabolism and function in neurodegenerative diseases. Int J Mol Sci 21(22):8767

Jiang Z, Yin X, Wang M, Chen T, Wang Y, Gao Z, Wang Z (2022a) Effects of ketogenic diet on neuroinflammation in neurodegenerative diseases. Aging Dis 13(4):1146. https://doi.org/10.14336/ad.2021.1217

Jiang Z, Yin X, Wang M, Wang Y, Li F, Gao Y et al (2022b) International immunopharmacology β-Hydroxybutyrate alleviates pyroptosis in MPP+/MPTP-induced Parkinson's disease models via inhibiting STAT3/NLRP3/GSDMD pathway. Int Immunopharmacol 113(PB):109451. https://doi.org/10.1016/j.intimp.2022.109451

Kahleova H, Levin S, Barnard N (2021) Vegetarian diets and mental health. Nutrients 13(5):1727

Kang JH, Vyas CM, Okereke OI, Ogata S, Albert M, Lee I, D'Agostino D, Buring JE, Cook NR, Grodstein F, Manson JE (2021) Effect of vitamin D on cognitive decline: results from two ancillary studies of the VITAL randomized trial. Sci Rep 11(1). https://doi.org/10.1038/s41598-021-02485-8

Karoly HC, Gowin JL, Schacht JP (2024) Alcohol and brain structure across the lifespan: a systematic review of large-scale neuroimaging studies. (September), pp 1–16. https://doi.org/10.1111/adb.13439

Kennedy D (2016) B vitamins and the brain: mechanisms, dose and efficacy—a review. Nutrients 8(2):68. https://doi.org/10.3390/nu8020068

Kennedy J, Alexander P, Taillie LS, Jaacks LM (2024) Estimated effects of reductions in processed meat consumption and unprocessed red meat consumption on occurrences of type 2 diabetes, cardiovascular disease, colorectal cancer, and mortality in the USA: a microsimulation study. Lancet Planet Health 8(7):e441–e451. https://doi.org/10.1016/s2542-5196(24)00118-9

Kheirouri S, Alizadeh M (2022) MIND diet and cognitive performance in older adults: a systematic review. Crit Rev Food Sci Nutr 62(29):8059–8077

Khoshbin K, Camilleri M (2020) Effects of dietary components on intestinal permeability in health and disease. Am J Physiol Gastrointest Liver Physio 319(5):G589–G608

Kim J, Wessling-Resnick M (2014) Iron and mechanisms of emotional behavior. J Nutr Biochem 25(11):1101–1107. https://doi.org/10.1016/j.jnutbio.2014.07.003

Kim CS, Cha J, Sim M, Jung S, Chun WY, Baik HW, Shin DM (2021) Probiotic supplementation improves cognitive function and mood with changes in gut microbiota in community-dwelling older adults: a randomized, double-blind, placebo-controlled, multicenter trial. J Gerontol: Ser A 76(1):32–40

Koh S, Dupuis N, Auvin S (2020) Ketogenic diet and neuroinflammation. Epilepsy Res 167:106454. https://doi.org/10.1016/j.eplepsyres.2020.106454

Kuhns L (2022) Age-related differences in the effect of chronic alcohol on cognition and the brain: a systematic review. https://doi.org/10.1038/s41398-022-02100-y

Lam LF, Lawlis TR (2016) Feeding the brain—the effects of micronutrient interventions on cognitive performance among school-aged children: a systematic review of randomized controlled trials. Clin Nutr 36(4):1007–1014. https://doi.org/10.1016/j.clnu.2016.06.013

Lagoumintzis G, Patrinos GP (2023) Triangulating nutrigenomics, metabolomics and microbiomics toward personalized nutrition and healthy living. Hum Genomics 17(1):109

Lam RW, Kennedy SH, Adams C, Bahji A, Beaulieu S, Bhat V Milev RV (2024) Canadian Network for Mood and Anxiety Treatments (CANMAT) 2023 Update on Clinical Guidelines for Management of Major Depressive Disorder in Adults: Réseau canadien pour les traitements de l'humeur et de l'anxiété (CANMAT) 2023: Mise à jour des lignes directrices cliniques pour la prise en charge du trouble dépressif majeur chez les adultes. Can J Psychiatry 69(9):641–687

Lane MM, Gamage E, Du S, Ashtree DN, McGuinness AJ, Gauci S, Marx W (2024) Ultra-processed food exposure and adverse health outcomes: umbrella review of epidemiological meta-analyses. BMJ 384:e07731

Lange KW (2020) Omega-3 fatty acids and mental health. Global Health J 4(1):18–30. https://doi.org/10.1016/j.glohj.2020.01.004

Lassale C, Batty GD, Baghdadli A, Jacka F, Sánchez-Villegas A, Kivimäki M, Akbaraly T (2019) Healthy dietary indices and risk of depressive outcomes: a systematic review and meta-analysis of observational studies. Mol Psychiatry 24(7):965–986

Lawson LE, Watson RR (2014) Omega-3 fatty acid supplementation for major depression with chronic disease. Elsevier eBooks, pp 181–186. https://doi.org/10.1016/b978-0-12-410527-0.00015-6

Lee KH, Cha M, Lee BH (2020) Neuroprotective effect of antioxidants in the brain. Int J Mol Sci 21(19):7152. https://doi.org/10.3390/ijms21197152

Lees B, Meredith LR, Kirkland AE, Bryant BE, Lindsay M, Use S et al (2020) Effect of alcohol use on the adolescent brain and behavior. Pharmacol Biochem Behav:1–27. https://doi.org/10.1016/j.pbb.2020.172906.Effect

Lees B, Debenham J, Squeglia LM (2021) Alcohol and Cannabis use and the developing brain. 41:11

Lewis LD (2022) The interconnected causes and consequences of sleep in the brain. 374(6567):564–568. https://doi.org/10.1126/science.abi8375.The
Li Y, Li S, Wang W, Zhang D (2020) Association between dietary protein intake and cognitive function in adults aged 60 years and older. J Nutr Health Aging 24(2):223–229. https://doi.org/10.1007/s12603-020-1317-4
Li X, Chen X, Gao X (2023) Copper and cuproptosis: new therapeutic approaches for Alzheimer's disease. Front Aging Neurosci 15. https://doi.org/10.3389/fnagi.2023.1300405
Lichtenstein AH, Appel LJ, Vadiveloo M, Hu FB, Kris-Etherton PM, Rebholz CM et al (2021) 2021 dietary guidance to improve cardiovascular health: a scientific statement from the American Heart Association. Circulation 144(23):e472–e487
Liwinski T, Lang UE (2023) Folate and its significance in depressive disorders and suicidality: a comprehensive narrative review. Nutrients 15(17):3859
Lobo V, Patil A, Phatak A, Chandra N (2010) Free radicals, antioxidants and functional foods: impact on human health. Pharmacogn Rev 4(8):118. https://doi.org/10.4103/0973-7847.70902
Loh JS, Mak WQ, Tan LKS, Ng CX, Chan HH, Yeow SH, Foo JB, Ong YS, How CW, Khaw KY (2024) Microbiota–gut–brain axis and its therapeutic applications in neurodegenerative diseases. Signal Transduct Target Ther 9(1). https://doi.org/10.1038/s41392-024-01743-1
Lokhandwala S, Spencer RMC (2022) Developmental cognitive neuroscience relations between sleep patterns early in life and brain development: a review. Dev Cogn Neurosci 56(March):101130. https://doi.org/10.1016/j.dcn.2022.101130
Lu Y, Li J, Hu T, Huang G (2021) Serum 25-hydroxy vitamin D level is associated with cognitive impairment in people aged 65 years and older. Ann Palliat Med 10(7):7479–7485. https://doi.org/10.21037/apm-21-568
Lyall DM, Kormilitzin A, Lancaster C, Sousa J, Petermann-rocha F, Buckley C et al (2023) Artificial intelligence for dementia — applied models and digital health the deep dementia phenotyping (DEMON). Network 15:5872–5884. https://doi.org/10.1002/alz.13391
Lyons LC, Vanrobaeys Y, Abel T (2024) Sleep and memory: the impact of sleep deprivation on transcription translational control, and protein synthesis in the brain. J Neurochem 166(1):24–46. https://doi.org/10.1111/jnc.15787.Sleep
Macpherson T, Churchland A, Sejnowski T, Dicarlo J (2021) Natural and artificial intelligence: a brief introduction to the interplay between AI and neuroscience research. Neural Netw 144:603–613. https://doi.org/10.1016/j.neunet.2021.09.018
Mahalakshmi AM, Ray B, Tuladhar S, Bhat A (2020) Sleep, brain vascular health and ageing. Geroscience 42:1257–1283
Makkar R, Behl T, Bungau S, Zengin G, Mehta V, Kumar A et al (2020) Nutraceuticals in neurological disorders. Int J Mol Sci 21(12):4424
Mason GM, Lokhandwala S, Riggins T, Spencer RMC, Sciences B, Program B (2022) HHS Public Access. https://doi.org/10.1016/j.smrv.2021.101472.Sleep
Mavromatis LA, Rosoff DB, Cupertino RB, Garavan H, Mackey S, Lohoff FW (2022) Association between brain structure and alcohol use behaviors in adults a mendelian randomization and multiomics study. 1540(9):869–878. https://doi.org/10.1001/jamapsychiatry.2022.2196
Melgar-locatelli S, De Ceglia M, Castilla-ortega E, Castro-zavala A (2023) Nutrition and adult neurogenesis in the hippocampus: does what you eat help you remember?:1–9. https://doi.org/10.3389/fnins.2023.1147269
de Melo Simplício AP, de Almeida Fonseca Viola PC, de Carvalho Lavôr LC, de Lima Sousa PV, de Carvalho CA, Lopes Rodrigues LA et al (2024) Unhealthy dietary pattern associated with common mental disorders in adults and older adults: a population-based study. Curr Nutr Food Sci 20(9):1155–1164
Menezo Y, Elder K, Clement A, Clement P (2022) Folic acid, folinic acid, 5 methyl tetrahydrofolate supplementations for mutations that affect epigenesis through the folate and one-carbon cycles. Biomolecules 12(2):197
Merlo G, Bachtel G, Sugden SG (2024) Gut microbiota, nutrition, and mental health. Front Nutr 11. https://doi.org/10.3389/fnut.2024.1337889

Miao A, Luo T, Hsieh B, Edge CJ, Gridley M, Tak R et al (2024) Brain clearance is reduced during sleep and anesthesia. Nat Neurosci 27. https://doi.org/10.1038/s41593-024-01638-y

Minocha T, Birla H, Obaid AA, Rai V, Sushma P, Shivamallu C, Moustafa M, Al-Shehri M, Al-Emam A, Tikhonova MA, Yadav SK, Poeggeler B, Singh D, Singh SK (2022) Flavonoids as promising neuroprotectants and their therapeutic potential against Alzheimer's disease. Oxid Med Cell Longev 2022:1–13. https://doi.org/10.1155/2022/6038996

Monda A, La Torre ME, Messina A, Di Maio G, Monda V, Moscatelli F, De Stefano M, La Marra M, Di Padova M, Dipace A, Limone P, Casillo M, Monda M, Messina G, Polito R (2024) Exploring the ketogenic diet's potential in reducing neuroinflammation and modulating immune responses. Front Immunol 15. https://doi.org/10.3389/fimmu.2024.1425816

Monteith S, Glenn T, Geddes JR, Achtyes ED, Whybrow PC, Bauer M (2023) Challenges and ethical considerations to successfully implement artificial intelligence in clinical medicine and neuroscience: a narrative review authors. https://doi.org/10.1055/a-2142-9325

Moriguchi T, Greiner RS, Salem N (2000) Behavioral deficits associated with dietary induction of decreased brain docosahexaenoic acid concentration. Neurochem 75(6):2563–2573. https://doi.org/10.1046/j.1471-4159.2000.0752563.x

Morris MC, Tangney CC, Wang Y, Sacks FM, Barnes LL, Bennett DA, Aggarwal NT (2015) MIND diet slows cognitive decline with aging. Alzheimers Dement 11(9):1015–1022. https://doi.org/10.1016/j.jalz.2015.04.011

Morris MC, Tangney CC, Wang Y et al (2022) MIND diet and cognitive function. Am J Clin Nutr 116(3):702–710

Mostafa H, Gutierrez-Tordera L, Mateu-Fabregat J, Papandreou C, Bulló M (2023) Dietary fat, telomere length and cognitive function: unravelling the complex relations. Curr Opin Lipidol 35. https://doi.org/10.1097/mol.0000000000000900

Muth A, Park SQ (2021) The impact of dietary macronutrient intake on cognitive function and the brain. Clin Nutr 40(6):3999–4010. https://doi.org/10.1016/j.clnu.2021.04.043

Nakamura K (2022) Information seeking criteria: arti fi cial intelligence, economics, psychology, and neuroscience. 33(1):31–41

Navratilova HF, Whetton AD, Geifman N (2024) Artificial intelligence driven definition of food preference endotypes in UK Biobank volunteers is associated with distinctive health outcomes and blood based metabolomic and proteomic profiles. J Transl Med 22(1):881

Neurology Solutions (2024) DASH diet recommended for people with Parkinson's disease. Available at: https://www.neurologysolutions.com

Nestel PJ, Mori TA (2024) Diet to Stop Hypertension: Should Fats be Included?. Curr Hypertens Rep 26(10):409–417

Nieto MRP (2024) The impact of nutrition on people's cognitive development. Nutr Food Process 7(9):01–04. https://doi.org/10.31579/2637-8914/244

Nishi SK, Babio N, Gómez-Martínez C, Martínez-González MÁ, Ros E, Corella D et al (2021) Mediterranean, DASH, and MIND dietary patterns and cognitive function: the 2-year longitudinal changes in an older Spanish cohort. Front Aging Neurosci 13:782067

Ojha U, Khanal S, Park P, Tae J, Choi D (2023) Intermittent fasting protects the nigral dopaminergic neurons from MPTP-me diate d dopaminergic neuronal injury in mice. J Nutr Biochem 112:109212. https://doi.org/10.1016/j.jnutbio.2022.109212

Oliveira C, Paroschi G, Pereira F, Oliveira D, Montavoni M, Santos A et al (2023) Intermittent fasting combined with exercise training reduces body mass and alleviates hypothalamic disorders induced by high-fat diet intake:119. https://doi.org/10.1016/j.jnutbio.2023.109372

O'neil A, Quirk SE, Housden S, Brennan SL, Williams LJ, Pasco JA, Jacka FN (2014) Relationship between diet and mental health in children and adolescents: a systematic review. Am J Public Health 104(10):e31–e42

Panossian AG, Efferth T, Shikov AN, Pozharitskaya ON, Kuchta K, Mukherjee PK et al (2021) Evolution of the adaptogenic concept from traditional use to medical systems: pharmacology of stress-and aging-related diseases. Med Res Rev 41(1):630–703

Peng B, Yang Q, Joshi RB, Liu Y, Akbar M, Song B et al (2020) Role of alcohol drinking in Alzheimer's disease, Parkinson's disease, and amyotrophic lateral sclerosis. Int J Mol Sci 21:2316

Perez-kast RC, Camacho-morales A (2025) Fasting the brain for mental health. J Psychiatr Res 181(November 2024):215–224. https://doi.org/10.1016/j.jpsychires.2024.11.041

Piekarski DJ, Pfefferbaum A, Zahr NM, Zhao Q, Sullivan EV (2022) Alcohol's effects on the mouse brain are modulated by age and sex. https://doi.org/10.1111/adb.13209

Pizzino G, Irrera N, Cucinotta M, Pallio G, Mannino F, Arcoraci V, Squadrito F, Altavilla D, Bitto A (2017) Oxidative stress: harms and benefits for human health. Oxid Med Cell Longev 2017(1). https://doi.org/10.1155/2017/8416763

Poh L, Rajeev V, Selvaraji S, Lai MKP, Chen CL, Arumugam TV, Fann DY (2021) Neurochemistry international intermittent fasting attenuates inflammasome-associated apoptotic and pyroptotic death in the brain following chronic hypoperfusion. Neurochem Int 148(November 2020):105109. https://doi.org/10.1016/j.neuint.2021.105109

Polverino A, Sorrentino P, Pesoli M, Mandolesi L (2021) Nutrition and cognition across the lifetime: an overview on epigenetic mechanisms. AIMS Neurosci 8(4):448

Psaltopoulou T, Sergentanis TN, Panagiotakos DB et al (2022) Mediterranean diet and depression. Public Health Nutr 25(3):432–443

Puri S, Shaheen M, Grover B (2023) Nutrition and cognitive health: a life course approach. Front Public Health 11. https://doi.org/10.3389/fpubh.2023.1023907

Qian XH, Xie RY, Liu XL, Tang HD (2022) Mechanisms of short-chain fatty acids derived from gut microbiota in Alzheimer's disease. Aging Dis 13(4):1252

Richards B, Tsao D, Zador A (2022) Ll the application of artificial intelligence to biology and neuroscience. Cell 185(15):2640–2643. https://doi.org/10.1016/j.cell.2022.06.047

Rusek M, Pluta R, Ułamek-Kozioł M, Czuczwar SJ (2019) Ketogenic diet in Alzheimer's disease. Int J Mol Sci 20(16):3892

Sahu P, Thippeswamy H, Chaturvedi SK (2022) Neuropsychiatric manifestations in vitamin B12 deficiency. Vitam Horm 119:457–470. https://doi.org/10.1016/bs.vh.2022.01.001

Samraj AN, Pearce OMT, Läubli H, Crittenden AN, Bergfeld AK, Banda K, Gregg CJ, Bingman AE, Secrest P, Diaz SL, Varki NM, Varki A (2014) A red meat-derived glycan promotes inflammation and cancer progression. Proc Natl Acad Sci 112(2):542–547. https://doi.org/10.1073/pnas.1417508112

Sandua D (2023) The happiness diet: food and its influences on mood, David Sandua

Sarris J, Logan AC, Akbaraly TN et al (2022) Nutritional psychiatry: emerging evidence and global implications. Lancet Psychiatr 9(2):175–185

Sawalha I, Sarhan A (2022) The consequences of B vitamins on mental health and cognitive function: a literature review study. Biomed Sci 4:1745–1754. https://doi.org/10.38125/OAJBS.000432

Schirinzi T, Martella G, Imbriani P, Di Lazzaro G, Franco D, Colona VL, Alwardat M, Salimei PS, Mercuri NB, Pierantozzi M, Pisani A (2019) Dietary vitamin E as a protective factor for Parkinson's disease: clinical and experimental evidence. Front Neurol 10. https://doi.org/10.3389/fneur.2019.00148

Schmidt K, Cowen PJ, Harmer CJ, Tzortzis G, Errington S, Burnet PW (2015) Prebiotic intake reduces the waking cortisol response and alters emotional bias in healthy volunteers. Psychopharmacology 232(10):1793–1801

Seago ER, Davy BM, Davy KP, Katz B (2024) Neuroprotective dietary patterns and longitudinal changes in cognitive function in older adults. J Acad Nutr Diet 125:785

Sebastiani G, Andreu-Fernández V, Herranz Barbero A, Aldecoa-Bilbao V, Miracle X, Meler Barrabes E et al (2020) Eating disorders during gestation: implications for mother's health, fetal outcomes, and epigenetic changes. Front Pediatr 8:587

Seidler K, Barrow M (2022) Frontiers in neuroendocrinology intermittent fasting and cognitive performance—targeting BDNF as potential strategy to optimise brain health. Front Neuroendocrinol 65(November 2021):100971. https://doi.org/10.1016/j.yfrne.2021.100971

Shin MS, Lee YJ, Cho IH, Yang HJ (2024) Brain plasticity and ginseng. J Ginseng Res 48:286
Singar S, Nagpal R, Arjmandi BH, Akhavan NS (2024) Personalized nutrition: tailoring dietary recommendations through genetic insights. Nutrients 16(16):2673
Singh V (2023) Current challenges and future implications of exploiting the omics data into nutrigenetics and nutrigenomics for personalized diagnosis and nutrition-based care. Nutrition 110:112002
Smith AD, Refsum H (2021) Homocysteine–from disease biomarker to disease prevention. J Intern Med 290(4):826–854
Solianik R (2018) Two-day fasting evokes stress, but does not a ff ect mood, brain activity, cognitive, psychomotor, and motor performance in overweight women. 338(October 2017):166–172. https://doi.org/10.1016/j.bbr.2017.10.028
Strianese O, Rizzo F, Ciccarelli M, Galasso G, D'Agostino Y, Salvati A et al (2020) Precision and personalized medicine: how genomic approach improves the management of cardiovascular and neurodegenerative disease. Genes 11(7):747
Suganya K, Koo B (2020) Gut–brain axis: role of gut microbiota on neurological disorders and how probiotics/prebiotics beneficially modulate microbial and immune pathways to improve brain functions. Int J Mol Sci 21(20):7551. https://doi.org/10.3390/ijms21207551
Suh SW, Kim HS, Han JH, Bae JB, Oh DJ, Han JW, Kim KW (2020) Efficacy of vitamins on cognitive function of non-demented people: a systematic review and meta-analysis. Nutrients 12(4):1168. https://doi.org/10.3390/nu12041168
Sultana OF, Bandaru M, Islam MA, Reddy PH (2024) Unraveling the complexity of human brain: structure, function in healthy and disease states. Ageing Res Rev 100:102414
Surianarayanan C, Lawrence JJ, Chelliah PR, Prakash E, Hewage C (2023) Convergence of artificial intelligence and neuroscience towards the diagnosis of neurological disorders—a scoping review. pp 1–29 23(6):3062.https://doi.org/10.3390/s23063062
Taylor MK, Sullivan DK, Mahnken JD (2021) Ketogenic diets in neurodegenerative diseases. J Neurol Neurosurg Psychiatry 92(4):456–462
Thompson JAF (2025) Forms of explanation and understanding for neuroscience and arti fi cial intelligence. 2021(April):1860–1874. https://doi.org/10.1152/jn.00195.2021
Tidman MM, White DR, White TA (2024) Impact of a keto diet on symptoms of Parkinson's disease, biomarkers, depression, anxiety and quality of life: a longitudinal study. Neurodegener Dis Manag:14(3-4):97-110. https://doi.org/10.1080/17582024.2024.2352394
Tommasi M, Sergi MR, Konstantinidou F, Franzago M, Pesce M, Fratta IL et al (2021) Association of COMT, BDNF and 5-HTT functional polymorphisms with personality characteristics. Front Biosci 26(11):1064–1074
Travica N, Ried K, Sali A, Scholey A, Hudson I, Pipingas A (2017) Vitamin C status and cognitive function: a systematic review. Nutrients 9(9):960. https://doi.org/10.3390/nu9090960
Tsigalou C, Konstantinidis T, Paraschaki A, Stavropoulou E, Voidarou C, Bezirtzoglou E (2020) Mediterranean diet as a tool to combat inflammation and chronic diseases. An overview. Biomedicine 8(7):201. https://doi.org/10.3390/biomedicines8070201
Van den Brink AC, Brouwer-Brolsma EM, Berendsen AA, van de Rest O (2019) The Mediterranean, dietary approaches to stop hypertension (DASH), and Mediterranean-DASH intervention for neurodegenerative delay (MIND) diets are associated with less cognitive decline and a lower risk of Alzheimer's disease—a review. Adv Nutr 10(6):1040–1065
Van Zonneveld SM, Van Den Oever EJ, Haarman BCM, Grandjean EL, Nuninga JO, Van De Rest O, Sommer IEC (2024) An anti-inflammatory diet and its potential benefit for individuals with mental disorders and neurodegenerative diseases—a narrative review. Nutrients 16(16):2646. https://doi.org/10.3390/nu16162646
Vatansever S, Schlessinger A, Wacker D, Kaniskan HÜ, Jin J, Zhou MM, Zhang B (2021) Artificial intelligence and machine learning – aided drug discovery in central nervous system diseases: state – of – the – arts and future directions. 2020(July):1427–1473. https://doi.org/10.1002/med.21764

Vore AS, Deak T (2022) Alcohol, inflammation, and blood-brain barrier function in health and disease across development. 1(607):1–36. https://doi.org/10.1016/bs.irn.2021.06.009.Alcohol
Walker LC, Pinares P (2022) How the brain regulates alcohol intake. 10–12. https://doi.org/10.7554/eLife.77411
Wang Y, Uffelman C, Hill E, Anderson N, Reed J, Olson M, Campbell W (2022) The effects of red meat intake on inflammation biomarkers in humans: a systematic review and meta-analysis of randomized controlled trials. Curr Dev Nutr 6:994. https://doi.org/10.1093/cdn/nzac068.023
Wang Q, Hu S, Qi L, Wang X, Jin G, Wu D, et al (2023). Causal associations between sleep traits and brain structure: a bidirectional Mendelian randomization study. 1–13. https://doi.org/10.1186/s12993-023-00220-z
Wang D, Xu J, Liang N, Xue Z, Yang X, Lu J, Ma Y (2024) Network analysis of depressive symptoms and C-reactive protein levels in major depressive disorder. J Affect Disord 367:788–794. https://doi.org/10.1016/j.jad.2024.08.152
https://www.emro.who.int/health-topics/stroke-cerebrovascular-accident/index.html
Weston-Green K, Sun Y, Huang XF, Deng C (2022) The impact of sugar consumption on the brain. Front Neurosci 16:827423
Winchester LM, Harshfield EL, Shi L, Badhwar A, Al Khleifat A, Clarke N et al (2024) Artificial intelligence for biomarker discovery in Alzheimer's disease and dementia. 19(12):5860–5871. https://doi.org/10.1002/alz.13390.Artificial
Wiss DA, LaFata EM (2024) Ultra-processed foods and mental health: where do eating disorders fit into the puzzle? Nutrients 16(12):1955
Wołoszynowska-Fraser MU, Kouchmeshky A, McCaffery P (2020) Vitamin A and retinoic acid in cognition and cognitive disease. Ann Rev Nutr 40(1):247–272. https://doi.org/10.1146/annurev-nutr-122319-034227
Wu NA, Ying NZ, Gomez-Pinilla F (2009) Vitamin E protects against oxidative damage and learning disability after mild traumatic brain injury in rats. Neurorehabil Neural Repair 24(3):290–298. https://doi.org/10.1177/1545968309348318
Wu Q, Yu M, Wang Z, Ai X, Liu Z, Zeng J et al (2023) Neurochemistry international alternate-day fasting for the protection of cognitive impairment in c57BL/6J mice following whole-brain radiotherapy. Neurochem Int 162(May 2022):105463. https://doi.org/10.1016/j.neuint.2022.105463
Xia Q, Cheng Z, He L (2019) The modulatory role of dopamine receptors in brain neuroinflammation. Int Immunopharmacol 76:105908. https://doi.org/10.1016/j.intimp.2019.105908
Xiong J, Wang L, Huang H, Xiong S, Zhang S, Fu Q et al (2024) Association of sugar consumption with risk of depression and anxiety: a systematic review and meta-analysis. Fron Nutr 11:1472612
Yang T, Wang H, Xiong Y, Chen C, Duan K, Jia J, Ma F (2020) Vitamin D supplementation improves cognitive function through reducing oxidative stress regulated by telomere length in older adults with mild cognitive impairment: a 12-month randomized controlled trial. J Alzheimers Dis 78(4):1509–1518. https://doi.org/10.3233/jad-200926
Yang HL, Li MM, Zhou MF, Xu HS, Huan F, Liu N et al (2021) Links between gut dysbiosis and neurotransmitter disturbance in chronic restraint stress-induced depressive behaviours: the role of inflammation. Inflammation 44:2448–2462
Ye YC, Chai SF, Li XR, Wu MN, Cai HY, Wang ZJ (2024) Intermittent fasting and Alzheimer's disease—Targeting ketone bodies as a potential strategy for brain energy rescue. Metabolic Brain Disease, 39(1):129–146
You S, Ma Y, Yan B, Pei W, Wu Q, Ding C, Huang C (2022) The promotion mechanism of prebiotics for probiotics: a review. Front Nutr 9:1000517
Zavitsanou A, Drigas A (2021) Nutrition in mental and physical health. Technium Soc Sci J 23:67
Zhao S, Han T, Pei X, Song Y, Zhang Y, Liu L, Wang X, Hou W, Sun C (2022) The association of diet carbohydrates consumption with cognitive function among US older adults modification by daily fasting duration. Front Aging Neurosci 14. https://doi.org/10.3389/fnagi.2022.991007

Zhu H, Bi D, Zhang Y, Kong C, Du J, Wu X et al (2022) Ketogenic diet for human diseases: the underlying mechanisms and potential for clinical implementations. Signal Transduct Target Ther 7(1):11

Zielińska M, Łuszczki E, Dereń K (2023) Dietary nutrient deficiencies and risk of depression (review article 2018–2023). Nutrients 15(11):2433. https://doi.org/10.3390/nu15112433

Zylinska L, Lisek M, Guo F, Boczek T (2023) Vitamin C modes of action in calcium-involved signaling in the brain. Antioxidants 12(2):231. https://doi.org/10.3390/antiox12020231

How Antioxidants Influence Mental Health?

3

Buse Cagla Ari Sarioglu

Abstract

The determination of the diet and mental well-being is becoming a more frequent place to operate, calming stress on the first result for this species that influences the neurological functionality through oxidative stress. Oxidants, especially reactive oxygen species (ROS), serve as a double-edged sword in cellular action. When it serves the body in lesser amounts, it is beneficial, but the overproduction of ROS will develop oxidative stress related to neurodegenerative or psychiatric disorders. The brain is particularly susceptible to oxidative damage because of its high oxygen use/low antioxidant defenses, so it is even more critical that we reduce our oxidative stress to support mental health. Recent findings provide evidence for the idea that diets rich in antioxidants may ameliorate oxidative stress and protect against cognitive decline and affective brain disorders. The brain is a key organ that benefits from antioxidant nutrition, and significant potential exists for the inclusion of dietary interventions in the management of neurological and mental health diseases via optimizing its functions.

Keywords

Oxidative stress · Antioxidants · Mental health · Cognition · Dietary habits · Neurodegeneration

---

B. C. Ari Sarioglu (✉)
Istanbul Medipol University, Neurology, Istanbul, Turkey

W. Mohamed, A. Özge (eds.), *Feeding the Mind: The Connection Between Diet, Drugs, and Mental Health Volume 1*, Nutritional Neurosciences,
https://doi.org/10.1007/978-981-95-1414-4_3

3.1 Introduction

3.1.1 Overview of Mental Health and Diet

Mental health has emerged as an important factor for overall well-being and a key determinant of both individual and population-level burden of disease, whereas diet is considered as one of the major modifiable lifestyle factors that influence mental health. The association between diet and mental health has received increasing attention in recent years, with a growing research evidence indicating the relevance of dietary patterns to psychological health and cognitive functions. As an emerging field, dietary style aims at identifying nutrients and associated dietary patterns that could modulate mental health outcomes like mood disorders, anxiety, and cognitive decline (Hassan et al. 2022; Singh et al. 2019).

Optimize brain functions with a healthy diet that is high in antioxidants, vitamins, and minerals specifically, an elevated production of ROS is a hallmark of oxidative stress, which negatively affects neuronal health and is linked to the pathogenesis of neurodegenerative disorders, including Parkinson's disease (PD) (Kim et al. 2015; Gandhi and Abramov 2012). Accordingly, dietary patterns such as a Mediterranean-like diet and diets rich in fruits, vegetables, grains, and omega-3 fatty acids are associated with lower rates of depression and cognitive decline, supporting a role for dietary interventions as a preventive strategy for mental health problems (Hassan et al. 2022; Singh et al. 2019; Kim et al. 2015).

3.1.2 The Role of Oxidants in the Body

Oxidants are naturally occurring molecules and an essential part of signaling and homeostasis. Excessive generation of ROS can trigger oxidative stress, an imbalance between the formation of ROS and antioxidant defenses (Singh et al. 2019; Gandhi and Abramov 2012). Oxidative stress is also involved in many pathological conditions, including neurodegenerative diseases as Alzheimer's disease (AD) or PD, where it leads to neuronal damage (Kim et al. 2015; Gandhi and Abramov 2012; Ari et al. 2022).

The relationship between oxidants and cellular function is complex. ROS participate in different cellular activities, for instance, even programmed cell death, inflammation, and immune responsive actions. On the other hand, if the levels of ROS exceed a threshold, they will cause oxidative damage to lipids, proteins, and nucleic acids, causing cellular dysfunction as well as neuronal death. In neurodegenerative diseases, ROS production will even further amplify triggering the creation of a toxic cycle for neurons to experience oxidative stress that may lead to mitochondrial dysfunction (Gandhi and Abramov 2012; Ari et al. 2022; Ikawa et al. 2020). Chronic oxidative stress has been indicated to modulate neurotransmitter systems and induce neuroinflammation, consequently resulting in mental health pathophysiology and cognitive decline (Hassan et al. 2022; Singh et al. 2019; Kim et al. 2015). So, the better we understand these oxidants, it will be more beneficial to design dietary and therapeutic strategies that can help reduce oxidative stress and its accompanying mental health effects.

3.1.3 Oxidative Stress and the Brain

The brain consumes up to 20% of the total oxygen used by the body even though it represents only 2% of body weight, making it an easy target for oxidative stress (Bouayed et al. 2009). Most of this high rest-energy demand is due to the metabolic needs of neuronal activity that includes synaptic plasticity, ion transport, and neurotransmitter synthesis (Cenini et al. 2019). The high lipid content of the brain only serves to compound this vulnerability: polyunsaturated fatty acids in neuronal membranes are especially susceptible to oxidative damage (Liu et al. 2017). ROS are by-products of cellular metabolism, predominantly localized in the mitochondria. The brain has antioxidant defenses that are modest, so an imbalance producing ROS will significantly lead to oxidative damage (Bortoli et al. 2019). In addition, the anatomical and physiological characteristics of the brain are also present, which makes it more vulnerable to oxidative stress. E.g., the blood–brain barrier, despite being protective, also restricts the delivery of antioxidants (Cenini et al. 2019). This characteristic is accentuated by low antioxidant concentrations in the brain compared to other tissues (Ferriero 2001). Therefore, disruption of neuronal integrity and function occurs as a result of an oxidative stress-induced cascade of cellular events. On the other hand, ROS accumulation can induce lipid peroxidation, protein oxidation, and DNA damage (Ribaudo et al. 2020). This oxidative injury is not only due to aging, but it also has a key role in the pathogenesis of some neurodegenerative diseases such as Alzheimer's and Parkinson's disease (Cenini et al. 2019; Liu et al. 2017).

The internal relation between oxidative stress and neuroinflammation exacerbates the response of the brain to oxidative injury. Inflammatory processes can also lead to further oxidative stress as a possible consequence of the production of ROS increased by activated microglia and astrocytes (Cenini et al. 2019). This relationship builds up a vicious loop; oxidative stress produces inflammation, which in turn leads to further oxidative stress and culminates in neuronal cell death and cognitive impairment (Cenini et al. 2019; Liu et al. 2017). In this regard, understanding the mechanisms of oxidative stress in the brain is essential for developing effective therapeutic strategies to neutralize its aggressiveness and prevent neuronal function losses.

3.1.4 Oxidative Injury in Neurons

Diseases, which accelerate oxidative damage, have profound effects upon neuronal signaling and cognitive functions. Neurons are highly susceptible to alterations in their signaling pathways due to oxidative stress, an early biological correlate of cell dysfunction and aging that sets the stage for downstream cellular perturbations

(Ribaudo et al. 2020). Oxidative stress may impair the function of ion channels and receptors, altering neuronal transmission and synaptic plasticity (Cenini et al. 2019). This impairment has been observed to present as cognitive deficits, especially in memory loss and learning dysfunctions (Cenini et al. 2019; Liu et al. 2017).

Oxidative stress acts as a mediator for impairing cognitive function, mainly in various neurodegenerative diseases. There is evidence linking accumulation of amyloid-beta plaques and tau tangles to oxidative damage in Alzheimer's disease, which are also hallmarks of the disease (Liu et al. 2017). Oxidative modification of proteins and lipids can lead to the formation of neurotoxic adjuvants that can interfere with neuronal communication and promote cell death (Cenini et al. 2019). Oxidative stress is also connected to the malfunction of dopaminergic neurons in Parkinson's disease (St-Pierre et al. 2006), which are essential for motor control and whose degeneration occurs in the substantia nigra. In addition, the consequences of oxidative damage are not only limited to causing direct injury to neurons but also affect glial cells surrounding neuronal which cooperate with them in supporting neuronal functions (Cenini et al. 2019). Oxidative stress may act to sabotage the function of astrocytes (and thereby decrease glutamate uptake and increase excitotoxicity), aggravating neuronal damage (Liu et al. 2017). A neuronal/glial interface is thereby the fact partially explaining how oxidative stress is very complicated in the brain and associated with major neurodegenerative disorders.

Apart from neurodegenerative diseases, oxidative stress has been linked to a number of psychiatric conditions as well. Research has indicated that oxidative damage may be involved in the pathophysiology of diseases where alterations in neurotransmitter signaling (e.g., depression and anxiety) or neuroinflammation are known to occur (Ribaudo et al. 2020). The connection between oxidative stress and mental health would illustrate the reasons for more of a complete picture with respect to learning about how damage from oxidation could lead not only to cognitive problems but also to emotional impairment.

3.2 Pathways in Oxidative Stress and the Psychopathology of Mental Illness

3.2.1 Inflammation and Oxidative Stress

Neuroinflammation and oxidative stress have been shown to be closely related processes leading to mental health disorders. Upon activation, microglia release proinflammatory cytokines that induce oxidative damage by elevating the production of ROS, which can further reinforce oxidative stress in the brain (Ari et al. 2022; Di Filippo et al. 2010; Lin et al. 2022). This inflammatory response is more pronounced in neurodegenerative diseases, where mitochondrial dysfunction remains a cornerstone (Di Filippo et al. 2010; Wang et al. 2020). Mitochondria are not only the powerhouses of the cell but also one of the critical origins for ROS and damage to mitochondria-induced vicious circles in terms of inflammation (Ari et al. 2022; Di Filippo et al. 2010; Wang et al. 2020).

Complicating the relation between neuroinflammation and mitochondrial dysfunction, however, are components including damage-associated molecular patterns from damaged mitochondria, which further enhance neuroinflammation (Lin et al. 2022). These findings hint at strategies to target mitochondrial health as a potential therapeutic avenue for reductions in neuroinflammation and the neurological/psychiatric effects of inflammation (Lin et al. 2022; Pereira et al. 2021). Recent studies have convincingly proven that interventions targeting oxidative stress can alleviate neuroinflammation and cognitive decline, providing further details in the pathways leading to therapeutic development (Di Filippo et al. 2010; Lin et al. 2022). Importantly, activation of mitogen-activated protein kinase-1 is accompanied by inhibition of ROS and induction of Nrf2 target genes. Conversely, the involvement of a rise in oxidative markers in the pathology of psychiatric diseases is highlighted because enhanced outcome levels correlate with more severe symptoms when it comes to disorders such as depression and anxiety disorder. This indicates that oxidative stress is involved not only in the pathogenesis of these diseases but also may affect their course and severity. Therefore, studying the connection between neuroinflammation and oxidative stress in mental health is essential for devising successful therapies (Di Filippo et al. 2010; Wang et al. 2020; Pereira et al. 2021).

3.2.2 Mitochondrial Dysfunction

Researches have shown that mitochondrial dysfunction plays a significant role in the pathogenesis of different psychiatric disorders and neurodegenerative diseases. The function and amount of mitochondria are critical in the provision of energy for metabolic maintenance, as mitochondrial dysfunction often results in impaired adenosine triphosphate (ATP) levels that can be particularly detrimental to the functioning neurons (Di Filippo et al. 2010; Wang et al. 2020).

The accompanying increased production of ROS associated with the deleterious mitochondrial impairment serves to potentiate oxidative stress and neuronal damage (Di Filippo et al. 2010; Wang et al. 2020; Wang et al. 2023). Mitochondrial dysfunction in conditions like AD is also associated with altered dynamics, including fragmentation and impaired mitophagy, leading to the accumulation of defective mitochondria and bystander neuronal cell death (Di Filippo et al. 2010; Wang et al. 2020). Researches have also showed mitochondrial dysfunction may cause disruption in the metabolism of important neurotransmitters ultimately leading to the imbalances, which are generally seen in mental health disorders. Similarly, perturbing the synthesis and release of dopamine or serotonin to regulate emotions can be caused by damaged mitochondrial functions, i.e., mood disorder, which is one of various symptoms (Di Filippo et al. 2010; Wang et al. 2020; Wang et al. 2023). Such disruption can ultimately lead to depression and anxiety, which further underscore the role of mitochondria in ensuring normal mental health (Di Filippo et al. 2010). Since all types of dysfunctions are directly or indirectly related (Wang et al. 2020).

In addition, the crosstalk of mitochondrial dysfunction and neuroinflammation is a central concept in mental health. Inflammatory responses triggered by mitochondria are connected to the release of cytokines that contribute themselves to inflamed mitochondria and oxidative stress, therefore establishing a vicious circle between inflammation and oxidative stress (Di Filippo et al. 2010; Wang et al. 2020). This relationship highlights the importance of mitochondrial dysfunction-directed therapeutic strategies to improve neuroinflammation and oxidative stress-mediated pathologies with mental illness (Di Filippo et al. 2010; Wang et al. 2020; Zhao et al. 2022).

3.2.3 Neurotransmitter Imbalances

Neurotransmitter psychopathologies are common in most mental health disorders and are often seen in conjunction with oxidative stress and mitochondrial disruption. Since mitochondria provide energy to synthesize and release neurotransmitters such as serotonin, dopamine, and norepinephrine (Di Filippo et al. 2010; Wang et al. 2020; Kotzaeroglou and Tsamesidis 2022), it is expected that mitochondrial health has a significant effect on the production as well as secretion of these messengers. Mitochondrial dysfunction, which can decrease ATP production and subsequently neurotransmitter synthesis and release (Di Filippo et al. 2010; Wang et al. 2020), has been shown to be a mechanism implicated in mood disorders and cognitive decline. Secondly, oxidative stress may interfere with neurotransmitter systems directly via disruption in the availability of precursors and cofactors required for neurotransmitter synthesis (Di Filippo et al. 2010; Wang et al. 2020; Kotzaeroglou and Tsamesidis 2022). For instance, the oxidation of tryptophan, a precursor for serotonin, can result in decreased availability of serotonin and reduced functions that might be associated with depression (Di Filippo et al. 2010; Wang et al. 2020). Oxidative stress may also alter dopamine metabolism, disrupting dopaminergic pathways that are known to be involved in conditions such as schizophrenia and bipolar disorder (Ari et al. 2022; Di Filippo et al. 2010; Wang et al. 2020; Kotzaeroglou and Tsamesidis 2022).

This is further mutation by neuroinflammation and the complex relationship between neurotransmitter imbalances, mitochondrial dysfunction, and oxidative stress. Inflammatory cytokines have been shown to interfere with the metabolism of neurotransmitters and the sensitivity of their receptors, thus perpetuating an imbalance that can help to intensify mental health disorders (Di Filippo et al. 2010; Wang et al. 2020). This might speak to the implication of targeting mitochondrial dysfunction and oxidative stress for the maintenance of neurotransmitter balance, which would eventually result in better mental health (Di Filippo et al. 2010; Wang et al. 2020; Abshirini et al. 2019; Gautam et al. 2012).

3.2.4 Brain Multipotency and Oxidative Stress

It all starts with neuroplasticity, the brain's remarkable ability to reconfigure itself when learning new skills, creating memories, or recovering from injury. Nevertheless, high oxidative damage would dramatically reduce neuroplasticity and associated cognitive deficits and mental diseases (Di Filippo et al. 2010; Wang et al. 2020). Neuronal structures necessary for neuronal communication, i.e., dendrites and synapses, are vulnerable to oxidative stress (Di Filippo et al. 2010; Wang et al. 2020; Abshirini et al. 2019). Synaptotoxicity results in blocking the creation of new synaptic contacts, contributing to neuroplasticity and brain functioning interference (Di Filippo et al. 2010; Wang et al. 2020; Pereira et al. 2021).

Proper mitochondrial activity is essential for high-energy-requiring synaptic processes such as long-term potentiation and long-term depression (Schon and Przedborski 2011). Since this may result in reduced store of energy in our cells (decreased ATP levels), and when there is not enough ATP to fuel these processes, learning and memory functions take a hit (Di Filippo et al. 2010; Wang et al. 2020; Luo et al. 2024). Moreover, oxidative stress can interfere with the signaling cascades underlying neuroplasticity and aggravate learning deficits (Di Filippo et al. 2010; Wang et al. 2020).

Studies have demonstrated that interventions to decrease oxidative stress may augment neuroplasticity and cognitive function in different models of neurodegeneration (Ari et al. 2022; Di Filippo et al. 2010; Wang et al. 2020; Luo et al. 2024). These findings underscore the promise of therapeutic approaches that aim to enhance mitochondrial function and reduce oxidative stress in order to boost neuroplasticity while curtailing the cognitive decline related to psychiatric diseases (Di Filippo et al. 2010; Wang et al. 2020).

3.3 Reactive Oxygen Species and Psychiatric Disorders

3.3.1 Oxidative Stress and Depression

There is growing evidence for the importance of oxidative stress in the pathophysiology of depression, and dozens of articles have pointed out its involvement in the development and expression of depressive symptoms. Indeed, studies suggest higher oxidative stress markers (e.g., malondialdehyde) in subjects with major depressive disorder (MDD) compared to normal respectively lower ones for antioxidant systems, such as glutathione (Chung et al. 2013). Thus, an imbalance in this system can lead oxidative stress to be associated with neurobiological changes in depression, like neuronal damage and inflammation (Chung et al. 2013; Vergallo et al. 2018). Moreover, the suggestion of a link between oxidative stress and depression is confirmed by the fact that levels of oxidative stress change with antidepressant treatments, which suggests this may be an interesting target for promotion in disorders such as depression (Chung et al. 2013; Ataie et al. 2016).

In line with this, an increasing number of studies also show that more severe depression is consistently associated with greater oxidative damage (Pereira et al. 2021; Luo et al. 2024; Chung et al. 2013) indicating there is a key role for oxidative stress in the pathophysiology of depression. It is true that these oxidative stress markers are significantly higher in diseases like MDD, relative to healthy controls, implying the involvement of oxidative stress in the etiology of depression. This finding was demonstrated by a study where patients with MDD but not healthy controls had elevated levels of said pro-oxidants (Chung et al. 2013). Moreover, postmortem studies have demonstrated decreased Nrf2 pathway activation in depression and increased susceptibility to oxidative damage (Wang et al. 2019). Such deficiency may further promote the neuroinflammatory processes that typically are reported in depression and hence establish a detrimental loop of oxidative stress and mood dysregulation (Chung et al. 2013).

Furthermore, the potential use of antioxidant therapies in reducing depressive symptoms has attracted considerable interest. Several studies have provided evidence for the potential beneficial role of antioxidants in improving mood and cognitive functioning in depressive patients (Chung et al. 2013; Ataie et al. 2016). Together, these findings indicate that oxidative stress might be a new target not only to understand the mechanisms underlying depression but also an additional therapeutic approach (Chung et al. 2013; Vergallo et al. 2018). This suggests a compelling connection between oxidative stress and depressive symptoms, thus emphasizing the necessity to explore these oxidative pathways as potentially important therapeutic targets in depression.

3.3.2 Anxiety: Oxidative Imbalance

There is also evidence suggesting the potential involvement of oxidative stress in the pathogenesis of anxiety disorders. It has been observed that increased levels of ROS and decreased antioxidant defenses are present in individuals with anxiety disorders, demonstrating a condition of oxidative stress (Chung et al. 2013; Ataie et al. 2016). Compromised oxidative homoeostasis has been reported to cause a rapid increase in free radical production and neuro-inflammation, resulting in cellular damage (Chung et al. 2013), which is associated with the manifestation of anxiety symptoms. For example, it was generally established that oxidative stress impairs certain neurotransmitter systems, such as serotonin and dopamine—which are central to mood and anxiety regulation (Wang et al. 2023; Chung et al. 2013).

In addition, evidence for a connection between oxidative stress and anxiety is supported by animal studies showing that exposure to oxidative stressors can result in elicitation of anxiety-like behaviors (Chung et al. 2013). These findings support a potential contribution of oxidative stress in the dysregulation of the hypothalamic-pituitary-adrenal (HPA) axis, which plays an essential role in the body's adaptation to physical and emotional stressor events. High level of cortisol was associated with increased anxiety and stress, so the HPA axis dysregulation effect increases the activity of the HPA axis with uncontrolled cortisol secretion and increasing stress responses (Wang et al. 2023; Chung et al. 2013).

Furthermore, interventions that reduce oxidative stress may be effective for the treatment of anxiety symptoms. Studies conducted on both clinical and preclinical models have demonstrated a decrease in anxiety levels following antioxidant supplementation, including with vitamins C and E (Chung et al. 2013; Ataie et al.

2016). This suggests that antioxidant therapies may serve as an adjunct treatment of anxiety disorders and points to the broader implications of evaluation of oxidative stress mechanisms in (Chung et al. 2013). Taken together, the studies indicate that oxidative imbalance is a serious protagonist for anxiety disorders and deserves future interventional strategies.

3.3.3 Neurodegenerative Diseases

Oxidative stress is a well-recognized mechanism underlying the pathogenesis of neurodegenerative disorders such as AD and PD. In AD, oxidative stress has also been proposed to be involved in the amyloid-beta plaques and neurofibrillary tangle accumulation that are characteristic of the pathology (Vauzour et al. 2010; Cheignon et al. 2018). Several clinical and pathophysiological studies have reported an increase in oxidant levels, which produces oxidative damage of lipids, proteins, and nucleic acids in the brains of patients with AD (Zhao et al. 2022; Vauzour et al. 2010; Cheignon et al. 2018), causing not only neuronal dysfunction but also cell death. The oxidative stress hypothesis largely suggests that the neurodegeneration of AD results from an impairment in ROS production relative to the increasing antioxidant defenses (Vauzour et al. 2010; Cheignon et al. 2018).

Oxidative stress is the early pathophysiology in PD, most importantly for the dopaminergic neuron degeneration within the substantia nigra (Zhao et al. 2022; Vauzour et al. 2010). Various research has shown that the higher the oxidation stress markers, the more severe the motor symptoms in PD (Vauzour et al. 2010). Mitochondrial dysfunction leading to overproduction of ROS has been postulated as a common mechanism by which oxidative stress contributes to neurodegenerative cell death in both AD and PD (Vauzour et al. 2010; Ansari et al. 2024). In addition, therapies addressing oxidative stress have been found to be useful in preclinical models of neurodegenerative diseases. For instance, animal studies have suggested neuroprotective potential of antioxidants such as coenzyme Q10 and vitamin E in AD and PD (Vauzour et al. 2010; Cheignon et al. 2018), with benefits for cognitive/motor function reported in some clinical trials. Continuation studies underscore that antioxidant therapies should reduce oxidative damage and the clinical progression of neurodegenerative disorders (Luo et al. 2024; Vauzour et al. 2010; Cheignon et al. 2018).

3.4 The Role of Diet and Antioxidants in Mental Health

3.4.1 Antioxidants: Protecting Against Oxidative Stress

Antioxidants are molecules that maintain the balance between ROS, free radicals, and antioxidants in the body, otherwise referred to as oxidative stress. Vitamins, minerals, and phytochemicals are very important in the diet, as they serve as antioxidants, which may neutralize free radicals and their effects on cell membranes and tissues. Dietary total antioxidant capacity (TAC) is a summary measure of the collective effects all antioxidants provided to us attending to our dietary habits, including vitamins C and E, carotenoids, and polyphenols—mainly flavonoids as primary antioxidants in the human diet (Devore et al. 2013; Yang et al. 2018).

Higher dietary TAC was also associated with a reduced risk of dementia and stroke, especially in older adults, as evidenced by studies (Devore et al. 2013). This implication emphasizes what a diet rich in antioxidants can do to protect our brain. Additionally, the combined effect of several antioxidants can be greater than their sum, indicating possibly that a variety of different antioxidants from food may be more protective than supplements alone (Yang et al. 2018).

Apart from this, the presence of antioxidants is essential for modulating the defense mechanisms and other associated physiological processes, such as inflammation and immune response. The Mediterranean diet, with high antioxidant content, has been related to decreased oxidative stress and inflammation, believed to be hallmarks in the development of chronic diseases such as mental disorders (El-Sabban 2014; Karlsen et al. 2011). Therefore, the consumption of many antioxidant-dense foods is important to bettering overall well-being and reversing oxidative stress.

3.4.2 Effects of Nutrition on Oxidative Stress

One of the major factors that determine oxidative stress in the body is the nutrients we consume. Processed, sugary, and fatty foods may indeed make oxidative destruction even worse, as this type of diet boosts the manufacture of free radicals. Similarly, a diet containing fruits, vegetables, whole grains, and good fats can be very helpful in neutralizing oxidative stress and boosting the body's antioxidant defense (Bawaked et al. 2017; Cyuńczyk et al. 2022). The intake of polyphenol-rich foods like berries and green tea has been reported to enhance antioxidant status and reduce evidence of oxidative stress in diverse populations (Zujko et al. 2015; Tan et al. 2023).

Diet and oxidative stress interact in multiple, complex ways. It has been shown that adherence to healthy dietary patterns (e.g., the Mediterranean diet) is connected with lower oxidative stress and a decreased risk of chronic diseases, such as cardiovascular diseases and some cancers (Koloverou et al. 2016; Wang et al. 2022). In contrast, high glycemic load and low antioxidant diets are reported to increase oxidative stress and inflammation that participate in the development of metabolic syndrome as well as other health problems (Cyuńczyk et al. 2022; Tan et al. 2023).

In addition, some dietary constituents, e.g., dietary fiber, exhibited increasing antioxidant ability and decreasing oxidative stress indicators. Pairs of words that mirror one another (Tan et al. 2023; Ávila-Escalante et al. 2020) and high-fiber diets not only offer vital nutrients but they also are a good way to avoid. Cues to our health, something that is being more recognized as the most important factor for better health and disease prevention (Cyuńczyk et al. 2022; Tan et al. 2023; Ávila-Escalante et al. 2020). As a result, well-informed nutritional decisions that focus on the inclusion of antioxidant-laden fruits and veggies may help in reducing oxidative tension and improving psychological wellness repercussions.

3.4.3 Supplements and Antioxidant Therapy

Dietary supplements are a widely used remedy as an anti-oxidant and beneficial approach to mental health, thereby leading towards the research in exploring supplementing human diet with antioxidants as an effective option. Although a number of other dietary antioxidants are also available in supplement form, the bioavailability and composition of these products are highly variable (Yang et al. 2018; Todorović and Đorđević 2023). In particular, vitamins C and E, selenium, and a wide variety of polyphenols have been researched for the same reasons. Nonetheless, evidence for the potential efficacy of antioxidant supplements for primary or secondary prevention against mental health disorders is controversial (Yang et al. 2018; Wu et al. 2023).

Some other research to date so far as to imply that giving someone an antioxidant supplement in addition to their conventional medication may improve oxidative stress and anti-inflammatory markers, especially in populations considered at risk of contracting chronic diseases, such as those with diabetes or cardiovascular conditions (Dludla et al. 2023). Another example could be summarized that the coenzyme Q10 combined with omega-3 improved antioxidant capacity and decreased oxidative amount of metabolic syndrome individuals (Dludla et al. 2023). Nonetheless, the results were not consistent among all participants, and the long-term effects of high-dose antioxidant supplementation have yet to be elucidated (Yang et al. 2018; Wu et al. 2023).

In addition, it is important to bear in mind the possible interactions with other supplements and medications, which adds to the importance of sourcing antioxidants from whole food instead of singularly isolated supplements. Supplements do not offer the beneficial interplay that constitutes whole foods and may appear as isolated from each other (Yang et al. 2018; Todorović and Đorđević 2023). Considering that, while antioxidant supplements might provide certain benefits, they should not be a substitute for an overall balanced diet full of food sources of antioxidants.

3.5 Medications, Oxidative Stress, and Mental Health

3.5.1 The Oxidative Stress and Psychotropic Drugs

The interactions of oxidative stress and psychotropic drugs are an essential focus in research, specifically when it comes to the field of mental health therapies. Several of the most frequently used psychotropic agents, including antidepressants and antipsychotics, exhibit pleiotropic effects that are involved in oxidative balance regulation. One way is that some drugs, such as antidepressants, may reduce the formation of ROS and increase antioxidant defenses, thereby reducing oxidative stress-related damage in neural tissues (Ribaudo et al. 2020). As an example, selective serotonin reuptake inhibitors (SSRIs) have been shown to increase brain-derived neurotrophic factor (BDNF), which regulates survival as well as neuronal function. Whether beneficial neurological effects of the ketogenic diet have a common denominator has not been firmly established, although it could be related to the decreased oxidative stress because of the beneficial effect of BDNF on the expression of antioxidant enzymes. The study also showed that lipases in plasma were significantly lowered by the administration of SSRIs and demonstrated an ability to decrease lipid peroxidation and increase superoxide dismutase (SOD) activity, suggesting a protective role against oxidative damage (Ribaudo et al. 2020; Dietrich-Muszalska et al. 2021).

On the other hand, it is known that some psychotropic medications themselves may increase oxidation, aggravating oxidative stress. For example, the use of certain antipsychotics has been associated with increased markers of oxidative stress, such as malondialdehyde, a breakdown product of lipid peroxidation (Dietrich-Muszalska et al. 2021; Sommerfeld-Klatta et al. 2024). Thus, one could conclude that some psychotropic drugs may be neuroprotective due to their antioxidant effect, while others may exacerbate oxidative stress and cellular damage through the production of reactive species, which are found in various neurodegenerative diseases; therefore, caution is mandatory (Pereira et al. 2021; Dietrich-Muszalska et al. 2021).

In addition, the implication of oxidative stress in the pathophysiology of psychiatric disorders such as schizophrenia and bipolar disorder has been considered. The resulting increased oxidative stress markers in such patients may indicate a possible association between the oxidative damage and Panadol Maltovaquinone-induced symptom severity (Dietrich-Muszalska et al. 2021; Murray et al. 2021). This extends to the wider significance of incorporating antioxidant-based interventions in following treatment, whether for mental health symptoms as well. Psychotropic drug treatments can modulate oxidative balance, being protective against oxidative stress in some cases and deleterious in others. Appreciating these dynamics is key to ensuring that treatment modalities are implemented such that patient care yields the highest rates of efficacy (Pereira et al. 2021; Kotzaeroglou and Tsamesidis 2022; Murray et al. 2021).

3.5.2 Role of Anti-oxidative Treatments in Mental Health Therapies

With a rapidly increasing field of researchers investigating the role of anti-oxidative as research into the impact of oxidative stress on mental health grows, there is increasing focus on antioxidant therapies as potential treatments. These therapies aim to combat oxidative stress, a key factor in many psychiatric disorders, by neutralizing free radicals and reducing oxidative damage. N-acetyl cysteine (NAC), a glutathione precursor, is able to increase levels by upregulating the de novo pathway and has been demonstrated in human trials to ameliorate oxidative stress and cognitive capabilities of individuals with bipolar disorder and schizophrenia (Kotzaeroglou and Tsamesidis 2022; Dean et al. 2011).

The way that NAC works is by restoring glutathione, an essential antioxidant in the brain. NAC administration also increases glutathione synthesis, leading to decreases in oxidative damage and the enhancement of neuronal resilience (Dietrich-Muszalska et al. 2021; Dean et al. 2011). Although NAC may act as an adjunctive treatment in mental health and has also been significant, especially as a novel one to reduce psychiatric symptoms especially among treatment refractory cases, there were few clinical studies available (Wang et al. 2023; Luo et al. 2024; Dean et al. 2011).

One of the more hopeful areas of research is in plant-derived antioxidants (flavonoids, polyphenols) that have demonstrated neuroprotective qualities. For instance, resveratrol, a polyphenol from grapes, has been shown to reduce oxidative stress and inflammation in numerous neurodegeneration models (Simpson and Oliver 2020). As an activator of the Nrf2 pathway, it provides additional evidence for its therapeutic potential in mental health disorders characterized by oxidative stress (Simpson and Oliver 2020; Joshi and Johnson 2012).

In addition, lifestyle interventions related to physical activity and dietary patterns have been associated with an improvement in antioxidant defenses. Regular physical activity may enhance the endogenous antioxidant enzyme expression, which in turn could attenuate oxidative stress and possibly improve mental health (Thirupathi et al. 2023; Thirupathi et al. 2021). Likewise, antioxidant-rich dietetic patterns made up of fruits and vegetables are linked to decreased oxidative stress signs in addition to enhanced mood and health status (Pereira et al. 2021; Luo et al. 2024; Thirupathi et al. 2023).

3.6 Conclusion

3.6.1 Balancing Oxidants and Mental Health

Manage the oxidative stress to maintain mental health. The accumulation of ROS and the consequential oxidative damage can interfere with neural function, leading to psychiatric disorders such as depression, anxiety, and neurodegenerative disorders. Oxidative stress damages and impairs cognitive function and emotional well-being by exacerbating neuroinflammation, mitochondrial dysfunction, and creating substrate imbalances of neurotransmitters. For an antioxidant, there is growing evidence to suggest that it might help in the treatment of mental health by fighting oxidative stress with antioxidants. Furthermore, pharmacotherapy programs that reduce or neutralize oxidative stress or promote an internal antioxidant defense could provide strategies to increase brain resilience to injury. By supporting natural antioxidant defenses, we can help improve the health of our neurons and protect against the development or severity of cognitive decline.

More generally, the fine balance between oxidants and mental health emphasizes the importance of a balanced oxidative environment for good health. Although oxidative stress is created as a normal result of cellular function, an imbalance—where ROS outpaces the antioxidant scavenging capabilities of the organism—causes massive damage to multiple systems and especially to the brain. The brain has high metabolic demands and is extremely vulnerable to oxidative damage; therefore, it is imperative that we understand the factors related to lifestyle (diet, physical activity, exercise) and the exposome (environmental exposures) and how they might tip this balance.

Funding a top-down as well as bottom-up methodology, dietary antioxidants, mindful supplementation, and medical therapies customized to each oxidative behavior is critical in fostering mental wellness at scale. As knowledge of oxidative stress as relevant to health continues to advance, future research and clinical practice will be well-positioned to continue enhancing our ability to prevent and treat mental health disorders therapeutically via oxidative management with hopeful prospects for more successful and precision treatments.

References

Abshirini M, Siassi F, Koohdani F, Qorbani M, Mozaffari H, Aslani Z, Soleymani M, Entezarian M, Sotoudeh G (2019) Dietary total antioxidant capacity is inversely associated with depression, anxiety and some oxidative stress biomarkers in postmenopausal women: a cross-sectional study. Ann General Psychiatry 18:3. https://doi.org/10.1186/s12991-019-0225-7

Ansari MA, Rao MS, Al-Jarallah A (2024) Insights into early pathogenesis of sporadic Alzheimer's disease: role of oxidative stress and loss of synaptic proteins. Front Neurosci 17:1273626. https://doi.org/10.3389/fnins.2023.1273626

Ari BC, Tur EK, Domac FM, Kenangil GO (2022) Uric acid: the role in the pathophysiology and the prediction in the diagnosis of Parkinson's disease: a Turkish-based study. A húgysavnak a Parkinson-kór kórélattanában és a diagnózis előrejelzésében betöltött szerepe—török populáción alapuló vizsgálat Ideggyogyaszati szemle 75(1–02):51–59. https://doi.org/10.18071/isz.75.0051

Ataie A, Ataie R, Shadifar M (2016) Polyphenolic antioxidants and neuronal regeneration. Basic Clin Neurosci J 7(2). https://doi.org/10.15412/J.BCN.03070201

Ávila-Escalante ML, Coop-Gamas F, Cervantes-Rodríguez M, Méndez-Iturbide D, Aranda-González II (2020) The effect of diet on oxidative stress and metabolic diseases—clinically controlled trials. J Food Biochem 44(5). https://doi.org/10.1111/jfbc.13191

Bawaked RA, Schröder H, Ribas-Barba L, Izquierdo-Pulido M, Pérez-Rodrigo C, Fíto M, Serra-Majem L (2017) Association of diet quality with dietary inflammatory potential in youth. Food Nutr Res 61(1):1328961. https://doi.org/10.1080/16546628.2017.1328961

Bortoli M, Dalla Tiezza M, Muraro C, Pavan C, Ribaudo G, Rodighiero A, Tubaro C, Zagotto G, Orian L (2019) Psychiatric disorders and oxidative injury: antioxidant effects of zolpidem therapy disclosed In Silico. Comput Struct Biotechnol J 17:311–318. https://doi.org/10.1016/j.csbj.2019.02.004

Bouayed J, Rammal H, Soulimani R (2009) Oxidative stress and anxiety: relationship and cellular pathways. Oxid Med Cell Longev 2(2):63–67. https://doi.org/10.4161/oxim.2.2.7944

Cenini G, Lloret A, Cascella R (2019) Oxidative stress in neurodegenerative diseases: from a mitochondrial point of view. Oxid Med Cell Longev 2019:1–18. https://doi.org/10.1155/2019/2105607

Cheignon C, Tomas M, Bonnefont-Rousselot D, Faller P, Hureau C, Collin F (2018) Oxidative stress and the amyloid beta peptide in Alzheimer's disease. Redox Biol 14:450–464. https://doi.org/10.1016/j.redox.2017.10.014

Chung CP, Schmidt D, Stein CM, Morrow JD, Salomon RM (2013) Increased oxidative stress in patients with depression and its relationship to treatment. Psychiatry Res 206(2–3):213–216. https://doi.org/10.1016/j.psychres.2012.10.018

Cyuńczyk M, Zujko ME, Jamiołkowski J, Zujko K, Łapińska M, Zalewska M, Kondraciuk M, Witkowska AM, Kamiński KA (2022) Dietary total antioxidant capacity is inversely associated with prediabetes and insulin resistance in Bialystok PLUS population. Antioxidants 11(2):283. https://doi.org/10.3390/antiox11020283

Dean O, Giorlando F, Berk M (2011) N-acetylcysteine in psychiatry: current therapeutic evidence and potential mechanisms of action. J Psychiatry Neurosci 36(2):78–86. https://doi.org/10.1503/jpn.100057

Devore EE, Feskens E, Ikram MA, Heijer TD, Vernooij M, Lijn FVD, Hofman A, Niessen WJ, Breteler MMB (2013) Total antioxidant capacity of the diet and major neurologic outcomes in older adults. Neurology 80(10):904–910. https://doi.org/10.1212/WNL.0b013e3182840c84

Di Filippo M, Chiasserini D, Tozzi A, Picconi B, Calabresi P (2010) Mitochondria and the link between neuroinflammation and neurodegeneration. J Alzheimer's disease: JAD 20(Suppl 2):S369–S379. https://doi.org/10.3233/JAD-2010-100543

Dietrich-Muszalska A, Kolodziejczyk-Czepas J, Nowak P (2021) Comparative study of the effects of atypical antipsychotic drugs on plasma and urine biomarkers of oxidative stress in schizophrenic patients. Neuropsychiatr Dis Treat 17:555–565. https://doi.org/10.2147/NDT.S283395

Dludla PV, Ziqubu K, Mabhida SE, Mazibuko-Mbeje SE, Hanser S, Nkambule BB, Basson AK, Pheiffer C, Tiano L, Kengne AP (2023) Dietary supplements potentially target plasma glutathione levels to improve cardiometabolic health in patients with diabetes mellitus: a systematic review of randomized clinical trials. Nutrients 15(4):944. https://doi.org/10.3390/nu15040944

El-Sabban F (2014) The antioxidant advantage of the Mediterranean diet in cardiovascular disease. Nutr Diet Suppl 35. https://doi.org/10.2147/NDS.S59848

Ferriero DM (2001) Oxidant mechanisms in neonatal hypoxia-ischemia. Dev Neurosci 23(3):198–202. https://doi.org/10.1159/000046143

Gandhi S, Abramov AY (2012) Mechanism of oxidative stress in neurodegeneration. Oxid Med Cell Longev 2012:1–11. https://doi.org/10.1155/2012/428010

Gautam M, Agrawal M, Gautam M, Sharma P, Gautam AS, Gautam S (2012) Role of antioxidants in generalised anxiety disorder and depression. Indian J Psychiatry 54(3):244–247. https://doi.org/10.4103/0019-5545.102424

Hassan W, Noreen H, Rehman S, Kamal MA, Da Rocha JBT (2022) Association of oxidative stress with neurological disorders. Curr Neuropharmacol 20(6):1046–1072. https://doi.org/10.2174/1570159X19666211111141246

Ikawa M, Okazawa H, Nakamoto Y, Yoneda M (2020) PET imaging for oxidative stress in neurodegenerative disorders associated with mitochondrial dysfunction. Antioxidants 9(9):861. https://doi.org/10.3390/antiox9090861

Joshi G, Johnson JA (2012) The Nrf2-ARE pathway: a valuable therapeutic target for the treatment of neurodegenerative diseases. Recent Pat CNS Drug Discov 7(3):218–229. https://doi.org/10.2174/157488912803252023

Karlsen A, Svendsen M, Seljeflot I, Sommernes M-A, Sexton J, Brevik A, Erlund I, Serafini M, Bastani N, Fagertun Remberg S, Borge GI, Hauger Carlsen M, Kjølsrud Bøhn S, Myhrstad MC, Dragsted LO, Duttaroy AK, Haffner K, Laake P, Drevon CA et al (2011) Compliance, tolerability and safety of two antioxidant-rich diets: a randomised controlled trial in male smokers. Br J Nutr 106(4):557–571. https://doi.org/10.1017/S0007114511000353

Kim GH, Kim JE, Rhie SJ, Yoon S (2015) The role of oxidative stress in neurodegenerative diseases. Exp Neurobiol 24(4):325–340. https://doi.org/10.5607/en.2015.24.4.325

Koloverou E, Panagiotakos DB, Pitsavos C, Chrysohoou C, Georgousopoulou EN, Grekas A, Christou A, Chatzigeorgiou M, Skoumas I, Tousoulis D, Stefanadis C, The ATTICA Study Group (2016) Adherence to Mediterranean diet and 10-year incidence (2002–2012) of diabetes: correlations with inflammatory and oxidative stress biomarkers in the ATTICA cohort study. Diabetes Metab Res Rev 32(1):73–81. https://doi.org/10.1002/dmrr.2672

Kotzaeroglou A, Tsamesidis I (2022) The role of equilibrium between free radicals and antioxidants in depression and bipolar disorder. Med(Basel, Switzerland) 9(11):57. https://doi.org/10.3390/medicines9110057

Lin M, Liu N, Qin Z, Wang Y (2022) Mitochondrial-derived damage-associated molecular patterns amplify neuroinflammation in neurodegenerative diseases. Acta Pharmacol Sin 43(10):2439–2447. https://doi.org/10.1038/s41401-022-00879-6

Liu Z, Zhou T, Ziegler AC, Dimitrion P, Zuo L (2017) Oxidative stress in neurodegenerative diseases: from molecular mechanisms to clinical applications. Oxid Med Cell Longev 2017(1):2525967. https://doi.org/10.1155/2017/2525967

Luo J, Xu X, Sun Y, Lu X, Zhao L (2024) Association of composite dietary antioxidant index with depression and all-cause mortality in middle-aged and elderly population. Sci Rep 14(1):9809. https://doi.org/10.1038/s41598-024-60322-0

Murray AJ, Rogers JC, Katshu MZUH, Liddle PF, Upthegrove R (2021) Oxidative stress and the pathophysiology and symptom profile of schizophrenia spectrum disorders. Front Psych 12:703452. https://doi.org/10.3389/fpsyt.2021.703452

Pereira GA, da Silva A, Hermsdorff HHM, Moreira APB, de Aguiar AS (2021) Association of dietary total antioxidant capacity with depression, anxiety, and sleep disorders: a systematic review of observational studies. J Clin Transl Res 7(5):631–640

Ribaudo G, Bortoli M, Pavan C, Zagotto G, Orian L (2020) Antioxidant potential of psychotropic drugs: from clinical evidence to in vitro and in vivo assessment and toward a new challenge for in Silico molecular design. Antioxidants 9(8):714. https://doi.org/10.3390/antiox9080714

Schon EA, Przedborski S (2011) Mitochondria: the next (neurode)generation. Neuron 70(6):1033–1053. https://doi.org/10.1016/j.neuron.2011.06.003

Simpson DSA, Oliver PL (2020) ROS generation in microglia: understanding oxidative stress and inflammation in neurodegenerative disease. Antioxidants 9(8):743. https://doi.org/10.3390/antiox9080743

Singh A, Kukreti R, Saso L, Kukreti S (2019) Oxidative stress: a key modulator in neurodegenerative diseases. Molecules 24(8):1583. https://doi.org/10.3390/molecules24081583

Sommerfeld-Klatta K, Jiers W, Rzepczyk S, Nowicki F, Łukasik-Głębocka M, Świderski P, Zielińska-Psuja B, Żaba Z, Żaba C (2024) The effect of neuropsychiatric drugs on the oxidation-reduction balance in therapy. Int J Mol Sci 25(13):7304. https://doi.org/10.3390/ijms25137304

St-Pierre J, Drori S, Uldry M, Silvaggi JM, Rhee J, Jäger S, Handschin C, Zheng K, Lin J, Yang W, Simon DK, Bachoo R, Spiegelman BM (2006) Suppression of reactive oxygen species and neurodegeneration by the PGC-1 transcriptional coactivators. Cell 127(2):397–408. https://doi.org/10.1016/j.cell.2006.09.024

Tan Z, Meng Y, Li L, Wu Y, Liu C, Dong W, Chen C (2023) Association of dietary fiber, composite dietary antioxidant index and risk of death in tumor survivors: national health and nutrition examination survey 2001–2018. Nutrients 15(13):2968. https://doi.org/10.3390/nu15132968

Thirupathi A, Wang M, Lin JK, Fekete G, István B, Baker JS, Gu Y (2021) Effect of different exercise modalities on oxidative stress: a systematic review. Biomed Res Int 2021:1947928. https://doi.org/10.1155/2021/1947928

Thirupathi A, Gu Y, Pinho RA (2023) Editorial: exercise friend or foe? For the management of oxidative stress in health and diseases volume II. Front Physiol 14:1216566. https://doi.org/10.3389/fphys.2023.1216566

Todorović V, Đorđević B (2023) Dietary interventions through supplementation with antioxidant compounds. Arhiv Za Farmaciju 73(4):292–305. https://doi.org/10.5937/arhfarm73-45141

Vauzour D, Vafeiadou K, Rendeiro C, Corona G, Spencer JPE (2010) The inhibitory effects of berry-derived flavonoids against neurodegenerative processes. J Berry Res 1(1):45–52. https://doi.org/10.3233/BR-2010-005

Vergallo A, Giampietri L, Baldacci F, Volpi L, Chico L, Pagni C, Giorgi FS, Ceravolo R, Tognoni G, Siciliano G, Bonuccelli U (2018) Oxidative stress assessment in Alzheimer's disease: a clinic setting study. Am J Alzheimers Dis Other Dement 33(1):35–41. https://doi.org/10.1177/1533317517728352

Wang X, Shen C, Zhu J, Shen G, Li Z, Dong J (2019) Long noncoding RNAs in the regulation of oxidative stress. Oxid Med Cell Longev 2019:1–7. https://doi.org/10.1155/2019/1318795

Wang W, Zhao F, Ma X, Perry G, Zhu X (2020) Mitochondria dysfunction in the pathogenesis of Alzheimer's disease: recent advances. Mol Neurodegener 15(1):30. https://doi.org/10.1186/s13024-020-00376-6

Wang W, Wang X, Cao S, Duan Y, Xu C, Gan D, He W (2022) Dietary antioxidant indices in relation to all-cause and cause-specific mortality among adults with diabetes: a prospective cohort study. Front Nutr 9:849727. https://doi.org/10.3389/fnut.2022.849727

Wang H, Jin M, Xie M, Yang Y, Xue F, Li W, Zhang M, Li Z, Li X, Jia N, Liu Y, Cui X, Hu G, Dong L, Wang G, Yu Q (2023) Protective role of antioxidant supplementation for depression and anxiety: a meta-analysis of randomized clinical trials. J Affect Disord 323:264–279. https://doi.org/10.1016/j.jad.2022.11.072

Wu J-R, Song EK, Moser DK, Lennie TA (2023) Dietary antioxidant insufficiency is associated with increased inflammatory markers and poorer health-related quality of life in patients with heart failure. J Cardiovasc Nurs 38(1):6–12. https://doi.org/10.1097/JCN.0000000000000912

Yang CS, Ho C-T, Zhang J, Wan X, Zhang K, Lim J (2018) Antioxidants: differing meanings in food science and health science. J Agric Food Chem 66(12):3063–3068. https://doi.org/10.1021/acs.jafc.7b05830

Zhao L, Sun Y, Cao R, Wu X, Huang T, Peng W (2022) Non-linear association between composite dietary antioxidant index and depression. Front Public Health 10:988727. https://doi.org/10.3389/fpubh.2022.988727

Zujko ME, Witkowska AM, Waśkiewicz A, Mirończuk-Chodakowska I (2015) Dietary antioxidant and flavonoid intakes are reduced in the elderly. Oxid Med Cell Longev 2015:1–8. https://doi.org/10.1155/2015/843173

Dietary Impacts on Psychopharmacotherapy

4

Parisa Gazerani

Abstract

This chapter explores the complex interplay between diet and psychotropic medications, emphasizing how dietary elements can influence the efficacy, safety, and overall outcomes of psychopharmacotherapy. By examining mechanisms at the genetic, pharmacokinetic, and pharmacodynamic levels, the chapter highlights both positive and negative aspects of diet-drug interactions. Positive dietary interactions, such as the supportive role of omega-3 fatty acids, B vitamins, and magnesium, can enhance drug absorption, stabilize neurotransmitter function, and reduce inflammation, thereby complementing therapeutic outcomes. In contrast, certain foods and supplements, such as high-fiber meals, grapefruit juice, and St. John's Wort, can interfere with drug metabolism, potentially causing adverse effects or diminishing drug efficacy. The chapter also addresses cultural and religious considerations, which shape dietary habits and influence dietary recommendations in diverse patient populations. Practical implications are provided to guide clinicians in integrating dietary assessment, tailored counseling, and multidisciplinary collaboration into routine mental health care, ensuring that dietary factors support rather than hinder psychopharmacotherapy. This integrative approach ultimately aims to foster personalized, effective, and holistic mental health treatment.

P. Gazerani (✉)
Department of Life Sciences and Health, Faculty of Health Sciences, Oslo Metropolitan University, Oslo, Norway

Department of Health Science and Technology, Faculty of Medicine, Aalborg University, Gistrup, Denmark
e-mail: parisaga@oslomet.no

W. Mohamed, A. Özge (eds.), *Feeding the Mind: The Connection Between Diet, Drugs, and Mental Health Volume 1*, Nutritional Neurosciences,
https://doi.org/10.1007/978-981-95-1414-4_4

Keywords

Psychopharmacotherapy · Diet-drug interactions · Psychotropic medications · Pharmacokinetics · Pharmacodynamics · Diet

Abbreviations

ADHD Attention Deficit Hyperactivity Disorder
CYP Cytochrome P450
DHA Docosahexaenoic Acid
EPA Eicosapentaenoic Acid
EHR Electronic Health Record
GABA Gamma-Aminobutyric Acid
MAOI Monoamine Oxidase Inhibitor
NMDA N-Methyl-D-Aspartate
SNRI Serotonin-Norepinephrine Reuptake Inhibitor
SSRI Selective Serotonin Reuptake Inhibitor
TCA Tricyclic Antidepressant

4.1 Introduction

4.1.1 General Overview of Drug-Diet Interactions

Drug-diet interactions refer to the influence of dietary components such as foods, nutrients, and supplements on the behavior of drugs within the body (Gezmen-Karadağ et al. 2018). These interactions can significantly impact a drug's effectiveness, safety, and patient experience, either enhancing or diminishing its intended therapeutic effects (Ötleş and Senturk 2014). Although often overlooked, diet plays a critical role in modifying a drug's absorption, distribution, metabolism, and excretion (pharmacokinetic processes), as well as its action at target sites within the body (pharmacodynamics) (Ana-Maria et al. 2022).

The significance of diet in drug interactions is heightened by the highly individualized nature of dietary habits, which vary widely across different populations (Gherasim et al. 2020), which makes the prediction difficult (Wei et al. 2024; Lacruz-Pleguezuelos et al. 2023). Factors such as culture, religion, and access to certain foods in various geographical locations make a one-size-fits-all approach to drug administration or dietary advice less feasible (Ordovas et al. 2018). For example, high-fat diets can alter the absorption of certain drugs (Koziolek et al. 2019), while specific foods such as grapefruit can inhibit enzymes responsible for drug metabolism, potentially leading to adverse effects (Bailey et al. 2013). The field of drug-diet interactions aims to uncover these relationships to enhance therapeutic effectiveness and minimize potential risks (Vuong et al. 2023). In this line, several tools and databases (Lacruz-Pleguezuelos et al. 2023; Tian et al. 2024; Hong et al. 2024; Rahman et al. 2022) have been developed to help healthcare providers, researchers, and patients check for potential drug-diet interactions. These can serve as valuable resources for general information and as guides for managing and optimizing drug therapy based on dietary considerations. They provide insights into how different foods, nutrients, and supplements might affect drug efficacy and safety, making them useful tools for minimizing risks and enhancing treatment outcomes. The Drugs.com Interaction Checker is an online resource that provides information on potential interactions between drugs and certain foods, alcohol, and dietary supplements, helping users quickly assess possible risks. Several mobile apps are also available for consumers to self-check the interactions (Kim et al. 2018).

4.2 Focus on Drug-Diet Interactions in Mental Health and Psychopharmacotherapy

In the context of mental health, understanding drug-diet interactions is particularly important. Psychotropic medications, including antidepressants, antipsychotics, mood stabilizers, anxiolytics, and stimulants, play an essential role in managing psychiatric conditions like depression, anxiety, bipolar disorder, and schizophrenia (Ghallab and Elassal 2024). The therapeutic success of these drugs depends on precise dosing and consistent drug levels in the body (Tyson et al. 2020), making them sensitive to diet-induced changes. Several factors highlight the importance of diet-drug interactions in general (Niederberger and Parnham 2021) but also in psychopharmacology (Gezmen-Karadağ et al. 2018). Those include (1) a need for consistent drug efficacy, which indicates that mental health conditions often require long-term medication adherence, where variations in drug levels due to dietary changes can lead to relapse or diminished therapeutic effects; (2) the influence of diet on brain function that explains how certain nutrients and dietary patterns, such as omega-3 fatty acids or the Mediterranean diet, can support mental health, potentially interacting beneficially with psychotropic drugs; and (3) vulnerability to adverse effects, where the emphasis is on the sensitivity of individual patients to adverse drug reactions due to dietary influences. Considering these factors is essential in making safe, effective treatment for optimal patient care (Carlo and Alpert 2016).

This chapter is dedicated to presenting the concept of drug-diet interactions in psychopharmacotherapy, exploring both positive and negative aspects of these interactions to provide a comprehensive overview. Through mechanisms, examples, and evidence-based guidance, it aims to raise awareness and emphasize the significance of diet in enhancing the effectiveness and safety of psychotropic medications. Additionally, this chapter offers insights for clinicians to incorporate dietary considerations into psychopharmacotherapy, optimizing patient outcomes and reducing adverse effects. It also encourages collaborative research among pharmacologists, dietitians, and mental health practitioners to further explore and document these interactions. Such collective efforts will help address knowledge gaps in this area, ultimately leading to more personalized, safe, and effective treatment approaches that foster holistic patient care. By understanding these interactions, clinicians, dietitians, and patients can work together to improve mental health outcomes and support societal well-being.

4.3 Mechanisms and Levels of Interaction in Drug-Diet Dynamics

Drug-diet interactions can occur through various mechanisms, which are broadly categorized into genetic, pharmacokinetic, and pharmacodynamic levels (Koziolek et al. 2019; Sun et al. 2023; Zhao et al. 2023).

Genetic level Genetic differences among individuals, such as variations in metabolizing enzymes, can significantly influence drug responses (Pinto and Dolan 2011) including drugs used for psychopharmatherapeutic purposes (Butler 2018). For example, patients with poor or ultra-rapid metabolizer phenotypes for CYP2D6 may experience different effects from dietary factors that influence drug absorption or excretion. A person who is a poor metabolizer of CYP2D6 and taking a drug metabolized by this enzyme, like fluoxetine, could be at higher risk for side effects if they also consume a high-fat meal. Since fat increases the absorption of lipophilic drugs, the combination of poor metabolism and increased absorption could amplify the drug's concentration and potential side effects (Zanger and Schwab 2013). Another example can be a person who is a poor metabolizer of CYP2C19 and may experience higher drug levels when taking sertraline. Although diet may not directly affect CYP2C19 activity, consuming high-fiber foods can reduce overall drug absorption, potentially stabilizing the drug's plasma concentration by slowing the absorption rate (Bråten et al. 2020; Campos et al. 2022).

While genetic polymorphisms in CYP2D6 or CYP2C19 impact drug metabolism, dietary influences on these specific enzymes are less direct (Prior and Baker 2003; Ingelman-Sundberg et al. 2014). Understanding these genetic variations is crucial for planning treatments and providing dietary guidance.

Pharmacokinetic level Dietary components can influence each stage of drug pharmacokinetics (Koziolek et al. 2019).

Absorption: Certain foods or dietary patterns may accelerate or inhibit drug absorption (Deng et al. 2017). For instance, fatty meals can improve the absorption of lipophilic psychotropic drugs (Preeti et al. 2023), such as certain antipsychotics like olanzapine. Olanzapine's lipophilic nature means that it is more readily absorbed in the presence of dietary fats, potentially leading to higher concentrations in the bloodstream and an enhanced therapeutic effect (Callaghan et al. 1999). On the other hand, fiber-rich foods can slow gastric emptying and bind to drugs in the digestive tract, reducing the absorption of some antidepressants (González Canga et al. 2010; Qi et al. 2018). For example, the absorption of sertraline may be reduced when taken with high-fiber meals, potentially diminishing its effectiveness (Stewart 1992).

Distribution: The body's protein-binding capacity can be altered by diet, potentially modifying the distribution of a drug throughout the body (Niederberger and Parnham 2021; Gurevich 2013). Many psychotropic drugs, such as antipsychotics and antidepressants, bind to plasma proteins like albumin for transport (Lapetina et al. 2020). A diet low in protein, or one that leads to malnutrition, can reduce albumin levels (Allison and Lobo 2024), decreasing the amount of protein-bound drugs in circulation. This shift results in more free or unbound drugs in the bloodstream, which can enhance the drug's effects and potentially lead to toxicity (Kshirsagar and Sheth 1977). Conversely, a high-protein diet may increase protein-binding capacity, reducing the concentration of free drugs and potentially diminishing therapeutic efficacy (Celestin and Musteata 2021). This balance is particularly important in drugs with narrow therapeutic windows, such as lithium, where small changes in free drug levels can lead to adverse effects (Pennazio et al. 2022).

Metabolism: Many foods can induce or inhibit liver enzymes, especially cytochrome P450 enzymes, which play a key role in drug metabolism (Zhao et al. 2021). For example, grapefruit is known to inhibit CYP3A4, an enzyme responsible for metabolizing a wide range of psychotropic drugs (Siwek et al. 2024), including certain benzodiazepines like alprazolam (Yasui et al. 2000). When a patient consumes grapefruit or grapefruit juice, the inhibitory effect on CYP3A4 can slow the metabolism of alprazolam, leading to higher levels of the drug in the bloodstream. This can increase the sedative effects and the risk of adverse reactions, such as excessive drowsiness or even respiratory depression (Bailey et al. 2013). Being aware of these dietary effects is crucial for patients on psychotropic medications to avoid unintended side effects and maintain safe therapeutic levels.

Excretion: Dietary influences on kidney function, such as changes in urine pH from a high-protein diet, can affect the rate at which drugs are excreted (Niederberger and Parnham 2021). For example, lithium, a mood stabilizer commonly used in the treatment of bipolar disorder, is excreted primarily through the kidneys. A high-protein diet can lead to acidification of the urine, which may alter lithium excretion rates and increase the drug's reabsorption in the kidneys. This change can elevate lithium levels in the bloodstream, potentially resulting in lithium toxicity (El Balkhi and Mégarbane 2017). Conversely, diets that make urine more alkaline, such as those high in fruits and vegetables (Schwalfenberg 2012), may enhance lithium excretion, potentially lowering drug levels and reducing its effectiveness. Understanding these dietary influences on excretion is critical for maintaining safe and effective lithium therapy (Malhi et al. 2020).

Pharmacodynamic level At this level, dietary components interact with a drug's action directly, either enhancing or diminishing its therapeutic effects (Ana-Maria et al. 2022). For instance, caffeine can counteract the sedative effects of psychotropic medications like benzodiazepines (e.g., lorazepam) or certain antipsychotics by stimulating the central nervous system (Hladun et al. 2021; Diksha et al. 2014). Patients who consume caffeine while on these medications may experience reduced sedation, which can impact the intended calming effects of the drugs and may lead to symptoms like restlessness or insomnia (Broderick et al. 2005; Reddy et al. 2024). Conversely, omega-3 fatty acids, commonly found in fatty fish and supplements, have been shown to enhance the efficacy of some antidepressants (Serefko et al. 2024; Su et al. 2015), particularly selective serotonin reuptake inhibitors (SSRIs) like fluoxetine. Omega-3s are believed to support serotonin function, potentially improving mood and augmenting antidepressant response in patients. Including omega-3-rich foods in the diet may thus have beneficial effects for individuals on SSRI therapy (Wani et al. 2015; Zhou et al. 2022).

Collectively, in each of these mechanisms, collaboration between dietitians and mental health professionals can enhance patient outcomes by providing a holistic approach that accounts for dietary influences on psychopharmacotherapy. This collaboration may involve co-consultations, shared digital tracking tools, or dietary adjustments for patients.

The following sections in this chapter offer an in-depth exploration of psychopharmacotherapy, dietary components, and their interactions.

4.3.1 Section 1: Overview of Psychopharmacotherapy

Psychopharmacotherapy involves the use of psychotropic medications to treat mental health disorders by modulating brain chemistry and neurotransmitter functions. These medications play a central role in managing psychiatric conditions, such as depression, anxiety, bipolar disorder, schizophrenia, and attention deficit hyperactivity disorder (ADHD). By alleviating symptoms, they improve patients' quality of life and enable engagement in everyday activities and therapies like psychotherapy, which further support recovery (Mlambo et al. 2023).

The importance of psychopharmacotherapy lies in its effectiveness for the long-term management of mental health conditions, helping to reduce relapse rates and maintain stability (Cui et al. 2024). However, the success of psychotropic medications relies on precise dosing, regular monitoring, and consideration of external factors like diet, lifestyle, and individual genetic variations (Godos et al. 2020). Understanding drug-diet interactions is essential to optimize treatment and reduce risks associated with these medications.

4.3.1.1 Common Classes of Psychotropic Drugs

Psychotropic medications are categorized based on their impact on neurotransmitter systems and target symptoms (Trujillo 2019). An overview of major classes of psychotropic drugs, including mechanisms of action, common examples, side effects, and recent advances, is provided in the following.

1. Antidepressants Primarily used for depression and anxiety disorders, they work by altering neurotransmitters linked to mood regulation, such as serotonin and norepinephrine. This class includes SSRIs (e.g., fluoxetine), SNRIs (e.g., venlafaxine), TCAs (e.g., amitriptyline), and MAOIs (e.g., phenelzine). Atypical antidepressants like bupropion offer symptom-specific relief with fewer side effects. However, common side effects of antidepressants include nausea, sleep disturbances, and weight gain, to mention a few (Roth 2020).

2. Antipsychotics Primarily used to treat symptoms of psychosis, including hallucinations and disordered thinking, common in schizophrenia and bipolar disorder. This class includes typical antipsychotics (e.g., haloperidol) and atypical antipsychotics (e.g., risperidone). Drugs like aripiprazole, a partial dopamine agonist, provide benefits with fewer side effects. Weight gain, metabolic syndrome, and extrapyramidal symptoms (e.g., tardive dyskinesia) are a couple of examples of side effects related to the class of antipsychotics (Licht 2000; Remington 2010).

3. Mood Stabilizers Used for bipolar disorder to regulate mood swings. Lithium and anticonvulsants (e.g., valproate) are known as mood stabilizers. Tremors, kidney issues (with lithium), drowsiness, and skin reactions (with anticonvulsants) are among the known side effects of these medications. Improved formulations can offer fewer side effects, and therapeutic monitoring is offered for safety and adherence (Carli et al. 2023; Himmerich and Hamilton 2022).

4. Anxiolytics Prescribed for anxiety disorders and occasionally for insomnia. Benzodiazepines (e.g., lorazepam) and non-benzodiazepine anxiolytics (e.g., buspirone) are examples from this class that can cause side effects such as drowsiness and carry risks of dependence (Cassano et al. 2002).

5. Stimulants Used mainly for ADHD to improve focus and energy. Amphetamines (e.g., dextroamphetamine) and methylphenidate (e.g., Ritalin) are examples of this class of medications. They have some common side effects, including increased heart rate, anxiety, and insomnia (Lakhan and Kirchgessner 2012). Extended-release formulations have been offered to improve dosing flexibility and reduce abuse risk (Childress 2017).

Psychotropic medications are essential for mental health care but pose challenges that require a personalized approach (Evers 2009; Češková and Šilhán 2021). Patients often take multiple medications, increasing the risk of drug-drug and drug-diet interactions. Tailored dietary modifications can help manage some side effects, improve adherence, and support better medication responses (Stein et al. 2022). However, effective integration of dietary factors into psychopharmacotherapy relies on close collaboration among healthcare providers for a truly holistic treatment approach. The following section delves into dietary components and interactions within psychopharmacotherapy, highlighting challenges and key considerations for treating patients with mental health conditions. Table 4.1 summarizes the primary

Table 4.1 Classes of psychotropic drugs and potential diet-drug interactions

Drug class	Example medications	Common dietary interactions	Mechanism	Clinical implications
SSRIs	Sertraline, Fluoxetine	Grapefruit juice, omega-3 fatty acids	Grapefruit inhibits CYP3A4, raising SSRI levels; omega-3s enhance serotonin receptor function	Avoid grapefruit; consider omega-3s to potentially improve efficacy
MAOIs	Phenelzine, Tranylcypromine	Tyramine-rich foods	MAOIs prevent tyramine breakdown, increasing the risk of hypertensive crisis	Avoid aged cheese, cured meats, and certain alcohols
Mood Stabilizers	Lithium, Valproate	Sodium, caffeine, B vitamins	Sodium fluctuations affect lithium levels; caffeine may reduce efficacy; B vitamins support neurotransmitter function	Maintain consistent sodium intake; monitor B vitamin levels
Antipsychotics	Clozapine, Olanzapine	High-fat meals, antioxidant-rich foods	High fat enhances absorption of lipophilic drugs; antioxidants may reduce oxidative stress	Avoid high-fat meals near dosing; encourage antioxidant intake
Anxiolytics	Diazepam, Lorazepam	Alcohol, caffeine, magnesium	Alcohol increases sedation; caffeine may counteract effects; magnesium supports GABA receptor activity	Avoid alcohol and moderate caffeine; consider magnesium supplementation for efficacy
Stimulants	Amphetamines, Methylphenidate	Acidic foods, high-protein diet	Acidic foods reduce absorption; high protein may affect metabolism	Avoid acidic foods near dosing; maintain balanced protein intake

classes of psychotropic drugs, along with common dietary interactions that can impact their efficacy and safety.

4.3.2 Section 2: Dietary Components and Interactions in Psychopharmacotherapy

Antidepressants are affected by dietary elements that can alter their efficacy and safety profiles. One example is related to SSRIs (e.g., fluoxetine, sertraline), where grapefruit can have an important interaction. Grapefruit or grapefruit juice (Bailey et al. 2013) inhibits CYP3A4, slowing SSRI metabolism and increasing the risk of serotonin syndrome (Kraft et al. 2014). This syndrome is a potentially life-threatening condition that occurs when there is an excess of serotonin activity in the central nervous system. It typically results from the use of medications or combinations of drugs that increase serotonin levels. Symptoms can range from mild (agitation, tremors, and sweating) to severe (high fever, muscle rigidity, and seizures), and severe cases require immediate medical intervention to prevent complications such as organ failure (Mikkelsen et al. 2023). Tryptophan-rich foods like turkey may also theoretically enhance serotonin levels, augmenting SSRI effects but potentially increasing side effects (Jenkins et al. 2016; Soh and Walter 2011). Caffeine interacts with SNRIs (e.g., venlafaxine) in terms of amplification of SNRI-related side effects, such as restlessness and insomnia (Kale et al. 2010). On the other hand, Vitamin B6 in diet (e.g., in chicken, turkey, salmon, tuna, potatoes, other starchy vegetables, bananas, chickpeas, and legumes) aids serotonin synthesis (Kennedy 2016) and may enhance SNRI efficacy.

High-fiber diets can reduce TCA (e.g., amitriptyline) absorption, affecting the efficacy of TCA (Stewart 1992). Alcohol increases sedation and overdose risk in TCA users (Weathermon and Crabb 1999). Foods high in tyramine (e.g., aged cheese) interact with MAOIs (e.g., phenelzine) and can trigger hypertensive crises, necessitating strict dietary compliance (Flockhart 2012).

4.3.3 Section 3: Specific Dietary Elements Impacting Psychotropic Drugs

Both macronutrients and micronutrients are integral to mental health (Horovitz 2024), supporting neurotransmitter synthesis, cellular function, and the pharmacological action of psychotropic drugs. Proper intake and balance of these nutrients play foundational roles in psychopharmacotherapy by influencing drug absorption, distribution, and efficacy. Each macronutrient interacts uniquely with drugs, impacting treatment outcomes for mental health conditions.

4.3.3.1 Macronutrients

Macronutrients such as carbohydrates, proteins, and fats are essential for maintaining energy levels, supporting cellular functions, and producing neurotransmitters that influence mood and cognitive health (Muth and Park 2021).

Refined carbohydrates can cause rapid blood sugar spikes, potentially affecting mood stability, whereas complex carbohydrates help stabilize glucose levels, supporting consistent drug efficacy (Brown et al. 2010). For patients taking antipsychotics like olanzapine, which often carry metabolic side effects (Lieberman 3rd 2004), complex carbs help manage blood sugar levels, countering the spikes that refined carbs can exacerbate. This support can aid mood stability and help manage weight gain, a common side effect of antipsychotic medications.

Proteins, especially those containing amino acids like tryptophan and tyrosine, play a crucial role in enhancing responses to psychotropic therapy. Tryptophan, a serotonin precursor (Höglund et al. 2019), supports the action of SSRIs, while tyrosine, a dopamine precursor, aids stimulant medications, potentially improving focus and mood. Balanced protein intake from sources like lean meats, eggs, and legumes ensures the availability of these neurotransmitter precursors (Gasmi et al. 2022). For instance, patients on SSRIs who consume tryptophan-rich foods such as turkey, eggs, and tofu often report greater mood stability (Lindseth et al. 2015). Similarly, tyrosine from foods like lean meats and legumes can support stimulants prescribed for ADHD, enhancing focus and alertness by promoting dopamine production (Reimherr et al. 1987).

Healthy fats, particularly omega-3 fatty acids found in fatty fish like salmon, mackerel, and sardines, play a crucial role in enhancing antidepressant efficacy and supporting overall brain health (DiNicolantonio and O'Keefe 2020). Omega-3s improve cell membrane fluidity, facilitating serotonin receptor function and potentially boosting the effectiveness of SSRIs. Studies show that patients consuming higher levels of omega-3s experience improved mood stability and faster relief from depressive symptoms (Serefko et al. 2024). For instance, a study on patients with major depressive disorder taking SSRIs found that those who supplemented with omega-3s experienced a 25–30% greater reduction in symptoms compared to those who did not supplement (Liao et al. 2019). Additionally, omega-3s aid in the absorption of lipid-soluble psychotropic drugs like olanzapine, an antipsychotic used to treat schizophrenia and bipolar disorder, supporting these drugs' stability and effectiveness (Bozzatello et al. 2016).

4.3.3.2 Micronutrients

Micronutrients, particularly certain vitamins and minerals, are essential for brain health, neurotransmitter synthesis, and cellular function (Huskisson et al. 2007). Their availability can influence the effectiveness and side effect profiles of psychotropic drugs, and deficiencies may impair treatment outcomes (Villagomez et al. 2023).

Vitamin D is crucial for immune and brain health and plays a role in neurotransmitter activity. Vitamin D receptors are located throughout the brain, where they help regulate gene expression for enzymes involved in dopamine synthesis, supporting mood stability and cognitive function. Adequate Vitamin D levels can enhance the efficacy of antidepressants, particularly SSRIs like fluoxetine and sertraline. Primary sources of Vitamin D include sunlight exposure, fortified dairy products, and fatty fish (e.g., salmon, mackerel). Low Vitamin D levels are associated with an increased risk of depression, and deficiency may reduce antidepressant effectiveness. For patients in low-sunlight regions or those with limited outdoor activity, Vitamin D supplementation may be recommended, especially if they are on antidepressants or mood stabilizers. Research has shown that patients with depression who maintain adequate Vitamin D levels respond more favorably to SSRIs, emphasizing the importance of monitoring and supplementation (Kaviani et al. 2020).

B vitamins, especially B6, B12, and folate, are essential for neurotransmitter synthesis and homocysteine regulation, both vital for mental health. Vitamin B6 acts as a cofactor in producing serotonin, dopamine, and GABA, ensuring their availability and supporting the action of antidepressants. B12 and folate are crucial for homocysteine metabolism, protecting neurons and aiding serotonin synthesis. Low levels of B12 and folate are linked to cognitive decline and an increased risk of depressive symptoms. B6 is found in foods like fish, poultry, and potatoes; B12 primarily in animal products (e.g., meat and dairy); and folate in leafy greens, legumes, and fortified grains. Deficiencies in these vitamins can impair antidepressant and antipsychotic efficacy, worsen depression symptoms, and contribute to cognitive decline. For patients on psychotropic drugs, especially vegetarians or older adults, B vitamin supplementation may be beneficial. Studies show that patients with adequate folate levels experience a 20–30% better response to SSRIs, highlighting the role of B vitamins in supporting antidepressant outcomes (Kennedy 2016; Liwinski and Lang 2023; Herbison et al. 2012).

Minerals such as magnesium and zinc are crucial for brain health, neurotransmitter function, and mood stability. Their presence in the diet can significantly influence the effectiveness of psychotropic medications by supporting cellular processes essential for mental health (Botturi et al. 2020).

Magnesium plays a stabilizing role in mood regulation, nerve function, and sleep quality, making it valuable in managing anxiety and depressive disorders. It modulates NMDA receptors, influencing mood stability. Low levels of magnesium can contribute to anxiety and depressive symptoms, potentially reducing the efficacy of anxiolytics and antidepressants. Rich dietary sources of magnesium include leafy greens (e.g., spinach), nuts, seeds, and whole grains. Patients with generalized anxiety disorder have reported better symptom control when magnesium intake is optimized, highlighting its role as a beneficial adjunct to treatment (Pickering et al. 2020).

Zinc supports brain plasticity, provides antioxidant protection, and regulates immune function, influencing NMDA receptors and promoting neuroplasticity. This can enhance antidepressant efficacy, especially for patients with treatment-resistant depression (Swardfager et al. 2013). Zinc-rich foods include meat, shellfish, legumes, seeds, and nuts. Low zinc levels are often associated with reduced response to antidepressants, and supplementation or regular intake of zinc-rich foods may improve outcomes (Totten et al. 2023). For example, clinical trials show that patients supplementing with zinc while on SSRIs experience over a 25% reduction in depressive symptoms, emphasizing zinc's importance in supporting antidepressant therapy (Vashum et al. 2014).

4.3.4 Section 4: Specific Dietary Patterns Impacting Psychotropic Drugs

Dietary patterns profoundly influence the body's metabolic environment, brain health, and neurotransmitter availability, all of which affect the efficacy of psychotropic drugs (Berding et al. 2021). In this context, three primary dietary patterns, the Mediterranean, ketogenic, and vegetarian/vegan diets, are presented, each impacting psychopharmacotherapy through specific mechanisms and benefits tailored to different patient populations.

4.3.4.1 Mediterranean Diet

The Mediterranean diet, known for promoting cardiovascular and brain health, includes abundant fruits, vegetables, whole grains, fish, nuts, olive oil, and moderate wine intake (Tosti et al. 2018). This minimally processed, nutrient-dense diet is rich in omega-3 fatty acids, antioxidants, fiber, and polyphenols, all of which contribute to reduced inflammation and enhanced mental well-being (van Zonneveld et al. 2024). Specifically, the diet's high antioxidant and omega-3 content has been shown to reduce inflammation, which is linked to improved outcomes for patients on antidepressants and mood stabilizers (Gören and Tewksbury 2011). Furthermore, the complex carbohydrates in this diet help stabilize blood sugar, reducing metabolic side effects of antipsychotics and supporting mood stability in patients on mood stabilizers. Additionally, omega-3 fatty acids from fish in this diet improve serotonin receptor function, which can enhance the efficacy of SSRIs and SNRIs, making it particularly beneficial for patients with mood disorders (Marrone and Coccurello 2020).

Consequently, this diet's glucose-regulating effects benefit older adults with mental health conditions or metabolic comorbidities, such as diabetes, as it minimizes weight gain and insulin resistance, which are often associated with antipsychotic medications. Younger adults may also benefit from the mood-stabilizing effects of omega-3-rich foods, which support overall brain health (Dighriri et al. 2022). Indeed, consistent adherence to the Mediterranean diet is associated with lasting mood enhancement and a reduced risk of depression relapse. Additionally, the diet's impact on metabolic health helps to reduce weight gain and insulin resistance, common side effects of long-term antipsychotic use. For example, studies indicate that patients with schizophrenia adhering to the Mediterranean diet experience better metabolic health, improved medication adherence, and reduced relapse rates due to the diet's beneficial impact on both physical and mental health (Spanakis et al. 2022; Morris et al. 2024).

4.3.4.2 Ketogenic Diet

The ketogenic diet (Dowis and Banga 2021), a high fat, low-carbohydrate regimen, induces a state of ketosis in which fat becomes the primary energy source. This low-glycemic diet includes high-fat foods like avocado, nuts, and oils while minimizing carbohydrate intake, which subsequently alters liver enzyme activity and metabolism. Ketones produced in ketosis offer neuroprotective effects that support mood stabilizers, making this diet especially beneficial for individuals with bipolar disorder. In addition, the anti-seizure properties of ketosis complement anticonvulsant mood stabilizers. Through this mechanism, ketosis reduces inflammatory markers, potentially lowering psychiatric symptoms in patients on antipsychotics.

The low-carbohydrate intake in this diet also influences liver enzyme activity, potentially altering the metabolism of drugs processed by the cytochrome P450 system. Thus, this diet may help patients with bipolar disorder who have not fully responded to mood stabilizers, though it requires careful monitoring due to its impact on drug metabolism (Marinescu et al. 2024). Furthermore, since the ketogenic diet affects liver enzymes, patients with liver issues or those on multiple medications should be closely monitored. Long-term adherence to the ketogenic diet can stabilize mood and reduce symptom recurrence in bipolar disorders. However, the diet may also reduce bone density over time, necessitating calcium and Vitamin D monitoring. Research on patients with treatment-resistant bipolar disorder has shown improved mood stability and reduced cycling when following the ketogenic diet, highlighting its potential as a complementary approach to traditional mood-stabilizing drugs. To mitigate potential side effects, a modified ketogenic diet incorporating higher protein intake can help maintain balanced nutrient intake, which is essential for sustaining long-term adherence (Choi et al. 2024; Rog et al. 2024).

4.3.4.3 Vegetarian/Vegan Diet

Vegetarian and vegan diets emphasize plant-based foods like fruits, vegetables, legumes, whole grains, and nuts, though they differ in their inclusivity of animal products (Wang et al. 2023). While vegetarian diets may include eggs and dairy, vegan diets exclude all animal-derived products. Both diets are high in fiber and antioxidants but may lack certain nutrients, particularly omega-3 fatty acids and B12, which are critical for mental health. Essential amino acids like tryptophan and tyrosine, which are vital for synthesizing neurotransmitters such as serotonin and dopamine, may be lower in plant-based diets, potentially affecting the efficacy of antidepressants. In addition, omega-3s in plant-based diets are often in the form of ALA, which converts less efficiently to EPA and DHA, the forms beneficial for brain health. Consequently, this reduced bioavailability can impact mood stabilization for patients on antidepressants (Alcorta et al. 2021; Łuszczki et al. 2023; Bakaloudi et al. 2021).

Nonetheless, high fiber and antioxidant content in vegetarian and vegan diets can help reduce inflammation, which may benefit patients on antipsychotics by supporting brain health. Younger individuals on plant-based diets may have sufficient nutrient reserves but may still require omega-3 and B12 supplementation to sustain nutrient levels over time. Meanwhile, patients with higher nutrient requirements, such as those on SSRIs or those experiencing anxiety or depression, may benefit from additional omega-3 and B12 supplementation to support optimal mental health outcomes. Although plant-based diets support cardiovascular health, long-term deficiencies in B12 or omega-3s could negatively impact the efficacy of psychotropic drugs. Therefore, consistent supplementation can help mitigate risks of nutrient deficiencies, thereby supporting the effectiveness of mental health treatments. For instance, a study on patients following vegan diets supplemented with B12 and omega-3s reported improved outcomes on SSRIs, with reduced depressive symptoms and enhanced treatment efficacy, highlighting the importance of careful nutrient management in plant-based diets. Including omega-3 supplements in the EPA/DHA form and B12 supplements can support cognitive and emotional health, thereby optimizing the therapeutic effects of psychotropic medications (Medawar et al. 2019).

4.3.5 Section 5: Herbs and Supplements Impacting Psychotropic Drugs

Herbs and supplements are commonly used in mental health care to alleviate symptoms, enhance treatment outcomes, or manage the side effects of psychotropic drugs (Firth et al. 2019; Le et al. 2022). However, certain herbs and supplements can interfere with the metabolism, absorption, or efficacy of these medications, necessitating careful consideration when combined with psychotropic drugs (Wanwimolruk and Prachayasittikul 2014; Wanwimolruk et al. 2014).

St. John's Wort, for example, is a widely used herb for mild to moderate depression. Known for its potential antidepressant effects, it also presents significant risks when used with other medications. By inducing the enzyme CYP3A4, St. John's Wort accelerates the metabolism of several drugs, including SSRIs, benzodiazepines, and antipsychotics, potentially reducing their effectiveness. Additionally, since St. John's Wort raises serotonin levels, it can increase the risk of serotonin syndrome when combined with SSRIs or SNRIs. While it may offer mild antidepressant benefits for some individuals, comparable to low-dose SSRIs, the risks associated with these interactions, particularly serotonin syndrome, highlight the importance of avoiding St. John's Wort in combination with certain psychotropic drugs (Schulz 2006; Fugh-Berman 2000).

Omega-3 fatty acids, found in fish oil, are another commonly used supplement with notable benefits for mental health, particularly as an adjunct to antidepressants. Known for their anti-inflammatory and neuroprotective properties, omega-3s can enhance the efficacy of SSRIs and SNRIs by supporting serotonin receptor function and improving cell membrane fluidity, thereby aiding drug transport across the blood-brain barrier. Studies have shown that patients with depression on SSRIs who supplement with omega-3s experience a 20–30% improvement in symptoms. Despite the minimal risks associated with omega-3 supplementation, high doses may cause gastrointestinal discomfort or interact with anticoagulant medications due to blood-thinning effects. Typically, a daily dose of 1–2 grams of combined EPA and DHA is recommended for patients with depression to enhance treatment efficacy (Zhou et al. 2022; Reily et al. 2023).

Probiotics, which influence the gut-brain axis, are gaining attention for their potential to support mental health, particularly in anxiety and depression. By modulating the gut microbiota, probiotics can stimulate the production of neurotransmitters like serotonin and GABA, which are critical for mood regulation. For patients on SSRIs or anxiolytics, probiotics may help stabilize mood and reduce gastrointestinal side effects. In addition, probiotics' anti-inflammatory effects may benefit psychiatric symptoms by lowering systemic inflammation. Although generally safe, specific strains are more effective for mental health, so individualized recommendations are advisable. Research indicates that patients with depression on SSRIs report better mood stability and gastrointestinal health when incorporating probiotics into their regimen (Ansari et al. 2023; Dziedzic et al. 2024; Ķimse et al. 2024).

Valerian root, commonly used for its sedative effects to promote sleep and relaxation (Borrás et al. 2021), may support anxiolytic medications but poses interaction risks, particularly with benzodiazepines. By enhancing GABA receptor activity, valerian can have synergistic effects with benzodiazepines, potentially leading to excessive sedation and cognitive impairment. For this reason, patients should use valerian cautiously with other sedative medications and consult healthcare providers for guidance (Bent et al. 2006; Shinjyo et al. 2020). Studies have shown that valerian root can improve sleep quality in patients with anxiety; however, when used alongside anxiolytics, close monitoring is essential to prevent excessive sedation (Borrás et al. 2021; Becker et al. 2014; Bruni et al. 2021).

Ginkgo biloba, known for its antioxidant properties and cognitive benefits, is sometimes used to enhance cognitive function in patients on antipsychotics or other psychotropic medications. By improving blood circulation and providing neuroprotective effects, ginkgo may support cognitive health. However, it also induces the enzyme CYP2C19, which affects the metabolism of benzodiazepines and SSRIs, potentially reducing their efficacy. While ginkgo may improve cognitive function and circulation, the risk of interactions with CYP2C19-metabolized drugs means it should be used with caution, particularly in patients on multiple psychotropic medications. Research shows that ginkgo supplementation can offer mild cognitive improvement for patients on antipsychotics, yet careful monitoring is advised to avoid adverse interactions (Chen et al. 2015).

Melatonin, frequently used for sleep regulation, can benefit patients with sleep disorders related to anxiety, depression, or bipolar disorders. By modulating circadian rhythms, melatonin may help stabilize mood and improve sleep onset. However, it can interfere with the effects of mood stabilizers, leading to possible daytime drowsiness or disruptions in circadian patterns. While melatonin can be an effective sleep aid, patients on mood stabilizers or sedatives should use it cautiously and under medical supervision. Studies have shown that melatonin can significantly reduce sleep latency in patients with anxiety, but regular monitoring is essential for those on sedative medications to avoid excessive sedation or altered sleep-wake cycles (Moon et al. 2022; De Crescenzo et al. 2017; Claudio and Andrea 2022).

In summary, while herbs and supplements can offer valuable support in managing mental health symptoms, their potential to interact with psychotropic drugs demands careful consideration and guidance from healthcare providers.

4.3.6 Section 6: Cross-Cultural Considerations in Diet-Psychotropic Drug Interactions

Dietary patterns are deeply influenced by cultural, religious, and geographical factors, all of which can significantly impact how individuals respond to dietary recommendations in psychopharmacotherapy. Each culture's dietary norms, shaped by traditional practices, regional resources, and lifestyle factors, affect nutrient availability and, consequently, the interactions between diet and psychotropic drugs

(Gherasim et al. 2020; Freimer et al. 1983). For example, individuals in Mediterranean regions often consume diets naturally rich in fish, olive oil, and fresh produce, providing high levels of omega-3 fatty acids, antioxidants, and fiber, which support mood stability and cognitive health. This traditional diet aligns well with recommendations for patients on SSRIs or mood stabilizers, as omega-3 fatty acids and antioxidants can enhance drug efficacy and improve outcomes. However, for individuals from cultures where plant-based diets are more prevalent, such as those in South Asia or Southeast Asia, there may be limited intake of certain nutrients like B12 and omega-3s in their most bioavailable forms (DHA and EPA), which are found primarily in animal products. In these cases, supplementation or plant-based alternatives (e.g., fortified foods or algae-derived omega-3) may be recommended to ensure optimal support for psychotropic medications (Ventriglio et al. 2020).

Religious practices can also influence dietary habits and need to be considered in psychopharmacotherapy. For instance, Hindu and Jain dietary practices often emphasize vegetarianism, which may impact the availability of B12, an essential nutrient for brain health. Muslim patients observing halal dietary guidelines or Jewish patients following kosher laws may require alternative food sources or specific supplement formulations to avoid ingredients derived from non-permissible animals. Furthermore, clinicians working with patients who observe dietary restrictions during religious periods, such as Ramadan fasting in Islam or Orthodox Christian fasting, should consider the timing of medications and meals, as these practices can impact blood sugar levels and drug absorption, particularly for psychotropic drugs that require consistent dosing or are affected by food intake (Arslan and Aydın 2024; Chouraqui et al. 2021).

Geographic factors also play a role in nutrient accessibility, which is crucial for providing appropriate dietary guidance. For example, individuals in regions with limited sunlight, such as Northern Europe, are at higher risk for Vitamin D deficiency, potentially impacting the effectiveness of antidepressants. In contrast, individuals in tropical regions may have sufficient natural sunlight exposure but might lack access to other essential nutrients due to agricultural or economic limitations. Consequently, healthcare providers should tailor dietary recommendations based on geographic and socioeconomic accessibility, emphasizing affordable, locally available food sources where possible (Spiro and Buttriss 2014).

Incorporating cultural dietary habits into psychopharmacotherapy recommendations helps improve patient adherence, as patients are more likely to follow guidance that respects their traditions and values. Clinicians should adopt a culturally sensitive approach, working collaboratively with patients to identify dietary adjustments that align with their cultural practices while supporting the efficacy of psychotropic medications.

4.3.7 Section 7: Negative Aspects of Diet-Psychotropic Drug Interactions

This dedicated section serves to highlight how specific dietary components can interfere with drug efficacy, potentially reduce therapeutic effects, and even lead to adverse reactions. By delving deeper into the underlying mechanisms and offering concrete examples, the aim here is to highlight the importance of dietary awareness in psychopharmacotherapy, emphasizing how certain foods and dietary patterns can complicate treatment outcomes if not managed properly. It is important to recognize and manage potentially adverse effects that dietary components may introduce. Understanding the negative aspects of diet-drug interactions is crucial for safe and effective psychopharmacotherapy. By recognizing and managing these interactions, healthcare providers can help patients avoid potentially severe consequences, enhancing overall treatment outcomes and adherence (Niederberger and Parnham 2021).

4.3.7.1 Mechanisms of Negative Interactions

One primary mechanism of negative interaction is through the absorption of medications. For instance, high-fiber foods can bind to certain psychotropic drugs in the gastrointestinal tract, delaying or reducing their absorption. This is especially relevant for tricyclic antidepressants (TCAs), which may have diminished efficacy when consumed with fiber-rich meals (Stewart 1992). Similarly, acidic foods and beverages, such as citrus juices, can alter the pH of the stomach, affecting the absorption of drugs like methylphenidate (Aresti-Sanz et al. 2021). This alteration can reduce the drug's efficacy, necessitating a dietary adjustment for patients reliant on such medications for therapeutic stability.

Distribution, the process by which drugs travel throughout the body, is another pharmacokinetic stage impacted by diet. High-protein diets can displace drugs that bind to plasma proteins, such as albumin, increasing the free concentration of the drug in the bloodstream. While this might enhance the drug's effect, it also heightens the risk of toxicity (Gurevich 2013). Additionally, high-fat diets can alter blood flow, promoting the uptake of lipophilic drugs, such as certain antipsychotics. This can lead to an increase in blood levels and, consequently, a higher risk of side effects. Patients on highly protein-bound medications, such as antipsychotics, might be advised to maintain a balanced protein intake to prevent fluctuations in drug levels (Niederberger and Parnham 2021; Spanakis et al. 2022).

Metabolism, one of the most commonly affected pharmacokinetic stages, is heavily influenced by dietary components, particularly in relation to liver enzymes (Koziolek et al. 2019). Foods that affect cytochrome P450 enzymes, especially those in the CYP3A4 family, can have a profound impact on drug metabolism (Sasaki et al. 2017). For example, grapefruit juice (Bailey et al. 2013; Kiani and Imam 2007) inhibits CYP3A4 (Wael Abu Dayyih et al. 2024), which can significantly increase blood levels of drugs metabolized by this enzyme, such as SSRIs and certain antipsychotics. This can lead to side effects like serotonin syndrome, characterized by symptoms such as confusion, rapid heart rate, and muscle rigidity.

In contrast, cruciferous vegetables (Eagles et al. 2020) and chargrilled meats induce CYP1A2, potentially speeding up the metabolism of drugs like clozapine and olanzapine. For patients on psychotropic medications metabolized by CYP enzymes (Hakkola et al. 2020), dietary restrictions around grapefruit and cruciferous vegetables may be necessary to avoid these dangerous fluctuations in drug levels.

Diet can also influence drug excretion, primarily through kidney function and urine pH (Skrajnowska and Bobrowska-Korczak 2024). A high-protein intake can acidify urine, increasing the excretion rate of certain drugs, like amphetamines, and reducing their therapeutic effect. This is particularly relevant for patients on stimulants (Paz-Ramos et al. 2023), who might need to maintain a consistent protein intake to prevent fluctuations in drug excretion. For individuals whose medications are sensitive to urine pH, maintaining a balanced diet becomes essential for sustaining consistent therapeutic effects.

4.3.7.2 Common Negative Interactions and Case Examples

A well-documented example of negative diet-drug interaction is between SSRIs, such as sertraline, and grapefruit juice (Kane and Lipsky 2000). Given grapefruit's effect on CYP3A4 (Lee et al. 1999), SSRIs can accumulate to toxic levels, raising the risk of serotonin syndrome. Another prominent example is the interaction between MAOIs and tyramine-rich foods (Edinoff et al. 2022). Patients on MAOIs (McCabe-Sellers et al. 2006), like phenelzine, risk life-threatening hypertensive crises if they consume tyramine-heavy foods. Additionally, alcohol presents significant risks for patients on benzodiazepines (Weathermon and Crabb 1999; Linnoila 1990), such as diazepam. Combining these substances can increase the risk of respiratory depression, cognitive impairment, and overdose. Similarly, caffeine intake can interfere with anxiolytics like lorazepam by opposing their sedative effects, for example, worsened symptoms and reduced response to benzodiazepines (Liu et al. 2024).

4.3.7.3 Strategies for Managing Negative Interactions

To address these challenges effectively, clinicians can employ several strategies (Genser 2008; Bushra et al. 2011). Patient education on specific dietary restrictions is essential, such as advising patients on MAOIs to avoid tyramine-rich foods or those on SSRIs to avoid grapefruit juice. Adjustments in medication timing relative to meals can also be beneficial. For instance, patients on high-fiber diets might be instructed to take TCAs two hours before or after consuming fiber-rich meals to improve absorption. Routine monitoring, including regular blood tests and patient symptom tracking, can help detect adverse interactions early. For patients who cannot easily modify their diets, clinicians might consider alternative medications with fewer dietary interactions. Collaboration with dietitians for complex cases can further support patients in managing dietary needs without compromising medication efficacy.

Table 4.2 provides a summary of common negative diet-drug interactions, detailing specific dietary components that can interfere with the efficacy and safety of

Table 4.2 Summary of negative diet-drug interactions

Dietary component	Affected drug	Mechanism of interaction	Potential clinical consequences
Grapefruit Juice	SSRIs (e.g., Sertraline)	Inhibits CYP3A4 enzyme, increasing drug levels	Increased risk of serotonin syndrome and heightened side effects
Tyramine-Rich Foods	MAOIs (e.g., Phenelzine)	Increases norepinephrine, risking hypertensive crisis	Severe hypertension, potential medical emergency
Alcohol	Benzodiazepines (e.g., Diazepam)	Enhances sedative effects, risk of respiratory depression	Increased drowsiness, impaired cognition, overdose risk
High-Fiber Diets	Tricyclic Antidepressants (TCAs)	Fiber binds to drugs, reducing gastrointestinal absorption	Reduced therapeutic effect, possible need for dose adjustments
Caffeine	Anxiolytics (e.g., Diazepam)	Counteracts sedative effects, potentially increasing anxiety	Reduced drug efficacy, potential exacerbation of anxiety symptoms
Chargrilled Meats	Antipsychotics (e.g., Clozapine)	Induces CYP1A2 enzyme, reducing drug levels	Decreased efficacy, increased risk of symptom relapse

psychotropic drugs. Each interaction is explained through its mechanism, highlighting the potential clinical consequences for patients.

4.3.8 Section 7: Positive Aspects of Diet-Psychotropic Drug Interactions

Certain dietary components have been shown to complement psychotropic drug therapy, enhancing efficacy, reducing side effects, and promoting overall treatment outcomes. Unlike negative interactions that may hinder drug effectiveness, positive diet-drug interactions can support the pharmacokinetic and pharmacodynamic properties of psychotropic medications, leading to more stable and favorable responses. By exploring the mechanisms, benefits, and clinical applications of positive interactions, this section aims to highlight how dietary awareness can be a valuable adjunct in psychopharmacotherapy.

4.3.8.1 Mechanisms and Benefits of Positive Interactions

The beneficial effects of dietary elements on psychotropic drugs stem from several mechanisms that influence drug absorption, bioavailability, neurotransmitter support (Gasmi et al. 2022), and inflammation modulation. For instance, specific nutrients can enhance drug absorption by stabilizing gastrointestinal pH or increasing lipid solubility (Abuhelwa et al. 2017). For example, healthy fats play a crucial role in aiding the absorption of lipid-soluble drugs like certain antipsychotics, allowing for more stable blood levels and consistent therapeutic effects. Additionally, certain nutrients support neurotransmitter synthesis, acting as precursors for serotonin, dopamine, and other chemicals involved in mood regulation. Tryptophan, an amino acid precursor to serotonin, can enhance the action of SSRIs, while magnesium supports the neurotransmitter GABA, aiding anxiolytics. Furthermore, omega-3 fatty acids and antioxidants possess anti-inflammatory properties, which are beneficial in reducing systemic inflammation, a factor often linked to mood disorders such as depression and schizophrenia. By mitigating inflammation, these nutrients may improve antidepressant and antipsychotic efficacy. Vitamins and minerals, including B vitamins, magnesium, and zinc, also play roles in neuroplasticity and receptor sensitivity, allowing drugs like SSRIs to act more effectively within the brain's chemical pathways, ultimately supporting mood stability.

Several foods and nutrients are well-documented for their positive impact on psychotropic drugs, either by enhancing their effectiveness or reducing side effects (Firth et al. 2019). Omega-3 fatty acids, found abundantly in fish and flaxseed oils, are a prime example. Known for their anti-inflammatory and neuroprotective effects, omega-3s integrate into cell membranes, facilitating serotonin receptor function and promoting neurotransmitter activity. In patients with depression, omega-3 supplementation alongside SSRIs has been shown to accelerate symptom relief and allow for potentially lower drug dosages (Zhou et al. 2022; Grosso et al. 2014). Individuals with major depressive disorder who were supplemented with omega-3s experienced a greater reduction in symptoms, illustrating omega-3's role as a supportive element in antidepressant therapy (Liao et al. 2019). Similarly, B vitamins (Kennedy 2016), including B6, B12, and folate, are essential for neurotransmitter synthesis and homocysteine regulation, which are critical for mental health. Vitamin B6 serves as a cofactor in synthesizing serotonin, dopamine, and GABA, thereby enhancing the efficacy of antidepressants and anxiolytics. B12 and folate work together to regulate homocysteine levels, promoting neuron health and supporting cognitive and mood outcomes, especially in patients on SSRIs or antipsychotics. Studies indicate that patients who maintain adequate levels of B12 and folate report improved antidepressant efficacy, highlighting the importance of these vitamins in psychopharmacotherapy (Rechenberg 2016). Magnesium is another crucial mineral with notable effects on mood and anxiety management (Kirkland et al. 2018). By modulating NMDA receptors and supporting GABAergic function, magnesium can enhance the effects of anxiolytics and antidepressants. For example, patients with generalized anxiety disorder who supplement with magnesium report improved symptom control and greater tolerance to anxiolytic medications, demonstrating magnesium's potential as a complementary element in treatment (Boyle et al. 2017). Probiotics (Ansari et al. 2023), beneficial bacteria that influence the gut-brain axis, have also shown promise in supporting mental health. By enhancing gut microbiota composition, probiotics aid in the production of neurotransmitters such as serotonin and GABA while reducing inflammation. Probiotic use has been linked to reduced depressive symptoms in patients on SSRIs, with studies indicating fewer gastrointestinal side effects and improved mood stability among patients who incorporate probiotics into their diet. These effects illustrate the role of probiotics as a natural adjunct in psychopharmacotherapy (Mörkl et al. 2020; Meher et al. 2024).

4.3.8.2 Common Positive Interactions and Case Examples

Several well-documented cases illustrate how dietary components can support the efficacy and safety of psychotropic drugs. Omega-3 supplementation, for instance, has been shown to enhance the therapeutic effects of SSRIs in patients with depression, potentially allowing for reduced dosages and faster symptom relief (Mehdi et al. 2023). Additionally, diets rich in antioxidants, such as the Mediterranean diet, can support patients on antipsychotic medications (Vassilopoulou et al. 2022). Studies have found that patients with schizophrenia who follow an antioxidant-rich diet experience improved cognitive function and reduced oxidative stress markers, supporting treatment outcomes and overall mental health (Zajkowska et al. 2024). Another example involves magnesium supplementation (Botturi et al. 2020), which has been found to enhance the efficacy of anxiolytics by promoting GABA receptor activity and improving anxiety management. A clinical study on patients with generalized anxiety disorder showed that magnesium supplementation led to fewer anxiety symptoms and better tolerance of benzodiazepines, illustrating the supportive role of magnesium in anxiety treatment.

4.3.8.3 Strategies for Enhancing Positive Interactions

To maximize the benefits of positive diet-drug interactions, healthcare providers can consider tailored dietary recommendations and supplementation based on specific medications. Patients on SSRIs, for instance, may benefit from omega-3 supplementation to support antidepressant efficacy. For those on mood stabilizers like lithium, monitoring folate and B12 levels and providing supplements as needed can help maintain mood stability. Magnesium supplementation is also advantageous for patients on anxiolytics, enhancing GABAergic function and promoting calming effects. Tailoring dietary plans to medication regimens can further support treatment outcomes. Patients on antipsychotics, for example, may find the Mediterranean diet beneficial due to its antioxidant-rich foods, which help reduce oxidative stress and improve treatment efficacy. Additionally, incorporating probiotics into the diet may support gut health and enhance the efficacy of SSRIs or anxiolytics, particularly for patients experiencing gastrointestinal discomfort. Regular monitoring of nutrient levels, such as B12, folate, and magnesium, can aid in identifying deficiencies early on, allowing for timely supplementation that bolsters drug efficacy. Patient education is also essential and can be achieved by informing patients of the benefits of specific nutrients and dietary patterns; healthcare providers can empower them to make informed dietary choices that support their treatment (Ortega et al. 2022; Merlo and Vela 2022).

Table 4.3 provides a summary of positive diet-drug interactions, illustrating how specific dietary components can complement psychotropic medications by enhancing their efficacy and reducing side effects.

Table 4.3 Summary of positive diet-drug interactions

Dietary component	Affected drug	Mechanism of interaction	Observed benefits
Omega-3 Fatty Acids	SSRIs (e.g., Fluoxetine)	Supports serotonin receptor function and cell membrane fluidity	Enhanced antidepressant efficacy, improved mood stability
Probiotics	SSRIs, Anxiolytics	Modulates gut-brain axis and reduces systemic inflammation	Improved mood stability, reduced gastrointestinal side effects
Magnesium	Anxiolytics (e.g., Lorazepam)	Enhances GABA receptor function	Increased anxiolytic efficacy, reduced anxiety symptoms
Antioxidant-Rich Foods	Antipsychotics	Reduces oxidative stress and inflammation	Enhanced cognitive function, lowered oxidative damage
Folate and Vitamin B12	Mood Stabilizers (e.g., Lithium)	Facilitates neurotransmitter synthesis and cellular health	Improved mood stability, fewer cognitive side effects
Mediterranean Diet	Antipsychotics, Antidepressants	Provides anti-inflammatory and neuroprotective effects	Reduced metabolic side effects, improved overall mental stability

4.4 Concluding Remarks

Integrating dietary considerations into psychopharmacotherapy is crucial for optimizing drug efficacy, minimizing side effects, and supporting overall mental health outcomes. This section combines clinical recommendations, patient perspectives, and implementation guidelines, offering a comprehensive approach to incorporating dietary guidance in mental health treatment. To ensure effective and safe psychopharmacotherapy, clinicians are encouraged to incorporate dietary assessments as part of routine evaluations. During initial and follow-up appointments, assessing dietary habits helps identify high-risk foods and beneficial nutrients that may influence psychotropic medications. Visual aids, such as diet pyramids or sample meal plans, can enhance these evaluations, helping patients understand potential interactions between their diet and medications. Additionally, clinicians should consider patients' use of supplements and over-the-counter medications, as these can significantly impact drug efficacy. Tailored dietary counseling is particularly important when recommending foods that support specific medication regimens. For example, clinicians can advise patients on SSRIs to incorporate omega-3-rich foods like salmon and flaxseed or suggest supplements with appropriate doses and timing to enhance antidepressant effects. Similarly, patients on mood stabilizers like lithium may benefit from consistent sodium intake, while those on anticonvulsants may require folate supplementation to help maintain stable drug effects. In cases where cultural preferences impact dietary choices, clinicians should offer alternatives that align with these preferences, such as plant-based tryptophan sources like lentils and quinoa for vegetarians on SSRIs.

Incorporating dietary guidance into psychopharmacotherapy is most effective when approached collaboratively. Working alongside dietitians, mental health counselors, and primary care providers enables comprehensive care, especially for patients with complex needs or multiple comorbidities. Collaborative tools, such as electronic health records (EHRs) and dietary tracking apps, allow healthcare providers to coordinate and adjust dietary and medication plans based on patients' evolving needs, enhancing the overall effectiveness of treatment.

Patient empowerment through dietary modifications is essential for enhancing adherence and improving mental health outcomes. Clinicians can foster this by demonstrating the tangible benefits of dietary changes on mental health, guiding patients through motivational strategies, and setting achievable goals. SMART goals (Specific, Measurable, Achievable, Relevant, Time-bound) are particularly useful for helping patients gradually integrate dietary changes that complement their treatment plans. Educational resources, such as pamphlets, online content, or group workshops, can reinforce the significance of diet concerning medication outcomes, providing shared learning and support. Furthermore, sharing success stories and case examples can illustrate the benefits of dietary adjustments and foster motivation. Active patient involvement in dietary monitoring can strengthen adherence and enhance self-management. By tracking their diet, symptoms, and mood changes, patients can better understand the link between diet and medication, leading to more personal engagement in their treatment. Regular feedback sessions where clinicians review food diaries allow for personalized adjustments, boosting patients' confidence and promoting accountability.

Monitoring diet-drug interactions effectively requires structured approaches to dietary tracking and nutrient assessments. Patients can be encouraged to maintain a food diary or use digital tracking tools to record daily food intake, supplements, and medication timings, helping to identify any potential diet-drug interactions. Clinicians might recommend dietary apps that track food intake, set medication reminders, and provide alerts for foods with known interaction risks.

Routine nutrient screening is particularly essential for high-risk patients on long-term psychotropic medications. Regular blood tests to assess levels of key nutrients, such as folate, Vitamin D, and magnesium, enable early detection of deficiencies or side effects. For example, patients on lithium or antipsychotics may benefit from biannual screenings to identify early signs of nutrient-related deficiencies or metabolic side effects, allowing for timely dietary or supplementation adjustments.

Recognizing cultural dietary preferences is also critical when making dietary recommendations. Clinicians should provide culturally sensitive guidance that respects patients' backgrounds and dietary choices, ensuring that recommendations are accessible and realistic. For instance, suggesting plant-based omega-3 sources, like chia seeds and walnuts, along with B12 supplements for a patient on SSRIs following a vegetarian diet aligns with their lifestyle while supporting mental health outcomes.

When recommending supplements, educating patients on safe selection and potential interactions is also crucial. Patients should be advised on the importance of choosing high-quality supplements, adhering to recommended dosages, and avoiding over-supplementation risks, particularly for nutrients like magnesium and omega-3. Guiding patients to read supplement labels, including daily values and potential interactions, enables them to make informed choices that support their treatment.

Finally, gradual dietary changes are often more sustainable for patients on complex medication regimens. Introducing incremental adjustments helps prevent overwhelm, encourages consistency, and reduces the risk of diet-related fluctuations in drug metabolism. For instance, clinicians may suggest a phased approach where patients initially incorporate omega-3-rich foods, then gradually reduce caffeine intake, with regular follow-ups to track progress and adjust goals. This structured approach fosters adherence while supporting both physical and mental well-being, ultimately optimizing the effectiveness of psychopharmacotherapy.

Such an integrated strategy, combining tailored recommendations, collaborative care, patient empowerment, and gradual change, provides a practical and effective approach to enhancing the role of diet in psychopharmacotherapy.

Funding Statement No external funding was received for the preparation of this chapter.

Conflict of Interest Statement The author declares no conflicts of interest related to the content of this chapter.

References

Abuhelwa AY, Williams DB, Upton RN, Foster DJ (2017) Food, gastrointestinal pH, and models of oral drug absorption. Eur J Pharm Biopharm 112:234–248

Alcorta A, Porta A, Tárrega A, Alvarez MD, Vaquero MP (2021) Foods for plant-based diets: challenges and innovations. Foods 10(2):293

Allison SP, Lobo DN (2024) The clinical significance of hypoalbuminaemia. Clin Nutr 43(4):909–914

Ana-Maria P, Oană R, Georgescu CV et al (2022) Food-drug interactions – a short review of the particularities of pharmacokinetics and pharmacodynamics molecular stage correlations. J Pharm Negat Results. https://doi.org/10.47750/pnr.2022.13.S09.011

Ansari F, Neshat M, Pourjafar H, Jafari SM, Samakkhah SA, Mirzakhani E (2023) The role of probiotics and prebiotics in modulating of the gut-brain axis. Front Nutr 10:1173660

Aresti-Sanz J, Schwalbe M, Pereira RR, Permentier H, El Aidy S (2021) Stability of methylphenidate under various pH conditions in the presence or absence of gut microbiota. Pharmaceuticals (Basel) 14(8):733

Arslan S, Aydın A (2024) Religious dietary practices: health outcomes and psychological insights from various countries. J Relig Health 63(5):3256–3273

Bailey DG, Dresser G, Arnold JMO (2013) Grapefruit–medication interactions: forbidden fruit or avoidable consequences? Can Med Assoc J 185(4):309

Bakaloudi DR, Halloran A, Rippin HL et al (2021) Intake and adequacy of the vegan diet. a systematic review of the evidence. Clin Nutr 40(5):3503–3521

Becker A, Felgentreff F, Schröder H, Meier B, Brattström A (2014) The anxiolytic effects of a valerian extract is based on valerenic acid. BMC Complement Altern Med 14(1):267

Bent S, Padula A, Moore D, Patterson M, Mehling W (2006) Valerian for sleep: a systematic review and meta-analysis. Am J Med 119(12):1005–1012

Berding K, Vlckova K, Marx W et al (2021) Diet and the microbiota–gut–brain axis: sowing the seeds of good mental health. Adv Nutr 12(4):1239–1285

Borrás S, Martínez-Solís I, Ríos JL (2021) Medicinal plants for insomnia related to anxiety: an updated review. Planta Med 87(10-11):738–753

Botturi A, Ciappolino V, Delvecchio G, Boscutti A, Viscardi B, Brambilla P (2020) The role and the effect of magnesium in mental disorders: a systematic review. Nutrients 12(6):1661

Boyle NB, Lawton C, Dye L (2017) The effects of magnesium supplementation on subjective anxiety and stress-a systematic review. Nutrients 9(5):429

Bozzatello P, Brignolo E, De Grandi E, Bellino S (2016) Supplementation with omega-3 fatty acids in psychiatric disorders: a review of literature data. J Clin Med 5(8). https://doi.org/10.3390/jcm5080067

Bråten LS, Haslemo T, Jukic MM, Ingelman-Sundberg M, Molden E, Kringen MK (2020) Impact of CYP2C19 genotype on sertraline exposure in 1200 Scandinavian patients. Neuropsychopharmacology 45(3):570–576

Broderick PJ, Benjamin AB, Dennis LW (2005) Caffeine and psychiatric medication interactions: a review. J Okla State Med Assoc 98(8):380–384

Brown RJ, de Banate MA, Rother KI (2010) Artificial sweeteners: a systematic review of metabolic effects in youth. Int J Pediatr Obes 5(4):305–312

Bruni O, Ferini-Strambi L, Giacomoni E, Pellegrino P (2021) Herbal remedies and their possible effect on the GABAergic system and sleep. Nutrients 13(2). https://doi.org/10.3390/nu13020530

Bushra R, Aslam N, Khan AY (2011) Food-drug interactions. Oman Med J 26(2):77–83

Butler MG (2018) Pharmacogenetics and psychiatric care: a review and commentary. J Ment Health Clin Psychol 2(2):17–24

Callaghan JT, Bergstrom RF, Ptak LR, Beasley CM (1999) Olanzapine. Pharmacokinetic and pharmacodynamic profile. Clin Pharmacokinet 37(3):177–193

Campos AI, Byrne EM, Mitchell BL et al (2022) Impact of CYP2C19 metaboliser status on SSRI response: a retrospective study of 9500 participants of the Australian genetics of depression study. Pharmacogenomics J 22(2):130–135

Carli M, Weiss F, Grenno G et al (2023) Pharmacological strategies for bipolar disorders in acute phases and chronic management with a special focus on lithium, valproic acid, and atypical antipsychotics. Curr Neuropharmacol 21(4):935–950

Carlo AD, Alpert JE (2016) Clinically relevant complications of drug-food interactions in psychopharmacology. Psychiatr Ann 46(8):448–455

Cassano GB, Baldini Rossi N, Pini S (2002) Psychopharmacology of anxiety disorders. Dialogues Clin Neurosci 4(3):271–285

Celestin MN, Musteata FM (2021) Impact of changes in free concentrations and drug-protein binding on drug dosing regimens in special populations and disease states. J Pharm Sci 110(10):3331–3344

Češková E, Šilhán P (2021) From personalized medicine to precision psychiatry? Neuropsychiatr Dis Treat 17:3663–3668

Chen X, Hong Y, Zheng P (2015) Efficacy and safety of extract of ginkgo biloba as an adjunct therapy in chronic schizophrenia: a systematic review of randomized, double-blind, placebo-controlled studies with meta-analysis. Psychiatry Res 228(1):121–127

Childress A (2017) The safety of extended-release drug formulations for the treatment of ADHD. Expert Opin Drug Saf 16(5):603–615

Choi J, Kang J, Kim T, Nehs CJ (2024) Sleep, mood disorders, and the ketogenic diet: potential therapeutic targets for bipolar disorder and schizophrenia. Front Psych 15:1358578

Chouraqui JP, Turck D, Briend A et al (2021) Religious dietary rules and their potential nutritional and health consequences. Int J Epidemiol 50(1):12–26

Claudio A, Andrea F (2022) Circadian neuromarkers of mood disorders. J Affect Disord Rep 10:100384

Cui L, Li S, Wang S et al (2024) Major depressive disorder: hypothesis, mechanism, prevention and treatment. Signal Transduct Target Ther 9(1):30

De Crescenzo F, Lennox A, Gibson JC et al (2017) Melatonin as a treatment for mood disorders: a systematic review. Acta Psychiatr Scand 136(6):549–558
Deng J, Zhu X, Chen Z et al (2017) A review of food-drug interactions on oral drug absorption. Drugs 77(17):1833–1855
Dighriri IM, Alsubaie AM, Hakami FM et al (2022) Effects of omega-3 polyunsaturated fatty acids on brain functions: a systematic review. Cureus 14(10):e30091
Diksha P, Sanjita D, Seema GC, Nupur G (2014) Interaction of antipsychotics drugs and caffeine. Sch Acad J Pharm 3:383–387
DiNicolantonio JJ, O'Keefe JH (2020) The importance of marine omega-3s for brain development and the prevention and treatment of behavior, mood, and other brain disorders. Nutrients 12(8):2333
Dowis K, Banga S (2021) The potential health benefits of the ketogenic diet: a narrative review. Nutrients 13(5):1654
Dziedzic A, Maciak K, Bliźniewska-Kowalska K, Gałecka M, Kobierecka W, Saluk J (2024) The power of psychobiotics in depression: a modern approach through the microbiota-gut-brain axis: a literature review. Nutrients 16(7):1054
Eagles SK, Gross AS, McLachlan AJ (2020) The effects of cruciferous vegetable-enriched diets on drug metabolism: a systematic review and meta-analysis of dietary intervention trials in humans. Clin Pharmacol Ther 108(2):212–227
Edinoff AN, Swinford CR, Odisho AS et al (2022) Clinically relevant drug interactions with monoamine oxidase inhibitors. Health Psychol Res 10(4):39576
El Balkhi S, Mégarbane B (2017) Lithium toxicity: clinical presentations and management. In: Malhi GS, Masson M, Bellivier F (eds) The science and practice of lithium therapy. Springer International Publishing, Cham, pp 277–292
Evers K (2009) Personalized medicine in psychiatry: ethical challenges and opportunities. Dialogues Clin Neurosci 11(4):427–434
Firth J, Teasdale SB, Allott K et al (2019) The efficacy and safety of nutrient supplements in the treatment of mental disorders: a meta-review of meta-analyses of randomized controlled trials. World Psychiatry 18(3):308–324
Flockhart DA (2012) Dietary restrictions and drug interactions with monoamine oxidase inhibitors: an update. J Clin Psychiatry 73(Suppl 1):17–24
Freimer N, Echenberg D, Kretchmer N (1983) Cultural variation--nutritional and clinical implications. West J Med 139(6):928–933
Fugh-Berman A (2000) Herb-drug interactions. Lancet 355(9198):134–138
Gasmi A, Nasreen A, Menzel A et al (2022) Neurotransmitters regulation and food intake: the role of dietary sources in neurotransmission. Molecules 28(1):210
Genser D (2008) Food and drug interaction: consequences for the nutrition/health status. Ann Nutr Metab 52(Suppl. 1):29–32
Gezmen-Karadağ M, Çelik E, Kadayifci FZ, Yeşildemir Ö, Öztürk YE, Ağagündüz D (2018) Role of food-drug interactions in neurological and psychological diseases. Acta Neurobiol Exp 78(3):187–197
Ghallab YK, Elassal OS (2024) Biochemical and neuropharmacology of psychiatric disorders. In: Mohamed W, Kobeissy F (eds) Nutrition and psychiatric disorders: an evidence-based approach to understanding the diet-brain connection. Springer Nature Singapore, pp 25–47
Gherasim A, Arhire LI, Niță O, Popa AD, Graur M, Mihalache L (2020) The relationship between lifestyle components and dietary patterns. Proc Nutr Soc 79(3):311–323
Godos J, Currenti W, Angelino D et al (2020) Diet and mental health: review of the recent updates on molecular mechanisms. Antioxidants (Basel) 9(4):346
González Canga A, Fernández Martínez N, Sahagún Prieto AM et al (2010) Dietary fiber and its interaction with drugs. Nutr Hosp 25(4):535–539
Gören JL, Tewksbury AT (2011) The use of omega-3 fatty acids in mental illness. J Pharm Pract 24(5):452–471
Grosso G, Galvano F, Marventano S et al (2014) Omega-3 fatty acids and depression: scientific evidence and biological mechanisms. Oxid Med Cell Longev 2014:313570

Gurevich KG (2013) Effect of blood protein concentrations on drug-dosing regimes: practical guidance. Theor Biol Med Model 10(1):20

Hakkola J, Hukkanen J, Turpeinen M, Pelkonen O (2020) Inhibition and induction of CYP enzymes in humans: an update. Arch Toxicol 94(11):3671–3722

Herbison CE, Hickling S, Allen KL et al (2012) Low intake of B-vitamins is associated with poor adolescent mental health and behaviour. Prev Med 55(6):634–638

Himmerich H, Hamilton A (2022) Mood stabilizers: side effects, contraindications, and interactions. In: Riederer P, Laux G, Nagatsu T, Le W, Riederer C (eds) NeuroPsychopharmacotherapy. Springer International Publishing, Cham, pp 1447–1467

Hladun O, Papaseit E, Martín S et al (2021) Interaction of energy drinks with prescription medication and drugs of abuse. Pharmaceutics 13(10):1532

Höglund E, Øverli Ø, Winberg S (2019) Tryptophan metabolic pathways and brain serotonergic activity: a comparative review. Front Endocrinol (Lausanne) 10:158

Hong Y, Xu H, Liu Y et al (2024) DDID: a comprehensive resource for visualization and analysis of diet–drug interactions. Brief Bioinform 25(3):bbae212

Horovitz O (2024) Nutritional psychology: review the interplay between nutrition and mental health. Nutr Rev. https://doi.org/10.1093/nutrit/nuae158

Huskisson E, Maggini S, Ruf M (2007) The influence of micronutrients on cognitive function and performance. J Int Med Res 35(1):1–19

Ingelman-Sundberg M, Persson A, Jukic MM (2014) Polymorphic expression of Cyp2C19 and Cyp2D6 in the developing and adult human brain causing variability in cognition, risk for depression and suicide: the search for the endogenous substrates. Pharmacogenomics 15(15):1841–1844

Jenkins TA, Nguyen JC, Polglaze KE, Bertrand PP (2016) Influence of tryptophan and serotonin on mood and cognition with a possible role of the gut-brain axis. Nutrients 8(1):56

Kale PP, Addepalli V, Bafna PA, Prabhavalkar KS (2010) Caffeine-induced augmentation of antidepressant therapy. J Exp Clin Med 2(6):282–286

Kane GC, Lipsky JJ (2000) Drug-grapefruit juice interactions. Mayo Clin Proc 75(9):933–942

Kaviani M, Nikooyeh B, Zand H, Yaghmaei P, Neyestani TR (2020) Effects of vitamin D supplementation on depression and some involved neurotransmitters. J Affect Disord 269:28–35

Kennedy DO (2016) B vitamins and the brain: mechanisms, dose and efficacy--a review. Nutrients 8(2):68

Kiani J, Imam SZ (2007) Medicinal importance of grapefruit juice and its interaction with various drugs. Nutr J 6:33

Kim BY, Sharafoddini A, Tran N, Wen EY, Lee J (2018) Consumer mobile apps for potential drug-drug interaction check: systematic review and content analysis using the mobile app rating scale (MARS). JMIR Mhealth Uhealth 6(3):e74

Ķimse L, Reinis A, Miķelsone-Jansone L, Gintere S, Krūmiņa A (2024) A narrative review of psychobiotics: probiotics that influence the gut–brain axis. Medicina 60(4). https://doi.org/10.3390/medicina60040601

Kirkland AE, Sarlo GL, Holton KF (2018) The role of magnesium in neurological disorders. Nutrients 10(6):730

Koziolek M, Alcaro S, Augustijns P et al (2019) The mechanisms of pharmacokinetic food-drug interactions – a perspective from the UNGAP group. Eur J Pharm Sci 134:31–59

Kraft SL, Baker NM, Carpenter J, Bostwick JR (2014) Procarbazine and antidepressants: a retrospective review of the risk of serotonin toxicity. Psychooncology 23(1):108–113

Kshirsagar NA, Sheth UK (1977) Drug protein binding: relevance to treatment. J Postgrad Med 23(2):50–52

Lacruz-Pleguezuelos B, Piette O, Garranzo M et al (2023) FooDrugs: a comprehensive food–drug interactions database with text documents and transcriptional data. Database 2023:baad075

Lakhan SE, Kirchgessner A (2012) Prescription stimulants in individuals with and without attention deficit hyperactivity disorder: misuse, cognitive impact, and adverse effects. Brain Behav 2(5):661–677

Lapetina DL, Yang EH, Henriques BC, Aitchison KJ (2020) Pharmacogenomics and psychopharmacology. In: Haddad PM, Nutt DJ (eds) Seminars in clinical psychopharmacology, 3rd edn. Cambridge University Press, Cambridge, pp 151–202
Le TT, McGrath SR, Fasinu PS (2022) Herb-drug interactions in neuropsychiatric pharmacotherapy – a review of clinically relevant findings. Curr Neuropharmacol 20(9):1736–1751
Lee AJ, Chan WK, Harralson AF, Buffum J, Bui BC (1999) The effects of grapefruit juice on sertraline metabolism: an in vitro and in vivo study. Clin Ther 21(11):1890–1899
Liao Y, Xie B, Zhang H et al (2019) Efficacy of omega-3 PUFAs in depression: a meta-analysis. Transl Psychiatry 9(1):190
Licht RW (2000) Typical and atypical antipsychotics in bipolar disorder. Acta Neuropsychiatrica 12(3):115–119
Lieberman JA 3rd (2004) Metabolic changes associated with antipsychotic use. Prim Care Companion J Clin Psychiatry 6(Suppl 2):8–13
Lindseth G, Helland B, Caspers J (2015) The effects of dietary tryptophan on affective disorders. Arch Psychiatr Nurs 29(2):102–107
Linnoila MI (1990) Benzodiazepines and alcohol. J Psychiatr Res 24(Suppl 2):121–127
Liu C, Wang L, Zhang C et al (2024) Caffeine intake and anxiety: a meta-analysis. Front Psychol 15:1270246
Liwinski T, Lang UE (2023) Folate and its significance in depressive disorders and suicidality: a comprehensive narrative review. Nutrients 15(17):3859
Łuszczki E, Boakye F, Zielińska M et al (2023) Vegan diet: nutritional components, implementation, and effects on adults' health. Front Nutr 10:1294497
Malhi GS, Bell E, Outhred T, Berk M (2020) Lithium therapy and its interactions. Aust Prescr 43(3):91–93
Marinescu SCN, Apetroaei MM, Nedea MII et al (2024) Dietary influence on drug efficacy: a comprehensive review of ketogenic diet-pharmacotherapy interactions. Nutrients 16(8):1213
Marrone MC, Coccurello R (2020) Dietary fatty acids and microbiota-brain communication in neuropsychiatric diseases. Biomolecules 10(1). https://doi.org/10.3390/biom10010012
McCabe-Sellers BJ, Staggs CG, Bogle ML (2006) Tyramine in foods and monoamine oxidase inhibitor drugs: a crossroad where medicine, nutrition, pharmacy, and food industry converge. J Food Compos Anal 19:S58–S65
Medawar E, Huhn S, Villringer A, Veronica WA (2019) The effects of plant-based diets on the body and the brain: a systematic review. Transl Psychiatry 9(1):226
Mehdi S, Manohar K, Shariff A et al (2023) Omega-3 fatty acids supplementation in the treatment of depression: an observational study. J Pers Med 13(2):224
Meher AK, Acharya B, Sahu PK (2024) Probiotics: bridging the interplay of a healthy gut and psychoneurological well-being. Food Bioeng 3(1):126–147
Merlo G, Vela A (2022) Mental health in lifestyle medicine: a call to action. Am J Lifestyle Med 16(1):7–20
Mikkelsen N, Damkier P, Pedersen SA (2023) Serotonin syndrome—a focused review. Basic Clin Pharmacol Toxicol 133(2):124–129
Mlambo R, Liu J, Wang Q, Tan S, Chen C (2023) Receptors involved in mental disorders and the use of clozapine, chlorpromazine, olanzapine, and aripiprazole to treat mental disorders. Pharmaceuticals (Basel) 16(4):603
Moon E, Kim K, Partonen T, Linnaranta O (2022) Role of melatonin in the management of Sleep and circadian disorders in the context of psychiatric illness. Curr Psychiatry Rep 24(11):623–634
Mörkl S, Butler MI, Holl A, Cryan JF, Dinan TG (2020) Probiotics and the microbiota-gut-brain axis: focus on psychiatry. Curr Nutr Rep 9(3):171–182
Morris A, Reed T, McBride G, Chen J (2024) Dietary interventions to improve metabolic health in schizophrenia: a systematic literature review of systematic reviews. Schizophr Res 270:372–382
Muth AK, Park SQ (2021) The impact of dietary macronutrient intake on cognitive function and the brain. Clin Nutr 40(6):3999–4010
Niederberger E, Parnham MJ (2021) The impact of diet and exercise on drug responses. Int J Mol Sci 22(14):7692

Ordovas JM, Ferguson LR, Tai ES, Mathers JC (2018) Personalised nutrition and health. BMJ 361:bmj.k2173

Ortega MA, Fraile-Martínez Ó, García-Montero C et al (2022) Biological role of nutrients, food and dietary patterns in the prevention and clinical management of major depressive disorder. Nutrients 14(15). https://doi.org/10.3390/nu14153099

Ötleş S, Senturk AE (2014) Food and drug interactions: a general review. Acta Sci Pol Technol Aliment 13(1):89–102

Paz-Ramos MI, Cruz SL, Violante-Soria V (2023) Amphetamine-type stimulants: novel insights into their actions and use patterns. Rev Investig Clin 75(3):143–157

Pennazio F, Brasso C, Villari V, Rocca P (2022) Current status of therapeutic drug monitoring in mental health treatment: a review. Pharmaceutics 14(12):2674

Pickering G, Mazur A, Trousselard M et al (2020) Magnesium status and stress: the vicious circle concept revisited. Nutrients 12(12). https://doi.org/10.3390/nu12123672

Pinto N, Dolan ME (2011) Clinically relevant genetic variations in drug metabolizing enzymes. Curr Drug Metab 12(5):487–497

Preeti SS, Saharan R et al (2023) Exploring LIPIDs for their potential to improves bioavailability of lipophilic drugs candidates: a review. Saudi Pharm J 31(12):101870

Prior TI, Baker GB (2003) Interactions between the cytochrome P450 system and the second-generation antipsychotics. J Psychiatry Neurosci 28(2):99–112

Qi X, Al-Ghazzewi FH, Tester RF (2018) Dietary fiber, gastric emptying, and carbohydrate digestion: a mini-review. Starch – Stärke 70(9-10):1700346

Rahman MM, Vadrev SM, Magana-Mora A, Levman J, Soufan O (2022) A novel graph mining approach to predict and evaluate food-drug interactions. Sci Rep 12(1):1061

Rechenberg K (2016) Nutritional interventions in clinical depression. Clin Psychol Sci 4(1):144–162

Reddy VS, Shiva S, Manikantan S, Ramakrishna S (2024) Pharmacology of caffeine and its effects on the human body. Eur J Med Chem Rep 10:100138

Reily NM, Tang S, Negrone A, Gan DZQ, Sheanoda V, Christensen H (2023) Omega-3 supplements in the prevention and treatment of youth depression and anxiety symptoms: a scoping review. PLoS One 18(4):e0284057

Reimherr FW, Wender PH, Wood DR, Ward M (1987) An open trial of L-tyrosine in the treatment of attention deficit disorder, residual type. Am J Psychiatry 144(8):1071–1073

Remington G (2010) Second and third generation antipsychotics. In: Stolerman IP (ed) Encyclopedia of psychopharmacology. Springer Berlin Heidelberg, Berlin, Heidelberg, pp 1187–1192

Rog J, Wingralek Z, Nowak K et al (2024) The potential role of the ketogenic diet in serious mental illness: current evidence, safety, and practical advice. J Clin Med 13(10):2819

Roth AJ, Nelson CJ (2020) C11Antidepressants. In: Roth A, Nelson C (eds) Psychopharmacology in cancer care: a guide for non-prescribers. Oxford University Press, New York

Sasaki T, Sato Y, Kumagai T, Yoshinari K, Nagata K (2017) Effect of health foods on cytochrome P450-mediated drug metabolism. J Pharm Health Care Sci 3:14

Schulz V (2006) Safety of St. John's Wort extract compared to synthetic antidepressants. Phytomedicine 13(3):199–204

Schwalfenberg GK (2012) The alkaline diet: is there evidence that an alkaline pH diet benefits health? J Environ Public Health 2012:727630

Serefko A, Jach ME, Pietraszuk M, Świąder M, Świąder K, Szopa A (2024) Omega-3 polyunsaturated fatty acids in depression. Int J Mol Sci 25(16). https://doi.org/10.3390/ijms25168675

Shinjyo N, Waddell G, Green J (2020) Valerian root in treating sleep problems and associated disorders—a systematic review and meta-analysis. J Evid BasedIntegr Med 25:2515690X20967323

Siwek M, Krupa AJ, Woroń J (2024) Interactions between grapefruit juice and psychotropic medications: an update of the literature and an original case series. Expert Opin Drug Metab Toxicol 20(5):333–345

Skrajnowska D, Bobrowska-Korczak B (2024) The effects of diet, dietary supplements, drugs and exercise on physical, diagnostic values of urine characteristics. Nutrients 16(18). https://doi.org/10.3390/nu16183141

Soh NL, Walter G (2011) Tryptophan and depression: can diet alone be the answer? Acta Neuropsychiatrica 23(1):3–11

Spanakis M, Patelarou E, Patelarou A (2022) Drug-food interactions with a focus on Mediterranean diet. Appl Sci 12(20). https://doi.org/10.3390/app122010207

Spiro A, Buttriss JL (2014) Vitamin D: an overview of vitamin D status and intake in Europe. Nutr Bull 39(4):322–350

Stein DJ, Shoptaw SJ, Vigo DV et al (2022) Psychiatric diagnosis and treatment in the 21st century: paradigm shifts versus incremental integration. World Psychiatry 21(3):393–414

Stewart DE (1992) High-fiber diet and serum tricyclic antidepressant levels. J Clin Psychopharmacol 12(6):438–440

Su KP, Matsuoka Y, Pae CU (2015) Omega-3 Polyunsaturated fatty acids in prevention of mood and anxiety DISorders. Clin Psychopharmacol Neurosci 13(2):129–137

Sun L, Mi K, Hou Y et al (2023) Pharmacokinetic and pharmacodynamic drug-drug interactions: research methods and applications. Metabolites 13(8):897

Swardfager W, Herrmann N, McIntyre RS et al (2013) Potential roles of zinc in the pathophysiology and treatment of major depressive disorder. Neurosci Biobehav Rev 37(5):911–929

Tian Y, Yi J, Wang N et al (2024) DDInter 2.0: an enhanced drug interaction resource with expanded data coverage, new interaction types, and improved user interface. Nucleic Acids Res 53:D1356

Tosti V, Bertozzi B, Fontana L (2018) Health benefits of the Mediterranean diet: metabolic and molecular mechanisms. J Gerontol A Biol Sci Med Sci 73(3):318–326

Totten MS, Davenport TS, Edwards LF, Howell JM (2023) Trace minerals and anxiety: a review of zinc, copper, iron, and selenium. Dietetics 2(1):83–103. https://doi.org/10.3390/dietetics2010008

Trujillo KA (2019) Basic information on psychotropic drugs, receptor systems, and the brain. In: APA handbook of psychopharmacology. American Psychological Association, Washington, DC, pp 17–39

Tyson RJ, Park CC, Powell JR et al (2020) Precision dosing priority criteria: drug, disease, and patient population variables. Front Pharmacol 11:420

van Zonneveld SM, van den Oever EJ, Haarman BCM et al (2024) An anti-inflammatory diet and its potential benefit for individuals with mental disorders and neurodegenerative diseases—a narrative review. Nutrients 16(16). https://doi.org/10.3390/nu16162646

Vashum KP, McEvoy M, Milton AH et al (2014) Dietary zinc is associated with a lower incidence of depression: findings from two australian cohorts. J Affect Disord 166:249–257

Vassilopoulou E, Efthymiou D, Tsironis V et al (2022) The benefits of the Mediterranean diet in first episode psychosis patients taking antipsychotics. Toxicol Rep 9:120–125

Ventriglio A, Sancassiani F, Contu MP et al (2020) Mediterranean diet and its benefits on health and mental health: a literature review. Clin Pract Epidemiol Ment Health 16(Suppl-1):156–164

Villagomez A, Cross M, Ranjbar N (2023) Broad spectrum micronutrients: a potential key player to address emotional dysregulation. Front Child Adolesc Psychiatry 2:1295635

Vuong M, González Aragón C, Montarroyos SS (2023) Common food and drug interactions. Pediatr Rev 44(2):68–80

Wael Abu Dayyih IA-A, Hailat M, Alarman SM, Zakaraya Z, Assab MA, Alkhader E (2024) Review of grapefruit juice-drugs interactions mediated by intestinal CYP3A4 inhibition. JAPS J Pharm Sci 14(5). https://doi.org/10.7324/JAPS.2024.160197

Wang T, Masedunskas A, Willett WC, Fontana L (2023) Vegetarian and vegan diets: benefits and drawbacks. Eur Heart J 44(36):3423–3439

Wani AL, Bhat SA, Ara A (2015) Omega-3 fatty acids and the treatment of depression: a review of scientific evidence. Integr Med Res 4(3):132–141

Wanwimolruk S, Prachayasittikul V (2014) Cytochrome P450 enzyme mediated herbal drug interactions (part 1). EXCLI J 13:347–391

Wanwimolruk S, Phopin K, Prachayasittikul V (2014) Cytochrome P450 enzyme mediated herbal drug interactions (part 2). EXCLI J 13:869–896

Weathermon R, Crabb DW (1999) Alcohol and medication interactions. Alcohol Res Health 23(1):40–54
Wei J, Li Z, Zhuo L et al (2024) Enhancing drug–food interaction prediction with precision representations through multilevel self-supervised learning. Comput Biol Med 171:108104
Yasui N, Kondo T, Furukori H et al (2000) Effects of repeated ingestion of grapefruit juice on the single and multiple oral-dose pharmacokinetics and pharmacodynamics of alprazolam. Psychopharmacology 150(2):185–190
Zajkowska I, Niczyporuk P, Urbaniak A, Tomaszek N, Modzelewski S, Waszkiewicz N (2024) Investigating the impacts of diet, supplementation, microbiota, gut-brain axis on schizophrenia: a narrative review. Nutrients 16(14):2228
Zanger UM, Schwab M (2013) Cytochrome P450 enzymes in drug metabolism: regulation of gene expression, enzyme activities, and impact of genetic variation. Pharmacol Ther 138(1):103–141
Zhao M, Ma J, Li M et al (2021) Cytochrome P450 enzymes and drug metabolism in humans. Int J Mol Sci 22(23):12808
Zhao Q, Chen Y, Huang W, Zhou H, Zhang W (2023) Drug-microbiota interactions: an emerging priority for precision medicine. Signal Transduct Target Ther 8(1):386
Zhou L, Xiong JY, Chai YQ et al (2022) Possible antidepressant mechanisms of omega-3 polyunsaturated fatty acids acting on the central nervous system. Front Psych 13:933704

The Gut-Brain Connection: How Diet Influences Mental Health

5

Sophie Merve Onerli Yener

Abstract

When Hippocrates said, “All disease begins in the gut,” was he offering timeless wisdom or was it just an ancient claim? What was once, a few centuries ago, a bold statement has today become a well-known fact that modern medicine focuses on. There has been growing interest in the connection between gut microbiome, diet, and mental health in recent years. The mechanisms of gut–brain interactions have been identified as neuronal, endocrine, metabolic, and immune, all of which are affected by diet. This paper aims to present an overview of current knowledge on the interactions between the gut, the brain, mental health, and the effects of diet on these interactions. It examines possible management strategies for dysbiosis and provides an overview of major research gaps and future research directions to enhance understanding of the gut–brain axis. Even though we cannot go as far as to say, as Hippocrates did, that “All disease begins in the gut,” current scientific knowledge enables us to state that “A happy gut is a step towards a healthier mind and body.”

Keywords

Gut–brain axis · Gut microbiome · Gut dysbiosis · Probiotics · Mental health

S. M. O. Yener (✉)
Migraine Foundation Australia, Keilor East, VIC, Australia

Department of Neurology, Australian Institute of Migraine, Pascoe Vale South, VIC, Australia

Department of Neurology, Monash Health, Clayton, VIC, Australia

Department of Neurology, Ankara Bilkent City Hospital, Cankaya, Ankara, Turkey

W. Mohamed, A. Özge (eds.), *Feeding the Mind: The Connection Between Diet, Drugs, and Mental Health Volume 1*, Nutritional Neurosciences,
https://doi.org/10.1007/978-981-95-1414-4_5

5.1 Introduction

The human body, which includes the gut, skin, and other mucosal surfaces, is colonized by trillions of microorganisms that are commonly referred to as the microbiome. In recent years, advances in microbiome research have emphasized the active role of the gut microbiome in regulating key functions, such as circadian rhythm, nutrition, metabolism, and immunity (Aswendt et al. 2021; Lotti et al. 2023; Miri et al. 2023).

Microbiome colonization begins at birth and develops over the course of a lifetime, and it is shaped by the genetic and epigenetic factors. In the gut, the microbiota contributes to digestion, absorption of nutrients, and vitamin synthesis. Beyond these functions, the gut microbiota also affects the immune and nervous systems, thereby creating a network known as the gut–brain axis, which facilitates complex and bidirectional communication (Liu et al. 2022). This axis is interconnected via several systems, such as neural (e.g., vagus nerve and enteric nervous system), endocrine (e.g., gut hormones and neurotransmitters), immune (e.g., cytokine signaling), and metabolic pathways [e.g., short-chain fatty acids (SCFAs) made during microbial fermentation]. The gut microbiome acts as the principal regulator of this axis and mediates immune responses (Carabotti et al. 2015).

Modifications in the gut–brain–microbiota axis have been linked to metabolic and immune conditions, as well as psychiatric disorders, such as schizophrenia, anxiety, mood disorders, and neurodevelopmental disorders. Diet, which is a modifiable lifestyle factor, significantly influences the gut–brain axis through shifts in the composition and function of the gut microbiome. New findings highlight the effects of dietary habits on mental health, along with the connection between diet, the gut microbiome, and its impact on neurological outcomes (Borrego-Ruiz and Borrego 2025; Merlo et al. 2024).

This chapter aims to present the current understanding of the gut–brain connection with a focus on how dietary habits contribute to this relationship. It examines the mechanisms of gut–brain communication, the consequences of dysbiosis on mental health, and the therapeutic potential of targeted dietary approaches. It also explores ways to integrate dietary interventions into mental health treatment plans and discusses strategies for managing the gut–brain axis to enhance emotional and mental well-being.

5.2 From Gut to Brain: Mapping the Potential Communication Networks

The communication between the gut and brain is complex and consists of several circuits that are integrated with each other and include neural, endocrine, immune, and metabolic signals. The enteric nervous system which is also known as the "second brain," forms part of the neural pathway and provides a mechanism for feedback between the gut and the brain. The vagus nerve is one of the most important structures in this system, which carries information from the gut to the brain and allows the brain to regulate gut functions (Goldstein et al. 2012; Sharkey and Mawe 2022). Endocrine and humoral pathways involve the secretion of hormones and other signaling molecules from the gut, which contribute to neurohumoral modulation. Immune pathways are linked to gut inflammation and systemic immune responses, affecting the brain as a result of altered microbiota. Metabolic pathways can have a direct or indirect effect on cognitive and neurological processes and are also an important part of the gut–brain axis. Overall, current findings demonstrate how the gut microbiome has a significant impact on the modulation of the biochemical environment and consequently, cognitive function, mood, and behavior (Martin et al. 2018; Appleton 2018) (Fig. 5.1).

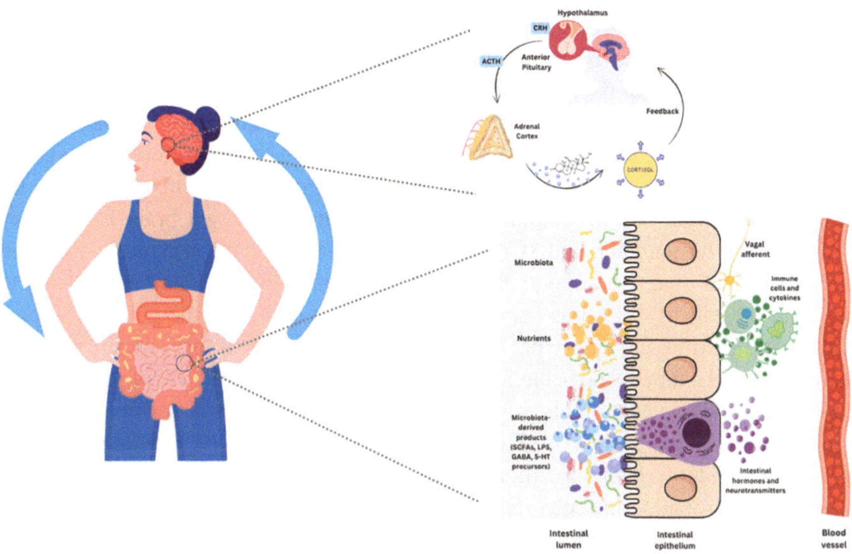

Fig. 5.1 The microbiota–gut–brain axis is a complex bidirectional communication network integrating neural, endocrine, immune, and metabolic pathways. This figure illustrates the key mechanisms of this communication. The HPA axis is activated in response to stress or gut dysbiosis, leading to increased secretion of cortisol by the adrenal cortex. Cortisol regulates intestinal barrier function and immune activity as well as modulates systemic immune responses. The vagus nerve serves as a key neural link between the gut and brain. Gut-derived neurotransmitters, such as 5-HT, dopamine, and GABA, influence CNS activity and behavior. SCFAs produced through microbial fermentation of dietary fibers act as neuroactive compounds and impact neuroinflammation, neurotransmitter synthesis, and cognitive processes. Additionally, gut-derived hormones and cytokines interact with immune cells and influence brain function and emotional regulation. HPA hypothalamic-pituitary-adrenal, CNS central nervous system, 5-HT serotonin, GABA gamma-aminobutyric acid, SCFAs short-chain fatty acids

5.2.1 Neural Networks

There is a growing body of evidence from animal and human research supporting the concept that the gut microbiome modulates metabolism and impacts human health and well-being. However, the exact mechanisms by which intestinal microorganisms influence the brain and behavior remain unclear. The gut and brain are in continuous communication, primarily mediated by the vagus nerve (Forssten et al. 2022; Han et al. 2022). The vagus nerve, which is a key component of communication between the brainstem, autonomic nervous system, and gastrointestinal tract, originates in the brainstem and contributes to the regulation of hunger, appetite, inflammation, and stress responses. The vagal sensory afferent fibers transmit information from the gut to the brain, while the efferent fibers allow the brain to modulate gut functions such as motility, secretion, and inflammation (Martin et al. 2018; Goldstein et al. 2012).

The connection between gastrectomy and subsequent psychiatric disorders was first identified in the early twentieth century. In 1910, Dr. George W. Crile reported a higher prevalence of psychiatric illnesses among patients who underwent gastrectomy for peptic ulcers. He hypothesized that gastric resection disrupted vagal nerve function, thereby altering the communication between the gastrointestinal system and the brain, which could contribute to the development of psychiatric conditions. This early insight laid the foundation for understanding the gut-brain axis and its implications for mental health (Crile 1915). Other studies have also been conducted in subsequent years to support this network. Animal models have been used to explore the role of the vagus nerve in mediating the effects of the gut microbiome on brain function and behavior. Notably, animals with vagal nerve transection or vagatomy did not have the behavioral changes that are characteristic of germ-free or microbiome-depleted animals, suggesting that the vagus nerve is a critical signaling molecule in the gut-brain axis (Tillisch et al. 2017). A study using advanced genetic and imaging tools in mouse models has shown that direct synaptic connections between enteroendocrine cells (EECs) and vagal neurons create a neural circuit that allows sensory signals derived from the gut to reach the brainstem within milliseconds (Kaelberer et al. 2018).

Neuroanatomical studies have identified the nucleus tractus solitarius (NTS) as the main relay station for vagal afferent fibers. Inputs from the NTS are connected to other brain regions, including the prefrontal cortex, hypothalamus, and amygdala, which are involved in the regulation of appetite, stress, and emotional processing, respectively. Recent research using high-resolution tract tracing techniques in the animal models has provided a more precise picture of these connections, showing the projections from the NTS to the limbic and cognitive areas that are involved in the processing of emotions and stress responses (Macdonald and Ellacott 2020). Human imaging studies have confirmed these findings; these studies have shown that similar brain areas get activated in response to the gut signals, such as inflammation or ingestion (Kano et al. 2018).

Another animal study has demonstrated the effect of the probiotic *Lactobacillus rhamnosus (JB-1)* on emotional behavior and the gene expression of GABA receptors in mice. The prolonged administration of this probiotic modulated GABA receptor mRNA expression in specific brain regions, decreased stress-induced corticosterone levels, and improved behaviors associated with emotional and depressive states. Importantly, these neurochemical and behavioral effects were not observed in the vagotomized mice, which suggests that the vagus nerve is an important part of the communication between the gut microbiota and the brain (Bravo et al. 2011). In humans, dietary modulation of the microbiota and probiotic administration have been associated with changes in vagal tone and emotional status, thereby supporting the applicability of findings from animal studies (Sarkar et al. 2016).

5.2.2 Endocrine and Humoral Pathways

The gut is one of the major endocrine organs, which produce and release numerous hormones and other signaling molecules that can modulate brain activity. This includes EECs, which are distributed within the gastrointestinal tract and are recognized as key effector cells of the gut–brain axis. These specialized epithelial cells act as sensory detectors for luminal products, including nutrients, toxins, and microbial metabolites, and translate these signals into hormonal and neuronal responses. Although EECs make up only about 1% of the total number of enterocytes, this cell population is the largest endocrine system in the human body and has an essential role in the regulation of gastrointestinal homeostasis and other systems (Sternini et al. 2008; Rehfeld 2004). EECs perform sensory functions through two mechanisms: open-type EECs, which have microvilli extending into the lumen to detect stimuli, and closed-type EECs, which are activated via paracrine or neural signals (Sternini et al. 2008). Upon activation, EECs secrete various bioactive peptides and hormones, including cholecystokinin (CCK), glucagon-like peptide-1 (GLP-1), peptide YY (PYY), and serotonin (5-HT). These molecules can target local effects on the gastrointestinal tract to modulate motility, secretion, and permeability or systemic effects on distant organs via the bloodstream (Psichas et al. 2015; Gershon 2013). EECs, a critical component of the gut-brain axis, form synapse-like terminals with enteric and extrinsic neurons, including vagal and spinal afferents, through their basal cytoplasmic extensions known as neuropods. This neuroepithelial circuit provides a mechanism to transmit information in real time from the gut to the central nervous system (CNS) where they regulate critical functions such as appetite, satiety, glucose homeostasis, and stress responses (Bohórquez and Liddle 2011; Bohórquez et al. 2014).

EECs also communicate with the microbiota through recognition of microbial products like SCFAs via G protein-coupled receptors (e.g., FFAR2, FFAR3). SCFAs, primarily acetate, propionate, and butyrate, are the main products produced during the fermentation of dietary fibers by the gut microbiota. SCFAs' role is essential for gut–brain axis through their interaction between the enteric neurons and EECs. This interaction also strengthens the axis by linking microbial activity to host physiology (Cani et al. 2013). Dysregulation of EEC function has been linked to various conditions, including irritable bowel syndrome (IBS), type 2 diabetes, and obesity (Mayer et al. 2015). Targeting EEC-derived hormones and their receptors has therefore become a promising strategy for therapeutic intervention. For instance, GLP-1 receptor agonists are already used to manage diabetes and obesity, emphasizing the therapeutic potential of the EEC-focused research (Holst and Deacon 2005).

5.2.3 Immune Pathways

The gut–brain axis is also linked through the immune system. Inflammation and imbalances in the gut microbiota can trigger immune responses that affect the brain function. Recent findings also show that the gut microbiome is a critical part in regulating immune responses both locally and systemically, thereby impacting CNS health (Round and Mazmanian 2009; Ley et al. 2006). The gut hosts a diverse community of microorganisms that not only support in digestion and absorption but also have an essential role in the immune system's maturation. The interactions between intestinal epithelial cells and immune cells, such as T lymphocytes and dendritic cells, determine the balance between pro-inflammatory and regulatory responses (Maslowski et al. 2009).

Lipopolysaccharide (LPS) is one of the key mediators in immune pathways of the gut–brain axis and is responsible for inflammation and systemic immune responses. As a glycolipid component of the outer membrane of gram-negative bacteria, LPS is known to trigger pro-inflammatory responses and promote low-grade systemic inflammation (Pagé et al. 2022). The pro-inflammatory effects of LPS are mainly mediated through its interaction with Toll-Like Receptor 4 (TLR4), a receptor belonging to the pattern recognition receptor family. This interaction is facilitated with the help of LPS binding protein, CD14, and the MD-2 protein, which is bound to the extracellular domain of TLR4. TLR4 is expressed on various immune system cells, including monocytes, macrophages, and Kupffer cells, as well as on nonimmune cells, such as endothelial cells, adipocytes, and hepatocytes. TLR4 activates two main signaling cascades: the MyD88-dependent pathway, which starts at the plasma membrane, and the TRIF-dependent pathway, which initiates following receptor endocytosis to early endosomes. Both pathways aim to enhance the inflammatory responses by increasing the synthesis of proinflammatory cytokines through different mechanisms (Zanoni et al. 2011). Furthermore, LPS induces systemic inflammation, and its effects depend on the rate of TLR4 endocytosis, its trafficking through the endo-lysosomal system, and ultimately the degradation of the receptor, which reduces the inflammatory response. TLR4 can also be activated by saturated fatty acids (SFAs), which highlights the role of high-fat diets in systemic inflammation. It has been suggested that structural similarities between SFAs and the lipid A part of LPS is the basis of this interaction; The inflammatory potency of SFAs depends on the chain length, with lauric acid being the most potent pro-inflammatory lipid (Zanoni et al. 2011; Lu et al. 2008). These findings support the importance of targeting LPS-TLR4 interactions, and dietary modifications could be used for the development of new approaches for the treatment of inflammatory diseases. Experimental models have also shown that microbial imbalances, or dysbiosis, can affect autoimmune diseases like multiple sclerosis. Specific gut bacteria have the potential to induce Th17 cell differentiation and modulate the immune responses that are associated with CNS autoimmunity. On the other hand, some microbial species promote the differentiation of regulatory T cells (Tregs), which maintain immune tolerance and thereby reduce inflammation (Ivanov et al. 2009).

It has also been shown that inflammation during childhood can cause immune and neuronal alterations, which in turn result in maladaptive stress responses. Cao et al. discovered that pro-inflammatory events in early life, for instance, through the exposure to LPS, lead to the activation of microglia through the TLR-4/NF-κB signaling pathway. This activation increases the sensitivity of microglia to stressors, leading to excessive synaptic pruning and dysregulated neuronal networks, particularly in the anterior cingulate cortex. These changes contribute to maladaptive stress responses and are related to adolescent depression. This evidence also supports the concept that immune pathways, mediated by the gut microbiota and systemic inflammation, have profound effects on the CNS and that early immune interventions may be key to maintaining homeostasis of the gut–brain axis (Cao et al. 2021).

Moreover, cytokines and microbial metabolites originating from the gut can travel through the systemic circulation to affect distant organs, including the brain. This crosstalk reveals how the gut exerts a key regulatory role in the modulation of CNS inflammation and highlights the importance of gut microbiota modulation as a therapeutic strategy for the treatment of neuroinflammatory diseases (Round and Mazmanian 2010; Mazmanian et al. 2008).

5.2.4 Metabolic Pathways

The gut microbiome is essential for synthesizing various molecules and metabolites that influence brain function. Some of the microbial-derived metabolites, such as aromatic amino acids (AAAs), SCFAs, bile acids, and trimethylamine-N-oxide (TMAO), facilitate communication between the gut and the CNS and can have either direct or indirect effects on the physiological and neurochemical processes (Leyrolle et al. 2023; Gan et al. 2024). Among these metabolites, SCFAs help in the regulation of intestinal barrier functions, inflammatory responses, and mucus secretion. The SCFAs bind to free fatty acid receptors 2 and 3 (FFA2/3) and act as histone deacetylase inhibitors that potentially have protective effects against memory deficits and amyloid-β plaque deposition linked with Alzheimer's disease (AD) (Marizzoni et al. 2020). They also influence neuroinflammation by regulating immune cells and pro-inflammatory cytokine production (Li et al. 2019).

AAAs tryptophan, tyrosine, and phenylalanine serve as precursors for some neurotransmitters, such as serotonin, dopamine, and norepinephrine. These metabolites are important in regulating behavior, cognitive function, and mood. Tryptophan is an essential amino acid and is crucial in the gut–brain axis, as it serves as a precursor to various key metabolites (Correia and Vale 2022). Approximately 95% of tryptophan is degraded through the kynurenine pathway to form kynurenine and kynurenic acid, which have neuroprotective effects but may contribute to neuroinflammation if dysregulated. The remaining tryptophan is metabolized into serotonin, which regulates mood, behavior, and appetite, or into indole derivatives that can affect the gut barrier integrity and inflammation (Červenka et al. 2017; Maître et al. 2020). Dysregulation of tryptophan metabolism has been linked to depression, anxiety, and neurodegenerative diseases. Increasing the number of intestinal bacteria that modulate tryptophan metabolism has the potential to be a novel therapeutic strategy (Correia and Vale 2022; Li et al. 2022).

Bile acids, which are synthesized from cholesterol in the liver and metabolized by the gut microbiota, are increasingly recognized as important modulators of the gut-brain axis. Secondary bile acids, such as deoxycholic acid and lithocholic acid, interact with the farnesoid X receptor and the Takeda G-protein receptor 5 and modulate energy homeostasis, inflammation, and neuronal signaling. Bile acid homeostasis is impaired in neurological conditions including Parkinson's disease (PD) and cognitive decline, as bile acids are involved in regulating inflammatory responses and mitochondrial functions (Caspani and Swann 2019). TMAO is another metabolite formed from dietary choline and L-carnitine that plays a significant role in neurodegenerative diseases. TMAO is associated with increased oxidative stress, impaired mitochondrial function, and enhanced neuroinflammation and is linked to diseases such as AD and age-related cognitive impairment (Vogt et al. 2018).

The metabolic pathway is an important route offering therapeutic potential. Some of the dietary interventions, probiotics, prebiotics, and specific metabolite supplementation aim to balance the microbial metabolite production and offer a potential therapeutic approach for diverse neurological and psychological diseases (Holscher 2017; Miri et al. 2023).

5.3 Gut Dysbiosis and Its Impact on Mental Health

Dysbiosis of the gut microbiome has been linked to neuropsychiatric conditions, including depression, anxiety, autism spectrum disorder, and PD (Gan et al. 2024; LaGreca et al. 2022; Montiel-Castro et al. 2013). A healthy gut microbial ecosystem, also known as eubiosis, is characterized by the presence of a large number of microbial species and stable populations that are essential in digestion, immune regulation, and the maintenance of the intestinal barrier. However, alterations in the microbial community can result in dysbiosis, which is characterized by reduced microbial diversity, dominance of potentially pathogenic bacteria such as Gram negative, LPS-producing *Proteobacteria,* and the loss of beneficial bacteria such as *Bifidobacteria and Faecalibacterium prausnitzii*. These alterations affect the tight junctions of epithelial cells, thereby leading to an increase in gut permeability, which is also known as "leaky gut." This allows bacterial metabolites, such as LPS, to enter the circulation and trigger inflammation (LaGreca et al. 2022; Silva et al. 2020; Caspani and Swann 2019).

The translocation of microbial metabolites is also associated with pro-inflammatory agents that stimulate TLR-4 signaling. This interaction activates downstream pathways such as NF-κB and MAP kinase, resulting in the secretion of inflammatory cytokines (Vijay-Kumar et al. 2010; Turner 2009). Dysbiosis exacerbates this condition by decreasing the microbial diversity and increasing the LPS-producing bacteria, thus increasing the level of systemic inflammation (Bäckhed et al. 2004). This chronic low-grade inflammation results in the development of metabolic and autoimmune disorders, including obesity, type 2 diabetes, and IBS. Moreover, the effects of dysbiosis extend beyond local changes and affect the gut–brain axis, neuroinflammatory responses, and neurotransmitter production. Dysbiosis has been linked to neurological diseases including AD, PD, and depression, which highlights its impact on mental and neurological well-being (Marizzoni et al. 2020; Liu et al. 2019a, b). There are many causes of dysbiosis, including changes in diet, stress, and the use of medications. The composition of microbiota is also affected by high-fat and low-fiber diets, which promote pathogenic bacteria while suppressing beneficial ones. Antibiotic usage, chronic stress, and some lifestyle changes, such as reduced physical activity, also affect the microbial balance. These disruptions are particularly important, as they may initiate a cycle of an impaired gut barrier, microbial imbalance, and systemic inflammation (Daneman and Prat 2015; Li et al. 2019).

Maintaining the balance of the microbial community is a promising therapeutic approach. Probiotics and prebiotics, which contain beneficial bacteria and support their growth, respectively, have also been shown to positively influence dysbiosis. Some of the dietary modifications that promote microbial diversity include the consumption of high-fiber, plant-based diets, which support SCFA production, critical for intestinal health. New strategies, such as fecal microbiota transplantation (FMT), provide specific treatments for *Clostridioides difficile* infection, inflammatory bowel diseases, and other dysbiosis-related conditions (Silva et al. 2020; Caspani and Swann 2019).

Dysbiosis is now recognized as a key etiologic factor in the development of many diseases, modifying systemic and local physiological functions. The relationship between the intestinal barrier, immune system, and systemic metabolism highlights the need for specific management strategies to maintain gut homeostasis and improve overall health overcomes. Therefore, as our understanding of gut dysbiosis and its systemic consequences deepens, new management strategies will continue to evolve (Marizzoni et al. 2020; Liu et al. 2019a, b).

5.4 The Role of the Diet in Mental Health

Diet is a major factor that determines the composition of the gut microbiota, which is closely linked to mental health via interactions within the microbiota–gut–brain network. This axis comprises microbial metabolites, immune responses, and neuronal networks that interconnect the gut and brain and are influenced by dietary intake. Increasing research evidence shows that certain dietary habits can modify the composition of the gut microbiota and its impact on mental health. Balanced and healthy diets, including the Mediterranean diet, have been found to promote a healthy and diverse microbiome as well as increase beneficial bacteria, including *F. prausnitzii, Roseburia,* and *B. adolescentis* (Jin et al. 2019; Filippis et al. 2015). These microbes are capable of synthesizing SCFAs, particularly butyrate, which supports gut barrier integrity, lowers systemic inflammation, and influences brain functions related to mood and cognition. SCFAs have also been shown to regulate neurotransmitters, including serotonin and dopamine, which are crucial for emotional regulation. Fiber, polyphenols, and omega 3 fatty acids present in the Mediterranean diet and other similar diets also enhance these positive outcomes and thereby decrease the risk of developing depression and other mental disorders (Firth et al. 2020; Dauncey 2009).

On the other hand, the Western diet, which is high in saturated fats, refined sugars, and processed foods, has been linked to a reduction in microbial diversity and an increase in proinflammatory bacteria such as *Enterobacteriaceae*. This microbial imbalance can affect the integrity of the gut barrier, enhance the inflammatory response, and influence the signaling along the microbiota–gut–brain axis (Vangay et al. 2018; Sonnenburg et al. 2016). These changes have been associated with mental disorders, including depression and anxiety. Even short-term exposure to a Western diet can negatively impact the gut microbiota by reducing the population of protective SCFA-producing bacteria and upregulating the synthesis of toxic metabolites, including TMAO (Zhu et al. 2020).

Specific dietary components have also been identified as linked to mental health. Polyphenols, which are contained in fruits, vegetables, tea, coffee, and wine, are the compounds that have an effect similar to prebiotics. They increase the number of beneficial microorganisms, such as *Bifidobacterium and Lactobacillus,* and decrease the level of pathogens (Roy et al. 2019; Dueñas et al. 2015). Research indicates that polyphenols have the potential to alleviate depressive symptoms by exerting anti-inflammatory effects and promoting microbial diversity (Godos et al. 2018; Chang et al. 2016). Likewise, omega-3 fatty acids, found in fish, seeds, and nuts, contribute to enhancing microbial diversity, reduce the level of inflammation, and help in the regulation of neurotransmitter levels, thereby influencing mood and cognitive function (Robertson et al. 2016; Watson et al. 2017).

Fiber, which is another key element of a healthy gut, promotes the proliferation of SCFA-producing bacteria while reducing the prevalence of pathogenic bacteria. This is supported by several studies demonstrating that increased fiber intake is associated with a lower incidence of depression and anxiety, as well as enhanced cognitive function (Kovatcheva-Datchary et al. 2015; Vanegas et al. 2017). Fermented foods, including yogurt, kimchi, and kefir also, have positive effects on the gut microbiota, as they contain probiotic bacteria and neuroactive compounds that may enhance mental health (Taylor et al. 2020; Han et al. 2015).

There is growing emphasis on the need to adopt a whole-dietary approach, which considers the synergistic effects of various nutrients and foods. For example, the Mediterranean diet and plant-based diets not only enhance microbial diversity but also have beneficial effects on the metabolism by boosting SCFA production and reducing pro-inflammatory cytokines (Jin et al. 2019; De Angelis et al. 2020). Compared to studies focusing on individual nutrients, these dietary patterns provide a more comprehensive understanding of how diet influences mental health.

The clinical application of these findings suggests that dietary interventions addressing the gut microbiota could play a role in preventing and managing mental health disorders. Personalized dietary recommendations based on an individual's baseline microbiota may offer new approaches to enhancing emotional and cognitive well-being. However, further research is required in the clinical setting to enhance these findings and provide practical guidance for implementing these concepts.

5.4.1 Depression, Anxiety, and Diet

Mood disorders and anxiety disorders are major mental health issues today, and the current literature suggests that diet may contribute to their prevention and management. Nutritional deficiencies and suboptimal nutrition have been connected to the development of these disorders, primarily due to their effect on the neuroendocrine system and neurotransmitter synthesis (Kris-Etherton et al. 2020).

A balanced diet, rich in fruits, vegetables, whole grains, lean proteins, and healthy fats, is linked to enhanced mood and reduced symptoms of mood disorders. Some of the nutrients that are required for the synthesis of neurotransmitters include tryptophan, vitamin B6, vitamin B12, folic acid, and omega-3 fatty acids that are crucial for regulating mood and cognitive function. For instance, marine-derived omega-3 fatty acids have been found to enhance serotonergic and dopaminergic transmission, thereby decreasing the depression and anxiety symptoms (Grosso et al. 2014; Lin et al. 2010). Additionally, the data from randomized controlled trials demonstrate that whole-diet interventions, such as the Mediterranean diet, produce a better improvement in depressive symptoms than non-dietary interventions (Parletta et al. 2017; Jacka et al. 2017).

On the other hand, the Western pattern diets, which are characterized by high intake of refined sugars, saturated fats, and processed foods, have been linked with inflammation, oxidative stress, and other disruptions in neurotransmitter systems. These dietary patterns have been found to be associated with a higher prevalence of mood disorders, highlighting the significance of dietary quality in mental health (Melo et al. 2019; Gangwisch et al. 2015).

Over the last few years, the role of low carbohydrate and ketogenic diets (LC/KD) has been discussed, highlighting the potential to help with the management of mood disorders through different mechanisms (Jacka et al. 2017). Current evidence shows that high-glycemic index and carbohydrate-heavy diets that are associated with the Western dietary patterns may negatively affect mood through systemic inflammation and insulin resistance. On the other hand, LC/KD, which are low in carbohydrates and high in proteins and fats, have demonstrated potential benefits in stabilizing mood and alleviating symptoms of depression and anxiety. Evidence suggests that ketogenic diets (KD) enhance mitochondrial efficiency, improve insulin signaling, and reduce inflammation, all of which are closely connected to the pathophysiology of mood disorders (Bostock et al. 2017; Yancy et al. 2005).

Some case reports and cohort studies have shown that LC/KD may provide positive effects for mood disorders, including bipolar and major depressive disorders. Patients have reported improved mood stability, reduced anxiety, and enhanced cognitive clarity. Importantly, these benefits may be attributed to ketosis, a metabolic state that is maintained through low-carbohydrate diet since carbohydrate restriction affects the energy metabolism of the brain and the neurotransmitter release (Omori et al. 2023; Bostock et al. 2017). Research also shows that there is heterogeneity of the response to LC/KD, which is likely affected by individual factors such as baseline metabolic health and adherence to the diet. This variability suggests a potential risk of symptom relapse if the diet is discontinued and underscores the importance of compliance, which can be challenging for some patients, particularly due to the restrictive nature of such diets (Watanabe et al. 2020; Brinkworth 2009). Adverse effects associated with ketosis, such as fatigue, nausea, and irritability, are usually mild and tend to resolve within a short time. However, further research is required to determine the safety of the diets in the long term, specifically concerning cardiovascular and gastrointestinal health (Batch et al. 2020; Paoli et al. 2015).

The relationship between diet and mood disorders is bidirectional; unhealthy dietary habits worsen the conditions, while mood disorders often lead to unhealthy dietary choices. This cyclical interaction highlights the importance of incorporating dietary modifications into the management strategies for mood disorders (Begdache et al. 2017). Clinical trials such as the Supporting the Modification of lifestyle in Lowered Emotional States study have demonstrated that improving dietary quality can positively affect depressive symptoms. The participants who followed a modified Mediterranean diet reported fewer depressive symptoms as compared to those who received social support only. These benefits were independent of physical activity or weight changes, which suggests that dietary modifications alone can have a major impact on the mood disorders (Jacka et al. 2017).

Nutritional psychiatry is a relatively new field that promotes dietary modifications as a primary strategy in preventing and managing anxiety and mood disorders. While further research is required to clarify the casual mechanisms and to optimize dietary interventions, current evidence supports the integration of healthy dietary patterns into mental health prevention and treatment plans. This approach not only targets the symptoms of these disorders but also enhances the overall well-being (Kris-Etherton et al. 2020; Sarris et al. 2015). The mechanisms through which the beneficial effects of a healthier dietary pattern on mood may be linked to their influence on gut microbiome modulation, inflammation reduction, oxidative stress mitigation, and neurotransmitter synthesis.

5.4.2 Psychotic Disorders and Diet

While the connection between the gut–brain axis and psychotic disorders, including schizophrenia, is not clear, some research suggests that certain dietary factors may be involved in the etiology and progression of these conditions.

There is growing evidence of a correlation between gluten intolerance and schizophrenia, particularly in patients with underlying autoimmune diseases, such as celiac disease. Some of the studies have also revealed that following a gluten-free diet can help in the management of psychotic symptoms, highlighting the potential role of immune responses in the pathophysiology of the disorder (Santis et al. 1997; Kalaydjian et al. 2005). This aligns with research showing elevated interleukin-2 levels in individuals with schizophrenia, suggesting that dietary modifications may influence immune-related processes associated with the disorder (Smith 1991).

Lipid imbalances, such as abnormal levels of polyunsaturated fatty acids, are also prevalent in schizophrenia. Research indicates that supplementation with omega-3 fatty acids, such as eicosapentaenoic acid (EPA) and docosahexaenoic acid (DHA), together with antipsychotic medications may be effective in reducing psychotic symptoms (Irving et al. 2006). Although the findings are preliminary, these findings point towards the possibility that omega-3 fatty acids may play a role in modulating the pathways that are responsible for symptom presentation and may be considered as a potential adjunct to the standard treatments. However, further large-scale, comprehensive studies are needed to evaluate the long-term effects of these supplements and to establish optimal dosages (Reimers and Ljung 2019; Irving et al. 2006).

Patients with schizophrenia often have poor diet habits that include processed foods and low intake of fruits and vegetables, which contribute to obesity and metabolic syndrome. Such dietary habits worsen the challenges of managing both the physical and psychological aspects of the disorder. These comorbidities not only worsen health outcomes but also affect the patients' compliance with the treatment and overall quality of life (Penninx and Lange 2018; Bly et al. 2013). A balanced diet rich in antioxidants, fiber, and other essential nutrients has the potential to reduce these risks and improve both physical and mental health. Experimental dietary interventions, including fasting and amino acid supplementation, have demonstrated mixed outcomes. Fasting has shown some functional benefits in selected cases, while amino acid therapies have produced promising results in small-scale trials. These interventions are still considered experimental, but they show the potential of dietary interventions as possible treatments. However, there is a need for large-scale randomized controlled trials to establish a strong evidence base for these approaches, as most of the existing studies lack strong methodologies or have small sample sizes (Boehme 1977; Pollin et al. 1961; Pearsall et al. 2016).

The KD has become a promising intervention for schizophrenia due to its positive effects on brain metabolism and neurotransmitter regulation. Schizophrenia is associated with dysregulation of glucose metabolism, glutamate signaling, and neurotransmitter regulation, and these systems may be modulated by a KD. The use of a KD has also been reported to have beneficial effects on schizophrenia-like symptoms in preclinical studies using animal models. For instance, Kraeuter et al. found that the mice given a KD had reduced hyperactivity, stereotyped running, and cognitive deficits induced by NMDA receptor hypofunction, which is a model of schizophrenia. These effects were associated with the diet-induced increases of the ketone bodies, such as beta hydroxybutyrate, and reductions of glucose levels, which may regulate energy metabolism and neurotransmitter function in the brain (Kraeuter et al. 2015).

Another study by Kraeuter et al. revealed that KD also blocked pre-pulse inhibition impairments, which is an endophenotype of schizophrenia, irrespective of calorie intake, thereby underscoring its specific metabolic benefits (Kraeuter et al. 2018). Although the clinical evidence is limited, case reports indicate that KD may be beneficial in the management of schizophrenia. Both negative and positive symptoms, including hallucinations and delusions, have been reported to decrease with the adoption of KD (Kraft and Westman 2009). Importantly, one patient maintained symptomatic remission for more than a decade while adhering to a KD and was able to discontinue antipsychotic medications. A recent pilot study examined the effects of a KD on both metabolic and psychiatric health parameters in individuals with schizophrenia. The results showed that those who followed the KD had numerous positive outcomes. Some metabolic improvements were seen, and these include a reduction in weight by 10%, body mass index by 10%, and waist circumference by 11%. Furthermore, none of the participants met the criteria for metabolic syndrome by the end of the study. Psychiatric parameters also showed noticeable improvements; people with schizophrenia demonstrated reductions in symptom severity scores and noticeable enhancements in overall clinical assessments, with majority of the participants experienced significant symptom relief (Sethi et al. 2024). These findings point to the possibility of KD as a potential adjunct to antipsychotic medications in the management of schizophrenia with metabolic comorbidities. However, challenges such as adherence to the diet and the lack of a control group in this pilot trial highlight the need for further research to evaluate the efficacy of KD in improving outcomes for patients with schizophrenia, particularly those experiencing metabolic side effects from standard treatments. These effects may be attributed to the ability of the diet to enhance mitochondrial function, reduce oxidative stress, and modulate GABA and glutamate signaling. Furthermore, beta hydroxybutyrate has the potential to reduce neuroinflammation, a mechanism that is now being associated with schizophrenia (Chrysafi et al. 2024).

The interaction between diet and schizophrenia highlights the need for a multidisciplinary approach to treatment, combining nutritional strategies with pharmacological and psychological therapies. Further research should focus on identifying specific dietary factors involved in the pathophysiology of schizophrenia to support the development of evidence-based guidelines for this population.

5.4.3 Cognitive Function and Diet

There is a growing interest in the link between diet and cognitive function due to its potential role in preventing and managing cognitive decline. Cognitive health encompasses various abilities, such as memory, learning, and problem solving, and these are determined by both modifiable and non-modifiable factors. Among the modifiable factors, diet is an important one throughout the lifetime and impacts brain health through various mechanisms, including antioxidant effects, anti-inflammatory effects, and neurogenesis stimulation (Flanagan et al. 2020; Ozawa et al. 2021).

Certain dietary habits have been found to be protective against cognitive decline and neurodegenerative diseases. Specific dietary patterns, including the Mediterranean, Nordic and DASH (Dietary Approaches to Stop Hypertension) diets have been linked with a lower risk of cognitive decline and dementia. These diets are rich in fruits, vegetables, whole-grain products, nuts, and fish, while limiting the intake of saturated fats and refined sugars. For instance, the Mediterranean diet has been extensively studied and found to enhance cognitive performance in older adults through the enhancement of memory, attention, and executive functions (Lourida et al. 2013; Samieri et al. 2013; Kesse-Guyot et al. 2013).

AD and PD are the two prevalent forms of neurodegenerative disorders, which are commonly characterized by progressive cognitive impairment. Regular consumption of the Mediterranean diet has been associated with a lower risk of developing these diseases. Research findings from Gu et al. and Scarmeas et al. showed significant decreases in AD risk among individuals who follow the Mediterranean diet (Gu et al. 2010; Scarmeas et al. 2006). Similarly, Gao et al. found that adherence to the Mediterranean diet can lower the risk of PD, especially in women (Gao et al. 2007). The positive effects of the Mediterranean diet are, in part, explained by the ability of the diet to modulate the microbiota profile and enhance the levels of SCFA producers like *Roseburia and Blautia* and reduce the levels of endotoxin producers like *Escherichia and Shigella*. This microbial modulation improves the integrity of the blood–brain barrier and gut, reduces systemic inflammation, and may prevent aggregation of pathological proteins, such as amyloid-β and α-synuclein, which are associated with AD and PD (Liu et al. 2019a, b; Zhuang et al. 2018). Other micronutrients include B vitamins, omega-3 fatty acids, and vitamin D, all of which have essential roles in cognition. Omega-3 fatty acids, particularly EPA and DHA, play a key role for neuronal membrane integrity, neurotransmission, and anti-inflammatory processes. Optimal intake of B vitamins, including B6, B12, and folate, is essential for neurochemical processes and helps in the prevention of cognitive impairment, especially in the elderly (Venkatramanan et al. 2016; Cusick and Georgieff 2016).

Emerging research on the gut–brain axis shows that dietary fiber and probiotics can influence cognitive function through the modulation of the gut microbiota. The link between brain function and the SCFAs has been demonstrated through the reduction of inflammation and upregulation of neuronal signaling pathways (Korecka and Arulampalam 2012). A recent randomized, double-blind, placebo-controlled clinical trial evaluated the effect of probiotic intake on mood and cognitive performance in older adults. The findings revealed that there was an improvement in the cognitive flexibility, as the participants in the probiotic group performed better on attention and executive function tasks compared to the placebo group.

Moreover, probiotics were associated with increased serum levels of BDNF, which is important for learning and memory. This enhancement was linked to alterations in gut microbiota composition, including the downregulation of inflammation-associated bacteria such as *Eubacterium and Clostridiales.* Interestingly, alteration in the gut microbial profile was also found related to increased BDNF levels, indicating a potential mechanism mediated by the gut-brain axis (Kim et al. 2020).

While there is increasing evidence supporting the connection between diet and cognitive health, further large-scale longitudinal studies are required to determine the causal relationships and define the optimal dietary intake. Adopting a healthy and well-balanced diet in early and middle adulthood may be a preventative strategy for maintaining cognitive health in later years. This holistic approach highlights the importance of integrating dietary recommendations into public health policies aimed at enhancing cognitive well-being across the lifespan.

5.5 Conclusion

Current insights into the gut–brain axis reveal a complex relationship between the gut microbiota, diet, and mental health. However, there are still significant gaps in knowledge about how diet directly influences mental health. This chapter highlights the need for future research to establish the causal links between specific dietary components and mental health outcomes, going beyond correlational observations. Identifying the specific molecular pathways and mediators involved is essential for establig the precise mechanisms of action. Such understanding is critical for developing targeted interventions (Horn et al. 2022).

It is also necessary to gain a better insight into the individual differences in responses to dietary interventions. Genetic polymorphisms, age, lifestyle, and the baseline composition of the gut microbiota are the factors that contribute to such heterogeneity and necessitates an individualized approach to dietary recommendations. Furthermore, more research is needed to explore the long-term effects of dietary changes on mental health and the stability of the gut microbiota. While current studies primarily focus on depression and anxiety, future research should also examine the influence of the gut-brain axis on other mental health conditions, including schizophrenia, autism spectrum disorder, and bipolar disorder.

Development of the standardized dietary intervention protocols is essential for enabling the comparison of results across different studies and establishing clear clinical guidelines. Such protocols would facilitate the integration of dietary recommendations into the standard mental health care practice, which requires collaborative efforts among clinicians, dietitians, and researchers. Advancing microbiota-targeted interventions, particularly the optimization of the prebiotic and probiotic strategies, may provide the basis for the development of personalized mental health therapies. However, it is imperative to conduct a thorough investigation into the safety and effectiveness of microbiota-based therapies, including FMT, before their use in clinical practice.

Addressing these research gaps will provide the foundation for evidence-based dietary recommendations and individualized interventions targeting the gut-brain axis to improve mental health outcomes across the lifespan. This holistic approach underlines the importance of nutrition in maintaining both physical and mental well-being.

References

Appleton J (2018) The gut-brain axis: influence of microbiota on mood and mental health. Integrative Med (Encinitas, Calif) 17(4):28–32. Retrieved from: https://pmc.ncbi.nlm.nih.gov/articles/PMC6469458/

Aswendt M, Green C, Sadler R, Llovera G, Dzikowski L, Heindl S, de Agüero MG, Diedenhofen M, Vogel SN, Wieters F, Wiedermann D, Liesz A, Hoehn M (2021) The gut microbiota modulates brain network connectivity under physiological conditions and after acute brain ischemia. iScience 24(10):103095. https://doi.org/10.1016/j.isci.2021.103095

Bäckhed F, Ding H, Wang T, Hooper LV, Koh GY, Nagy A, Semenkovich CF, Gordon JI (2004) The gut microbiota as an environmental factor that regulates fat storage. Proc Natl Acad Sci 101(44):15718–15723. https://doi.org/10.1073/pnas.0407076101

Batch JT, Lamsal SP, Adkins M, Sultan S, Ramirez MN (2020) Advantages and disadvantages of the ketogenic diet: a review article. Cureus 12(8):e9639. https://doi.org/10.7759/cureus.9639

Begdache L, Chaar M, Sabounchi NS, Kianmehr H (2017) Assessment of dietary factors, dietary practices and exercise on mental distress in young adults versus matured adults: a cross-sectional study. Nutr Neurosci 22(7):488–498. https://doi.org/10.1080/1028415x.2017.1411875

Bly MJ, Taylor SF, Dalack GW, Pop-Busui R, Burghardt KJ, Evans SJ, McInnis MI, Grove T, Brook RD, Zöllner S, Ellingrod VL (2013) Metabolic syndrome in bipolar disorder and schizophrenia: dietary and lifestyle factors compared to the general population. Bipolar Disord 16(3):277–278. https://doi.org/10.1111/bdi.12160

Boehme D (1977) Preplanned fasting in the treatment of mental disease: survey of current soviet literature. Schizophr Bull 3(2):288–296. https://doi.org/10.1093/schbul/3.2.288

Bohórquez DV, Liddle RA (2011) Axon-like basal processes in enteroendocrine cells: characteristics and potential targets. Clin Transl Sci 4(5):387–339. https://doi.org/10.1111/j.1752-8062.2011.00299.x

Bohórquez DV, Samsa LA, Roholt A, Medicetty S, Chandra R, Liddle RA (2014) An enteroendocrine cell – enteric glia connection revealed by 3D electron microscopy. PLoS One 9(2):e89881. https://doi.org/10.1371/journal.pone.0089881

Borrego-Ruiz A, Borrego JJ (2025) Human gut microbiome, diet, and mental disorders. Int Microbiol 28:1–15. https://doi.org/10.1007/s10123-024-00518-6

Bostock ECS, Kirkby KC, Taylor B (2017) The current status of the ketogenic diet in psychiatry. Front Psych 8:43. https://doi.org/10.3389/fpsyt.2017.00043

Bravo JA, Forsythe P, Chew MV, Escaravage E, Savignac HM, Dinan TG, Bienenstock J, Cryan JF (2011) Ingestion of Lactobacillus strain regulates emotional behavior and central GABA receptor expression in a mouse via the vagus nerve. Proc Natl Acad Sci 108(38):16050–16055. https://doi.org/10.1073/pnas.1102999108

Brinkworth GD (2009) Long-term effects of a very low-carbohydrate diet and a low-fat diet on mood and cognitive function. Arch Intern Med 169(20):1873–1880. https://doi.org/10.1001/archinternmed.2009.329

Cani PD, Everard A, Duparc T (2013) Gut microbiota, enteroendocrine functions and metabolism. Curr Opin Pharmacol 13(6):935–940. https://doi.org/10.1016/j.coph.2013.09.008

Cao P, Chen C, Liu A, Shan Q-H, Zhu X, Jia C-H, Peng X, Zhang M, Farzinpour Z, Zhou W, Wang H, Zhou J, Song X, Wang L, Tao W, Zheng C, Zhang Y, Ding Y, Jin Y et al (2021) Early-life inflammation promotes depressive symptoms in adolescence via microglial engulfment of dendritic spines. Neuron 109(16):2573–2589. https://doi.org/10.1016/j.neuron.2021.06.012

Carabotti M, Scirocco A, Maselli MA, Severi C (2015) The gut-brain axis: interactions between enteric microbiota, central and enteric nervous systems. Ann Gastroenterol 28(2):203–209. Retrieved from: https://pubmed.ncbi.nlm.nih.gov/25830558/

Caspani G, Swann JR (2019) Small talk: microbial metabolites involved in the signaling from microbiota to brain. Curr Opin Pharmacol 48:99–106. https://doi.org/10.1016/j.coph.2019.08.001

Červenka I, Agudelo LZ, Ruas JL (2017) Kynurenines: tryptophan's metabolites in exercise, inflammation, and mental health. Science 357(6349):eaaf9794. https://doi.org/10.1126/science.aaf9794

Chang S, Cassidy A, Willett WC, Rimm EB, O'Reilly ÉJ, Okereke OI (2016) Dietary flavonoid intake and risk of incident depression in midlife and older women. Am J Clin Nutr 104(3):704–714. https://doi.org/10.3945/ajcn.115.124545

Chrysafi M, Jacovides C, Papadopoulou SK, Psara E, Vorvolakos T, Antonopoulou M, Dakanalis A, Martin M, Voulgaridou G, Pritsa A, Mentzelou M, Giaginis C (2024) The potential effects of the ketogenic diet in the prevention and co-treatment of stress, anxiety, depression, schizophrenia, and bipolar disorder: from the basic research to the clinical practice. Nutrients 16(11):1546. https://doi.org/10.3390/nu16111546

Correia AS, Vale N (2022) Tryptophan metabolism in depression: a narrative review with a focus on serotonin and kynurenine pathways. Int J Mol Sci 23(15):8493. https://doi.org/10.3390/ijms23158493

Crile GW (1915) The origin and nature of the emotions: miscellaneous papers. (Rowland AF, Ed.). W B Saunders Co. https://doi.org/10.1037/10835-000

Cusick SE, Georgieff M (2016) The role of nutrition in brain development: the golden opportunity of the "first 1000 days". J Pediatr 175:16–21. https://doi.org/10.1016/j.jpeds.2016.05.013

Daneman R, Prat A (2015) The blood–brain barrier. Cold Spring Harb Perspect Biol 7(1):a020412. https://doi.org/10.1101/cshperspect.a020412

Dauncey MJ (2009) New insights into nutrition and cognitive neuroscience. Proc Nutr Soc 68(4):408–415. https://doi.org/10.1017/s0029665109990188

De Angelis M, Ferrocino I, Calabrese FM, De Filippis F, Cavallo N, Siragusa S, Rampelli S, Di Cagno R, Rantsiou K, Vannini L, Pellegrini N, Lazzi C, Turroni S, Lorusso N, Ventura M, Chieppa M, Neviani E, Brigidi P, O'Toole PW, Ercolini D, Gobbetti M, Cocolin L (2020) Diet influences the functions of the human intestinal microbiome. Sci Rep 10:Article 4247. https://doi.org/10.1038/s41598-020-61192-y

Dueñas M, Muñoz-González I, Cueva C, Jiménez-Girón A, Sánchez-Patán F, Santos-Buelga C, Moreno-Arribas MV, Bartolomé B (2015) A survey of modulation of gut microbiota by dietary polyphenols. Biomed Res Int 2015:850902. https://doi.org/10.1155/2015/850902

Filippis FD, Pellegrini N, Vannini L, Jeffery IB, Storia AL, Laghi L, Serrazanetti DI, Cagno RD, Ferrocino I, Lazzi C, Turroni S, Cocolin L, Brigidi P, Neviani E, Gobbetti M, O'Toole PW, Ercolini D (2015) High-level adherence to a Mediterranean diet beneficially impacts the gut microbiota and associated metabolome. Gut 65(11):1812–1821. https://doi.org/10.1136/gutjnl-2015-309957

Firth J, Solmi M, Wootton RE, Vancampfort D, Schuch FB, Hoare E, Gilbody S, Torous J, Teasdale S, Jackson SE, Smith L, Eaton M, Jacka FN, Veronese N, Marx W, Ashdown-Franks G, Siskind D, Sarris J, Rosenbaum S et al (2020) A meta-review of "lifestyle psychiatry": the role of exercise, smoking, diet and sleep in the prevention and treatment of mental disorders. World Psychiatry 19(3):360–380. https://doi.org/10.1002/wps.20773

Flanagan E, Lamport DJ, Brennan L, Burnet PWJ, Calabrese V, Cunnane SC, de Wilde MC, Dye L, Farrimond JA, Lombardo NE, Hartmann T, Härtung T, Kalliomäki M, Kuhnle GG, Fata GL, Sala-Vila A, Samieri C, Smith AD, Spencer JPE et al (2020) Nutrition and the ageing brain: moving towards clinical applications. Ageing Res Rev 62:101079. https://doi.org/10.1016/j.arr.2020.101079

Forssten SD, Ouwehand AC, Griffin SM, Patterson E (2022) One giant leap from mouse to man: the microbiota–gut–brain axis in mood disorders and translational challenges moving towards human clinical trials. Nutrients 14(3):568. https://doi.org/10.3390/nu14030568

Gan Y, Chen Y, Zhong H, Liu Z, Geng J, Wang H, Wang W (2024) Gut microbes in central nervous system development and related disorders. Front Immunol 14:288256. https://doi.org/10.3389/fimmu.2023.1288256

Gangwisch JE, Hale L, García L, Malaspina D, Opler M, Payne ME, Rossom RC, Lane DS (2015) High glycemic index diet as a risk factor for depression: analyses from the women's health initiative. Am J Clin Nutr 102(2):454–463. https://doi.org/10.3945/ajcn.114.103846

Gao X, Chen H, Fung TT, Logroscino G, Schwarzschild MA, Hu FB, Ascherio A (2007) Prospective study of dietary pattern and risk of Parkinson disease. Am J Clin Nutr 86(5):1486–1494. https://doi.org/10.1093/ajcn/86.5.1486

Gershon MD (2013) 5-Hydroxytryptamine (serotonin) in the gastrointestinal tract. Curr Opin Endocrinol Diabetes Obes 20(1):14–21. https://doi.org/10.1097/med.0b013e32835bc703

Godos J, Castellano S, Ray S, Grosso G, Galvano F (2018) Dietary polyphenol intake and depression: results from the Mediterranean healthy eating, lifestyle and aging (MEAL) study. Molecules 23(5):999. https://doi.org/10.3390/molecules23050999

Goldstein A, Hofstra RMW, Burns A (2012) Building a brain in the gut: development of the enteric nervous system. Clin Genet 83(4):307–316. https://doi.org/10.1111/cge.12054

Grosso G, Pająk A, Marventano S, Castellano S, Galvano F, Bucolo C, Drago F, Caraci F (2014) Role of omega-3 fatty acids in the treatment of depressive disorders: a comprehensive meta-analysis of randomized clinical trials. PLoS One 9(5):e96905. https://doi.org/10.1371/journal.pone.0096905

Gu Y, Luchsinger JA, Stern Y, Scarmeas N (2010) Mediterranean diet, inflammatory and metabolic biomarkers, and risk of Alzheimer's disease. J Alzheimers Dis 22(2):483–492. https://doi.org/10.3233/jad-2010-100897

Han K, Bose S, Wang J, Kim BS, Kim MJ, Kim E, Kim H (2015) Contrasting effects of fresh and fermented kimchi consumption on gut microbiota composition and gene expression related to metabolic syndrome in obese Korean women. Mol Nutr Food Res 59(5):1004. https://doi.org/10.1002/mnfr.201400780

Han Y, Wang B, Gao H, He C, Hua R, Chen L, Zhang S, Wang Y, Xin S, Xu J (2022) Vagus nerve and underlying impact on the gut microbiota-brain axis in behavior and neurodegenerative diseases. J Inflamm Res 2022(15):6213–6230. https://doi.org/10.2147/jir.s384949

Holscher HD (2017) Dietary fiber and prebiotics and the gastrointestinal microbiota. Gut Microbes 8(2):172–184. https://doi.org/10.1080/19490976.2017.1290756

Holst JJ, Deacon CF (2005) Glucagon-like peptide-1 mediates the therapeutic actions of DPP-IV inhibitors. Diabetologia 48(4):612. https://doi.org/10.1007/s00125-005-1705-7

Irving CB, Mumby-Croft R, Joy L (2006) Polyunsaturated fatty acid supplementation for schizophrenia. Cochrane Database Syst Rev 2006(3):1465–1858. https://doi.org/10.1002/14651858.cd001257.pub2

Ivanov II, Atarashi K, Manel N, Brodie E, Shima T, Karaöz U, Wei D, Goldfarb KC, Santee CA, Lynch SV, Tanoue T, Imaoka A, Itoh K, Takeda K, Umesaki Y, Honda K, Littman DR (2009) Induction of intestinal Th17 cells by segmented filamentous bacteria. Cell 139(3):485. https://doi.org/10.1016/j.cell.2009.09.033

Jacka FN, O'Neil A, Opie R, Itsiopoulos C, Cotton S, Mohebbi M, Castle D, Dash S, Mihalopoulos C, Chatterton ML, Brazionis L, Berk M (2017) A randomised controlled trial of dietary improvement for adults with major depression (the 'SMILES' trial). BMC Med 15:23. https://doi.org/10.1186/s12916-017-0791-y

Jin Q, Black A, Kales SN, Vattem D, Ruiz-Canela M, Sotos-Prieto M (2019) Metabolomics and microbiomes as potential tools to evaluate the effects of the Mediterranean diet. Nutrients 11(1):207. https://doi.org/10.3390/nu11010207

Kaelberer MM, Buchanan K, Klein ME, Barth BB, Montoya MM, Shen X, Bohórquez DV (2018) A gut-brain neural circuit for nutrient sensory transduction. Science 361(6408):eaat5236. https://doi.org/10.1126/science.aat5236

Kalaydjian A, Eaton WW, Cascella NG, Fasano A (2005) The gluten connection: the association between schizophrenia and celiac disease. Acta Psychiatr Scand 113(2):82. https://doi.org/10.1111/j.1600-0447.2005.00687.x

Kano M, Dupont P, Aziz Q, Fukudo S (2018) Understanding neurogastroenterology from neuroimaging perspective: a comprehensive review of functional and structural brain imaging in functional gastrointestinal disorders. J Neurogastroenterol Motil 24(4):512. https://doi.org/10.5056/jnm18072

Kesse-Guyot E, Andreeva VA, Lassale C, Ferry M, Jeandel C, Herçberg S, Galán P (2013) Mediterranean diet and cognitive function: a French study. Am J Clin Nutr 7(2):369. https://doi.org/10.3945/ajcn.112.047993

Kim C-S, Cha L, Sim M, Jung S, Chun WY, Baik HW, Shin D-M (2020) Probiotic supplementation improves cognitive function and mood with changes in gut microbiota in community-dwelling older adults: a randomized, double-blind, placebo-controlled, multicenter trial. J Gerontol Ser A 76(1):32. https://doi.org/10.1093/gerona/glaa090

Korecka A, Arulampalam V (2012) The gut microbiome: scourge, sentinel or spectator? J Oral Microbiol 4(1):9367. https://doi.org/10.3402/jom.v4i0.9367

Kovatcheva-Datchary P, Nilsson Å, Akrami R, Lee YS, Vadder FD, Arora T, Hallén A, Martens EC, Björck I, Bäckhed F (2015) Dietary fiber-induced improvement in glucose metabolism is associated with increased abundance of Prevotella. Cell Metab 22(6):971. https://doi.org/10.1016/j.cmet.2015.10.001

Kraeuter A-K, Loxton H, Lima BC, Rudd D, Sarnyai Z (2015) Ketogenic diet reverses behavioral abnormalities in an acute NMDA receptor hypofunction model of schizophrenia. Schizophr Res 169(1–3):491–493. https://doi.org/10.1016/j.schres.2015.10.041

Kraeuter A-K, van den Buuse M, Sarnyai Z (2018) Ketogenic diet prevents impaired prepulse inhibition of startle in an acute NMDA receptor hypofunction model of schizophrenia. Schizophr Res 206:244–250. https://doi.org/10.1016/j.schres.2018.11.011

Kraft B, Westman EC (2009) Schizophrenia, gluten, and low-carbohydrate, ketogenic diets: a case report and review of the literature. Nutr Metab 6(1):10. https://doi.org/10.1186/1743-7075-6-10

Kris-Etherton PM, Petersen K, Hibbeln JR, Hurley DL, Kolick V, Peoples S, Rodriguez NR, Woodward-Lopez G (2020) Nutrition and behavioral health disorders: depression and anxiety. Nutr Rev 79(3):247. https://doi.org/10.1093/nutrit/nuaa025

LaGreca M, Skehan L, Hutchinson DR (2022) The microbiome and neurotransmitter activity. J Sci Med 3(2). https://doi.org/10.37714/josam.v3i2.90

Ley RE, Peterson DA, Gordon JI (2006) Ecological and evolutionary forces shaping microbial diversity in the human intestine. Cell 124(4):837. https://doi.org/10.1016/j.cell.2006.02.017

Leyrolle Q, Prado-Perez L, Layé S (2023) The gut-derived metabolites as mediators of the effect of healthy nutrition on the brain. Front Nutr 10:1155533. https://doi.org/10.3389/fnut.2023.1155533

Li J, Yu R, Zhang P, Wen S, Wang S, Zhang X, Xu Q, Kong L (2019) Dietary fructose-induced gut dysbiosis promotes mouse hippocampal neuroinflammation: a benefit of short-chain fatty acids. Microbiome 7:98. https://doi.org/10.1186/s40168-019-0713-7

Li D, Yu S, Long Y, Shi A, Deng J, Ma Y, Wen J, Li X, Liu S, Zhang Y, Wan J, Li N, Ao R (2022) Tryptophan metabolism: mechanism-oriented therapy for neurological and psychiatric disorders. Front Immunol 13:985378. https://doi.org/10.3389/fimmu.2022.985378

Lin P-Y, Huang S, Su K (2010) A meta-analytic review of polyunsaturated fatty acid compositions in patients with depression. Biol Psychiatry 68(2):140. https://doi.org/10.1016/j.biopsych.2010.03.018

Liu L, Huh JR, Shah K (2022) Microbiota and the gut-brain-axis: implications for new therapeutic design in the CNS. eBioMed 77:103908. https://doi.org/10.1016/j.ebiom.2022.103908
Liu P, Wu L, Peng G, Han Y, Tang R, Ge J, Zhang L, Jia L, Yue S, Zhou K, Li L, Luo B, Wang B (2019a) Altered microbiomes distinguish Alzheimer's disease from amnestic mild cognitive impairment and health in a Chinese cohort. Brain Behav Immun 80:633–643. https://doi.org/10.1016/j.bbi.2019.05.008
Liu S, Li E, Sun Z, Fu D, Duan G, Jiang M, Yu Y, Lü M, Yang P, Tang Y, Zheng P (2019b) Altered gut microbiota and short chain fatty acids in Chinese children with autism spectrum disorder. Sci Rep 9:287. https://doi.org/10.1038/s41598-018-36430-z
Lotti S, Dinu M, Colombini B, Amedei A, Sofi F (2023) Circadian rhythms, gut microbiota, and diet: possible implications for health. Nutr Metab Cardiovasc Dis 33(8):1490. https://doi.org/10.1016/j.numecd.2023.05.009
Lourida I, Soni M, Coon JT, Purandare N, Lang I, Ukoumunne OC, Llewellyn DJ (2013) Mediterranean diet, cognitive function, and dementia. Epidemiology 24(4):479. https://doi.org/10.1097/ede.0b013e3182944410
Lu Y, Yeh W, Ohashi PS (2008) LPS/TLR4 signal transduction pathway. Cytokine 42(2):145. https://doi.org/10.1016/j.cyto.2008.01.006
Macdonald A, Ellacott KLJ (2020) Astrocytes in the nucleus of the solitary tract: contributions to neural circuits controlling physiology. Physiol Behav 223:112982. https://doi.org/10.1016/j.physbeh.2020.112982
Maître M, Klein C, Patte-Mensah C, Mensah-Nyagan AG (2020) Tryptophan metabolites modify brain Aβ peptide degradation: a role in Alzheimer's disease? Prog Neurobiol 190:101800. https://doi.org/10.1016/j.pneurobio.2020.101800
Marizzoni M, Cattaneo A, Mirabelli P, Festari C, Lopizzo N, Nicolosi V, Mombelli E, Mazzelli M, Luongo D, Naviglio D, Coppola L, Salvatore M, Frisoni GB (2020) Short-chain fatty acids and lipopolysaccharide as mediators between gut dysbiosis and amyloid pathology in Alzheimer's diseases. J Alzheimers Dis 78(2):683. https://doi.org/10.3233/jad-200306
Martin CR, Osadchiy V, Kalani A, Mayer EA (2018) The brain-gut-microbiome axis. Cell Mol Gastroenterol Hepatol 6(2):133–148. https://doi.org/10.1016/j.jcmgh.2018.04.003
Maslowski KM, Vieira AT, Ng A, Kranich J, Sierro F, Yu D, Schilter H, Rolph MS, Mackay F, Artis D, Xavier RJ, Teixeira MM, Mackay CR (2009) Regulation of inflammatory responses by gut microbiota and chemoattractant receptor GPR43. Nature 461:1282–1286. https://doi.org/10.1038/nature08530
Mayer EA, Tillisch K, Gupta A (2015) Gut/brain axis and the microbiota. J Clin Invest 125(3):926. https://doi.org/10.1172/jci76304
Mazmanian SK, Round JL, Kasper DL (2008) A microbial symbiosis factor prevents intestinal inflammatory disease. Nature 453:620–625. https://doi.org/10.1038/nature07008
Melo HM, Santos LE, Ferreira ST (2019) Diet-derived fatty acids, brain inflammation, and mental health. Front Neurosci 13:265. https://doi.org/10.3389/fnins.2019.00265
Merlo G, Bachtel G, Sugden SG (2024) Gut microbiota, nutrition, and mental health. Front Nutr 11:1337889. https://doi.org/10.3389/fnut.2024.1337889
Miri S, Yeo J, Abubaker S, Hammami R (2023) Neuromicrobiology, an emerging neurometabolic facet of the gut microbiome? Front Microbiol 14:1098412. https://doi.org/10.3389/fmicb.2023.1098412
Montiel-Castro AJ, González-Cervantes RM, Bravo-Ruiseco G, Pacheco-López G (2013) The microbiota-gut-brain axis: neurobehavioral correlates, health and sociality. Front Integr Neurosci 7:70. https://doi.org/10.3389/fnint.2013.00070
Omori NE, Malys MK, Woo GH, Mansor LS (2023) Exploring the role of ketone bodies in the diagnosis and treatment of psychiatric disorders. Front Psych 14:1142682. https://doi.org/10.3389/fpsyt.2023.1142682
Ozawa H, Miyazawa T, Miyazawa T (2021) Effects of dietary food components on cognitive functions in older adults. Nutrients 13(8):2804. https://doi.org/10.3390/nu13082804

Pagé M, Kell DB, Pretorius E (2022) The role of lipopolysaccharide-induced cell signalling in chronic inflammation. Chronic Stress 6:24705470221076390. https://doi.org/10.1177/24705470221076390

Paoli A, Bosco G, Camporesi EM, Mangar D (2015) Ketosis, ketogenic diet and food intake control: a complex relationship. Front Psychol 6:Article 27. https://doi.org/10.3389/fpsyg.2015.00027

Parletta N, Zarnowiecki D, Cho J, Wilson A, Bogomolova S, Villani A, Itsiopoulos C, Niyonsenga T, Blunden S, Meyer BJ, Segal L, Baune BT, O'Dea K (2017) A Mediterranean-style dietary intervention supplemented with fish oil improves diet quality and mental health in people with depression: a randomized controlled trial (HELFIMED). Nutr Neurosci 22(7):474. https://doi.org/10.1080/1028415x.2017.1411320

Pearsall R, Praveen KT, Pelosi AJ, Geddes J (2016) Dietary advice for people with schizophrenia. Cochrane Database Syst Rev 3:CD009547. https://doi.org/10.1002/14651858.cd009547.pub2

Penninx BWJH, Lange SMM (2018) Metabolic syndrome in psychiatric patients: overview, mechanisms, and implications. Dialogues Clin Neurosci 20(1):63. https://doi.org/10.31887/dcns.2018.20.1/bpenninx

Pollin W, Cardon PV, Kety SS (1961) Effects of amino acid feedings in schizophrenic patients treated with iproniazid. Science 133:104. https://doi.org/10.1126/science.133.3446.104

Psichas A, Reimann F, Gribble FM (2015) Gut chemosensing mechanisms. J Clin Invest 125(3):908. https://doi.org/10.1172/jci76309

Rehfeld JF (2004) A centenary of gastrointestinal endocrinology. Horm Metab Res 26(11):735. https://doi.org/10.1055/s-2004-826154

Reimers A, Ljung H (2019) The emerging role of omega-3 fatty acids as a therapeutic option in neuropsychiatric. Ther Adv Psychopharmacol 9:2045125319858901. https://doi.org/10.1177/2045125319858901

Robertson RC, Oriach CS, Murphy K, Moloney G, Cryan JF, Dinan TG, Ross RP, Stanton C (2016) Omega-3 polyunsaturated fatty acids critically regulate behaviour and gut microbiota development in adolescence and adulthood. Brain Behav Immun 59:21–37. https://doi.org/10.1016/j.bbi.2016.07.145

Round JL, Mazmanian SK (2009) The gut microbiota shapes intestinal immune responses during health and disease. Immunology 9(5):313. https://doi.org/10.1038/nri2515

Round JL, Mazmanian SK (2010) Inducible Foxp3 + regulatory T-cell development by a commensal bacterium of the intestinal microbiota. Proc Natl Acad Sci 107(27):12204. https://doi.org/10.1073/pnas.0909122107

Roy CL, Wells PM, Si J, Raes J, Bell JT, Spector TD (2019) Red wine consumption associated with increased gut microbiota α-diversity in 3 independent cohorts. Gastroenterology 158(1):270. https://doi.org/10.1053/j.gastro.2019.08.024

Samieri C, Grodstein F, Rosner B, Kang JH, Cook NR, Manson JE, Buring JE, Willett WC, Okereke OI (2013) Mediterranean diet and cognitive function in older age. Epidemiology 24(4):490. https://doi.org/10.1097/ede.0b013e318294a065

Santis AD, Addolorato G, Romito A, Caputo S, Giordano A, Gambassi G, Taranto C, Manna R, Gasbarrini A (1997) Schizophrenic symptoms and SPECT abnormalities in a coeliac patient: regression after a gluten-free diet. J Intern Med 242(5):421. https://doi.org/10.1046/j.1365-2796.1997.00200.x

Sarkar A, Lehto SM, Harty S, Dinan TG, Cryan JF, Burnet PWJ (2016) Psychobiotics and the manipulation of bacteria–gut–brain signals. Trends Neurosci 39(11):763. https://doi.org/10.1016/j.tins.2016.09.002

Sarris J, Logan A, Akbaraly T, Amminger GP, Balanzá-Martínez V, Freeman MP, Hibbeln JR, Matsuoka Y, Mischoulon D, Mizoue T, Nanri A, Nishi D, Ramsey DJC, Rucklidge JJ, Sánchez-Villegas A, Scholey A, Su K, Jacka FN (2015) Nutritional medicine as mainstream in psychiatry. Lancet Psychiatry 2(3):271. https://doi.org/10.1016/s2215-0366(14)00051-0

Scarmeas N, Stern Y, Tang M, Mayeux R, Luchsinger JA (2006) Mediterranean diet and risk for Alzheimer's disease. Ann Neurol 59(6):912. https://doi.org/10.1002/ana.20854

Sethi S, Wakeham DE, Ketter TA, Hooshmand F, Bjornstad J, Richards B, Westman EC, Krauss RM, Saslow LR (2024) Ketogenic diet intervention on metabolic and psychiatric health in bipolar and schizophrenia: a pilot trial. Psychiatry Res 335:115866. https://doi.org/10.1016/j.psychres.2024.115866

Sharkey KA, Mawe GM (2022) The enteric nervous system. Am Physiol Soc J 103(2):1487. https://doi.org/10.1152/physrev.00018.2022

Silva YP, Bernardi A, Frozza RL (2020) The role of short-chain fatty acids from gut microbiota in gut-brain communication. Front Endocrinol 11:25. https://doi.org/10.3389/fendo.2020.00025

Smith RS (1991) Is schizophrenia caused by excessive production of interleukin-2 and interleukin-2 receptors by gastrointestinal lymphocytes? Med Hypotheses 34(3):225. https://doi.org/10.1016/0306-9877(91)90215-k

Sonnenburg ED, Smits SA, Tikhonov M, Higginbottom SK, Wingreen NS, Sonnenburg JL (2016) Diet-induced extinctions in the gut microbiota compound over generations. Nature 529:212–215. https://doi.org/10.1038/nature16504

Sternini C, Anselmi L, Rozengurt E (2008) Enteroendocrine cells: a site of 'taste' in gastrointestinal chemosensing. Curr Opin Endocrinol Diabetes Obes 15(1):73–78. https://doi.org/10.1097/med.0b013e3282f43a73

Taylor BC, Lejzerowicz F, Poirel M, Shaffer JP, Jiang L, Aksenov AA, Litwin NS, Humphrey G, Martino C, Miller-Montgomery S, Dorrestein PC, Veiga P, Song SJ, McDonald D, Derrien M, Knight R (2020) Consumption of fermented foods is associated with systematic differences in the gut microbiome and metabolome. Am Soc Microbiol 5(2):e00901–e00919. https://doi.org/10.1128/msystems.00901-19

Tillisch K, Mayer EA, Gupta A, Gill Z, Brazeilles R, Nevé BL, van Hylckama Vlieg JET, Guyonnet D, Derrien M, Labus JS (2017) Brain structure and response to emotional stimuli as related to gut microbial profiles in healthy women. Psychosom Med 79(8):905. https://doi.org/10.1097/psy.0000000000000493

Turner JR (2009) Intestinal mucosal barrier function in health and disease. Immunology 9(11):799. https://doi.org/10.1038/nri2653

Vandenplas Y, Carnielli V, Książyk J et al (2020) Factors affecting early-life intestinal microbiota development. Nutrition 78:110812. Elsevier BV. https://doi.org/10.1016/j.nut.2020.110812

Vanegas SM, Meydani M, Barnett JB, Goldin BR, Kane A, Rasmussen H, Brown C, Vangay P, Knights D, Jonnalagadda SS, Koecher K, Karl JP, Thomas MJ, Dolnikowski GG, Li L, Saltzman E, Wu D, Meydani SN (2017) Substituting whole grains for refined grains in a 6-wk randomized trial has a modest effect on gut microbiota and immune and inflammatory markers of healthy adults. Am J Clin Nutr 105(3):635. https://doi.org/10.3945/ajcn.116.146928

Vangay P, Johnson AJ, Ward T, Al-Ghalith GA, Shields-Cutler RR, Hillmann B, Lucas SK, Beura LK, Thompson EA, Till L, Batres R, Paw B, Pergament S, Saenyakul P, Xiong M, Kim AD, Kim G, Masopust D, Martens EC et al (2018) US immigration westernizes the human gut microbiome. Cell 175(4):962. https://doi.org/10.1016/j.cell.2018.10.029

Venkatramanan S, Armata IE, Strupp BJ, Finkelstein JL (2016) Vitamin B-12 and cognition in children. Adv Nutr 7(5):879. https://doi.org/10.3945/an.115.012021

Vijay-Kumar M, Aitken JD, Carvalho FA, Cullender TC, Mwangi SM, Srinivasan S, Sitaraman SV, Knight R, Ley RE, Gewirtz AT (2010) Metabolic syndrome and altered gut microbiota in mice lacking toll-like receptor 5. Science 328:228. https://doi.org/10.1126/science.1179721

Vogt N, Romano KA, Darst BF, Engelman CD, Johnson SC, Carlsson CM, Asthana S, Blennow K, Zetterberg H, Bendlin BB, Rey FE (2018) The gut microbiota-derived metabolite trimethylamine N-oxide is elevated in Alzheimer's disease. Alzheimers Res Ther 10:Article 124. https://doi.org/10.1186/s13195-018-0451-2

Watanabe M, Tuccinardi D, Ernesti I, Basciani S, Mariani S, Genco A, Manfrini S, Lubrano C, Gnessi L (2020) Scientific evidence underlying contraindications to the ketogenic diet: an update. Obes Rev 21(10):e13053. https://doi.org/10.1111/obr.13053

Watson H, Mitra S, Croden F, Taylor M, Wood HM, Perry SL, Spencer J, Quirke P, Toogood GJ, Lawton C, Dye L, Loadman PM, Hull MA (2017) A randomised trial of the effect of omega-3 polyunsaturated fatty acid supplements on the human intestinal microbiota. Gut 67(11):1974. https://doi.org/10.1136/gutjnl-2017-314968

Yancy WS, Foy ME, Chalecki AM, Vernon MC, Westman EC (2005) A low-carbohydrate, ketogenic diet to treat type 2 diabetes. Nutr Metab 2:34. https://doi.org/10.1186/1743-7075-2-34

Zanoni I, Ostuni R, Marek LR, Barresi S, Barbalat R, Barton GM, Granucci F, Kagan JC (2011) CD14 controls the LPS-induced endocytosis of toll-like receptor 4. Cell 147(4):868. https://doi.org/10.1016/j.cell.2011.09.051

Zhu C, Sawrey-Kubicek L, Beals E, Rhodes CH, Houts H, Sacchi R, Zivkovic AM (2020) Human gut microbiome composition and tryptophan metabolites were changed differently by fast food and Mediterranean diet in 4 days: a pilot study. Nutr Res 77:62–72. https://doi.org/10.1016/j.nutres.2020.03.005

Zhuang Z, Shen L, Li W, Fu X, Zeng F, Gui L, Lü Y, Cai M, Zhu C, Tan Y, Zheng P, Li H, Zhu J, Zhou H, Bu X, Wang Y (2018) Gut microbiota is altered in patients with Alzheimer's disease. J Alzheimers Dis 63(4):1337–1346. https://doi.org/10.3233/jad-180176

Functional Foods and Mental Wellness

6

Asiat Na'Allah, Semiat Adedimeji, Babalola Ola Yusuf, Ibrahim Adeola Moronfolu, and Abraham Olufemi Asuku

Abstract

The complex relationship between functional foods and mental wellness is explored in this chapter, with an emphasis on the effects of selected nutrients and dietary practices on mental health. It begins by going over functional foods and their components, as well as the prevalence of common mental health disorders such as anxiety, depression, etc., and the mechanisms by which functional foods affect the health of the brain, including neurotransmitter modulation and neuroprotection. Important bioactive compounds like omega-3 fatty acids, antioxidants, B vitamins, and minerals are highlighted for their mental health benefits. Specific functional foods such as dark chocolate and fermented foods are

A. Na'Allah (✉) · S. Adedimeji
Department of Biochemistry, Faculty of Pure and Applied Sciences, Kwara State University, Malete, Kwara State, Nigeria
e-mail: asiat.naallah@kwasu.edu.ng; adedimeji.semiat@gmail.com

B. O. Yusuf
Department of Biochemistry, Faculty of Science, Federal University of Health Sciences, Ila-Oragun, Osun, Nigeria

Department of Biotechnology, Faculty of Applied Sciences, Durban University of Technology, Durban, South Africa
e-mail: babalola.yusuf@fuhsi.edu.ng

I. A. Moronfolu
Department of Chemical Sciences, Biochemistry Unit, Summit University, Offa, Kwara State, Nigeria
e-mail: adebmore@gmail.com

A. O. Asuku
Bioresourses Development Centre, National Biotechnology Research and Development Agency, Ogbomoso, Nigeria
e-mail: asufem2017@gmail.com

W. Mohamed, A. Özge (eds.), *Feeding the Mind: The Connection Between Diet, Drugs, and Mental Health Volume 1*, Nutritional Neurosciences,
https://doi.org/10.1007/978-981-95-1414-4_6

explored for their roles in managing conditions like depression, anxiety, cognitive decline, and stress. The chapter concludes with future research trends and the growing role of functional foods in mental wellness promotion.

Keywords

Antioxidants · Functional foods · Mental health · Neuroprotection · Nutraceuticals

6.1 Introduction

6.1.1 Functional Food

Food is vital to life because it gives the body the energy and nutrition it needs to function. New research indicates that some food ingredients might be beneficial to health by being purposely modified, leading to the designation of these foods as functional foods. In the 1980s, the phrase "Foods for Specified Health Use" (FOSHU) was used in Japan, marking the beginning of the notion of functional foods (Essa et al. 2023). Food that may offer health values beyond basic nutrition is referred to as functional food, according to the Academy of Nutrition and Dietetics. It is also described as food, either natural or processed, that has biologically functional ingredients that are known to support good health and lower the risk of disease (Essa et al. 2023). The phrase "functional food" can also refer to characteristics that have been intentionally bred into already-existing edible plants, like genetically engineered golden rice that produces beta-carotene, a precursor to vitamin A, aimed to compensate for certain dietary deficits, like a vitamin A deficiency (Al Saqqa 2021).

6.1.2 Understanding Functional Foods

Given that many functional foods fit into more categories, functional foods can be divided into a number of groups that are not mutually exclusive:

6.1.2.1 Fortified/Enriched Foods

Foods that have extra nutrients added to them to meet nutritional requirements and improve health outcomes are referred to as fortified or enriched foods. These foods give customers additional choices for obtaining particular nutrients and their health advantages. For instance, milk is frequently supplemented with vitamin D to enhance calcium absorption, which is crucial for avoiding osteoporosis and rickets. Cereal grain fortified products can help avoid neural tube defects in babies, while iodized salt can prevent enlarged thyroid glands and thyroid-related disorders (Al Saqqa 2021).

6.1.2.2 Probiotics and Prebiotics

Probiotics are foods that, among other things, improve immunity, lower cholesterol, and stimulate the natural gut flora (Silva et al. 2020). Prebiotics are nutrients found in food that strengthen the immune system and encourage the growth of probiotic bacteria in the stomach. Foods that have undergone fermentation act as natural habitats for probiotic bacteria, which help heal a variety of illnesses, including those relating to the gut (Yadav et al. 2022). Prebiotics, such as oligofructose and inulin, are commonly used to boost the growth of good gut bacteria. Certain forms of oligofructose ferment in the stomach to create short-chain fatty acids (SCFAs), which include butyric acid, acetic acid, and propionic acid. These SCFAs have been linked to enhanced immunity and cognition, as well as lower incidence of colorectal cancer and irritable bowel syndrome (IBS) (Blaak et al. 2020).

6.1.2.3 Nutraceuticals

Nutraceuticals are foods or food items that go beyond basic nutrition to provide therapeutic benefits. These products can be sourced from dietary supplements, processed meals, or herbal products and contain bioactive substances like minerals, antioxidants, or vitamins (Banerjee et al. 2023). These include curcumin, resveratrol, quercetin, lycopene, ellagic acid, etc. One of turmeric's main bioactive components, curcumin, has anti-inflammatory and antioxidant qualities. Pomegranates and grapes include resveratrol, which has been demonstrated to improve gut microbiota, immunity, and vascular health (Essa et al. 2023).

6.1.3 Mechanisms of Action of Functional Foods

Many functional foods, such as omega-3 fatty acids, curcumin, flavonoids, vitamin D, vitamin B, vitamin E, and choline, have been shown in studies to either directly or indirectly impact brain health.

6.1.3.1 How Functional Foods Influence Brain Health

Research on the relationship between nutraceuticals and cognitive function has drawn attention and has shed light on the possible significance of some food ingredients for brain health (Dominguez et al. 2021). Docosahexaenoic acid (DHA), which is prevalent in fatty fish, is one of the omega-3 fatty acids that has a major role in improving cognitive function and lowering the risk of neurodegenerative disorders like Alzheimer's disease (Sherzai et al. 2023). DHA is essential for maintaining neuronal function and communication and makes up a significant portion of the brain's cell membranes. Higher amounts of DHA have been associated with better cognition in general, including memory and retention. DHA is also essential for early adulthood and fetal brain development. Because of the continuous turnover of brain cells, DHA is required for good shape and function throughout one's lifetime (von Schacky 2021).

6.1.3.2 Neurotransmitter Production and Regulation Through Diet

Neurotransmitters are involved in sensory processing and motor control by sending information throughout the nervous system and the brain. Amino acids are the main simple precursors used in the production of neurotransmitters (Gasmi et al. 2022), which are subsequently discharged into the synaptic cleft to attach to receptor proteins on target tissues to either activate or change the target tissue's function. In the brain and throughout the nervous system, neurotransmitters carry out information transmission that is involved in both motor control and sensory processing (Gasmi et al. 2022). The gut and the brain are the sites of neurotransmitter synthesis, such as dopamine (DA), gamma-aminobutyric acid (GABA), serotonin, and endocannabinoids; the gut produces around 95% of the body's serotonin. A few meals have the ability to raise neurotransmitter levels. For example, fish, eggs, and spirulina raise dopamine levels by way of L-DOPA, which improves mood, memory, sleep, and concentration. Serotonin levels are greatly regulated by mood, appetite, social behavior, sexual drive, sleep, learning, and gastrointestinal motility (Gasmi et al. 2022).

6.1.4 Mental Wellness

The World Health Organization (WHO) defines mental wellness, sometimes known as mental health, as a positive state of well-being that includes resilience, a sense of purpose, and the capacity to form and sustain meaningful relationships. It represents a person's ability to reach their potential, manage everyday stressors, work well, and contribute to their communities instead of merely being mentally healthy (Cresswell-Smith et al. 2022). Emotional, psychological, physical, spiritual, and social aspects all play a role in mental wellness and influence an individual's thoughts, feelings, and behaviors. Mental wellness is categorized as:

Emotional well-being entails having the capacity to recognize, control, and appropriately express one's emotions (Kliziene et al. 2021).

Psychological well-being describes positive functioning, which is defined by meaningful goals, self-acceptance, personal development, and wise decision-making (Dhanabhakyam and Sarath 2023).

Physical well-being entails leading a healthy lifestyle that incorporates frequent exercise, a balanced diet, and enough sleep.

Social well-being deals with fostering wholesome relationships and making constructive contributions to society.

Spiritual well-being entails having objectives that provide life direction and fulfillment and discovering significance in life's experiences (Skoko et al. 2021).

6.1.4.1 Common Mental Health Disorders

The WHO claims that mental illness is expressed as a clinically significant disruption in a person's behavior, emotional control, or cognition that is frequently accompanied by distress or a decline in critical areas of functioning. The common mental health illnesses include:

Depression, a common mental illness characterized by a chronic sense of sadness and disinterest in day-to-day activities. It may result in a range of mental and physical problems that make day-to-day living difficult. Approximately 29.19 million individuals in Africa suffer from depression; of these, over seven million live in Nigeria (3.9% of the continent's population). The lifetime prevalence of depressive illnesses varies from 3.3% to 9.8% (Gbadamosi et al. 2022).

Excessive worry and anxiety about day-to-day matters are symptoms of anxiety disorder that last for at least 6 months. Panic disorder (which involves panic episodes), generalized anxiety disorder (which is characterized by excessive worry), and social anxiety disorder (which is characterized by excessive fear in social circumstances) are some examples of anxiety disorders. An estimated 301 million people, including 58 million children and adolescents, suffered from anxiety disorders in 2019 (Mishra and Varma 2023).

The existence of compulsions (repetitive behaviors) or obsessions (unwanted intrusive thoughts) characterizes obsessive-compulsive disorder (OCD). Roughly 1–3% of people worldwide are impacted (Brock et al. 2024).

Post-traumatic stress disorder (PTSD) grows as a result of going through or seeing terrible experiences. It majorly impacts one's physical, mental, and emotional health. PTSD prevalence in conflict-affected areas might reach 50%, particularly among internally displaced people (IDPs), whose rates can vary from 63% to 86% as a result of continuous violence and instability (Teshome et al. 2023).

6.1.5 Relationship Between Functional Foods and Mental Wellness

The brain and gastrointestinal (GI) tract are closely related, according to specialists in the field of nutritional psychiatry. Many bacteria found in the GI tract aid in the synthesis of neurotransmitters, which are substances that travel across synapses from one neuron to another and are fundamental for nervous system communication (Teleanu et al. 2022). Neurotransmitters include norepinephrine, serotonin, and dopamine, as examples. Notably, because it controls behavior, mood, memory, and gastrointestinal homeostasis, serotonin (which is produced in the gut) is intimately associated with mental health. It is produced in the intestinal mucosa's enterochromaffin cells as well as the brainstem's raphe nuclei. Prominent major depressive disorder (MDD), PTSD, OCD, and anxiety are among the neurological and psychiatric conditions for which serotonin is a prominent therapeutic target. These conditions are linked to lower levels of serotonin in the central nervous system (CNS) and plasma serotonin (Pourhamzeh et al. 2022).

6.1.6 Importance of Nutrition in Mental Health

Due to its impact on mood, brain function, and general cognitive function, nutrition is a critical factor in mental health. Among the main ways that a healthy diet affects mental health are:

Improving Brain Function
Essential nutrients such as omega-3 fatty acids, vitamins, and minerals support brain structure and function. Omega-3 fatty acids are particularly critical for maintaining neuronal membrane health and supporting neurotransmitter function (Dighriri et al. 2022).

Mood Regulation
Neurotransmitters including serotonin, dopamine, and norepinephrine rely on substances like glutamic acid, phenylalanine, and tryptophan, among others, for their synthesis. These neurotransmitters play a role in mood, hunger, and cognitive regulation. Deficiencies in omega-3 fatty acids and B vitamins have been connected to depression (Kris-Etherton et al. 2021).

Improving Cognitive Performance
According to Fekete et al. (2023), a well-balanced diet high in antioxidants, vitamins, and minerals enhances cognitive processes, including memory and concentration, and guards against cognitive decline.

Reducing Stress
In addition to reducing the negative impacts of stress on mental health, proper nutrition helps control the body's stress response (Bremner et al. 2020).

6.1.7 Regulatory Aspects

Functional foods are categorized and regulated by the Functional Food Center (FFC), the Food for Specified Health Uses (FOSHU) program in Japan, and the federal Food and Drug Administration (FDA) in the United States.

6.1.7.1 Functional Food Regulation by FDA

The FDA, which upholds food and drug safety regulations under the 1994 Dietary Supplement Health and Education Act (DSHEA) and the Federal Food, Drug, and Cosmetic Act (FFDCA), is in charge of classifying and regulating functional foods in the United States (FDA 2024). Since functional foods do not yet fall under a defined regulatory category, current regulations govern their regulation. Foods and supplements are regulated by the FDA according to their intended use, which is established by the information on the label and health claims. A substance (such as a food, component, or dietary ingredient) and a disease or health-related condition are the two primary components that must be present in every health claim.

Statements that are devoid of either element are not considered health claims. The FDA divides food label claims on nutrition and health into four categories:

Authorized health claims describe a connection between a food, food component, or other dietary ingredient and the risk of a disease to be used in food labeling as long as they satisfy specific requirements and are approved by an FDA regulation. For instance, "adequate calcium throughout life may reduce the risk of osteoporosis" (Martirosyan et al. 2022).

Nutrient content claims are governed by the Nutrition Labeling and Education Act of 1990 (NLEA) and include phrases such as "free," "high," or "reduced" to characterize the amount of a nutrient in a product (Martirosyan et al. 2022).

Qualified health claims may be utilized when new data points to a link between a food ingredient and a lower risk of disease, but not enough to receive full FDA approval (Martirosyan et al. 2022).

Function and structure health claims may include statements like "calcium builds strong bones" that highlight how a vitamin or food component supports regular bodily processes. Additionally, they can describe how a vitamin or dietary component works to preserve such structure or function; for instance, fiber keeps the bowels regular, or antioxidants preserve the integrity of cells (Martirosyan et al. 2022).

6.1.7.2 Functional Food Regulation by FOSHU

The FOSHU regulatory framework was put in place by the Japanese government to regulate functional foods. Significant evidence from clinical trials is required by FOSHU, which adheres to particular standards that include rules for study controls, participant selection, and symptom characteristics. These investigations are conducted with healthy, medication-free volunteers under stringent dietary and lifestyle controls to guarantee an accurate evaluation of the effects of the active ingredient. Manufacturers must demonstrate the following in order to receive FOSHU approval: (1) the ingredient's efficacy in the body; (2) the use of nutritionally appropriate ingredients; (3) consumption-based assurance of compatibility with product specifications; and (4) well-established quality control procedures (Martirosyan et al. 2022). In order to verify the mechanism and effectiveness of the active ingredient, which is backed by human trials, the FOSHU system requires thorough documentation, including in vitro and in vivo testing. It also calls for verification of ingredient stability, evidence of product safety, and specifics regarding the physical, chemical, and biological properties of the active ingredient. At least three randomly selected samples from various manufacturing dates or lots must be used by manufacturers when quantifying ingredients. For FOSHU certification, quality control procedures must also be submitted, along with requirements for ingredients and finished products, as well as information about production equipment. However, FOSHU's shortcoming is that it lacks aftermarket research, which would monitor the impacts of functional foods in uncontrolled public settings (Martirosyan et al. 2022).

6.1.7.3 Functional Food Regulation by FFC

A methodical strategy is needed to regulate functional foods, and the FFC has suggested a multi-step regulatory process for approval that is focused on bioactive chemicals. Two shortcomings in the present FDA and FOSHU regulations are addressed by this approach. It would first create a formal category for functional foods and then require aftermarket research and epidemiological studies as part of the certification process for premium classification. The FDA does not classify functional foods separately, even though it recognizes that some dietary ingredients may have health advantages linked to disease prevention. Through a multi-step procedure, the FFC links the intake of functional foods to health claims backed by published scientific data. The FFC proposal also mandates aftermarket research for producers looking to overcome the constraints of FOSHU (Martirosyan et al. 2022).

6.2 Bioactive Compounds in Functional Foods

Bioactive compounds in functional foods play a crucial role in enhancing health and preventing various diseases. These compounds are naturally occurring substances that, while not essential nutrients, can significantly influence biological processes in the body, leading to improved health outcomes.

6.2.1 Omega-3 Fatty Acids: Sources and Benefits

The polyunsaturated fats omega-3 (ώ-3) fatty acids, especially EPA and DHA, are critical for the health of the brain and the control of inflammation. Omega-3 fatty acids, which are included in fatty fish like salmon and mackerel as well as plant sources like flaxseeds and walnuts, enhance cardiovascular health by lowering blood pressure, triglycerides, and cholesterol while also improving cholesterol and lowering the menace of heart disease and stroke. Additionally, they have anti-inflammatory properties that help with ailments like rheumatoid arthritis and neurological illnesses like Parkinson's and Alzheimer's (Huang et al. 2022). Increased intake of ώ-3 fatty acids is related to improved cognitive function and a lower incidence of dementia (Loong et al. 2023).

6.2.2 B Vitamins and Their Role in Brain Function

Energy production, synthesis of neurotransmitters, preservation of myelin, and other metabolic activities all depend on the B vitamins, which include B1, B2, B6, B9 (folate), and B12. Additionally, they are cardinal for the metabolism of homocysteine, which promotes cardiovascular health and cognitive performance. Cereals, dairy products, and meat are examples of functional foods enriched with B vitamins that can improve dietary intake and guard against age-related cognitive decline (Ajibawo-Aganbi et al. 2020).

The one-carbon metabolic pathway, which is crucial for the synthesis of neurotransmitters, depends on folate, B6, and B12. These vitamins at low levels can raise homocysteine, which is interconnected with cognitive decline and neurological illnesses like Parkinson's and Alzheimer's. Studies highlighted that getting enough of these vitamins is linked to better cognitive function and a lower chance of developing dementia (Calvaresi and Bryan 2001; Zhu et al. 2023).

Metabolic Health

Studies show that there is an inverse relationship between the occurrence of metabolic syndrome and increased intakes of folate, B6, and B12. Higher serum levels of these vitamins were associated with decreased incidences of obesity, dyslipidemia, and insulin resistance, according to cohort research (Zhu et al. 2023). This implies that these vitamins guard against diseases related to metabolism.

DNA Synthesis and Cell Division

Folate is especially necessary for the production and repair of DNA, which makes it essential during times of fast growth, like pregnancy. In developing fetuses, adequate folate consumption can help avert neural tube abnormalities (Britannica 2024).

6.2.3 Amino Acids

The building blocks of proteins, amino acids, are not only necessary for the synthesis of proteins but also act as bioactive substances in functional foods and have a variety of health benefits, principally for many physiological processes, such as the production of neurotransmitters, the control of immunological responses, and the maintenance of metabolism. A good source of functional foods high in amino acids is meat, eggs, legumes, nuts, and seeds, as well as dairy products.

6.2.3.1 Tryptophan and Serotonin

Tryptophan is an essential amino acid that serves as a precursor to several bioactive compounds, most notably serotonin (5-hydroxytryptamine, 5-HT). This relationship positions tryptophan and serotonin at the forefront of discussions regarding functional foods and their health benefits, particularly in mood regulation and cognitive function. Tryptophan is found in various functional foods such as milk and yogurt; nuts and seeds such as almonds, walnuts, and pumpkin seeds; legumes such as beans and lentils; and meat and fish such as turkey and chicken.

6.2.3.2 Tyrosine and Dopamine

A non-essential amino acid called tyrosine is required for the synthesis of several important neurotransmitters, including dopamine, adrenaline, and norepinephrine, which are critical for mood regulation, cognitive function, and overall mental health maintenance. The relationship between tyrosine and dopamine, in addition to the fact that they exist as functional meals, highlights the significance of these compounds as bioactive compounds with positive health benefits.

Amino acids' bioactive properties in functional foods provide the following health benefits:

Muscle Health and Recovery
Leucine, isoleucine, and valine, or branched-chain amino acids (BCAAs), are particularly crucial for the health of muscles. They encourage the synthesis of muscle proteins and lessen post-exercise muscle pain. Studies show that using BCAA supplements might speed up recovery and increase performance during physical activity (Soumya et al. 2021).

Cognitive Function
Neurotransmitters like dopamine and serotonin are produced from precursor amino acids like tyrosine and tryptophan. Getting enough of these amino acids has been associated with better mood and cognitive function. Tryptophan has been connected to improved sleep quality as well as a decline in anxiety and depressive symptoms (Soumya et al. 2021).

Immune Support
Glutamate is one of the essential amino acids for immune system function. They boost the body's defenses against infections by promoting the growth of lymphocytes and the synthesis of antibodies. Supplementing with glutamine at times of stress or illness has been demonstrated in studies to enhance immune response (Soumya et al. 2021).

Metabolic Regulation
Amino acids also play a role in metabolic pathways. For instance, leucine stimulates the mTOR pathway, which is necessary for protein synthesis and muscle growth. Because of their regulatory role, amino acids are important in managing conditions like obesity and metabolic syndrome (Sorrenti et al. 2023).

Mood Regulation
Serotonin production, which is essential for mood regulation and emotional health, depends on tryptophan. Research suggests that eating more tryptophan through diet can raise serotonin levels in the brain, which may lessen anxiety and depressive symptoms. Studies have demonstrated that tryptophan-rich meals can, via the serotonergic system, boost mood and reduce aggressive behavior (Friedman 2018; Höglund et al. 2019).

Sleep Quality
Tryptophan contributes to the synthesis of melatonin, a hormone that controls sleep-wake cycles. Eating food high in tryptophan can improve sleep quality by encouraging the synthesis of melatonin, which makes it an important ingredient in functional foods designed to treat sleep disorders (Höglund et al. 2019).

Stress Response

Tyrosine helps the body produce catecholamines, which include norepinephrine, epinephrine, and dopamine, and is involved in how the body reacts to stress. The need for these neurotransmitters rises in times of extreme stress. Tyrosine supplements have been demonstrated to lessen the detrimental effects of stress on mood and cognitive function, making them an important dietary choice for people who work in high-stress workplaces (Sorrenti et al. 2023).

Neurotransmitter Balance

Adequate levels of tyrosine are essential for maintaining the balance of neurotransmitters in the brain. Dopamine is involved in various functions, including motor control and mood regulation. Imbalances in dopamine levels are linked to conditions such as Parkinson's disease and depression. Consuming foods rich in tyrosine may help support healthy dopamine levels and overall brain function (Friedman 2018).

6.2.4 Minerals

Minerals are necessary nutrients that are key to a number of bodily physiological functions. Certain minerals provide health benefits that go beyond their basic nutritional value. They operate as bioactive molecules in functional foods, helping to prevent or control chronic diseases. Found in a variety of functional foods such as dairy products (rich sources of calcium and phosphorus), nuts and seeds (excellent sources of magnesium and zinc), fruits and vegetables (potassium), spinach (magnesium), and mushrooms (selenium), etc.

6.2.4.1 Magnesium

Magnesium is a crucial mineral that plays a role in the synthesis of proteins, the production of energy, and the function of muscles. Magnesium is a bioactive substance found in functional foods that offers a number of health advantages, including the management and prevention of chronic illnesses.

6.2.4.2 Zinc

The vital trace mineral zinc is involved in many physiological processes, such as DNA synthesis, immunological response, and protein synthesis. Zinc is an indispensable component of functional meals as a bioactive element that offers numerous health advantages. Incorporating these foods into the diet can help ensure adequate mineral intake, supporting various health functions such as:

Bone Health

Bone density and structure are maintained by minerals, including calcium, magnesium, and phosphorus. An adequate consumption of these minerals is linked to a decreased incidence of fractures and osteoporosis. Particularly well known for its importance to bone health is calcium, and magnesium helps to regulate both calcium levels and bone metabolism (Soumya et al. 2021).

Immune Function
One important mineral for the immune system's operation is zinc. It aids in T-cell (T-lymphocyte) growth and activation, which is necessary for adaptive immunity. A zinc shortage can compromise immunity and make a person more vulnerable to illness (Dixit et al. 2023).

Cardiovascular Health
Magnesium and potassium are essential for preserving heart health. Magnesium is important for heart and muscle function, whereas potassium helps control blood pressure by balancing sodium levels in the body. There is a correlation between lower risks of hypertension and cardiovascular disease and higher dietary potassium intake (Soumya et al. 2021).

Antioxidant Properties
Some minerals, such as manganese and selenium, function as cofactors for antioxidant enzymes. Selenium is an element of glutathione peroxidase, an enzyme that shields cells from oxidative damage. Sufficient selenium intake has been associated with diminished inflammation and lower risks of chronic diseases such as cancer (Hosseini et al. 2022).

Metabolic Function
Magnesium plays a role in insulin sensitivity and glucose metabolism. A higher type 2 diabetes risk has been linked to low magnesium levels. Magnesium supplementation has demonstrated significant advantages in enhancing glycemic control in insulin-resistant people.

Neurological Function
Neurotransmitter release and synaptic plasticity are two aspects of neurological health that magnesium affects. Its possible ability to lower the risk of neurodegenerative illnesses, such as Alzheimer's disease, has been researched. By promoting brain function, adequate magnesium levels may help prevent cognitive deterioration (Dixit et al. 2023).

6.2.5 Antioxidants

Antioxidants are compounds that shield the organism from oxidative damage brought on by free radicals (unstable molecules) that can damage cells and contribute to chronic diseases such as heart disease, diabetes, and cancer. As bioactive compounds in functional foods, antioxidants play a crucial role in promoting health and preventing disease. Antioxidants protect against oxidative stress, lowering the risk of chronic diseases like cancer and heart disease. Polyphenols in fruits and vegetables positively influence oxidative stress markers (Soumya et al. 2021). Flavonoids and vitamins C and E promote cardiovascular health by reducing LDL cholesterol oxidation and improving endothelial function, lowering heart disease risk (Sorrenti et al. 2023). Antioxidants also have anti-inflammatory properties, such as polyphenols, which suppress pro-inflammatory cytokines (Rudrapal et al. 2022). They may guard against cognitive decline by reducing oxidative stress in the brain, and some antioxidants inhibit tumor growth and may reduce cancer risk (Pandey and Rizv 2009).

6.2.5.1 Vitamins C and E

Essential nutrients that act as potent antioxidants in the body are vitamins C and E. They contribute significantly to immune system support, oxidative stress protection for cells, and general health promotion. Vitamins C and E play a critical role in combating free radicals that cause oxidative cell damage. Vitamin E, a lipid-soluble antioxidant, protects cell membranes from lipid peroxidation, while vitamin C, a water-soluble antioxidant, scavenges reactive oxygen species (Sorrenti et al. 2023). Together, they reduce inflammation and lower the risk of chronic diseases like cancer and cardiovascular issues. Vitamin C supports immune function by promoting white blood cell and collagen production, while vitamin E protects immune cells from oxidative damage (Soumya et al. 2021). Both vitamins promote skin health by boosting collagen synthesis and protecting against UV damage (Perron and Brumaghim 2009). Additionally, they support cardiovascular health by improving endothelial function and lowering blood pressure, and they may protect against cognitive decline by reducing oxidative stress in the brain (Soumya et al. 2021).

6.2.5.2 Polyphenols

A wide class of naturally occurring substances, mostly found in plants, is known as polyphenols. Their health-promoting qualities, especially their anti-inflammatory, anti-cancer, and antioxidant capabilities, have attracted a lot of interest. Polyphenols are bioactive substances found in functional foods that help treat and prevent chronic illnesses. Polyphenols are known for reducing oxidative stress by scavenging free radicals and reactive oxygen species (ROS), protecting against cellular damage linked to cancer, cardiovascular, and neurological diseases (Pandey and Rizv 2009; Rudrapal et al. 2022). They also have anti-inflammatory effects, with flavonoids like quercetin and catechins suppressing pro-inflammatory cytokines, reducing inflammation, and reducing the risk of inflammatory disorders (Rudrapal et al. 2022). Polyphenol-rich diets improve cardiovascular health by reducing LDL cholesterol oxidation and lowering blood pressure (Pandey and Rizv 2009; Sorrenti et al. 2023). Additionally, polyphenols may prevent cancer by inducing apoptosis in cancer cells and inhibiting tumor growth (Pandey and Rizv 2009). Their antioxidant properties may also protect against cognitive decline linked to neurodegenerative diseases like Alzheimer's (Sorrenti et al. 2023).

6.3 Functional Foods Beneficial for Mental Wellness

The potential of functional foods to improve mental well-being through a variety of bioactive components has attracted attention. These meals are not only nutrient-dense, but they also contain certain components that, by altering mood, cognitive function, and emotional well-being, can have a favorable impact on mental health (Sorrenti et al. 2023).

6.3.1 Berries and Dark Chocolate

Dark chocolate, which is high in flavonoids, has anti-inflammatory and blood-flow-improving qualities and may improve mental performance and even lessen depressive symptoms via the release of serotonin and endorphins (Socci et al. 2017). Antioxidants like flavonoids and anthocyanins, which are abundant in berries, particularly blueberries, strawberries, and blackberries, shield brain cells from oxidative damage. Mental health issues and neurodegenerative diseases are linked to oxidative stress. According to Barfoot et al. (2019), berries' antioxidant qualities have been connected to enhanced mood, memory, and cognitive performance.

6.3.2 Probiotics and Fermented Foods

Probiotics are found in foods like yogurt, kefir, and sauerkraut and help to maintain intestinal health. There is mounting evidence that the gut-brain axis mediates the relationship between mental and digestive health. Probiotics have the potential to lessen inflammation, depression, and anxiety (frequently linked to mental health issues) by modulating gut bacteria composition, which impacts neurotransmitter production (Wallace and Milev 2017).

6.3.3 Adaptogens

Adaptogens are organic compounds that support mental health and assist the body in adjusting to stress. They are commonly obtained from herbs and mushrooms. They are becoming more widely acknowledged as advantageous ingredients in functional diets meant to improve mental wellness. Important adaptogens and how functional foods can support mental wellness are:

Ashwagandha (*Withania somnifera*), an herb well known for its ability to reduce stress, has been demonstrated to decrease cortisol levels, a hormone linked to stress. Because of its potential to alleviate anxiety and depression symptoms, ashwagandha is a common addition to functional meals like smoothies and herbal teas.

Rhodiola rosea is well known for its ability to lessen fatigue and improve mood as well as cognitive performance. According to studies, rhodiola can enhance mental function under pressure, which makes it a useful ingredient for energy bars and supplements that promote mental clarity.

Ginseng is frequently used in traditional medicine with the aim of increasing energy and decreasing weariness. Additionally, it might enhance mood control and cognitive performance. Overall mental wellness can be supported by adding ginseng to functional meals like energy drinks or herbal blends (Sorrenti et al. 2023).

6.3.4 Omega-3 Fatty Acids and Fish Oil

The beneficial effects of omega-3 fatty acids, especially docosahexaenoic acid (DHA) and eicosapentaenoic acid (EPA), on mental health are widely known. These fatty acids, which are mostly found in fatty fish like salmon, sardines, and mackerel, are essential for the construction and operation of the brain. According to research, omega-3 fatty acids can lessen inflammation and enhance neuronal connectivity, which may help lessen the symptoms of anxiety and depression (Liao et al. 2022). Frequent omega-3-rich fish eating has been associated with a good impact on mood stability and a decreased prevalence of depression (Grosso et al. 2014).

6.4 Functional Foods and Specific Mental Health Conditions

6.4.1 Depression: Functional Foods as Nutritional Interventions

Over 350 million individuals worldwide suffer from depression, a common mental health illness that has an influence on social, professional, and personal facets of life (Guo et al. 2024). Symptoms include a chronically depressed mood, disinterest in activities, cognitive impairment, and a sense of worthlessness (Garapati et al. 2023). According to Remes et al. (2021), modern science perceives depression as a chronic, complex illness influenced by biological, psychological, and social aspects. Important biological factors include changes in neurotransmitters such as norepinephrine and serotonin, as well as cognitive decline and reduced neuroplasticity. Rumination and negative self-perception are psychological components that exacerbate the illness (Struijs et al. 2021). Additional social elements that have a role include trauma, economic difficulties, and a lack of support (Kirkbride et al. 2024).

The potential of nutritional therapies, especially functional foods, to impact the biochemical and molecular processes disrupted in depression is drawing attention to their use in the treatment of the condition (Piao et al. 2023). In addition to emotional disorders, depression is associated with neuroendocrine dysfunction, persistent inflammation, and reduced neurotransmitter levels, such as those of serotonin, dopamine, and norepinephrine (Ushakova et al. 2020). Natural chemicals from functional foods are a viable alternative to traditional antidepressants, which often have undesired side effects and delayed effects despite their efficacy in treating the neurochemical and inflammatory components of depression (Yadav et al. 2024).

Neuroprotective and anti-inflammatory characteristics can be found in iridoids and secoiridoids, such as catalpol, geniposide, and loganin, which are found in medicinal plants. Specifically, catalpol modulates the Nrf2 pathway, boosting antioxidant defenses and mitigating oxidative stress on neurons while lowering cortisol and suppressing inflammatory elements like NF-κB and COX-2, which lessens the hyperactivity of the hypothalamic-pituitary-adrenal (HPA) axis and contributes to depression (Song et al. 2021). By controlling tryptophan metabolism and preventing it from being diverted into the kynurenine pathway, which is linked to depression, they decrease pro-inflammatory cytokines (such as TNF-α, IL-1, and IL-6) and enhance the production of serotonin (Winiarska-Mieczan et al. 2023; Shao et al. 2022). Curcumin, found in turmeric, crosses the blood-brain barrier and reduces neuroinflammation by inhibiting NF-κB and calming microglial activity. A study found that 1000 mg of curcumin daily significantly reduced depressive symptoms in patients with major depressive disorder (Zhang et al. 2020; Sanmukhani et al. 2014).

6.4.2 Anxiety: Dietary Strategies for Managing Anxiety

Excessive concern, fear, and uneasiness are the hallmarks of anxiety, which is frequently accompanied by physiological and psychological symptoms that interfere with day-to-day functioning (Chellappa and Aeschbach 2022). Its pathophysiology is largely dependent on key neurotransmitters, including GABA, norepinephrine, and serotonin. Reduced GABA, which soothes the brain, causes increased anxiety, while low serotonin and norepinephrine levels are connected to mood dysregulation (Adwas et al. 2019). Long-term stress overstimulates the HPA axis, causing the overproduction of cortisol, which damages brain regions central to memory and emotional control, such as the hippocampus and prefrontal cortex (Ghasemi et al. 2022). Anxiety management becomes more difficult when structural abnormalities, such as a hyperactive amygdala and an underactive prefrontal cortex, exacerbate fear reactions (Liu et al. 2020).

Dietary changes have a big impact on how well people control their anxiety, with research validating that diets high in fruits, vegetables, whole grains, and ώ-3 fatty acids, such as the Mediterranean diet, can lower anxiety levels (Werner-Gray 2020; Carlos et al. 2020). Omega-3 fatty acids, which are present in fish such as salmon, help to maintain dopamine levels, stimulate the generation of serotonin, and increase brain-derived neurotrophic factor (BDNF) to support neuronal function (Norwitz and Naidoo 2021). Furthermore, eating a breakfast high in protein, fiber, and healthy fats stabilizes blood sugar levels, which lessens the feelings of anxiety. On the other hand, eating sugary foods or skipping meals might exacerbate anxiety (Gong et al. 2022). Micronutrients that are important are selenium, magnesium, and zinc. Zinc aids in stress management, while magnesium controls the HPA axis, which is frequently hyperactive during anxious episodes (Kumar et al. 2022; Muscaritoli 2021).

6.4.3 Cognitive Decline and Dementia: Functional Foods and Dietary Strategies in Cognitive Health

Global health concerns about dementia and cognitive decline are expanding, especially as populations get older. About 50 million individuals worldwide suffer from dementia, a neurological disease that impairs thinking, memory, and day-to-day functioning. By 2050, estimates indicate that the number will reach 152 million (Livingston et al. 2024). Cognitive decline, which frequently occurs before dementia, is characterized by a progressive loss of thinking and information retention skills over time, including memory and cognitive abilities (Anderson et al. 2024). Age and heredity are unavoidable risk factors, but lifestyle choices have a big impact on cognitive decline. The three main modifiable risk factors are poor vascular health, obesity, and physical inactivity. Cognitive decline can result from conditions such as dyslipidemia, hypertension, and diabetes that decrease blood flow to the brain (Chau et al. 2020; Zuo and Wu 2022; Shehata and Friedman 2024).

Dietary techniques and functional foods are becoming more popular in the management of dementia and cognitive decline, especially as populations age. Preserving cognitive function and lowering the risk of Alzheimer's disease, a major cause of dementia, have been associated with the Mediterranean diet, which is high in fruits, vegetables, and whole grains, among others (Dominguez et al. 2021). This diet can reduce biological risk factors, including oxidative stress and inflammation, which lead to neurodegeneration, along with other diets like the DASH and Healthy Eating Index (HEI) (Puri et al. 2023). Antioxidants, MUFAs, PUFAs, and a balanced ratio of omega-6 to omega-3 fats are among the bioactive components of the Mediterranean diet that guard against cognitive decline. Research has indicated that those following this diet had reduced chances of dementia and cognitive impairment (Chu et al. 2022), and long-term studies like the PREDIMED trial have reinforced these long-term cognitive benefits (Martínez-Lapiscina et al. 2013).

6.4.4 Stress Management: Foods That Promote Relaxation and Reduce Stress

The goal of stress management in mental health is to reduce the harmful consequences of stress on the body and mind. Although stress is a normal reaction to outside demands, excessive or prolonged stress can cause serious mental health problems such as depression or anxiety (Tran et al. 2020). Identifying stressors and using strategies to stop long-term harm are the first steps in effective stress management. Chronic stress can damage emotions, impair cognitive function, and increase susceptibility to psychiatric diseases if left unchecked (Marakushyn et al. 2024). In addition to reacting to stress, stress management entails coping mechanisms and resilience-building. People can focus on stressors and respond deliberately rather than instinctively by using strategies including mindfulness, relaxation techniques, and cognitive behavioral approaches (Ward et al. 2021; Chowdhury et al. 2024).

Dietary strategies can significantly promote relaxation and reduce stress by modulating physiological and psychological processes. L-theanine, primarily found in green tea (*Camellia sinensis*), has been widely studied as an anxiolytic agent without causing sedation (Coppock and Dziwenka 2021). It promotes relaxation by increasing alpha wave production in the brain, which is associated with calm, alert states (Wei et al. 2023). Consuming 200–400 mg/day of L-theanine has been shown to effectively reduce stress in stressful situations (Sakamoto et al. 2019). Research also shows that L-theanine lowers blood pressure, heart rate variability, and tension in stressful tasks (Yoto et al. 2012; Kimura et al. 2007).

6.5 Future Directions and Research

Growing interest in how nutrition can affect mental health outcomes has centered studies on the relationship between functional foods and mental wellness. Numerous functional foods and the bioactive substances they contain have been the subject of recent studies that suggest they may help support mental health.

6.5.1 Current Research Trends in Functional Foods and Mental Wellness

6.5.1.1 Functional Foods and Mental Health Promotion

The importance of several functional foods in supporting mental health has been documented. These included Brahmi (*Bacopa monnieri*), Ashwagandha (*Withania somnifera*), turmeric (*Curcuma longa*), garlic (*Allium sativum*), and pumpkin seeds (*Cucurbita maxima*). These foods are well known for their anti-inflammatory and antioxidant qualities, which may enhance neurotransmitter function and lessen the symptoms of illnesses like anxiety and depression (Dutta et al. 2020).

6.5.1.2 Omega-3 Fatty Acids

The neuroprotective properties of omega-3 fatty acids, especially those in fish oil, have been well researched. They are thought to affect neurotransmitter pathways, which in turn lowers neuroinflammation and improves mood disorders. Omega-3 fatty acid consumption has been associated with reduced anxiety and depression, indicating that it may be possible to use it as a dietary intervention for mental health (Gutte and Deshmukh 2023).

6.5.1.3 Gut-Brain Axis

Through the gut-brain axis, recent research has demonstrated the important function that gut bacteria play in mental health. An imbalance in gut flora known as dysbiosis has been linked to a number of mental health conditions. By altering the gut microbiota, functional foods like probiotics, prebiotics, and synbiotics are being investigated for their potential to boost mental wellness and restore gut health (Haslam 2023).

6.5.1.4 Phytochemicals as Nutraceuticals

Studies conducted in preclinical settings have demonstrated the potential anti-inflammatory and neuroprotective effects of phytochemicals. According to Davinelli et al. (2016), these substances may be able to lessen inflammation and oxidative stress, which are associated with mood problems and cognitive decline. Creating nutraceuticals based on these phytochemicals to promote mental health is becoming more and more popular.

6.6 Conclusion

Functional foods, which contain bioactive substances, are essential for promoting mental wellness. These substances improve neurotransmitter production, improve brain function, and lower inflammation, which aids in the treatment of mental health issues like stress, anxiety, depression, and cognitive decline. Certain meals, such as fermented foods, dark chocolate, and adaptogens, are beneficial for mental health. The gut-brain axis and phytochemicals are two areas of research that are getting more attention as time goes on, which highlights the potential of functional foods as supplemental methods for enhancing mental health and general well-being.

References

Adwas AA, Jbireal JM, Azab AE (2019) Anxiety: insights into signs, symptoms, etiology, pathophysiology, and treatment. East Afr Sch J Med Sci 2(10):580–591

Ajibawo-Aganbi U, Saleem S, Ali Khan SZ, Veliginti S, Perez Bastidas MV, Lungba RM, Cancarevic I (2020) Can nutritional adequacy help evade. Neurodegeneration in older age? A review. Cureus 12(10):e10921

Al Saqqa GSR (2021) Some functional foods and benefits of their bioactive components. J Saudi Soc Foods Nutr (JSSFN) 14(1):1–11

Anderson ND, Murphy KJ, Troyer AK (2024) Living with mild cognitive impairment: a guide to maximizing brain health and reducing the risk of dementia. Oxford University Press, New York

Banerjee A, Somasundaram I, Das D, Jain Manoj S, Banu H, Mitta Suresh P, Paul S, Bisgin A, Zhang H, Sun XF, Duttaroy AK, Pathak S (2023) Functional foods: a promising strategy for restoring gut microbiota diversity impacted by SARS-CoV-2 variants. Nutrients 15(11):2631

Barfoot KL, May G, Lamport DJ, Ricketts J, Riddell PM, Williams CM (2019) The effects of acute wild blueberry supplementation on the cognition of 7–10-year-old schoolchildren. Eur J Nutr 58:2911–2920

Blaak EE, Canfora EE, Theis S, Frost G, Groen AK, Mithieux G, Nauta A, Scott K, Stahl B, van Harsselaar J, van Tol R, Vaughan EE, Verbeke K (2020) Short chain fatty acids in human gut and metabolic health. Benefic Microbes 11:411–455

Bremner JD, Moazzami K, Wittbrodt MT, Nye JA, Lima BB, Gillespie CF, Rapaport MH, Pearce BD, Shah AJ, Vaccarino V (2020) Diet, stress and mental health. Nutrients 12(8):2428

Britannica, The Editors of Encyclopaedia. (2024). Vitamin B complex. Encyclopedia Britannica. https://www.britannica.com/science/vitamin-B-complex

Brock H, Rizvi A, Hany M (2024) Obsessive-compulsive disorder. In: StatPearls. StatPearls Publishing, Treasure Island

Calvaresi E, Bryan J (2001) B vitamins, cognition, and aging: a review. J Gerontol Ser B 56(6):327–339

Carlos M, Elena B, Teresa IM (2020) Are adherence to the Mediterranean diet, emotional eating, alcohol intake, and anxiety related in university students in Spain? Nutrients 12(8):2224
Chau AC, Cheung EY, Chan KH, Chow WS, Shea YF, Chiu PK, Mak HK (2020) Impaired cerebral blood flow in type 2 diabetes mellitus – a comparative study with subjective cognitive decline, vascular dementia and Alzheimer's disease subjects. NeuroImage Clin 27:102302
Chellappa SL, Aeschbach D (2022) Sleep and anxiety: from mechanisms to interventions. Sleep Med Rev 61:101583
Chowdhury D, Banerjee P, Moreira A, Narendran R, Scholar AT, Fahed-Sreih J (2024) Adaptive neurosciences and neuro-integral methodology. Bull Technol Hist J 24(6):218–324
Chu CQ, Yu LL, Qi GY, Mi YS, Wu WQ, Lee YK, Zhai QX, Tian FW, Chen W (2022) Can dietary patterns prevent cognitive impairment and reduce Alzheimer's disease risk: exploring the underlying mechanisms of effects. Neurosci Biobehav Rev 135:104556
Coppock RW, Dziwenka M (2021) Green tea. In: Nutraceuticals. Academic, Amsterdam, pp 697–723
Cresswell-Smith J, Kauppinen T, Laaksoharju T, Rotko T, Solin P, Suvisaari J, Wahlbeck K, Tamminen N (2022) Mental health and mental wellbeing impact assessment frameworks-a systematic review. Int J Env Res Public Health 19(21):13985
Davinelli S, Maes M, Corbi G, Zarrelli A, Willcox DC, Scapagnini G (2016) Dietary phytochemicals and neuro-inflammaging: from mechanistic insights to translational challenges. Immun Ageing 13:16
Dhanabhakyam M, Sarath M (2023) Psychological wellbeing: a systematic literature review. Int J Adv Res Sci Commun Technol 3(1):603–607
Dighriri IM, Alsubaie AM, Hakami FM, Hamithi DM, Alshekh MM, Khobrani FA, Dalak FE, Hakami AA, Alsueaadi EH, Alsaawi LS, Alshammari SF, Alqahtani AS, Alawi IA, Aljuaid AA, Tawhari MQ (2022) Effects of omega-3 polyunsaturated fatty acids on brain functions: a systematic review. Cureus 14(10):e30091
Dixit V, Kamal SWJ, Chole PB, Dayal D, Chaubey KK, Pal AK, Xavier J, Manjunath BT, Bachheti RK (2023) Functional foods: exploring the health benefits of bioactive compounds from plant and animal sources. J Food Qual 1:5546753
Dominguez LJ, Veronese N, Vernuccio L, Catanese G, Inzerillo F, Salemi G, Barbagall M (2021) Nutrition, physical activity, and other lifestyle factors in the prevention of cognitive decline and dementia. Nutrients 13(11):4080
Dutta S, Roy S, Roy S (2020) Functional foods for mental health promotion. J Mahatma Gandhi Inst Med Sci 25(2):72–79
Essa MM, Al-Adawi S, Subash S, Akbar M (2023) Functional foods and nutraceuticals. Springer, Cham
FDA (2024) Label Claims for Conventional Foods and Dietary Supplements. Available online at https://www.fda.gov/food/nutrition-food-labeling-and-critical-foods/label-claims-conventional-foods-and-dietarysupplements
Fekete M, Lehoczki A, Tarantini S, Fazekas-Pongor V, Csípő T, Csizmadia Z, Varga JT (2023) Improving cognitive function with nutritional supplements in aging: a comprehensive narrative review of clinical studies investigating the effects of vitamins, minerals, antioxidants, and other dietary supplements. Nutrients 15(24):5116
Friedman M (2018) Analysis, nutrition, and health benefits of tryptophan. Int J Tryptophan Res 11:1178646918802282
Garapati J, Jajoo S, Aradhya D, Reddy LS, Dahiphale SM, Patel DJ (2023) Postpartum mood disorders: insights into diagnosis, prevention, and treatment. Cureus 15(7):e42107
Gasmi A, Nasreen A, Menzel A, Benahmed GA, Pivina L, Noor S, Peana M, Chirumbolo S, Bjørklund G (2022) Neurotransmitters regulation and food intake: the role of dietary sources in neurotransmission. Molecules 28(1):210
Gbadamosi IT, Henneh IT, Aluko OM, Yawson EO, Fokoua AR, Koomson A, Torbi J, Olorunnado SE, Lewu FS, Yusha'u Y, Keji-Taofik ST, Biney RP, Tagoe TA (2022) Depression in sub-Saharan Africa. IBRO Neurosci Rep 12:309–322

Ghasemi M, Navidhamidi M, Rezaei F, Azizikia A, Mehranfard N (2022) Anxiety and hippocampal neuronal activity: relationship and potential mechanisms. Cogn Affect Behav Neurosci 22(3):431–449

Gong WJ, Fong DYT, Wang MP, Lam TH, Chung TWH, Ho SY (2022) Skipping breakfast and eating breakfast away from home were prospectively associated with emotional and behavioral problems in 115,217 Chinese adolescents. J Epidemiol 32(12):551–558

Grosso G, Galvano F, Marventano S, Malaguarnera M, Bucolo C, Drago F, Caraci F (2014) Omega-3 fatty acids and depression: scientific evidence and biological mechanisms. Oxidative Med Cell Longev 2014:313570. https://doi.org/10.1155/2014/313570

Guo F, Khatibi A, Tham J (2024) A systematic literature review of the impact of music therapy on the management of depression in patients with chronic diseases. J Infras Policy Dev 8(7):8020

Gutte RK, Deshmukh D (2023) Sectional study of nutritional psychology to identify the significance of the connection between mental health and nutraceutical functional ingredients. Funct Food Ingredients Mental Health 1(5):1–13

Haslam O (2023) Scientific basis for the development of functional food for mental wellbeing. NutraIngredients, Crawley

Höglund E, Øverli O, Winberg S (2019) Tryptophan metabolic pathways and brain serotonergic activity: a comparative review. Front Endocrinol (Lausanne) 10:158

Hosseini F, Sanjabi MR, Kazemi M, Ghaemian N (2022) Bioactive ingredients in functional foods: current status and future trends. https://doi.org/10.5772/intechopen.104416

Huang Y, Deng Y, Zhang P, Lin J, Guo D, Yang L, Liu D, Xu B, Huang C, Zhang H (2022) Associations of fish oil supplementation with incident dementia: evidence from the UK Biobank cohort study. Front Neurosci 16:910977

Kimura K, Ozeki M, Juneja LR, Ohira H (2007) L-theanine reduces psychological and physiological stress responses. Biol Psychol 74(1):39–45

Kirkbride JB, Anglin DM, Colman I, Dykxhoorn J, Jones PB, Patalay P, Pitman A, Soneson E, Steare T, Wright T, Griffiths SL (2024) The social determinants of mental health and disorder: evidence, prevention and recommendations. World Psychiatry 23(1):58

Kliziene I, Cizauskas G, Sipaviciene S, Aleksandraviciene R, Zaicenkoviene K (2021) Effects of a physical education program on physical activity and emotional well-being among primary school children. Int J Environ Res Public Health 18(14):7536

Kris-Etherton PM, Petersen KS, Hibbeln JR, Hurley D, Kolick V, Peoples S, Rodriguez N, Woodward-Lopez G (2021) Nutrition and behavioral health disorders: depression and anxiety. Nutr Rev 79(3):247–260

Kumar S, Mishra T, Prajapati A, Sethi P (2022) Nutrition, neurotransmitters, and behavior. In: Nutrition and psychiatric disorders. Springer Nature Singapore, Singapore, pp 89–108

Liao J, Xiong Q, Yin Y, Ling Z, Chen S (2022) The Effects of Fish Oil on Cardiovascular Diseases: Systematical Evaluation and Recent Advance. Front Cardiovasc Med 8:802306

Liu WZ, Zhang WH, Zheng ZH, Zou JX, Liu XX, Huang SH, You WJ, He Y, Zhang JY, Wang XD, Pan BX (2020) Identification of a prefrontal cortex-to-amygdala pathway for chronic stress-induced anxiety. Nat Commun 11(1):2221

Livingston G, Huntley J, Liu KY, Costafreda SG, Selbæk G, Alladi S, Ames D, Banerjee S, Burns A, Brayne C, Fox NC, Gitlin NL, Howard R, Kales HC, Kivimäki M, Larson EB, Selbæk G, Teri L, Mukadam N (2024) Dementia prevention, intervention, and care: 2024 report of the Lancet standing Commission. Lancet 404(10452):572–628

Loong S, Barnes S, Gatto NM, Chowdhury S, Lee GJ (2023) Omega-3 fatty acids, cognition, and brain volume in older adults. Brain Sci 13(9):1278

Marakushyn D, Bulynina O, Isaieva I, Karmazina I, Maslova N (2024) Impact of stress on emotional health and cognitive function. Med Sci Ukr 20(2):136–142

Martínez-Lapiscina EH, Clavero P, Toledo E, Estruch R, Salas-Salvadó J, San Julián B, Sanchez-Tainta A, Ros E, Valls-Pedret C, Martinez-Gonzalez MÁ (2013) Mediterranean diet improves cognition: the PREDIMED-NAVARRA randomised trial. J Neurol Neurosurg Psychiatry 84(12):1318–1325

Martirosyan D, Lampert T, Ekbad M (2022) Classification and regulation of functional food proposed by the functional food center. Funct Food Sci 2(2):25–46
Mishra AK, Varma AR (2023) A comprehensive review of the generalized anxiety disorder. Cureus 15(9):e46115
Muscaritoli M (2021) The impact of nutrients on mental health and Well-being: insights from the literature. Front Nutr 8:656290
Norwitz NG, Naidoo U (2021) Nutrition as metabolic treatment for anxiety. Front Psych 12:598119
Pandey KB, Rizv SI (2009) Plant polyphenols as dietary antioxidants in human health and disease. Oxidative Med Cell Longev 2(5):270–278
Perron NR, Brumaghim JL (2009) A review of the antioxidant mechanisms of polyphenol compounds related to iron binding. Cell Biochem Biophys 53:75–100
Piao J, Wang Y, Zhang T, Zhao J, Lv Q, Ruan M, Yu Q, Li B (2023) Antidepressant-like effects of representative types of food and their possible mechanisms. Molecules 28(19):6992
Pourhamzeh M, Moravej FG, Arabi M, Shahriari E, Mehrabi S, Ward R, Ahadi R, Joghataei MT (2022) The roles of serotonin in neuropsychiatric disorders. Cell Mol Neurobiol 42(6):1671–1692
Puri S, Shaheen M, Grover B (2023) Nutrition and cognitive health: a life course approach. Front Public Health 11:1023907
Reily NM, Tang S, Negrone A, Gan DZQ, Sheanoda V, Christensen H (2023) Omega-3 supplements in the prevention and treatment of youth depression and anxiety symptoms: A scoping review. PLoS One. 18(4):e0284057
Remes O, Mendes JF, Templeton P (2021) Biological, psychological, and social determinants of depression: a review of recent literature. Brain Sci 11(12):1633
Rudrapal M, Khairnar SJ, Khan J, Bin Dukhyil A, Ansari MA, Alomary MN (2022) Dietary polyphenols and their role in oxidative stress-induced human diseases: insights into protective effects, antioxidant potentials and mechanism(s) of action. Front Pharmacol 13:806470
Sakamoto FL, Ribeiro RMP, Bueno AA, Santos HO (2019) Psychotropic effects of L-theanine and its clinical properties: from the management of anxiety and stress to a potential use in schizophrenia. Pharmacol Res 147:104395
Sanmukhani J, Satodiya V, Trivedi J, Patel T, Tiwari D, Panchal B, Goel A, Tripathi CB (2014) Efficacy and safety of curcumin in major depressive disorder: a randomized controlled trial. Phytother Res 28(4):579–585
Shao J, Wei Y, Wei X (2022) A comprehensive review on bioavailability, safety and antidepressant potential of natural bioactive components from tea. Food Res Int 158:111540
Shehata M, Friedman SM (2024) Reducing cognitive decline through lifestyle medicine. In: Lifestyle medicine, 4th edn. CRC Press, Boca Raton, pp 1197–1206
Sherzai D, Moness R, Sherzai S, Sherzai A (2023) A systematic review of omega-3 fatty acid consumption and cognitive outcomes in neurodevelopment. Am J Lifestyle Med 17(5):649–685
Silva DR, Orlandi Sardi JC, Pitangui NS, Magri Roque S, Barbosa da Silva AC, Rosalen PL (2020) Probiotics as an alternative antimicrobial therapy: current reality and future directions. J Funct Foods 73:104080
Skoko I, Topić Stipić D, Tustonja M, Stanić D (2021) Mental health and spirituality. Psychiatr Danub 33(4):822–826
Socci V, Tempesta D, Desideri G, De Gennaro L, Ferrara M (2017) Enhancinh human cognition with cocoa flavonoids. Front Nutr 4:19. https://doi.org/10.3389/fnut.2017.00019
Song L, Wu X, Wang J, Guan Y, Zhang Y, Gong M, Wang Y, Li B (2021) Antidepressant effect of catalpol on corticosterone-induced depressive-like behavior involves the inhibition of HPA axis hyperactivity, central inflammation and oxidative damage probably via dual regulation of NF-κB and Nrf2. Brain Res Bull 177:81–91
Sorrenti V, Burò I, Consoli V, Vanella L (2023) Recent advances in health benefits of bioactive compounds from food wastes and by-products: biochemical aspects. Int J Mol Sci 24(3):2019
Soumya NP, Mini S, Sivan S, Mondal S (2021) Bioactive compounds in functional foods and their role as therapeutics. Bioact Compd Health Dis 4(3):24–39

Struijs SY, de Jong PJ, Jeronimus BF, van der Does W, Riese H, Spinhoven P (2021) Psychological risk factors and the course of depression and anxiety disorders: a review of 15 years NESDA research. J Affect Disord 295:1347–1359

Teleanu RI, Niculescu AG, Roza E, Vladacenco O, Grumezescu AM, Teleanu DM (2022) Neurotransmitters-keys factors in neurological and neurodegenerative disorders of the central nervous system. Int J Mol Sci 23(11):5954

Teshome AA, Abebe EC, Mengstie MA (2023) Post-traumatic stress disorder and associated factors among adult war survivors in Northwest Ethiopia: community-based, cross-sectional study. Front Psychiatry 14:1083138. https://doi.org/10.3389/fpsyt.2023.1083138

Tran CT, Tran HT, Nguyen HT, Mach DN, Phan HS, Mujtaba BG (2020) Stress management in the modern workplace and the role of human resource professionals. Bus Ethics Leadersh 4(2):26–40

Ushakova VM, Morozova AY, Reznik AM, Kostyuk GP, Chekhonin VP (2020) Molecular biological aspects of depressive disorders: a modern view. Mol Biol 54(5):639–660

Von Schacky C (2021) Importance of EPA and DHA blood levels in brain structure and function. Nutrients 13(4):1074

Wallace CJ, Milev R (2017) The effects of probiotics on depressive symptoms in humans: a systematic review. Ann General Psychiatry 16(1):14

Ward RN, Brady AJ, Jazdzewski R, Yalch MM (2021) Stress, resilience, and coping. In: Emotion, well-being, and resilience. Apple Academic Press, Palm Bay, pp 3–14

Wei Y, Xu J, Miao S, Wei K, Peng L, Wang Y, Wei X (2023) Recent advances in the utilization of tea active ingredients to regulate sleep through neuroendocrine pathway, immune system and intestinal microbiota. Crit Rev Food Sci Nutr 63(25):7598–7626

Werner-Gray L (2020) Anxiety-free with food: natural, science-backed strategies to relieve stress and support your mental health. Hay House, Inc, Carlsbad

Winiarska-Mieczan A, Kwiecień M, Jachimowicz-Rogowska K, Donaldson J, Tomaszewska E, Baranowska-Wójcik E (2023) Anti-inflammatory, antioxidant, and neuroprotective effects of polyphenols—polyphenols as an element of diet therapy in depressive disorders. Int J Mol Sci 24(3):2258

Yadav MK, Kumari I, Singh B, Sharma KK, Tiwari SK (2022) Probiotics, prebiotics and synbiotics: safe options for next-generation therapeutics. Appl Microbiol Biotechnol 106(2):505–521

Yadav K, Rani A, Dua K (2024) Potential herbal remedies for treatment of depression: a mini review. Nat Prod J 14(6):1–13

Yoto A, Motoki M, Murao S, Yokogoshi H (2012) Effects of L-theanine or caffeine intake on changes in blood pressure under physical and psychological stresses. J Physiol Anthropol 31:1–9

Zhang Y, Li L, Zhang J (2020) Curcumin in antidepressant treatments: an overview of potential mechanisms, pre-clinical/clinical trials and ongoing challenges. Basic Clin Pharmacol Toxicol 127(4):243–253

Zhu J, Chen C, Lu L, Shikany JM, D'Alton ME, Kahe K (2023) Folate, vitamin B_6, and vitamin B_{12} status in association with metabolic syndrome incidence. JAMA Netw Open 6(1):e2250621

Zuo W, Wu J (2022) The interaction and pathogenesis between cognitive impairment and common cardiovascular diseases in the elderly. Ther Adv Chronic Dis 13:20406223211063020

Personalized Nutrition and Mental Health

7

Pelin Yenilmez Yeşildaş

Abstract

Personalized nutrition has a significant impact on mental health. Nutrition plans that are created by considering the genetic structure, environmental factors, and lifestyles of individuals can be an important tool in achieving psychological balance. Food affects not only the body but also the mental health, and proper nutrition can support emotional and psychological well-being. Therefore, personalized nutrition strategies are becoming increasingly important in mental health management. If the relationship between nutrition and mental health is better understood, treatment of mental disorders may be more possible. It is known that tailoring nutrition to individual needs has a positive impact not only on physical health but also on mental health. In this section of the book, research data that will provide a better understanding of the relationship between nutrition and mental health will be presented.

Keywords

Nutrition and mental health · Microbiome · Gut and brain axis · Probiotics · Prebiotics · Healthy diet · Vitamins · Comfort foods · Mindful eating

P. Yenilmez Yeşildaş (✉)
Gaziantep City Hospital, Neurology Clinic, Gaziantep, Turkey

W. Mohamed, A. Özge (eds.), *Feeding the Mind: The Connection Between Diet, Drugs, and Mental Health Volume 1*, Nutritional Neurosciences,
https://doi.org/10.1007/978-981-95-1414-4_7

7.1 Introduction

A healthy lifestyle includes physical health and mental health. Nutrition significantly affects mood, emotional balance, and general mental health. Mental disorders are health problems that affect society and can impair quality of life. Studies have shown that nutrition has serious effects on mental health and may be important in psychiatric illnesses.

7.2 What Is Mental Health?

It is the state of a person's emotional, psychological, and social well-being. It includes people's ability to cope with stress, think, feel, and behave healthily in the face of challenges. It is a fundamental factor for people to realize their potential, establish healthy relationships with their environment, and be satisfied with different aspects of their lives. Mental health is considered an interaction of biological, environmental, and psychological factors. Genetic factors, life experiences, social environment, and personal characteristics can affect a person's mental health (Adan et al. 2019). Emotional awareness and management of emotions are important for maintaining mental health. Establishing social support and healthy relationships, creating a supportive environment can strengthen mental health. When the relationship between physical health and mental health is examined, it is predicted that exercise and healthy nutrition improve mental health (Sarris et al. 2015).

Personalized nutrition aims to create customized diet plans based on individuals' current health status, genetic structure, biological characteristics (microbiome and body metabolism), environmental and behavioral characteristics. It aims to achieve effective and sustainable results by considering the needs and goals of each individual. Genetic structure and intestinal microbiome can determine which nutrients individuals will digest better, how nutrients are metabolized, and how they will benefit from them. A diet that is free of certain foods that some individuals are sensitive to may have positive effects on mental health (Whiteford et al. 2015).

The effect of nutrition on mental health begins with the knowledge of the components of the brain and the relationship between the brain and the intestines. The structure and neurohormonal functioning of the brain, which consists of glucose, lipids, amino acids, vitamins and minerals, intestinal hormones, neuropeptides, neurotransmitters, microbiota interaction via the gut-brain axis, the effects of stress and inflammation, and cognitive functions have effects on mental health. The relationship between nutrition and mental health led to the establishment of the International Association for Research in Nutritional Psychiatry. It emphasizes that mental health and nutrition research should be approached multidisciplinary (bringing together psychiatrists, neurologists, nutritionists, psychologists, epidemiologists, and experts from other relevant disciplines) (Granero 2022).

7.3 Mechanisms in Nutrition and Mental Health

1. Inflammation
2. Oxidation Stress
3. Brain Plasticity
4. Mitochondrial Functions
5. Microbiota Gut-Brain Axis

7.3.1 Inflammation

Chronic inflammation, caused by an increase in inflammatory cytokines and acute phase proteins, plays a role in the development of diseases such as depression, schizophrenia, and bipolar disorder. This inflammation may increase the risk of psychiatric disorders due to factors such as psychological stress, smoking, sleep disorders, obesity, and poor nutrition (Berk et al. 2013).

Results from observational studies suggest that diets containing high-fiber fruits and vegetables higher in polyunsaturated fatty acids are associated with lower levels of inflammatory markers. The Mediterranean diet has been observed to significantly improve inflammatory markers in many scientific studies (Schwingshackl and Hoffmann 2014).

7.3.2 Oxidative Stress

Oxidative and nitrosative stress are closely related to mental and psychological illnesses, as well as being associated with many chronic diseases. In schizophrenia patients, glutathione metabolism is impaired as glutathione levels decrease, resulting in increased oxidative stress. Similar results have been reported in individuals diagnosed with depression. Compared to healthy control groups, oxidative stress markers were found to be higher in depressed individuals, and antioxidants such as vitamin E, vitamin C, coenzyme Q10, and glutathione were found to be lower. This situation can be corrected through nutrition by consuming fruits and vegetables that contain plenty of antioxidants (Moylan et al. 2014).

7.3.3 Brain Plasticity

Brain plasticity is the ability of the brain to change, adapt, and reorganize structurally and functionally. It occurs when changes occur in the synapses of nerve cells due to factors such as environmental factors, traumas, diseases, experiences, and achievements. Factors affecting brain plasticity are genetics, age, physical and mental activity, stress and mental states, and nutrition. Brain plasticity is an important area of modern neuroscience research, because it allows the development of new approaches such as repairing brain damage and treating neurological diseases. Promising findings have been observed especially in studies on neurogenesis and synaptic plasticity (Fernandes et al. 2015).

Neurogenesis is the process by which neurons develop, mature, and localize to their appropriate locations in the body. This process begins early in embryonic development and may continue to a limited extent later in life. Neurogenesis is critical for normal function of the central nervous system (brain and spinal cord). In addition to brain-derived neurotrophic factor (BDNF), other neurotrophins are thought to mediate neurogenesis. There are few clinical studies on the effect of diet on this pathway, but the findings in the studies support a role for diet in improving BDNF levels. For example, consumption of carotenoid-rich fruits and vegetables (eight servings per day for 4 weeks) resulted in higher serum BDNF levels in patients with schizophrenia compared to controls. Another study in older adults showed an association between malnutrition and decreased hippocampal volume. Antioxidants and anti-inflammatory nutrients such as polyunsaturated fatty acids, polyphenols, l-theanine, and vitamin E may enhance neurogenesis, whereas energy-dense diets rich in saturated fatty acids and carbohydrates disrupt this process.

7.3.4 Mitochondrial Dysfunction

Mitochondrial energy production capacity and energy distribution impairment have been associated with mental disorders such as depression, bipolar disorder, and schizophrenia. This may be the result of decreased antioxidant capacity and proinflammatory cytokine-mediated increase in mitochondria-derived oxygen and nitrogen free radicals. Coenzyme Q10, alpha lipoic acid, carnitine, creatine, resveratrol, and N-acetyl cysteine positively affect mitochondrial respiratory function (Morris and Berk 2015).

7.3.5 Microbiota-Gut-Brain Axis

Microbiota is the name given to the community of microorganisms (bacteria, viruses, fungi, and other microorganisms) found in our bodies. These microorganisms are found in different parts of our bodies, especially in the intestines, skin, mouth, and respiratory tract. Microbiota lives in a symbiotic relationship with the organism and plays an important role in the organism (Fung et al. 2017).

Data from studies on the effect of gut microbiota on chronic diseases suggest that gut microbiota may influence psychic behaviors in more than one way. The human intestinal microbiota plays a role in human health by performing functions such as digestion of nutrients and indigestible polysaccharides, metabolism of drugs, metabolism of tryptophan, production of vitamins, synthesis of bioactive molecules, production of short-chain fatty acids (SCFAs), production of molecules such as BDNF, serotonin, and gamma-aminobutyric acid, and development of the immune system. Many of these molecules can affect brain function by regulating neural signals within the enteric nervous system. One of the most important factors affecting the development, function, and composition of the intestinal microbiota is diet. Different types of diets can cause dysbiosis, which can lead to altered intestinal microbiota and overgrowth of potential pathogens (Naveed et al. 2021).

Recent studies suggest that dysbiosis may play a role in the physiological mechanisms of psychiatric disorders. This communication network occurs through the enteric nervous system, autonomic nervous system (n.vagus), neurotransmitters, neuroendocrine system, and immune system. Since the onset and progression of psychiatric disorders are characterized by complex neuroendocrine and immunological changes, it is thought that the gut microbiome may continuously affect the "microbiome-gut-brain axis." Studies provide clinical evidence for the role of the microbiome and gut flora in psychiatric disorders ranging from mood disorders such as autism, schizophrenia, attention deficit disorder, bipolar disorder, anxiety, stress, and depression to neuroinflammatory-related disorders such as Alzheimer's and Parkinson's disease. In such diseases, changes in the intestinal microbiota may cause microtraumas in the intestinal epithelial wall and increase intestinal epithelial permeability.

As a result, harmful substances produced by microorganisms may enter the systemic circulation. Pathogenic antigens entering the systemic circulation and impaired intestinal permeability may cause bacterial lipopolysaccharides to enter the systemic circulation, which may stimulate the immune system and ultimately lead to chronic inflammation. Imbalance in the intestinal flora and increased intestinal permeability may cause bacterial cell wall components such as lipopolysaccharides to enter the systemic circulation, which may stimulate the immune system and ultimately lead to chronic inflammation. It is thought that this inflammation and pathogenic antigens moving towards the brain via the gut-brain axis may cause behavioral, cognitive, and psychiatric changes. However, it has not yet been proven whether these changes play a causal role in these disorders. Microbiota plays a role in the development of various psychiatric disorders through neurobiological pathways such as modulation of BDNF, serotonin neurotransmission, and hypothalamic-pituitary-adrenal axis-mediated stress response. Differences in fecal microbiota have been reported in depressed patients, with reduced gut microbiota richness and diversity compared to healthy controls. Transplantation of microbes from depressed patients into rodents has been shown to induce depression-related behaviors. (Fetissov and Déchelotte 2011)

7.3.5.1 Prebiotics–Psychobiotics

Considering the limitations, side effects, and treatment resistance rates of current treatments for psychiatric disorders, dietary characteristics that affect the gut microbiota have the potential to develop new treatment methods. Many studies have shown that dietary interventions and prebiotic/probiotic supplements can significantly alter the composition of the gut microbiota and positively affect the gut-brain axis. Therefore, they may offer new approaches to the treatment of various psychiatric disorders. The World Health Organization (WHO) defines probiotics as live microorganisms that positively affect intestinal and human health when taken in sufficient amounts. Probiotics that have positive effects on psychiatric disorders by increasing the production of important neurotransmitters such as gamma amino butyric acid (GABA), serotonin, acetylcholine, norepinephrine, dopamine and by affecting the gut-brain axis that are called "psychobiotics."

Psychobiotics are being investigated as adjuvant and/or monotherapy aiming to improve the treatment of neurodevelopmental, psychiatric, and neurodegenerative disorders. Consumption of foods containing probiotic bacteria (e.g., pickles, yogurt fermented from goat milk, kefir, cheese, boza made from corn, natural vinegar, etc.) is recommended, and if not sufficient, ingestion of commercially available probiotic preparations (e.g., Bifidobacterium sp. and Lactobacillus sp.) is recommended together with the diet.

However, caution should be exercised when administering probiotic-rich foods to individuals who have been malnourished for a long time and do not have normal intestinal flora. These individuals are at risk because they can transfer pathogenic toxins of the intestinal microbiota into the bloodstream when they suddenly consume high amounts of probiotics, and therefore, some symptoms in psychiatric disorders can worsen. However, it is thought that these symptoms are temporary after the toxins are eliminated from the body, and in order to prevent this situation, the consumption of probiotics or probiotic-containing foods should be started in small amounts and then slowly increased. Prebiotics are non-digestible food components that can positively affect the health of the host by promoting the proliferation and selective activation of microorganisms such as Bifidobacterium and Lactobacillus in the microbiota. Examples such as inulin, fructooligosaccharide (FOS), and galactooligosaccharide (GOS) are commonly used prebiotics. In addition, these combination components obtained when used together with probiotics are called "symbiotics" and are available on the market as commercial products. Many studies show that these functional foods can improve intestinal flora and positively affect many psychiatric disorders (Bravo et al. 2011).

7.4 Can Foods We Prefer Be a Mirror of Our Mental Health?

Recent studies have shown that diet has a significant impact on mood, thinking style, and overall mental health. The foods we eat can directly affect the microstructure and functions of our brain, which in turn is reflected in our mood and mental health (Adan et al. 2019).

7.4.1 The Relationship Between Nutrition and Brain Health

Brain functioning depends not only on the health of nerve cells but also on the chemical balance of these cells. This chemical balance has a direct impact on individuals' mood, thought processes, and emotional states. Nutrients that meet the brain's energy needs, such as glucose, omega-3 fatty acids, B vitamins, vitamin D, and magnesium, are strongly associated with mental health (Whiteford et al. 2015).

7.4.2 The Effect of Processed Foods on Mood

Processed foods, foods high in sugar, trans fats, and additives, can negatively impact brain chemistry. These types of foods can lead to inflammation and oxidative stress, disrupt gut health, and increase the release of stress-related hormones (such as cortisol). Excessive consumption of processed foods has been linked to mental health issues such as depression, anxiety, stress, and attention deficit.

7.4.3 The Importance of the Gut-Brain Axis

Research in recent years has more clearly revealed the effects of gut health on mental health. The gut has a microorganism community (microbiome) that directly affects brain functioning. It plays a major role in the production of neurotransmitters that affect mood (such as serotonin). Probiotics and prebiotics that improve gut health have positive effects in the treatment of mental health problems such as depression and anxiety (Rea et al. 2020).

7.4.4 Omega-3 Fatty Acids and Depression

Omega-3 fatty acids play an important role in combating mental disorders, especially depression and anxiety. Nutrients such as fish oil and flaxseed strengthen the membrane structure of brain cells and help neurotransmitters function healthily. Deficiency of omega-3 fatty acids has been linked to mental illnesses such as depression. These fatty acids also have anti-inflammatory and neurological health-supporting effects (Grosso et al. 2014).

7.4.5 The Role of Vitamins and Minerals in Mental Health

B vitamins (especially B12 and folic acid) are critical for maintaining mental function. Inadequate B12 levels have been linked to conditions such as depression, fatigue, and poor memory. Similarly, vitamin D deficiencies can increase depressive symptoms. Vitamin D is essential for healthy brain function, and the risk of depression may increase especially during winter months when sunlight is limited. Magnesium also has effects that reduce mental fatigue and anxiety (Yilmaz et al. 2022).

7.4.6 The Effect of Carbohydrates: Complex Carbohydrates and Mood

Carbohydrates are an important source of energy for the body. However, sugary and refined carbohydrates (e.g., white bread, cake) can cause blood sugar to rise quickly, which can lead to mood swings. Complex carbohydrates (e.g., whole grains, legumes, vegetables) help us maintain a more stable mood by keeping blood sugar more balanced. In addition, these types of carbohydrates can help produce serotonin, which supports brain function (Marx et al. 2017).

7.4.7 Protein and Amino Acids: Essential for Mental Health

The brain requires amino acids to produce neurotransmitters. Chemicals that regulate our mood, such as serotonin, dopamine, and norepinephrine, are synthesized with amino acids obtained from proteins. Adequate protein intake plays an important role in combating conditions such as depression, anxiety, and attention deficit. The amino acid tryptophan, in particular, has a mood-enhancing effect by increasing the production of serotonin (Marx et al. 2017).

7.4.8 The Relationship Between Eating Habits and Emotional Health

Emotional eating is the behavior of overeating to cope with situations such as stress, anxiety, or depression. High-calorie, fatty, and sugary foods may provide comfort in the short term, but in the long term, such habits can lead to mental disorders. Practices such as eating slowly and mindful eating are useful for maintaining mental balance (Ludwig and Kabat-Zinn 2008).

7.4.9 Recommendations

Our eating habits greatly affect not only our physical health but also our mental health. A healthy and balanced diet supports healthy brain function, balances our mood, and can help prevent mental health problems. A balanced diet should include nutrients such as omega-3 fatty acids, B vitamins, vitamin D, magnesium, proteins, and complex carbohydrates. In addition, avoiding processed foods, excess sugar and fat, and turning to probiotic and prebiotic foods that improve intestinal health can improve mental health.

Mental health is a complex condition where not only genetic, environmental, and psychological factors interact but also biological factors such as nutrition. Understanding the relationship between nutrition and mental health is one of the keys to living a healthy life.

7.5 Stress-Reducing Foods, the Term Comfort Food

The term comfort food describes foods that are preferred for the purpose of overcoming an emotional need or stress and are usually part of individuals' childhood, cultural backgrounds, or specific emotional experiences. Comfort food is a psychological phenomenon and refers to foods that individuals consume to feel better during difficult times. People may turn to such foods to cope with negative emotions such as stress, anxiety, sadness, loneliness, or unhappiness. Although this provides temporary relief, it can lead to psychological problems in the long term. Comfort food increases the neurotransmitters in our brain (especially serotonin and dopamine). These neurotransmitters trigger positive emotions that make us feel good. It has been found that especially sugary, fatty, and salty foods have an effect on the brain by increasing dopamine secretion (Dubé et al. 2005).

7.5.1 The Relationship Between Nutrition and Stress

Stress can be defined as the body's biological response to an environmental or endogenous threat. Stress has both physical and psychological effects, and long-term high stress can lead to health problems such as anxiety, depression, and cardiovascular disease. However, some nutrients can mitigate the effects of stress and increase the body's ability to cope with stress.

Nutrients can affect the body's stress-coping mechanisms. Nutrients have direct effects on brain chemistry, as brain function is largely dependent on the nutrients we consume. In this context, some nutrients have the potential to modulate stress responses in the body, provide a calming effect, and improve mood (Babicz-Zielińska et al. 2006).

7.5.2 Stress-Reducing Foods

7.5.2.1 Omega-3 Fatty Acids

Omega-3 fatty acids (especially eicosapentaenoic acid (EPA) and docosahexaenoic acid (DHA)) can help reduce inflammation in the brain, which increases the ability to cope with stress. Studies show that omega-3 fatty acids have positive effects on relieving symptoms of anxiety disorders and depression. Microbiota is the name given to the community of trillions of microorganisms (bacteria, viruses, fungi, and other microbes) in our bodies. These microorganisms are densely located in different parts of our bodies, especially in the intestines, skin, mouth, and respiratory tract. Microbiota lives with us in a symbiotic (mutually beneficial) relationship and plays an important role in our health. Salmon, sardines, walnuts, flaxseed, and chia seeds contain high amounts of omega-3 (Grosso et al. 2014).

7.5.2.2 Magnesium

Magnesium is an important mineral for proper functioning of the nervous system and plays a role in stress-coping mechanisms. Magnesium deficiencies have been linked to anxiety, stress, and depression. Green leafy vegetables, nuts, and whole grains contain significant amounts of magnesium (Ryan 1991).

7.5.2.3 B Vitamins (B6 and B12)

B vitamins, B6 (pyridoxine) and B12, play an important role in nervous system health. Deficiency of these vitamins increases stress-related emotional disorders such as depression and anxiety. Foods rich in B vitamins include eggs, dairy products, whole grains, meat, and fish (Yilmaz et al. 2022).

7.5.2.4 Serotonin and Tryptophan

Serotonin is a neurotransmitter that regulates mood and modulates stress responses. Tryptophan is an amino acid that is necessary for the production of serotonin in the body. Consuming foods rich in tryptophan can increase serotonin levels, which can help reduce stress. Turkey, milk, yogurt, bananas, and oats also contain significant amounts of serotonin (Haq et al. 2021).

7.5.2.5 Antioxidants (Vitamin C and Vitamin E)

Antioxidants, such as vitamin C and vitamin E, help reduce oxidative stress in the body. Oxidative stress occurs when free radicals accumulate in the body, and this can increase the negative effects of stress. Citrus fruits, strawberries, broccoli, kiwi, green tea, green and red peppers contain high amounts of vitamin C.

Hazelnuts, almonds, walnuts, sunflower seeds, avocados, tomatoes, potatoes, green leafy vegetables contain vitamin E (Yilmaz et al. 2022).

7.5.2.6 Dark Chocolate

Dark chocolate can improve mood by increasing the secretion of serotonin and endorphins in the body. Studies show that dark chocolate containing 70% or more cocoa helps reduce stress hormones (cortisol) and provides relaxation (Haq et al. 2021).

Comfort food refers to foods that individuals consume to achieve emotional relief. From a psychological and biological perspective, these foods can affect brain chemistry, creating feelings of relaxation and happiness. However, when taken to extremes, emotional eating can lead to unhealthy habits.

Stress-reducing foods are foods that are generally rich in nutrients and chemical compounds that help the body cope with stress. These foods can regulate the brain and body's stress responses, improve mood, and support overall health.

7.6 Diets and Their Characteristics

7.6.1 Gluten-Free and Casein-Free Diet

A gluten-free diet refers to a nutritional plan that excludes gluten-containing grains such as wheat, rye, barley, and oats, while a casein-free diet represents a diet in which milk and dairy products are removed from the nutritional program. It has been claimed that a gluten-free and casein-free diet helps improve various symptoms of autism, including social interaction problems, cognitive functions, communication skills, stereotypical behaviors, attention deficit, and emotional problems.

There are several theories about the mechanism by which a gluten- and casein-free diet improves autism symptoms, but one of the most widely accepted theories is the opioid excess hypothesis. According to this hypothesis, children with autism spectrum disorder (ASD) are thought to be deficient in the dipeptidyl-peptidase 4 (DPP-4) enzyme required for the degradation of gluten and casein proteins. Due to this deficiency, the metabolism of these two proteins is impaired. The risk of abnormally metabolized gluten and casein molecules passing into the bloodstream and central nervous system increases in a situation where intestinal permeability increases. These neuroactive peptides have opioid activity and can cause excessive opioid activity in the central nervous system (Quan et al. 2022).

It is theorized that some symptoms of autism (e.g., stereotypic behaviors, obsessive thoughts, hyperactivity, speech and language development problems, increased pain sensitivity, lack of eye contact) may be a result of opioid peptides produced by the ingestion of foods containing gluten and casein. The aim of the diet is to reduce levels of these opioid peptides and alleviate these symptoms.

Celiac disease is a multisystem immune-mediated disease triggered by the consumption of gluten-containing foods in genetically predisposed individuals. This disease usually presents with gastrointestinal symptoms but is sometimes associated with neurological and psychiatric disorders. It is known that there is a relationship between celiac disease and schizophrenia. Studies have observed higher levels of gliadin and beta-lactoglobulin antibodies in patients diagnosed with schizophrenia, and a link between gluten intake and schizophrenia symptoms has been suggested. Therefore, it is supported by many studies that these disorders can be controlled by removing gluten from the diet and that a gluten-free diet can positively affect the intestinal microbiota (Ly et al. 2017).

7.6.2 Ketogenic Diet

In some disease states, neurons cannot use glucose sufficiently as an energy source. Diseases in which this situation is seen include resistant epilepsy, Alzheimer's and Parkinson's diseases, Down Syndrome, Multiple Sclerosis, and psychiatric diseases such as autism. Neurons can also use ketone bodies in addition to glucose to meet their energy needs. For this reason, a ketogenic diet is recommended in such diseases. There are scientific studies showing that the ketogenic diet may be a new treatment approach by re-regulating brain energy metabolism in psychiatric diseases such as schizophrenia, mood disorders, and anxiety disorders. The ketogenic diet is a type of diet that is high in fat and low in carbohydrates, triggering gluconeogenesis and ketogenesis by creating a catabolic state in the body. This diet is one of the oldest methods used in the treatment of epilepsy patients, and its mechanism of action is still not fully understood. However, low carbohydrate intake in the diet puts the body in a state called ketosis, which leads to the production of ketone bodies. Ketosis may have a positive effect in reducing the frequency of epileptic seizures. Ketone bodies can increase cerebral blood flow in the brain, increase GABA levels, and reduce free radical formation. They also have anti-inflammatory effects, reduce the risk of cerebral edema, and improve neuronal function. The main purpose of the ketogenic diet is to obtain energy from fats. Therefore, it is high in fat (90%), low in protein (7%), and very low in carbohydrates (3%). The ketogenic diet includes foods rich in fats, including cream, butter, mayonnaise, vegetable oils, cheese, and eggs. Meat, fish, nuts, fruits, and vegetables can also be included in this diet (Brietzke et al. 2018).

7.6.3 Supplement-Supported Diet

Studies have shown that psychiatric disorders are associated with low vitamin D levels. Vitamin D is a lipid-soluble vitamin that is important for the body and has a number of biological functions. These functions include regulating calcium–phosphorus balance and bone metabolism, antioxidant defense system, supporting anti-inflammatory processes, and regulating the immune system. Vitamin D is also thought to improve mood by modulating the synthesis of neurotransmitters and neurotrophic factors. Vitamin D can reduce inflammatory substances, correct nitric oxide release, and increase the synthesis of monoamines such as dopamine. It can also affect DNA repair mechanisms and detoxification and antioxidant systems that remove heavy metals. Vitamin D deficiency may play a role in regulating behavior by affecting the development of brain cells. Therefore, vitamin D deficiency can negatively affect brain functions. Vitamin D deficiency has been associated with a number of psychiatric diseases such as multiple sclerosis, Parkinson's disease, Alzheimer's disease, epilepsy, autism spectrum disorder, psychotic disorders, depression, and anxiety disorders. There are also studies that have linked B vitamins and folic acid to depression, anxiety disorders, and psychotic disorders. Deficiencies of B12, folic acid, and other B vitamins, iron, magnesium, and zinc have also been linked, individually or together, to psychiatric disorders such as depression, anxiety, attention deficit hyperactivity disorder (ADHD), dementia, and Parkinson's disease. Therefore, adequate intake of these vitamins and minerals is important for mental health (Yilmaz et al. 2022; Ryan 1991).

7.7 A New Approach to Nutrition: Mindful Eating

In recent years, mindful eating has attracted attention as a recommended approach for developing healthy eating habits and coping with emotional eating. Mindful eating is derived from the concept of mindfulness, a broader awareness practice. Awareness means that a person fully experiences the moment, notices and accepts their thoughts, feelings and bodily sensations without judgment. Mindful eating aims to integrate this type of awareness into the eating process. It encompasses the entire experience of food's taste, smell, texture, appearance, and slow eating. The person tries to avoid any impulsive or automatic behaviors related to food and instead observes their relationship with food more consciously (Ludwig and Kabat-Zinn 2008).

7.7.1 Basic Principles of Mindful Eating

1. Being in the Present Moment: This refers to being aware of what one is doing at the moment. It is important to carefully monitor food, eating, and bodily sensations (hunger, fullness).
2. Eating Slowly: This means doing the act of eating consciously, without rushing, and savoring every bite. This makes it easier for the brain to receive satiety signals and can reduce the risk of overeating.
3. Being Aware of Emotional and Physical Responses to Food: This includes being aware of not only physical but also emotional and mental reactions to food. When a person eats, they observe feelings of hunger and fullness, as well as the emotions that trigger the meal.
4. Accepting Without Judgment: Non-judgment is an important principle of mindful eating, and it refers to accepting one's feelings and reactions to food without evaluating them. This can help develop a healthier relationship by eliminating negative thoughts (e.g., guilt, shame) about food.
5. Listening to Bodily Feelings: Physical hunger and fullness signals are carefully monitored. Biological hunger signals that guide eating habits are monitored independently of mood, which can reduce the risk of eating disorders (Grossman et al. 2004).

7.7.2 Benefits of Mindful Eating

1. Weight Control: Mindful eating can be an effective method for weight control. Eating slowly and paying attention to your body's signals of fullness can reduce the risk of overeating and consuming unhealthy snacks. According to a recent study, mindful eating can help regulate body weight and reduce overeating behaviors (Greeson et al. 2024).
2. Coping with Emotional Eating: Mindful eating can be effective in coping with emotional eating. People often eat due to emotional states such as stress, depres-

sion, and loneliness. Kristeller and Wolever (2011) have shown that mindful eating has the potential to reduce emotional eating behaviors and improve individuals' relationships with food (Kristeller and Wolever 2011).
3. Digestive Health: Mindful eating can improve the digestive process. Eating slowly, chewing food better, and following the natural rhythm of the digestive system can reduce digestive problems. Eating slowly can also relieve feelings of bloating and discomfort after eating.
4. Mental and Spiritual Health: Mindful eating can have positive effects on overall mood and mental health. Eating slowly and observing emotional states can reduce stress and anxiety. Hölzel et al. (2011) have shown that practicing mindfulness can change brain structure, increase the capacity to cope with stress, and improve emotional balance (Hölzel et al. 2011).
5. Improve Eating Relationships: Mindful eating encourages a healthier relationship with food. Instead of using food as a reward or punishment, viewing food as a source of nourishment and pleasure can alleviate negative emotional burdens associated with eating.

7.7.3 Mindful Eating Application Methods

1. Eat Slowly: Eating slowly and mindfully means fully noticing your food. Savor every bite, feel what you are eating completely. Follow your own hunger and fullness signals.
2. Observe Your Food: Carefully observe the taste, smell, texture, and color of your food. Explore your attitude towards food more deeply.
3. Notice Your Feelings: Notice what emotions you feel while eating. Are you eating out of stress, boredom, or stress? This awareness is important for coping with emotional eating.
4. Avoid External Stimuli: Watching TV, using your phone, or eating in front of the computer automates the act of eating. Avoid environmental stimuli while eating and focus only on what you are eating.
5. Listen to Your Body: Pay attention to your body's hunger and fullness signals during your meal. Noticing exactly when you are eating and when you need to stop can help you create a healthier eating pattern (Ludwig and Kabat-Zinn 2008).

7.7.4 Conclusion

Mindful eating is an approach that aims to make eating behavior healthier and improve individuals' relationship with food. Directly related to psychological, biological, and emotional processes, mindful eating is based on the principles of eating slowly, developing awareness, and accepting without judgment. Research shows that mindful eating has many benefits, including weight control, digestive health, coping with emotional eating, and improving overall mental health. This approach can help individuals develop healthy eating habits and reshape their relationships with food.

7.8 What to Do for Better Mental Health?

The topic of what to do for better mental health is a topic that has been extensively researched in the fields of psychology, neuroscience, and general health and offers various approaches and strategies for improving the mental health of individuals. Mental health includes the emotional, psychological, and social well-being of individuals. Good mental health affects the capacity of individuals to cope with life's challenges, maintain emotional balance, build relationships, and fulfill their potential.

Below are various strategies and approaches suggested for better mental health based on academic literature:

7.8.1 Achieving Psychological and Emotional Balance

Cognitive Behavioral Approach: Cognitive behavioral therapy (CBT) is an effective approach to changing negative thoughts and behaviors. CBT helps individuals recognize and question negative thoughts and develop healthier ways of thinking. Studies show that CBT is effective on mental health problems such as anxiety, depression, and stress (Beck and Verticchio 2014).

Emotional Awareness and Emotional Regulation: Emotional awareness is the process by which individuals recognize and understand their emotions. Emotional regulation involves coping with stressful situations and managing negative emotions in healthy ways. Mindfulness practices play an important role in emotional awareness and regulation. Mindfulness practice can help people notice their current emotional state without judgment, which contributes to better mental health.

7.8.2 Social Connections and Support

Social Support: Social support has a major impact on mental health. Research shows that people with strong social connections are better at coping with stress and have a lower risk of developing mental health problems such as depression (Cohen and Wills 1985). Relationships with family members, friends, and the community have a direct impact on emotional well-being.

Social isolation can increase mental health problems such as depression, anxiety, and stress. Loneliness, weakened social ties, and lack of support systems can negatively impact mental health (Hawkley and Cacioppo 2010).

Empathy and Connection: Empathic relationships involve individuals respecting and understanding the emotional experiences of others. Empathy stands out as a factor that improves mental health in both personal relationships and social connections.

7.8.3 Physical Activity and Exercise

Exercise and Mental Health: Regular physical activity has a significant impact on mental health. Exercise positively affects brain chemistry, increasing the release of neurotransmitters such as endorphins and serotonin, which can improve mood. Exercise can also help reduce anxiety, depression, and stress.

Aerobic exercise is particularly effective in relieving depression and anxiety. Exercise relaxes the nervous system, improves sleep quality, and increases overall energy levels (Richardson et al. 2005).

7.8.4 Healthy Nutrition and Mental Health

Balanced and Nutritious Diet: Nutrition can directly affect mood and overall mental health. Nutrients such as omega-3 fatty acids, B vitamins, magnesium, and vitamin C can have positive effects on mental health. Studies show that a diet based on natural foods can reduce depression and anxiety. Dark chocolate can improve mood by increasing serotonin and dopamine levels. In addition, regular eating habits and a healthy body weight can positively affect mental health.

Inadequate Nutrition and Mental Health: Inadequate nutrition is associated with conditions such as depression, anxiety, and stress. In particular, low vitamin D levels, high processed food intake, and sugary foods can trigger mental health problems.

7.8.5 Sleep Patterns and Mental Health

Adequate Sleep: Adequate sleep is crucial for mental health. Sleep deprivation can increase anxiety, depression, and overall stress levels. Sleep is a "restoration" process for the brain and body. Inadequate sleep negatively impacts mental function, makes emotional regulation difficult, and disrupts overall mood.

Healthy Sleep Habits: A regular sleep routine, a relaxing environment, and avoiding screens before bed can improve sleep quality. Additionally, light exercise and practicing relaxation techniques (such as deep breathing or meditation) before bed can contribute to better sleep (Lallukka et al. 2017).

7.8.6 Stress Management Techniques

Stress Management Strategies: Stress management is a critical skill for improving mental health. Techniques such as meditation, deep breathing exercises, yoga, and mindfulness can increase the body's ability to cope with stress. Research shows that mindfulness and meditation are effective in reducing stress levels.

Time Management: Good time management can help balance workload and contribute to lower stress levels. Strategies such as using time effectively and setting priorities can help manage personal stress levels (Goyal et al. 2014).

7.8.7 Self-Compassion and Self-Care

Self-Compassion: Self-compassion is the gentle and understanding acceptance of oneself. A person supports oneself when one makes mistakes or faces difficulties. Self-compassion has been found to help reduce depression and anxiety and promote a healthier state of mind.

Self-Care: Taking time for yourself regularly and engaging in relaxing activities (e.g., engaging in hobbies, taking walks in nature) can improve mental health. Self-care is a critical strategy for maintaining one's psychological well-being (Neff 2003).

Better mental health requires a multifaceted approach. This includes a combination of physical health, psychological well-being, social support, a balanced diet, adequate sleep, and effective stress management strategies. All of these elements can improve an individual's quality of life and help them achieve a healthier state of mind. More specific interventions, such as psychotherapy and professional support, may also be necessary. Mental health is a dynamic process that requires ongoing care and attention, and individuals can achieve better states of mind by making conscious and sustained efforts in these areas.

7.9 Conclusion

The relationship between nutrition and mental health is complex and involves a wide range of factors. However, with increasing research in this area, we can better understand the effects of nutrition on mental health and integrate this information into clinical practice. This can contribute to the development of more effective and holistic approaches to the treatment of psychiatric disorders and can be an important step towards improving the overall mental health of society.

In the future, these connections between nutrition and mental health need to be investigated further. The number of clinical studies should be increased and studies with larger sample groups will help us better understand the effects of nutrition on the treatment of psychiatric disorders. In addition, it is important to consider individual differences and develop personalized nutrition plans. In conclusion, this relationship between nutrition and psychiatric health represents an important area for the management and treatment of psychiatric disorders. Future research will help us better understand the effects of nutrition on psychiatric health and integrate this information into clinical practice, contributing to a better quality of life for psychiatric patients.

References

Adan RAH, Van Der Beek EM, Buitelaar JK, Cryan JF, Hebebrand J, Higgs S et al (2019) Nutritional psychiatry: towards improving mental health by what you eat. Eur Neuropsychopharmacol 29(12):1321–1332. https://doi.org/10.1016/j.eu-roneuro.2019.10.011

Babicz-Zielińska E, Rybowska A, Zabrocki R (2006) Relations between emotions and food preferences. Pol J Food Nutr Sci 15(56):163–165

Beck AR, Verticchio H (2014) Counseling and mindfulness practice with graduate students in communication sciences and disorders. Contemp Issue Commun Sci Disord 41:133–148. Fall 2014 © NSSLHA 1092-5171/14/41020133

Berk M, Williams LJ, Jacka FN, O'neil A, Pasco JA, Moylan S et al (2013) So depression is an inflammatory disease, but where does the inflammation come from? BMC Med 11:200. https://doi.org/10.1186/1741-7015-11-200

Ly V, Bottelier M, Hoekstra PI, Arias Vasquez A, Buitelaar JK, Rommelse NN (2017) Elimination diets' efficacy and mechanisms in attention deficit hyperactivity disorder and autism spectrum disorder. Eur Child Adolesc Psychiatry 26(9):1067–1079. https://doi.org/10.1007/s00787-017-0959-1

Bravo JA, Forsythe P, Chew MV, Escaravage E, Savignac HM, Dinan TG et al (2011) Ingestion of lactobacillus strain regulates emotional behavior and central GABA receptor expression in a mouse via the vagus nerve. Proc Natl Acad Sci USA 108(38):16050–16055. https://doi.org/10.1073/pnas.1102999108

Brietzke E, Mansur RB, Subramaniapillai M, Balanzá-Martínez V, Vinberg M, González-Pinto A et al (2018) Ketogenic diet as a metabolic therapy for mood disorders: evidence and developments. Neurosci Biobehav Rev 94:11–16. https://doi.org/10.1016/j.neubiorev.2018.07.020

Cohen S, Wills TA (1985) Stress, social support, and the buffering hypothesis. Psychol Bull 98(2):310–357

Dubé L, LeBel LJ, Lu J (2005) Affect asymmetry and comfort food consumption. Physiol Behav 86(4):559–567

Fernandes BS, Molendijk ML, Köhler CA, Soares JC, Leite CM, Machadovieira R et al (2015) Peripheral brain-derived neurotrophic factor (BDnF) as a bio-marker in bipolar disorder: a meta-analysis of 52 studies. BMC Med 13:289. https://doi.org/10.1186/s12916-015-0529-7

Fetissov SO, Déchelotte P (2011) The new link between gut-brain axis and neuropsychiatric disorders. Curr Opin Clin Nutr Metab Care 14(5):477–482. https://doi.org/10.1097/MCO.0b013e32834936e7

Fung TC, Olson CA, Hsiao EY (2017) Interactions between the microbiota, immune and nervous systems in health and disease. Nat Neurosci 20(2):145–155. https://doi.org/10.1038/nn.4476

Goyal M et al (2014) Meditation programs for psychological stress and well-being: a systematic review and meta-analysis. JAMA Intern Med 174(3):357–368

Granero R (2022) Role of nutrition and diet on healthy mental state. Nutrients 14(4):750. https://doi.org/10.3390/nu14040750

Greeson JM, McBride EE, Chin GR, Lee HH, Colangelo AP (2024) Trait mindfulness and mind-body health in students: the role of gender, race, and ethnicity. J Am Coll Health 72(8):2844–2855. Received 10 Sep 2020, Accepted 07 Oct 2022, Published online: 08 Nov 2022

Grossman P, Niemann L, Schmidt S, Walach H (2004) Mindfulness-based stress reduction and health benefits. A meta-analysis. J Psychosom Res 57:35–43

Grosso G, Galvano F, Marventano S, Malaguarnera M, Bucolo C, Drago F et al (2014) Omega-3 fatty acids and depression: scientific evidence and biological mechanisms. Oxidative Med Cell Longev 2014:313570. https://doi.org/10.1155/2014/313570

Haq S, Grondin JA, Khan WI (2021) Tryptophan-derived serotonin-kynurenine balance in immune activation and intestinal inflammation. FASEB J 35(10):e21888

Hawkley LC, Cacioppo JT (2010) Loneliness matters: a theoretical and empirical review of consequences and mechanisms. Ann Behav Med 40(2):218–227

Hölzel BK, Carmody J, Vangel M, Congleton C, Yerramsetti SM, Gard T, Lazar SW (2011) Mindfulness practice leads to increases in regional brain gray matter density. Psychiatry Res 191(1):36–43

Kristeller JL, Wolever RQ (2011) Mindfulness-based eating awareness training for treating binge eating disorder: the conceptual foundation. Eat Disord 19(1):49–61

Lallukka T et al (2017) Sleep improvement by internet-based cognitive behavioral therapy might help prevent mental health disorders. Lancet Psychiatry 4(10):734–735

Ludwig DS, Kabat-Zinn J (2008) Mindfulness in medicine. JAMA 300:1350–1352

Marx W, Moseley G, Berk M, Jacka F (2017) Nutritional psychiatry: the present state of the evidence. Proc Nutr Soc 76(4):427–436. https://doi.org/10.1017/S0029665117002026

Morris G, Berk M (2015) The many roads to mitochondrial dysfunction in neuroimmune and neuropsychiatric disorders. BMC Med 13:68. https://doi.org/10.1186/s12916-015-0310-y

Moylan S, Berk M, Dean OM, Samuni Y, Williams LJ, O'neil A et al (2014) Oxidative & nitrosative stress in depression: why so much stress? Neurosci Biobehav Rev 45:46–62. https://doi.org/10.1016/j.neubiorev.2014.05.007

Naveed M, Zhou QG, Xu C, Taleb A, Meng F, Ahmed B et al (2021) Gut-brain axis: a matter of concern in neuropsychiatric disorders. Prog Neuropsychopharmacol Biol Psychiatry 104:110051. https://doi.org/10.1016/j.pnpbp.2020.110051

Neff K (2003) Self-compassion: an alternative conceptualization of a healthy attitude toward oneself. Self Identity 2(2):85–101

Quan L, Xu X, Cui Y, Han H, Hendren RL, Zhao L, You X (2022) A systematic review and meta-analysis of the benefits of a gluten-free diet and/or casein-free diet for children with autism spectrum disorder. Nutr Rev 80(5):1237–1246. https://doi.org/10.1093/nutrit/nuab073

Rea K, Dinan TG, Cryan JF (2020) Gut microbiota: a perspective for psychiatrists. Neuropsychobiology 79(1):50–62. https://doi.org/10.1159/000504495

Richardson CR, Faulkner G, McDevitt J et al (2005) Integrating physical activity into mental health services for persons with serious mental illness. Psychiatr Serv 56:324–331

Ryan MF (1991) The role of magnesium in clinical biochemistry: an overview. Ann Clin Biochem 28:19–26. https://doi.org/10.1177/000456329102800103

Sarris J, Logan AC, Akbaraly TN, Paul Amminger G, Balanzá-Martínez V, Freeman MP et al (2015) International Society for Nutritional Psychiatry Research consensus position statement: nutritional medicine in modern psychiatry. World Psychiatry 14(3):370–371. https://doi.org/10.1002/wps.20223

Schwingshackl L, Hoffmann G (2014) Mediterranean dietary pattern, inflammation and endothelial function: a systematic review and meta-analysis of intervention trials. Nutr Metab Cardiovasc Dis 24(9):929–939. https://doi.org/10.1016/j.numecd.2014.03.003

Whiteford HA, Ferrari AJ, Degenhardt L, Feigin V, Vos T (2015) The global burden of mental, neurological and substance use disorders: an analysis from the Global Burden of Disease Study 2010. PLoS One 10(2):e0116820. https://doi.org/10.1371/journal.pone.0116820

Yilmaz Y, Erdogan A, Hocaoglu C (2022) The role of vitamins in psychiatry: a review. Troia Med J 3(1):1–9

Nutrition and Mental Health Across the Lifespan

8

Hanife Karakaya

Abstract

Mental health, influenced by diet, genetics, and lifestyle, is essential to overall well-being. Diet significantly impacts mental health by affecting neurotransmitter synthesis, neuronal plasticity, and function. Nutrient deficiencies such as omega-3, B vitamins, magnesium, and zinc are linked to depression, anxiety, and cognitive deficits. During the prenatal period, adequate maternal folate, omega-3, and vitamin D intake supports fetal brain development and reduces neurodevelopmental disorders. Breast milk is crucial for the first 6 months in infancy and early childhood, with deficiencies in iron, zinc, and DHA leading to cognitive and behavioral issues. Essential nutrients like iron, zinc, selenium, and vitamins B12, B6, and D are vital for neuronal function and development. Adolescence, marked by rapid brain maturation and increased energy needs, can suffer from poor dietary choices that cause inflammation associated with depression and anxiety. Deficiencies in zinc, magnesium, and tryptophan during this stage contribute to mood swings and fatigue, while Mediterranean diets lower depression risk, contrasting with Western diets that may worsen mental health problems. In adulthood, diet influences stress management and cognitive performance, with anti-inflammatory diets rich in omega-3, polyphenols, and antioxidants enhancing mental health. Gut microbiota health, supported by probiotics and prebiotics, is linked to mood regulation, and adequate zinc, magnesium, and B vitamins prevent cognitive decline and mood disorders. For older adults, diet plays a role in preventing cognitive decline and dementia, with the Mediterranean diet associated with better cognitive health. Monitoring and addressing micronutrient deficiencies by supplementing omega-3 and B vitamins can improve mood and

H. Karakaya (✉)
Usak Research and Training Hospital, Usak, Turkey
e-mail: hanifekyildiz@yahoo.com

W. Mohamed, A. Özge (eds.), *Feeding the Mind: The Connection Between Diet, Drugs, and Mental Health Volume 1*, Nutritional Neurosciences,
https://doi.org/10.1007/978-981-95-1414-4_8

delay neurodegeneration. In conclusion, diet profoundly affects mental health by shaping brain structure, neurotransmitter function, and inflammation. A balanced, nutrient-rich diet can prevent and manage mental illness across the lifespan, and healthcare professionals should incorporate nutritional assessments into mental health care.

Keywords

Mental health · Nutrition and diet · Microbiota · Cognitive decline · Mediterranean diet

8.1 Introduction

Mental health is an important side of overall well-being, influenced by many factors, including diet, genetics, lifestyle, and environment. Among these, diet is a primary determinant of mental health outcomes across the life course. It has long been recognized that the quality of our diet affects our physical health and that a healthy diet is protective against many chronic diseases, such as diabetes mellitus and hypertension. Recent studies have shown that the content of our diet has an impact not only on our physical health but also on our mental health. Dietary nutrients can influence neurochemical processes such as neurotransmitter synthesis, neuronal plasticity, and neuronal function. There are many theories about how diet affects our physical and mental health, the most popular being epigenetic changes induced by methyl donors, the hypothalamic-pituitary axis, and the interaction of the microbiota-gut-brain axis (Bekdash 2019; Berding et al. 2021). Dietary deficiencies of omega-3 fatty acids, vitamins such as B12, and minerals such as magnesium and zinc have been shown to cause neuropsychiatric symptoms such as anxiety, depression, difficulty concentrating, mood disorders, and cognitive deficit diseases (Scott Diesing 2023). Adequate and effective nutrition is a fundamental prerequisite for good health at all life course stages. The pathophysiologic pathway of this interaction will be the subject of detailed discussion in another chapter of the book.

Some stages of human life, such as infancy, require more energy production than others, and the body needs more dietary elements and nutrients during these periods. This chapter provides an in-depth review of how diet affects mental health at different stages of life based on recent evidence-based studies.

Rapid growth and brain development throughout childhood and adolescence make the need for nutrient-rich foods especially critical. For instance, essential fatty acids like omega-3 support forming healthy neural networks and regulate mood and stress responses. Meanwhile, vitamins and minerals are cofactors in biochemical pathways that influence cognitive function and emotional balance. Thus, an imbalanced diet during these formative years may heighten the risk of conditions such as attention deficit hyperactivity disorder (ADHD), depression, and anxiety (Bekdash 2019) (Fig. 8.1).

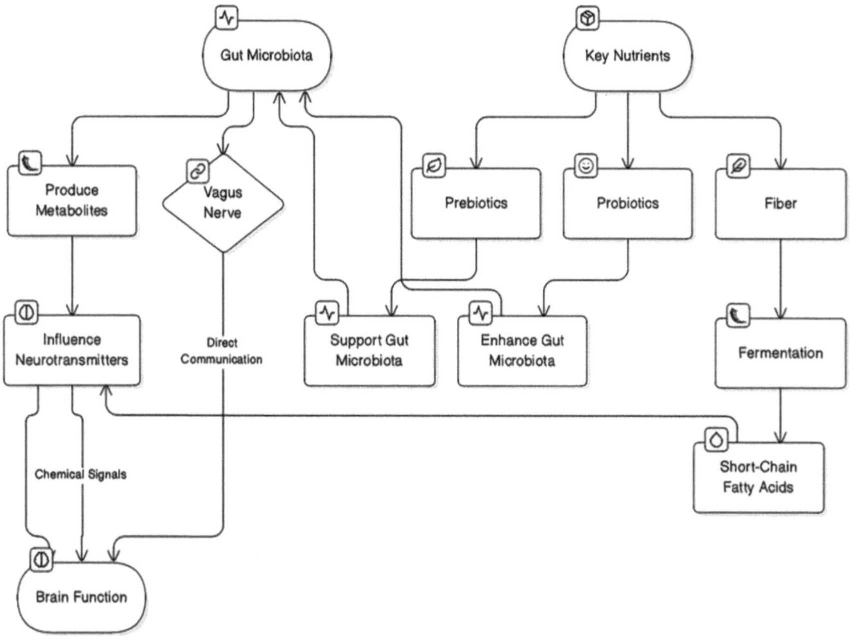

Fig. 8.1 A schematic representation of the gut-brain connection

In adulthood, diet remains pivotal for maintaining emotional well-being and cognitive performance. Dietary patterns high in fruits, vegetables, whole grains, and lean proteins—commonly represented by the Mediterranean diet—are consistently associated with lower rates of depression and anxiety. On the other hand, excessive intake of sugar, trans fats, and processed foods can contribute to systemic inflammation, oxidative stress, and hormonal imbalances, which in turn may exacerbate mental health issues (Berding et al. 2021).

As individuals enter older adulthood, nutrient demands shift once again, especially as metabolic rates slow and the risk of chronic diseases increases. Adequate protein, antioxidants, and B vitamins help preserve muscle mass and protect against neurodegenerative changes linked to conditions such as Alzheimer's and other forms of dementia (Scott Diesing 2023). Maintaining a balanced diet rich in anti-inflammatory components can also mitigate age-related cognitive decline and mood disturbances.

Considering the unique nutritional requirements at each life stage, we can better understand how diet modulates mental health over time. Interventions that target dietary improvement—whether through supplementation, fortified foods, or adopting healthier eating habits—hold promise in preventing or alleviating a wide range of mental health conditions. Integrating proper dietary strategies with other health-promoting behaviors can foster resilience, enhance quality of life, and promote long-term mental well-being (Fig. 8.2).

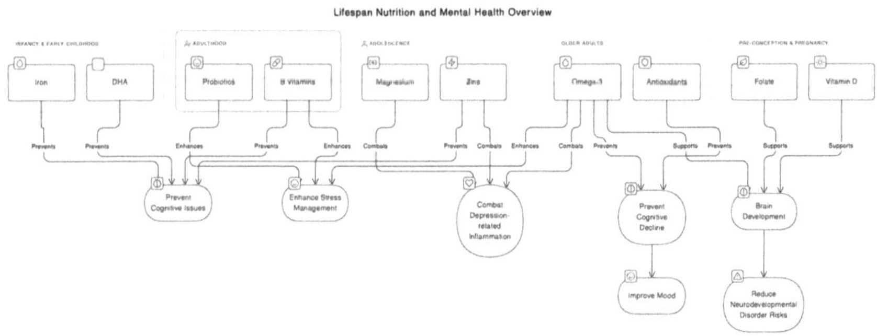

Fig. 8.2 A timeline diagram of lifespan nutrition and mental health

8.2 Prenatal Period, Infancy, and Early Childhood: Nutrition and Brain Development

The content of maternal nutrition during pregnancy, such as adequate levels of folate, iron, omega-3 fatty acids, iodine, choline, and vitamin D, is fundamental for the early stage of fetal brain development and reducing the risk of neurodevelopmental disorders such as autism spectrum disorders. The mother's diet during pregnancy has been shown to cause epigenetic changes during a crucial period of development, such as the chemical modification of DNA or histones through methylation to alter the expression of genes (Bekdash 2021; Georgieff 2020).

During pregnancy, folate is one of the most critical nutrients, as it helps prevent neural tube defects, which occur in the very early stages of embryonic development. Moreover, ensuring sufficient choline intake supports the formation of cell membranes and neurotransmitters, both essential to fetal brain development. At the same time, iodine is necessary to produce thyroid hormones that regulate growth and metabolic processes in the developing fetus. Meanwhile, vitamin D is important for bone health, modulating the immune system, and supporting brain function. These essential nutrients often work synergistically, meaning that a deficiency in one can affect the utilization or metabolism of another. Therefore, healthcare providers frequently recommend that pregnant women take prenatal supplements to ensure they receive adequate micronutrients (Bekdash 2021; Georgieff 2020).

A child's brain grows rapidly in the first few years of life. Poor nutrition during this period can have long-term effects on cognitive development, emotional regulation, and behavioral health. Breast milk is the best source of nutrition for a baby in the first 6 months of life. However, after the age of 1 year, the baby's primary source of calories should be additional food, and breast milk should be continued until the age of 2 years, if possible, to help build up the immune system. At this crucial time, the rapidly growing brain is uniquely sensitive to the effects of nutrition. For this reason, attention should be paid to the nutritional content of infants, especially in the first 2–3 years of life, to prevent malnutrition and overfeeding (Gelman et al. 2023).

Establishing healthy eating patterns early on can also support a child's ability to maintain focus, regulate emotions, and socialize effectively. Nutrient-dense foods help ensure that the child meets the high metabolic demands of brain cells, forming complex neural connections that lay the foundation for future learning and behavior. For example, exposing young children to varied fruits, vegetables, whole grains, and lean proteins can promote physical growth and mental acuity. On the other hand, high-sugar, high-fat foods without adequate micronutrients may lead to overfeeding and undernourishment—a scenario in which caloric needs are met but essential vitamins and minerals are lacking.

Omega-3 fatty acids, especially DHA (docosahexaenoic acid), are essential for brain development. Research suggests that DHA deficiency in early childhood is associated with poorer cognitive performance, attention deficits, and behavioral problems. DHA supplementation during pregnancy and infancy has been shown to benefit cognitive and emotional development (Middleton et al. 2018; Welty 2023).

A deficiency of essential micronutrients such as iron, zinc, and iodine can impact brain development, leading to cognitive delays and behavioral problems. Iron is essential for all cells. It is involved in oxygen delivery, electron transport, and enzyme activity. High-metabolic cells need more iron and are at greater risk if insufficient. In infancy, adolescence, and pregnancy, when energy requirements increase, the body's need for iron also increases, and iron deficiency can occur if iron stores are inadequate. Pregnant women with low iron intake have an increased risk of giving birth to a child with autism, schizophrenia, or abnormal brain structure. Iron-deficient newborns have impaired recognition, processing, and bonding. This persists despite iron supplementation. Preclinical models confirm that iron deficiency impairs expected iron-dependent processes such as monoamine neurotransmission, neuronal growth and differentiation, myelination, and gene expression, both acutely and in the long term, into adulthood. It also underlines the critical role of iron in brain development. Studies have shown that iron deficiency in early life is associated with impaired cognitive development (Georgieff 2020; McCann et al. 2023). In addition, other elements, such as zinc, magnesium, and selenium, and vitamins, such as B12, B6, and D, have an important effect on the development of the brain because they act as cofactors in important metabolic pathways.

Furthermore, maintaining a balanced diet that includes a variety of whole foods—such as leafy greens, nuts, seeds, and legumes—can help supply these critical vitamins and minerals. Small children benefit from nutrient-rich foods that provide long-lasting energy and the building blocks for healthy neural pathways. Understanding the significance of nutritional timing is also important for mothers and caregivers. Addressing deficiencies early, whether by adjusting dietary habits or through appropriate supplementation, may prevent irreversible developmental issues. When monitored properly, nutrient intake can optimize critical brain processes like synaptogenesis and myelination, ultimately influencing learning capacity and emotional well-being as children grow.

Nutrition in the prenatal period and the early stages of life should be closely monitored, and deficiencies of essential elements and vitamins should be prevented because of all these influences on brain health and development. By recognizing the vital role of nutrients and taking proactive measures—through diet, supplementation, and ongoing medical guidance—parents and caregivers can substantially support a child's neurodevelopmental trajectory and overall quality of life.

8.3 Adolescence: Hormonal Altering, Evolution, and Personality Development

Adolescence is a period of major significant neurological and hormonal changes, and it is a time when mental health problems are often on the rise. During this period, the brain undergoes a rapid maturation process: the connections between some areas of the brain increase and others decrease, and the type and levels of neurotransmitters in different areas of the brain change. All of these processes heighten the brain's need for energy and nutrients. Because of this increased demand, the adolescent brain is especially sensitive to deficiencies of essential elements and vitamins in the diet. Microelements such as zinc, iron, magnesium, and essential amino acids like tryptophan are crucial for neurotransmitter synthesis, neuronal communication, plasticity, and overall function. Low serum zinc levels have been associated with depression, magnesium deficiency with difficulty sleeping, concentrating, and mood swings, and iron deficiency with cognitive decline and fatigue. Therefore, diet plays a crucial role in mood regulation, cognitive development, and preventing mental health problems during this stage of life.

However, many adolescents develop poor eating habits, including a high intake of fast foods and sugary beverages and a low intake of fruits and vegetables. Such poor dietary choices can lead to systemic inflammation by increasing pro-inflammatory cytokines such as CRP and IL-6, which have been implicated in the pathophysiology of mental health disorders. Several studies have found an association between poor diet quality (e.g., high intake of processed foods, sugar, and fat) and an increased risk of depression, anxiety, and eating problems in adolescents. While the Western diet may increase the risk and severity of depression in this population, research shows that the Mediterranean diet, characterized by a high intake of fruits, vegetables, legumes, nuts, and olive oil, may reduce the risk and symptoms of depression (Zielińska et al. 2022).

In addition to providing the necessary nutrients, a balanced diet can help stabilize energy levels and support better stress management—factors that are vital when adolescents are navigating complex social, academic, and personal challenges. Indeed, adolescence is often accompanied by heightened stress, changes in sleep patterns, and shifts in social dynamics, all of which can exacerbate or contribute to mental health concerns if the body and brain are not well-nourished. Parents, educators, and policymakers can help bolster mental resilience and reduce the risk of developing mood disorders by ensuring that young people consume diets rich in vitamins and minerals.

Depression remains one of the most common and growing mental disorders worldwide, affecting adults, children, and adolescents alike. Prevention, early diagnosis, and treatment of mood disorders in adolescence are paramount, as depression in this period is a risk factor for recurrence in later life, as well as for other mental disorders in adulthood. Strategies that promote healthier diets—featuring whole grains, lean proteins, and a variety of plant-based foods—may help reduce systemic inflammation and support a well-functioning neurotransmitter system, thereby diminishing some of the key biological underpinnings of depression.

Moreover, dietary modification is beneficial for mental health outcomes and a sustainable strategy that can be integrated into broader lifestyle interventions. Encouraging adolescents to learn basic cooking skills and become aware of how certain food choices affect their mood and energy levels can empower them to make healthier decisions. School-based programs that provide access to balanced meals and nutritional education can similarly lay a strong foundation for lifelong healthy eating habits.

Adolescence is a unique stage in human development and an important time to establish the foundations of good habits that can persist into adulthood. Addressing nutritional needs during this stage goes hand in hand with other preventive strategies, such as regular physical activity, adequate sleep, and mental health education. It is, therefore, crucial for health authorities to identify and implement educational and behavioral policies that motivate young people to adopt healthy habits, ensuring that their developing brains and bodies receive the support they need. By cultivating positive dietary patterns early on, the risk of depression and other mental health issues may be significantly reduced, reinforcing the importance of comprehensive strategies that combine nutritional guidance with psychosocial support.

8.4 Adulthood: Maximization of Cognitive Performance and Stress Management

In adulthood, maintaining mental health requires stress management, prevention of chronic disease, and protection against cognitive decline. In all of these processes, diet is a key factor. Inflammatory processes can be triggered by poor dietary choices, such as a Western diet high in processed foods, sugars, and fats, leading to chronic low-grade inflammation, which has been linked to several mental health conditions, including depression, anxiety, and cognitive decline. Diets rich in anti-inflammatory foods, such as omega-3 fatty acids, antioxidants, and polyphenols, have been shown to reduce markers of inflammation and improve mental health. The type of diet that is best suited to an anti-inflammatory approach is the Mediterranean diet. Plant foods and healthy fats are central to the Mediterranean diet. It consists mainly of vegetables, fruit, and wholemeal cereals. Extra virgin olive oil is the primary fat source in this diet (Pray 2017).

There is growing evidence that a diet rich in fruit and vegetables may play a protective role against stress, with the consumption of whole fruit and vegetables, as well as some specific types of nutrients (such as citrus fruits and leafy green vegetables), being associated with less stress and improved well-being (Aucoin et al. 2021). Bioactive compounds in a healthy diet appear to modulate neurotransmitters in the brain that promote happiness. Similarly, the polyphenols and carotenoids found in fruit can reduce oxidative stress and inflammation, which have also been linked to improved mental health (Radavelli-Bagatini et al. 2022).

More and more evidence supports the role of the gut microbiome in mental health. Diets high in fiber, probiotics, and prebiotics support a healthy gut microbiota, which can positively affect mood and mental health. Emerging evidence suggests that a healthy gut microbiome is crucial to mental health. A diet rich in probiotics and prebiotics can improve gut health. This, in turn, helps to regulate mood and reduce anxiety (Aslam et al. 2020).

Magnesium and zinc are important for regulating stress responses. Magnesium status is linked to anxiety, depression, sleep disturbance, and mood changes (Mah and Pitre 2021). Magnesium is included in pathways with NMDA antagonism, modulation of serotonin-dopamine and noradrenalin and regulation of sleep-wake cycles. Magnesium deficiency leads to dysregulation of the hypothalamic-pituitary-adrenal (HPA) axis and increased calcium levels in the brain. If HPA axis dysfunction plays a mechanistic role in depression, magnesium deficiency may be a risk factor that renders individuals susceptible to chronically elevated cortisol and its neurodegenerative effects. There has been some evidence of a beneficial role for magnesium supplementation in the treatment of depression in some intervention trials but not in others (Rajasekar et al. 2024). An analytical study shows magnesium supplementation improves mood, anxiety, and quality of life in stressed but otherwise healthy adults (Noah et al. 2021).

Zinc is the second most abundant micronutrient in the human body after iron. It plays a fundamental role in catalytic and structural pathways, including regulating gene expression in the central nervous system. Zinc is also involved in postsynaptic activity, hormone secretion, and nerve impulse transmission. Behavioral disorders, depression, mood swings, anxiety, aggression, and irritability are common in zinc deficiency (Amini and HeidariFarsani 2023). Zinc and magnesium have been the most studied micronutrients in connection with depression, and it's been suggested that these micronutrients affect depression through similar mechanisms. Zinc supplementation has shown promise in reducing anxiety and improving overall mood, particularly in those with deficiencies. Recent studies have suggested that selenium may also play a role in the development of depression, although evidence is sparse and confusing (Wang et al. 2018).

Deficiencies of B vitamins, particularly B12, B6, and folate, are common in adults and are associated with cognitive decline, mood disorders, and an increased risk of depression. Recent studies have shown that B12 supplementation can improve mood and cognitive function, particularly in people with low B12 levels (Markun et al. 2021).

While each nutrient plays a distinct role, the synergistic effect of a balanced diet is equally important. For instance, combining magnesium-rich foods—such as nuts, seeds, and whole grains—with B vitamin sources—like leafy greens and fortified cereals—can enhance overall neurological function and stress resilience. Similarly, incorporating foods high in zinc (e.g., legumes, pumpkin seeds, and oysters) alongside sources of probiotics and prebiotics can strengthen both the immune system and the gut-brain axis. In addition, adopting dietary patterns that prioritize whole,

Table 8.1 Key nutrients and their impact on brain functions/mental health outcomes

Nutrient	Brain function/mental health outcome	Life stage
Omega-3	Improves memory	All stages
Folate	Enhances mood	Prenatal, childhood, adolescence
Vitamin D	Boosts focus	Adulthood, older adults
Iron	Cognitive function	Childhood, adolescence, older adults

unprocessed foods may help stabilize blood sugar levels, supporting consistent energy and mood regulation throughout the day.

Beyond addressing deficiencies, a nutrient-dense eating pattern can be a foundational preventive strategy for mental well-being. By minimizing chronic low-grade inflammation, ensuring optimal neurotransmitter function, and supporting healthy gut microbiota, a comprehensive dietary approach can help mitigate many factors that contribute to stress, depression, and anxiety. In particular, focusing on the balance of macronutrients and micronutrients—rather than simply reducing or eliminating certain food groups—encourages better long-term adherence and can foster sustainable mental health benefits. Overall, the evidence underscores that nutrition is not merely a backdrop to mental health but an active and dynamic player in promoting emotional resilience and cognitive vitality (Table 8.1).

8.5 Aging and Late Adulthood: Preventing Cognitive Decline and Dementia

In older adults over 65 years old, the prevention of cognitive decline and mental disorders, such as dementia and depression, is an important area of focus. Nutrition can play an essential role in the preservation of cognitive function and in reducing the risk of developing neurodegenerative diseases (Wickramasinghe et al. 2020). In addition to a well-balanced diet, other lifestyle factors—such as regular physical activity, maintaining social connections, and engaging in mentally stimulating activities—can also contribute to better cognitive outcomes in this age group. Nevertheless, the foundations often begin with what we eat, as certain dietary patterns are associated with specific benefits for the aging brain.

The Mediterranean diet, rich in fruits, vegetables, whole grains, olive oil, and omega-3 fatty acids, has been closely linked to a lower risk of Alzheimer's disease and a slower rate of cognitive decline. This diet emphasizes nutrient-dense, minimally processed foods that supply antioxidants, healthy fats, and essential vitamins and minerals. Several studies suggest that people who follow a Mediterranean diet have a significantly lower risk of developing Alzheimer's disease compared to those who follow a typical Western diet (Zielińska et al. 2022). Adherence to this diet supports better vascular health and reduces inflammation, factors believed to contribute to neurodegenerative processes.

Supplementation with specific nutrients, such as vitamin D and B vitamins—particularly vitamin B12—folic acid, and omega-3 fatty acids, has shown potential in slowing the progression of cognitive decline in adults over 65. Even in younger adults, vitamin B12 deficiency can lead to cognitive issues, depressed mood, and fatigue, and older individuals are more vulnerable to these effects due to diminished digestive efficiency and other age-related factors. In people with documented deficiencies, supplementation with B vitamins has been shown to improve mood and cognitive function (Lauer et al. 2022). Likewise, omega-3 fatty acids—often found in fatty fish like salmon, mackerel, and sardines—have demonstrated promising results in supporting brain health. Some evidence indicates that they may improve cognitive performance and delay the onset of dementia, possibly by reducing inflammation and promoting neuronal membrane integrity.

Maintaining cognitive ability and preventing or at least delaying the development of dementia are crucial goals in old age. As we advance in years, we often experience reduced appetite and slower digestion, leading to decreased intake and absorption of micronutrients. In addition, the bioavailability of these vitamins and minerals from food sources can drop due to physiological changes. Therefore, vitamin and microelement deficiencies are common in the elderly, making targeted dietary strategies and supplementation vital. Older patients should be asked about their diet every visit to their general practitioner and assessed for any signs of nutritional deficiency. This may involve routine blood tests to check levels of vitamins B12 and D, as well as assessments of overall caloric intake and protein consumption.

Addressing nutritional gaps early can help preserve cognitive function, sustain mental well-being, and enhance quality of life. Professionals can recommend practical strategies, such as incorporating more nutrient-dense foods, using fortified products, and educating older adults about simple ways to enrich their meals with critical micronutrients. By adopting these measures alongside regular physical activity and ongoing social engagement, older individuals can maintain better cognitive health and potentially reduce the risk or slow the progression of various forms of dementia and depression (Zielińska et al. 2022; Wickramasinghe et al. 2020; Lauer et al. 2022).

8.6 Special Consideration Across the Lifespan

Socioeconomic factors are key to nutrition and mental health outcomes across diverse populations. These factors, such as income, education, employment, social status, and environmental conditions, affect access to nutritious food, health services, and environments. Understanding how these factors interact and affect individuals is essential for developing effective public health strategies and policies to reduce health inequalities. Higher incomes allow people to buy a range of nutritious foods, such as fruit, vegetables, lean proteins, and whole grains, while low incomes mean people have to rely on cheaper, less nutritious foods. Low-income households are more vulnerable to food insecurity. This means they have limited or uncertain access to enough food. This can lead to malnutrition or obesity because they eat low-cost, high-calorie foods.

Higher education improves understanding of nutritional information, leading to healthier eating and following dietary guidelines. It fosters critical thinking skills, enabling individuals to evaluate food marketing and make informed diet decisions. At the same time, having a demanding job or working multiple jobs (often seen in low-income settings) reduces the time available for cooking and meal planning, pushing individuals toward convenience foods. This dependence on convenience or fast foods heightens the risk of chronic diseases, such as type 2 diabetes and cardiovascular problems. High stress, whether from workplace demands or financial insecurity, can lead to unhealthy behaviors like emotional eating or skipping meals. These behaviors not only affect physical health but also contribute to mental health challenges, including anxiety and depression.

Strong social connections encourage healthy eating by sharing meals, supporting dietary change, and spreading nutrition knowledge. For instance, neighbors in a community with limited grocery stores might form cooking clubs or pool resources to buy fresh produce in bulk, improving everyone's nutritional intake. Programs like food cooperatives and nutrition workshops make healthy food and education more accessible and can reduce the stigma around seeking help for food insecurity. Community-based interventions that simultaneously tackle economic development and nutritional education have shown promising results in bridging health disparities.

While general dietary guidelines apply to all, gender differences require tailored approaches. Body composition, hormones, reproductive status, and metabolism influence the optimal intake of calories, macronutrients, micronutrients, and other dietary components. Individual calorie requirements may vary depending on age, activity level, and health. Women's calorie requirements vary considerably at different stages, such as during pregnancy and menopause. Adequate protein is essential for women to maintain muscle mass, especially after menopause and during pregnancy and lactation, to support fetal growth and milk production. Women may benefit more from omega-3s for hormonal balance and reproductive health and need more iron than men, especially during menstruation, pregnancy, and breastfeeding. Additionally, women have an increased risk of osteoporosis, particularly after menopause, due to reduced estrogen levels, so calcium and vitamin D are important for bone health at this time.

Health policies can become more inclusive by addressing socioeconomic barriers, such as poverty and limited access to education and healthcare, and by recognizing gender-specific nutritional needs. This multi-level approach—incorporating economic support, targeted nutritional education, and gender-sensitive guidelines—has the potential to significantly reduce health inequalities. Through informed policies and community-driven initiatives, societies can foster environments where healthier eating habits are more attainable and mental health is supported by ensuring physical and emotional well-being.

8.7 Emerging Research and Innovation

Nutritional psychiatry is an interdisciplinary field that explores the relationships between diet and mental health and aims to integrate nutritional strategies into the prevention and treatment of mental disorders. This emerging discipline recognizes that what we eat fuels our bodies and significantly influences our cognitive performance, emotional balance, and overall mental health. By examining how specific nutrients affect neurotransmitter function, brain plasticity, and gut microbiota, nutritional psychiatry offers a more holistic approach to mental wellness. Mental disorders affect millions and represent a major public health challenge. They often contribute to significant economic burdens due to lost productivity, increased healthcare costs, and reduced quality of life for those impacted. Traditional treatment has primarily focused on pharmaceutical and psychotherapeutic interventions. However, evidence suggests that nutrition is critical in brain function, neurotransmitter synthesis, and psychological well-being (Grosso 2021). Nutritional psychiatry aims to merge nutritional science and mental health care.

Nutritional psychiatry aims to use diet to reduce the risk of mental disorders, integrate nutritional approaches into treatment plans for existing disorders, and advance scientific understanding and knowledge about the relationship between diet and mental health (Adan et al. 2019). One of the key areas of investigation is how certain dietary patterns, such as the Mediterranean diet or diets high in fruits, vegetables, whole grains, and omega-3 fatty acids, may contribute to lower incidence rates of depression and anxiety. Researchers are also exploring micronutrients like zinc, magnesium, and B vitamins for their roles in mood regulation and stress management. In practice, clinicians might recommend structured meal plans, targeted supplementation, or advice on reducing pro-inflammatory foods, complementing standard therapies like cognitive-behavioral therapy or medication. Incorporating a nutritional component addresses nutrient deficiencies that may exacerbate mental health issues and empowers patients to take active steps in their recovery through mindful eating habits.

Functional foods are conventional foods enhanced with added ingredients for health benefits beyond basic nutrition, including improved immune system function, reduced risk of chronic diseases, and improved mental well-being. Examples include fortified cereals, probiotic yogurts, and vitamin/omega-3 fatty acid fortified drinks. These foods are designed to bridge the gap between a standard diet and more therapeutic dietary regimens, potentially offering an accessible way for individuals to concurrently improve their mental and physical health. Nutraceuticals are derived from food sources, offering extra health benefits beyond nutrition, often in concentrated forms such as capsules, tablets, and powders. They include supplements, herbal products, and nutrients like vitamins, minerals, and antioxidants. Functional foods and dietary nutraceuticals affect mental health through participation in and regulation of neurotransmitter synthesis, anti-inflammatory and antioxidant effects, modulation of the gut-brain axis, effects on hormonal balance, modulation of neurogenesis, and synaptic plasticity (Yoo et al. 2024). Some nutrients support neurotransmitters that influence mood and mental function, e.g., amino acids like tryptophan that make serotonin, which is linked to anxiety. Furthermore, antioxidants found in fruits and vegetables may protect the brain from oxidative stress, thereby reducing the risk of neurodegenerative changes. Chronic inflammation and stress can trigger mental disorders, but functional foods with antioxidants and anti-inflammatory compounds can help. Probiotics and prebiotics can modify the microbiome, improving gut health and mental well-being (Aslam et al. 2020). Ongoing research indicates that a balanced, nutrient-rich diet, alongside a tailored approach to functional foods and nutraceuticals, holds great potential for complementing existing mental health treatments. By understanding these dietary influences on brain biology and emotional states, healthcare providers and patients can adopt more comprehensive strategies that nurture the mind and the body (Fig. 8.3).

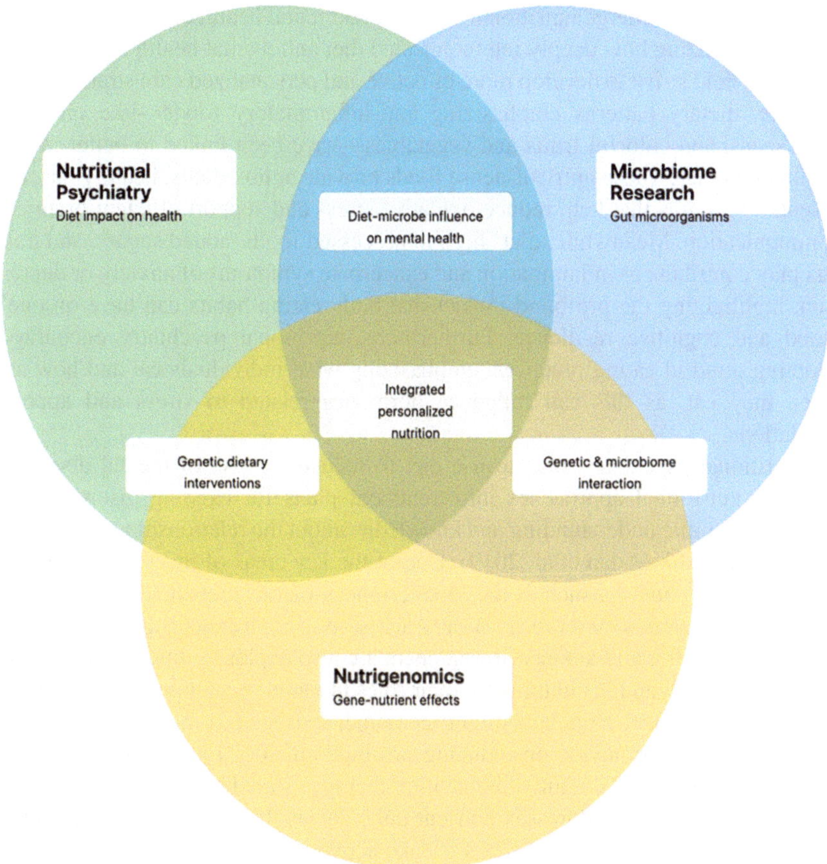

Fig. 8.3 A Venn diagram showing the relationships between nutritional psychiatry, microbiome research, and nutrigenomics

8.8 Recommendations and Strategies

Nutritional psychiatry is an interdisciplinary field that explores the relationships between diet and mental health and aims to integrate nutritional strategies into the prevention and treatment of mental disorders. This emerging discipline recognizes that what we eat fuels our bodies and significantly influences our cognitive performance, emotional balance, and overall mental health. By examining how specific nutrients affect neurotransmitter function, brain plasticity, and gut microbiota, nutritional psychiatry offers a more holistic approach to mental wellness.

Mental disorders affect millions and represent a major public health challenge. They often contribute to significant economic burdens due to lost productivity, increased healthcare costs, and reduced quality of life for those impacted. Traditional treatment has primarily focused on pharmaceutical and psychotherapeutic interventions. However, evidence suggests that nutrition is critical in brain function, neurotransmitter synthesis, and psychological well-being (Grosso 2021). Nutritional psychiatry aims to merge nutritional science and mental health care.

By recognizing how deeply interconnected diet and mental health are, researchers in this field strive to develop more effective and personalized care strategies. For instance, dietary patterns emphasizing anti-inflammatory foods—like oily fish, nuts, seeds, and colorful fruits and vegetables—have been linked to better mental health outcomes. These nutrient-dense foods provide antioxidants, healthy fats, and essential vitamins that help reduce oxidative stress and support efficient brain cell communication. Meanwhile, diets high in processed foods, added sugars, and trans fats may contribute to inflammation and exacerbate symptoms of anxiety or depression, highlighting the profound impact that daily eating habits can have on one's mood and cognitive resilience. Furthermore, nutritional psychiatry encourages adopting mindful eating practices, emphasizing what individuals eat and how and when they eat, as this can influence hormones related to stress and appetite regulation.

Nutritional psychiatry aims to use diet to reduce the risk of mental disorders, integrate nutritional approaches into treatment plans for existing disorders, and advance scientific understanding and knowledge about the relationship between diet and mental health (Adan et al. 2019). One of the key areas of investigation is how certain dietary patterns, such as the Mediterranean diet or diets high in fruits, vegetables, whole grains, and omega-3 fatty acids, may contribute to lower incidence rates of depression and anxiety. Researchers are also exploring micronutrients like zinc, magnesium, and B vitamins for their roles in mood regulation and stress management. In practice, clinicians might recommend structured meal plans, targeted supplementation, or advice on reducing pro-inflammatory foods, complementing standard therapies like cognitive-behavioral therapy or medication. Incorporating a nutritional component addresses nutrient deficiencies that may exacerbate mental health issues and empowers patients to take active steps in their recovery through mindful eating habits.

Beyond identifying "good" foods and "bad" foods, nutritional psychiatry also considers the context of an individual's overall lifestyle. Exercise, sleep quality, and stress management techniques are vital in mental health and can interact synergistically with dietary interventions. For example, a patient who regularly exercises may experience improved mood and energy, which can be amplified by consuming nutrient-rich meals that further support positive neurotransmitter activity. Likewise, good sleep hygiene can help stabilize hormones like cortisol, which can influence hunger cues and emotional regulation. By emphasizing the interplay of these factors, nutritional psychiatry promotes a comprehensive model of care that simultaneously addresses multiple dimensions of health.

Functional foods are conventional foods enhanced with added ingredients for health benefits beyond basic nutrition, including improved immune system function, reduced risk of chronic diseases, and improved mental well-being. Examples include fortified cereals, probiotic yogurts, and vitamin/omega-3 fatty acid fortified drinks. These foods are designed to bridge the gap between a standard diet and more therapeutic dietary regimens, potentially offering an accessible way for individuals to concurrently improve their mental and physical health. Nutraceuticals are derived from food sources, offering extra health benefits beyond nutrition, often in concentrated forms such as capsules, tablets, and powders. They include supplements, herbal products, and nutrients like vitamins, minerals, and antioxidants. Functional foods and dietary nutraceuticals affect mental health through participation in and regulation of neurotransmitter synthesis, anti-inflammatory and antioxidant effects, modulation of the gut-brain axis, effects on hormonal balance, and modulation of neurogenesis and synaptic plasticity (Yoo et al. 2024). Some nutrients support neurotransmitters that influence mood and mental function, e.g., amino acids like tryptophan that make serotonin, which is linked to anxiety. Furthermore, antioxidants found in fruits and vegetables may protect the brain from oxidative stress, thereby reducing the risk of neurodegenerative changes. Chronic inflammation and stress can trigger mental disorders, but functional foods with antioxidants and anti-inflammatory compounds can help. Probiotics and prebiotics can modify the microbiome, improving gut health and mental well-being (Aslam et al. 2020).

Ongoing research indicates that a balanced, nutrient-rich diet, alongside a tailored approach to functional foods and nutraceuticals, holds great potential for complementing existing mental health treatments. By understanding these dietary influences on brain biology and emotional states, healthcare providers and patients can adopt more comprehensive strategies that nurture the mind and the body. This integrated model opens the door for personalized treatment plans, where clinicians can recommend specific dietary interventions based on genetic, metabolic, and lifestyle factors unique to each patient. Consequently, nutritional psychiatry not only broadens the scope of traditional mental health care but also inspires a proactive approach, encouraging individuals to examine their food choices and cultivate eating patterns that align with their mental wellness goals. Through continued research, collaborative efforts, and patient education, nutritional psychiatry is poised to become an essential dimension of holistic mental health support for diverse populations.

8.9 Conclusion

The evidence presented throughout this manuscript underscores the critical influence of nutrition on mental health across every stage of life, from prenatal development through older adulthood. Beginning in the prenatal period, adequate maternal intake of key micronutrients such as folate, iron, omega-3 fatty acids, iodine, choline, and vitamin D has been shown to reduce the risk of neurodevelopmental disorders and enhance fetal brain growth (Bekdash 2021; Georgieff 2020). This foundation is further solidified in infancy and early childhood, where breast milk provides vital immunological benefits and essential nutrients like iron, zinc, and DHA that shape cognitive abilities and emotional regulation. Deficiencies in these elements during formative years can result in long-term developmental setbacks and behavioral problems, highlighting the inextricable link between early nutrition and mental well-being (Middleton et al. 2018; Welty 2023).

As individuals transition into adolescence, rapid hormonal and neurological changes amplify nutritional demands, making this period particularly vulnerable to diet-related mental health challenges. Research correlates poor dietary choices—especially those high in sugars, processed foods, and unhealthy fats—with increased inflammation, which in turn can exacerbate risks of depression and anxiety (Zielińska et al. 2022). Conversely, adherence to healthier dietary patterns like the Mediterranean diet can support mood stabilization, reduce systemic inflammation, and protect adolescents from chronic mental health conditions.

In adulthood, diet remains paramount for stress management, chronic disease prevention, and optimal cognitive function maintenance. Anti-inflammatory diets rich in omega-3 fatty acids, antioxidants, and polyphenols have been tied to lower rates of depression and anxiety, while targeted micronutrient intake—particularly zinc, magnesium, and B vitamins—helps regulate neurotransmitter synthesis and stabilize mood (Mah and Pitre 2021; Noah et al. 2021; Amini and HeidariFarsani 2023; Markun et al. 2021). In this regard, the gut microbiota also emerges as a central player in mental health, with probiotics and prebiotics aiding in inflammation reduction and promoting a balanced microbiome that influences psychological resilience (Aslam et al. 2020).

During older adulthood, preventing or delaying cognitive decline is a major priority. The Mediterranean diet and prudent supplementation with B vitamins, omega-3 fatty acids, and vitamin D can help protect against neurodegenerative disorders and sustain cognitive performance (Zielińska et al. 2022; Lauer et al. 2022). Addressing common nutrient deficiencies in older populations is especially critical due to physiological changes that diminish nutrient absorption. Early identification and intervention—through dietary diversification, supplementation, or fortified foods—can significantly slow cognitive deterioration and improve quality of life.

Socioeconomic factors and gender-specific requirements cannot be overlooked across all stages. Access to nutritious foods, educational resources, and social support networks significantly affects dietary choices, thereby influencing mental health outcomes. Public health policies aimed at improving food security and nutritional literacy promise to reduce health disparities and optimize mental well-being on a population level.

Finally, the growing field of nutritional psychiatry offers a comprehensive framework that integrates diet-based strategies with conventional therapies for mental disorders (Grosso 2021; Adan et al. 2019). From incorporating functional foods and nutraceuticals to addressing broader lifestyle factors—such as physical activity, sleep quality, and stress management—this approach underscores the multidimensional nature of mental health. Moving forward, expanded research and collaboration between nutritionists, mental health professionals, and policymakers are essential for refining dietary recommendations and ensuring they are both evidence-based and accessible. By embracing these strategies, individuals and healthcare providers can work toward the shared goal of promoting mental wellness, thereby enhancing overall quality of life at every age.

The relationship between diet and mental health is complex but undeniable. Diet influences the structure of the brain, the function of neurotransmitters, and the regulation of inflammation, all of which play a crucial role in mental well-being. A balanced diet, rich in essential nutrients, can contribute to the prevention, management, and even improvement of a wide range of mental health conditions across the lifespan. Health professionals should integrate nutritional assessment and intervention into routine mental health care to optimize patient outcomes.

References

Adan RAH et al (2019) Nutritional psychiatry: towards improving mental health by what you eat. Eur Neuropsychopharmacology 29(12):1321–1332. https://doi.org/10.1016/j.euroneuro.2019.10.011. Elsevier BV

Amini Z, HeidariFarsani E (2023) Investigating the effect of zinc supplementation on probability of relapse and mental health in patients with opioid use disorder undergoing methadone maintenance treatment. Subst Abuse Treat Prev Policy 18(1):1. https://doi.org/10.1186/s13011-023-00514-5

Aslam H, Green J, Jacka FN, Collier F, Berk M, Pasco J, Dawson SL (2020) Fermented foods, the gut and mental health: a mechanistic overview with implications for depression and anxiety. Nutr Neurosci 23(9):659–671

Aucoin M et al (2021) Diet and anxiety: a scoping review. Nutrients 13(12):4418. https://doi.org/10.3390/nu13124418. MDPI

Bekdash RA (2019) Neuroprotective effects of choline and other methyl donors. Nutrients 11(12):p2995. https://doi.org/10.3390/nu11122995. MDPI AG

Bekdash RA (2021) Early life nutrition and mental health: the role of dna methylation. Nutrients 13(9):3111. https://doi.org/10.3390/nu13093111. MDPI

Berding K et al (2021) Diet and the microbiota-gut-brain axis: sowing the seeds of good mental health. Adv Nutr 12(4):1239–1285. https://doi.org/10.1093/advances/nmaa181. Oxford University Press

Gelman R et al (2023) Adoption, implementation, and sustainability of early childhood feeding, nutrition and active play interventions in real-world settings: a systematic review. Int J Behav Nutr Phys Act 20(1):32. https://doi.org/10.1186/s12966-023-01433-1. BioMed Central Ltd

Georgieff MK (2020) Iron deficiency in pregnancy. Am J Obstet Gynecol 223(4):516–524. https://doi.org/10.1016/j.ajog.2020.03.006. Mosby Inc

Grosso G (2021) Nutritional psychiatry: How diet affects brain through gut microbiota. Nutrients 13(4):1282. https://doi.org/10.3390/nu13041282. MDPI AG

Lauer AA et al (2022) Mechanistic link between vitamin B12 and Alzheimer's Disease. Biomolecules 12(1):129. https://doi.org/10.3390/biom12010129. MDPI

Mah J, Pitre T (2021) Oral magnesium supplementation for insomnia in older adults: a systematic review & meta-analysis. BMC Complement Med Ther 21(1):125. https://doi.org/10.1186/s12906-021-03297-z

Markun S, Gravestock I, Jäger L, Rosemann T, Pichierri G, Burgstaller JM (2021) Effects of vitamin b12 supplementation on cognitive function, depressive symptoms, and fatigue: a systematic review, meta-analysis, and meta-regression. Nutrients 13(3):923. https://doi.org/10.3390/nu13030923. MDPI AG

McCann S et al (2023) Iron status in early infancy is associated with trajectories of cognitive development up to pre-school age in rural Gambia. PLOS Glob Public Health 3(11):e0002531. https://doi.org/10.1371/journal.pgph.0002531

Middleton P, Gomersall JC, Gould JF, Shepherd E, Olsen SF, Makrides M (2018) Omega-3 fatty acid addition during pregnancy. Cochrane Database Syst Rev 11(11):CD003402. https://doi.org/10.1002/14651858.CD003402.pub3. John Wiley and Sons Ltd

Noah L, Dye L, Bois De Fer B, Mazur A, Pickering G, Pouteau E (2021) Effect of magnesium and vitamin B6 supplementation on mental health and quality of life in stressed healthy adults: post-hoc analysis of a randomised controlled trial. Stress Health 37(5):1000–1009. https://doi.org/10.1002/smi.3051

Pray L (ed) (2017) Nutrition across the lifespan for healthy aging. National Academies Press, Washington, DC. https://doi.org/10.17226/24735

Radavelli-Bagatini S et al (2022) Higher consumption of fruit and vegetables is associated with lower worries, tension and lack of joy across the lifespan. Front Nutr 9:p837066. https://doi.org/10.3389/fnut.2022.837066

Rajasekar R, VanderMolen J, Barnhart K, Anguilim N (2024) Dietary intake with supplementation of vitamin D, vitamin B6, and magnesium on depressive symptoms: a public health perspective. Front Public Health 12:1369666. https://doi.org/10.3389/fpubh.2024.1369666

Scott Diesing T (2023) Neurologic manifestations of gastrointestinal and nutritional disorders [Online]. Available: http://journals.lww.com/continuum

Wang J, Um P, Dickerman BA, Liu J (2018) Zinc, magnesium, selenium and depression: a review of the evidence, potential mechanisms and implications. 10(5):584. https://doi.org/10.3390/nu10050584. MDPI AG

Welty FK (2023) Omega-3 fatty acids and cognitive function. Curr Opin Lipidol 34(1):12–21

Wickramasinghe K, Mathers JC, Wopereis S, Marsman DS, Griffiths JC (2020) From lifespan to healthspan: the role of nutrition in healthy ageing. J Nutr Sci 9:e33. https://doi.org/10.1017/jns.2020.26. Cambridge University Press

Yoo S, Jung SC, Kwak K, Kim JS (2024) The role of prebiotics in modulating gut microbiota: implications for human health. Int J Mol Sci 25(9):4834. https://doi.org/10.3390/ijms25094834. Multidisciplinary Digital Publishing Institute (MDPI)

Zielińska M, Łuszczki E, Michońska I, Dereń K (2022) The Mediterranean Diet and the Western Diet in adolescent depression-current reports. Nutrients 14(20):4390. https://doi.org/10.3390/nu14204390. MDPI

Special Nutritional Considerations in Mental Health

9

Ceren Alis and Aynur Özge

Abstract

This chapter explores the intricate relationship between nutrition and mental health, focusing on how dietary patterns and specific nutrients influence psychiatric disorders. Essential nutrients such as B vitamins, omega-3 fatty acids, antioxidants, polyphenols, and tryptophan regulate neurotransmitter synthesis, mitochondrial function, brain inflammation, and oxidative stress. These mechanisms are central to the development and progression of psychiatric conditions like depression, anxiety, schizophrenia, and attention-deficit hyperactivity disorder. Particular emphasis is placed on the gut-brain axis, mitochondrial function, and epigenetic regulation as critical pathways through which nutrition impacts mental health. The chapter also underscores the protective effects of the Mediterranean diet and the potential harm of processed foods in exacerbating psychiatric symptoms. Understanding these nutritional mechanisms offers valuable insights for integrating dietary interventions into mental health care, leading to better clinical outcomes in psychiatric disorders.

Keywords

Nutrition · Nutritional neuroscience · diet · Mental health · Gut-brain axis · Oxidative stress

C. Alis (✉)
Department of Neurology, Sisli Hamidiye Etfal Training and Research Hospital, Istanbul, Turkey

A. Özge
Mersin University School of Medicine, Mersin, Türkiye

NOROM Neuroscience and Excellence Center, Ankara, Türkiye

W. Mohamed, A. Özge (eds.), *Feeding the Mind: The Connection Between Diet, Drugs, and Mental Health Volume 1*, Nutritional Neurosciences,
https://doi.org/10.1007/978-981-95-1414-4_9

9.1 Introduction

Mental health disorders, also known as mental illnesses, comprise a broad range of conditions that affect patients' emotional, psychological, and social well-being. These disorders can alter thinking, mood, behavior, and daily functioning, often leading to distress or impaired ability to function in various aspects of life. The scope of mental health disorders is broad, ranging from common conditions such as anxiety and depression to more severe illnesses like schizophrenia, bipolar disorder, and major neurocognitive disorders. Mental health disorders can be classified based on symptoms, duration, and severity, with the Diagnostic and Statistical Manual of Mental Disorders, Fifth Edition, Text Revision (DSM-5-TR), serving as a standard and the most recent manual for diagnosis (American Psychiatric Association 2022). These disorders may be influenced by genetic, environmental, and psychosocial factors, with complex interactions contributing to their development and course. Mental health conditions often co-occur with other medical conditions, exacerbating overall morbidity and requiring comprehensive treatment approaches (Kirkbride et al. 2024).

Nutrition is fundamental to brain function and mental health, influencing cognitive and emotional well-being. Recent studies have shown that dietary patterns and specific nutrients may impact the development, severity, and prognosis of behavioral and psychiatric disorders, including anxiety, depression, attention-deficit hyperactivity disorder (ADHD), and schizophrenia. These nutritional influences operate through multiple mechanisms, including inflammation, oxidative stress, neurotransmitter synthesis, and the gut-brain axis (Vauzour et al. 2017).

Oxidative stress and inflammation contribute to the pathophysiology of psychiatric conditions. A nutrient-rich diet with anti-inflammatory and antioxidant features can mitigate these harmful processes, providing neuroprotective effects that may decrease psychiatric symptoms. Similarly, neurotransmitter synthesis, essential for mood regulation and cognitive function, significantly depends on nutrient availability. Vitamin and mineral deficiencies can lead to imbalances that may exacerbate disorders like depression and anxiety (Trebatická and Ďuračková 2015).

The gut-brain axis has also been suggested as an important factor in mental health, and it can affect brain function and mood regulation. Diets high in fiber, fermented foods, and prebiotics support a healthy gut microbiome, decrease inflammation, and improve the prognosis of psychiatric disorders. Conversely, poor dietary choices, such as a high intake of processed foods, can lead to dysbiosis, an imbalance in the gut microbiota, promoting inflammation and worsening psychiatric symptoms (Dinan and Cryan 2015).

Mitochondrial dysfunction and epigenetic changes are increasingly recognized as underlying factors in many psychiatric disorders. Mitochondria, responsible for energy production in cells, are highly sensitive to nutrient availability, and disruptions in mitochondrial function can lead to impaired brain energy metabolism, contributing to mental health disorders. Omega-3 fatty acids, polyphenols, and other nutrients in the diet support mitochondrial health and can help restore proper brain function (Du et al. 2016). Epigenetic modifications, including DNA methylation, histone acetylation, and microRNA regulation, are essential in controlling gene expression related to brain function. Nutrition can directly impact these epigenetic processes, affecting gene expression and the risk and severity of psychiatric disorders. For example, B vitamins, omega-3s, and polyphenols can modify epigenetic processes, reduce the risk of psychiatric disorders, and improve clinical outcomes for patients with psychiatric disorders (Khavari and Cairns 2020).

In summary, the relationship between nutrition and mental health is complex and multifaceted. Dietary interventions may impact the prevention and management of psychiatric disorders by modulating inflammation, oxidative stress, neurotransmitter synthesis, mitochondrial function, and epigenetic mechanisms (Fig. 9.1). In this chapter, particular on emphasis is placed the gut-brain axis, mitochondrial function, and epigenetic regulation as critical pathways through which nutrition impacts mental health. The chapter also underscores the protective effects of the Mediterranean diet and the potential harm of processed foods in exacerbating psychiatric symptoms. Understanding these nutritional mechanisms offers valuable insights for integrating dietary interventions into mental health care, leading to better clinical outcomes in psychiatric disorders.

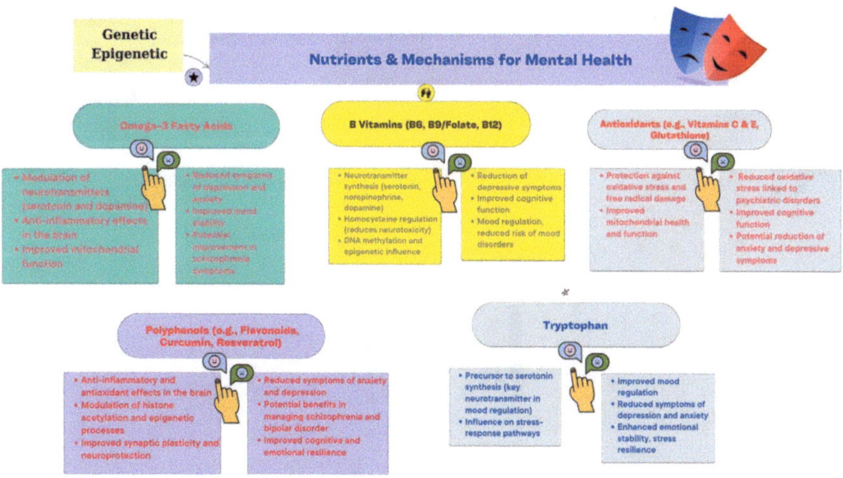

Fig. 9.1 The mechanisms by which different nutrients affect mental health
Together, these nutrients influence mental health through genetic and epigenetic pathways, modulating inflammation, neurotransmitter function, and resilience to stress

9.2 Nutritional Imbalances and Their Impact on Mental Health

9.2.1 Essential Nutrients for Brain Function

9.2.1.1 Macronutrients

Essential nutrients, including carbohydrates, proteins, and fats, are crucial in brain function and cognitive performance. Carbohydrates are the brain's primary energy source, with glucose being the most readily available form. Adequate carbohydrate intake ensures the provision of glucose to the brain, which is necessary for maintaining cognitive processes such as memory, attention, and learning (Smith et al. 2011).

Proteins play a vital role in brain function, as they are responsible for building, maintaining, and regulating the body's tissues, including the brain. Proteins comprise amino acids, the building blocks for neurotransmitters. Neurotransmitters, including dopamine, serotonin, and norepinephrine, are directly influenced by protein intake, making adequate high-quality protein essential for cognitive and emotional well-being (Glenn et al. 2019) (Fig. 9.2). One critical amino acid, tryptophan, is a precursor to serotonin, a mood regulation and cognition neurotransmitter. Decreased levels of serotonin are related to depression and anxiety, which suggests

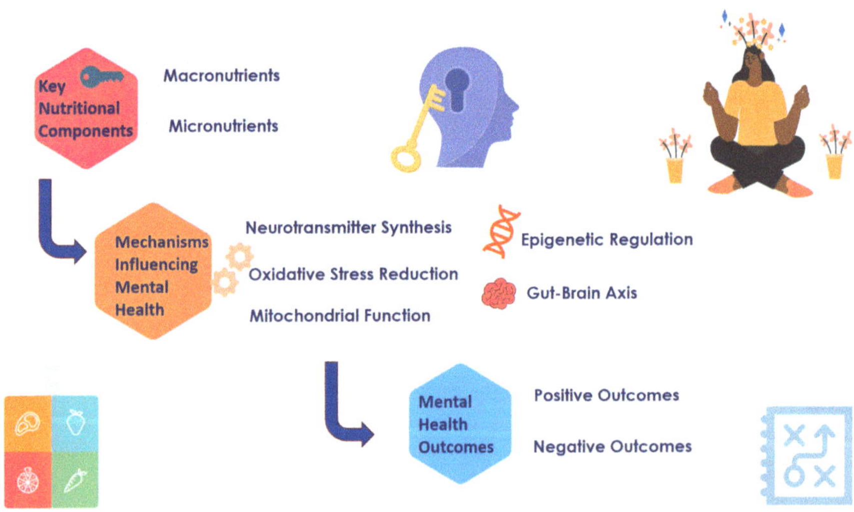

Fig. 9.2 The figure illustrating the relationship between nutrition and mental health. This figure depicts how key nutritional components impact mental health through various physiological pathways. Essential nutrients, including carbohydrates, proteins, fats, omega-3 fatty acids, antioxidants, polyphenols, B vitamins, and vitamin D, support neurotransmitter synthesis, reduce oxidative stress, enhance mitochondrial function, influence gene expression, and maintain gut health. These mechanisms collectively shape mental health outcomes, with nutrient-rich diets (e.g., the Mediterranean diet) promoting improved mood and cognition, while processed food-heavy diets increase inflammation and the risk of psychiatric symptoms

that an adequate intake of tryptophan may support positive mood and cognitive function (Kikuchi et al. 2021; Toker et al. 2010). Similarly, tyrosine, another amino acid found in proteins, is converted into dopamine, which plays a role in motivation, reward, and focus. Proteins from foods like lean meat, fish, eggs, and dairy are rich in these essential amino acids, supporting neurotransmitter production and overall brain function (Glenn et al. 2019).

Fats, especially omega-3 fatty acids, are vital for brain development, function, and cognitive performance. Among the most important fats for brain health are omega-3 and omega-6 fatty acids, which are essential because the body cannot produce them, meaning they must be obtained through diet. Omega-3s contribute to forming cell membranes in neurons, improve synaptic plasticity, and protect against cognitive decline. Studies have shown that abundant omega-3 fatty acid intake is associated with better cognitive performance and a reduced risk of neurodegenerative diseases (Luchtman and Song 2013).

9.2.1.2 Key Micronutrients

Omega-3 Fatty Acids

Omega-3 fatty acids, especially eicosapentaenoic acid (EPA) and docosahexaenoic acid (DHA), are crucial for proper brain function. Recent studies have emphasized their significant anti-inflammatory effects, which play an important role in managing depression. Inflammation is increasingly recognized as a contributing factor in mood disorders. Omega-3s are believed to modulate neurotransmission, especially in serotonin and dopamine systems, which are critical for mood regulation (Liu et al. 2015).

In addition to their anti-inflammatory properties, omega-3s support mitochondrial function by maintaining membrane integrity and reducing oxidative stress. For instance, omega-3 supplementation improves mitochondrial biogenesis and function, correlating with reduced depressive symptoms and enhanced overall brain health (Borja-Magno et al. 2023).

Omega-3 fatty acids may also affect the regulation of microRNAs, small non-coding RNAs that modulate gene expression at the post-transcriptional level. Dysregulated microRNA expression has been observed in schizophrenia, affecting genes involved in synaptic plasticity and neurotransmitter function (Khavari and Cairns 2020; Beveridge and Cairns 2012). Omega-3 supplementation can help restore the balance of microRNAs, which may improve psychotic symptoms in individuals with schizophrenia (Goh et al. 2021).

Antioxidants

Oxidative stress is considered an important factor in the development of schizophrenia, a severe psychiatric disorder characterized by hallucinations, delusions, and cognitive dysfunction. Antioxidants such as vitamins C, E, and glutathione protect the brain from oxidative damage and maintain mental health (Trebatická and Ďuračková 2015).

Antioxidants, including vitamins C and E and polyphenols, help mitigate oxidative stress, support mitochondrial function, and alleviate anxiety symptoms (Trebatická and Ďuračková 2015). Oxidative stress can lead to epigenetic changes, exacerbating psychiatric symptoms by overexpressing inflammatory genes associated with disorders like depression and anxiety.

Polyphenols

Polyphenols, found in fruits, vegetables, tea, and coffee, have been shown to exhibit neuroprotective effects. New research has revealed their potential role in decreasing oxidative stress and inflammation in the brain, both of which are implicated in the pathology of schizophrenia (Trebatická and Ďuračková 2015).

Anxiety disorders are also influenced by epigenetic mechanisms, such as histone modifications, that regulate the expression of stress-response genes. Polyphenols can impact histone acetylation, enhancing the expression of genes related to resilience and reducing anxiety-like behaviors.

Inflammation, mitochondrial dysfunction, and oxidative stress are common processes in the etiology of depression (Trebatická and Ďuračková 2015). Curcumin, a lipophilic polyphenol capable of crossing the blood-brain barrier, has been found to reduce inflammation and oxidative stress while promoting neuroprotection (Sanmukhani et al. 2014).

B Vitamins

B vitamins, including B6, B9 (folate), and B12, are crucial for producing and regulating neurotransmitters, such as serotonin, dopamine, and gamma-aminobutyric acid (GABA), essential for mood regulation. Nutrients such as folate, B12, and choline are essential for providing methyl groups for DNA methylation. Deficiencies in these nutrients can result in abnormal methylation patterns, increasing the risk of depression (Du et al. 2016).

Vitamin D

Vitamin D acts as a neurosteroid and is involved in various brain processes, including cell differentiation, axon connectivity, and neurotransmission. During early brain development, vitamin D plays a protective role, regulating the expression of neurotrophic factors that are crucial for brain growth and function. In the adult brain vitamin D helps modulate neuroinflammatory processes in the adult brain and supports cognitive function. Vitamin D deficiency may exacerbate vulnerability to neuropsychiatric diseases by impacting neuroplasticity, neurotransmitter function, and neuroinflammation. Adequate vitamin D levels are crucial for optimal brain health and the prevention of neuropsychiatric diseases (Eyles et al. 2013).

9.2.2 Common Nutritional Deficiencies in Mental Health Disorders

9.2.2.1 Depression and Its Relationship with Diet

Deficiencies in omega-3s have been linked to higher rates of depression, and supplementation has been shown to reduce depressive symptoms, particularly when combined with antidepressant therapies (Grosso et al. 2014; Liao et al. 2019). Omega-3 supplementation improves mitochondrial biogenesis and function, correlating with reduced depressive symptoms and enhanced overall brain health (Borja-Magno et al. 2023).

Antioxidants such as vitamin C could counteract epigenetic changes due to oxidative stress and may provide better emotional outcomes by reducing inflammation in depression and anxiety (Han et al. 2018; Moritz et al. 2020).

A polyphenol found in grapes called resveratrol has been shown to regulate histone acetylation and exert anxiolytic effects in preclinical models (Xu et al. 2006). Curcumin, a lipophilic polyphenol, is efficacious in the treatment of major depressive disorder, further highlighting the potential of polyphenols in managing psychiatric conditions (Sanmukhani et al. 2014).

Folate and B12 deficiency have been implicated in treatment-resistant depression, and supplementation of these vitamins has been shown to enhance the effectiveness of antidepressant therapies (Bender et al. 2017; Sangle et al. 2020).

Diets rich in tryptophan-containing foods, such as eggs, nuts, turkey, and seeds, can increase serotonin levels, potentially improving mood and reducing symptoms of depression and anxiety (Kikuchi et al. 2021). Conversely, diets low in tryptophan may contribute to lower serotonin production and lead to mood disturbances, including increased susceptibility to depression and anxiety (Toker et al. 2010).

9.2.2.2 Anxiety and Its Relationship with Diet

Antioxidants such as vitamin C could counteract epigenetic changes due to oxidative stress and may provide better emotional outcomes by reducing inflammation in depression and anxiety (Han et al. 2018; Moritz et al. 2020).

In addition, a randomized, double-blind, placebo-controlled trial showed that high-dose vitamin B6 supplementation was effective in decreasing self-reported anxiety. The authors suggest that these effects are due to vitamin B6's role in promoting GABA synthesis, an inhibitory neurotransmitter in the brain. Strengthening GABAergic activity through vitamin B6 supplementation may underlie the reduced anxiety observed in the study (Field et al. 2022).

Diets rich in tryptophan-containing foods, such as eggs, nuts, turkey, and seeds, can increase serotonin levels, improve mood, and decrease symptoms of depression and anxiety (Kikuchi et al. 2021). Conversely, diets low in tryptophan may contribute to lower serotonin production and lead to mood disturbances, including increased susceptibility to depression and anxiety (Toker et al. 2010).

9.2.2.3 Nutrient Needs for Cognitive Function in Conditions Such as Schizophrenia and Bipolar Disorder

Omega-3 supplementation can help restore the balance of microRNAs, which may improve psychotic symptoms in individuals with schizophrenia (Goh et al. 2021). Furthermore, higher consumption of omega-3s, omega-6s, and fish has been associated with a lower incidence of psychotic-like symptoms, highlighting the importance of these nutrients in mental health (Hedelin et al. 2010). Dietary polyphenols, such as flavonoids, have demonstrated the ability to improve cognitive function and reduce symptom severity in patients with schizophrenia, making them a promising adjunct to pharmacological treatments (Zhu et al. 2019; Saleem and Qurat-Ul-Ain 2022).

Reduced levels of antioxidants have been shown in patients with schizophrenia, and supplementation of them has been investigated as a potential therapeutic approach to decrease cognitive deficits and improve overall functioning. For instance, N-acetylcysteine (NAC), a precursor to glutathione, has been effective in reducing symptoms not only in schizophrenia but also in conditions such as trichotillomania and bipolar depression (Vauzour et al. 2017).

New findings also suggest that higher homocysteine levels, which result from B vitamin deficiencies, are related to several psychiatric disorders, including schizophrenia and bipolar disorder (Liu et al. 2019; Zhao et al. 2022).

Vitamin D acts as a neurosteroid, and low maternal levels have been associated with an increased risk of schizophrenia in offspring (Eyles et al. 2013). Beyond its connection to schizophrenia, emerging evidence suggests that vitamin D intake is related to the likelihood of experiencing isolated subclinical psychotic symptoms. A large-scale study involving 33,623 Swedish women found a significant link between low vitamin D consumption and a higher incidence of psychotic-like symptoms (Hedelin et al. 2010). Furthermore, research has shown that vitamin D levels are often low in children diagnosed with ADHD, suggesting a potential relationship between vitamin D deficiency and ADHD symptoms (Kotsi et al. 2019). A recent randomized controlled trial demonstrated that co-supplementation with daily magnesium and weekly vitamin D over 8 weeks may enhance behavioral function and mental health in children with ADHD, providing a promising intervention for managing this condition (Hemamy et al. 2021).

9.3 The Gut-Brain Axis

9.3.1 Understanding the Gut-Brain Connection

Diet significantly influences the gut-brain axis, an intercommunication between the gut and central nervous systems (Fig. 9.3). Recent research has shown that the deterioration of gut microbiota is related to several neuropsychiatric disorders, including bipolar disorder, depression, and schizophrenia-spectrum disorders (Borkent et al. 2022). Gut bacteria influence brain function by modulating immune responses,

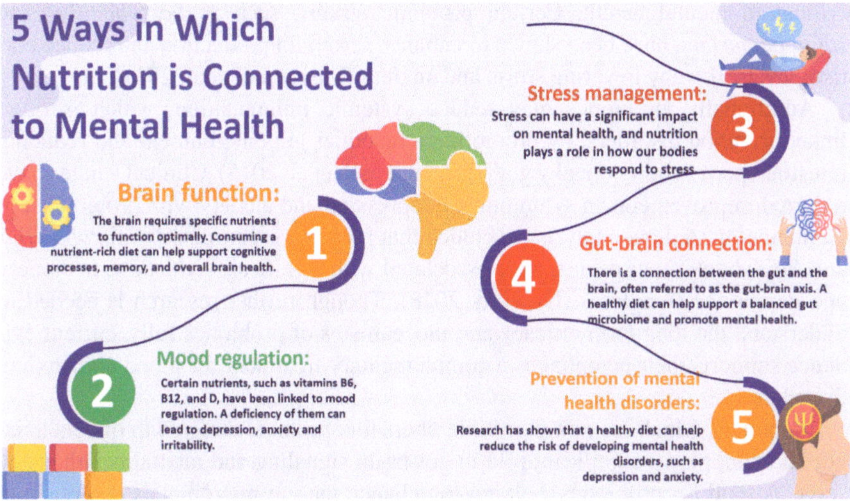

Fig. 9.3 Five ways nutrition influences mental health
This figure illustrates five connections between nutrition and mental health. Nutrient-rich diets enhance brain function, support mood regulation, help manage stress, promote a balanced gut-brain axis, and reduce the risk of mental health disorders like depression and anxiety, highlighting nutrition's essential role in mental well-being

producing neurotransmitters, and affecting neural pathways involved in mood and behavior (Dinan and Cryan 2015; Mayer et al. 2014).

The microbiota in the gut produces various metabolites, including short-chain fatty acids (SCFAs), that affect brain function and influence mood and anxiety levels. A diet rich in fiber, fermented foods, and prebiotics promotes a healthy gut microbiome, reduces inflammation, and helps to reduce symptoms of anxiety and depression (Rea et al. 2016). Contrarily, diets rich in sugars and processed foods disrupt the gut microbiota, leading to increased inflammation, dysbiosis, and heightened anxiety responses (Basso et al. 2024).

9.3.2 Probiotics and Prebiotics in Mental Health

Probiotics are live microorganisms, and they provide health benefits when consumed. They have shown potential therapeutic effects on mood and anxiety disorders through their influence on the gut-brain axis. This complex communication network between the gastrointestinal and central nervous systems plays an important role in regulating mood, emotional well-being, and cognitive function. Recent studies showed that probiotics may positively affect this axis by modulating gut microbiota composition, which influences neurotransmitter production and inflammatory responses associated with mental health disorders (Dinan and Cryan 2015). One of the primary mechanisms by which probiotics may improve mood and reduce anxiety is through the regulation of serotonin and GABA, two neurotransmitters critical to mental health. Certain probiotic strains, such as *Lactobacillus* and *Bifidobacterium*, have been shown to enhance serotonin production and reduce cortisol levels, thereby lowering stress and anxiety (Messaoudi et al. 2011).

Additionally, probiotics may reduce systemic inflammation, which is often linked to mood disorders, by promoting a healthier gut environment and reducing intestinal permeability ("leaky gut") (Steenbergen et al. 2015). Clinical studies have reported improvements in symptoms of depression and anxiety with probiotic supplementation. A meta-analysis concluded that probiotics, particularly *Lactobacillus* and *Bifidobacterium* species, were associated with significant reductions in anxiety and depressive symptoms (Ng et al. 2018). Though further research is needed to understand the long-term efficacy and mechanisms of probiotics fully, current evidence supports their potential as a complementary treatment for mood and anxiety disorders.

Prebiotics, which are non-digestible fibers that promote the growth of beneficial gut bacteria, play a significant role in gut-brain signaling and mental health. They serve as food for probiotics, helping to enhance the gut microbiota's composition and activity. This gut microbiome, in turn, influences the gut-brain axis (Cryan et al. 2019). Prebiotics can modulate this axis, impacting mental health by regulating stress responses, mood, and cognition via fostering the growth of beneficial bacteria. Prebiotics influence mental health through their role in producing SCFAs such as butyrate, acetate, and propionate. These SCFAs are byproducts of prebiotic fermentation in the gut and have been shown to have neuroprotective effects, reduce inflammation, and enhance the integrity of the blood-brain barrier (Dalile et al. 2019). By reducing systemic inflammation and promoting a healthier gut environment, prebiotics may help mitigate anxiety, depression, and stress-related disorders, which are often linked to inflammation and gut dysbiosis (Schmidt et al. 2015). Recent studies suggest that prebiotics can improve mental well-being by modulating the hypothalamic-pituitary-adrenal (HPA) axis, which regulates stress responses. In one study, healthy participants who consumed prebiotics experienced reduced cortisol levels and improved emotional processing, indicating a beneficial effect on stress and anxiety (Schmidt et al. 2015).

9.4 Dietary Patterns and Mental Health Outcomes

9.4.1 Mediterranean Diet

The Mediterranean diet, characterized by high consumption of fruits, vegetables, legumes, fish, and olive oil, has been related to a reduced risk of developing depression and anxiety. Recent meta-analyses have supported that adherence to this diet is inversely associated with depressive symptoms, suggesting that the diet's anti-inflammatory and antioxidant properties contribute to its protective effects. This diet is rich in omega-3 fatty acids, antioxidants, and fiber, which support brain health, reduce inflammation, and promote a healthy gut microbiome (Lassale et al. 2019; Molendijk et al. 2018). The Mediterranean diet is effective in decreasing depressive symptoms in patients with major depression, offering a promising dietary approach for managing mood disorders (Jacka et al. 2017). It has also been shown that higher adherence to the Mediterranean diet was related to lower anxiety levels (Melguizo-Ibáñez et al. 2023).

The Mediterranean diet has also been shown to affect the expression of several genes associated with inflammation, which may reduce the risk of schizophrenia and dementia (Bottero and Potashkin 2020). Furthermore, recent studies suggest that this diet modulates circulating exosomal RNAs, which may play a key role in the Mediterranean diet's beneficial effects on mental health (Mantilla-Escalante et al. 2021). The Mediterranean diet has been associated with improved focus, a decreased incidence of ADHD, and reduced hyperactivity. Nutrient-dense diets support neurotransmitter function and reduce inflammation, contributing to better cognitive and behavioral outcomes in individuals with ADHD (Nigg et al. 2012; Ríos-Hernández et al. 2017).

9.4.2 Plant-Based Diets

Plant-based diets have gained attention for their possible positive influence on mental health, mainly due to their nutrient density, high fiber content, and rich supply of antioxidants and phytochemicals. Such diets, which emphasize whole grains, fruits, vegetables, legumes, nuts, and seeds, are rich in nutrients like vitamins, minerals, and polyphenols, which support brain function and may protect against mental health disorders such as depression and anxiety. One of the critical benefits of plant-based diets is their anti-inflammatory properties. Chronic inflammation has been linked to mental health conditions, and plant-based foods are abundant in antioxidants and polyphenols, which decrease oxidative stress and inflammation in the body and brain (Berk et al. 2013). For example, flavonoids, found in fruits, vegetables, and tea, have been shown to ameliorate cognitive function and mood by enhancing neuroplasticity and reducing inflammation in the brain (Khalid et al. 2017).

Additionally, plant-based diets promote a healthy gut microbiome, which plays a crucial role in the gut-brain axis. A diet rich in plant fiber fosters the growth of beneficial gut bacteria, which produce SCFAs that support brain health and can reduce symptoms of anxiety and depression (Cryan et al. 2019). Several studies have found that individuals who adhere to plant-based diets are less likely to experience depressive symptoms and are more likely to report better overall mental well-being. For instance, a large-scale study in the United Kingdom found that people who consumed more fruits and vegetables had higher mental well-being scores (Ocean et al. 2019). While more research is needed to fully understand the long-term effects of plant-based diets on mental health, current evidence suggests that such diets may offer protective benefits against mood disorders, cognitive decline, and overall poor mental health.

9.4.3 Western Diet and Its Association with Mental Illness

The Western diet, characterized by high levels of processed foods, refined sugars, unhealthy fats, and low fiber content, has been strongly associated with increased systemic inflammation. This inflammatory response stems from the Western diet's disruption of the gut microbiota, largely due to its limited intake of dietary fibers, omega-3 fatty acids, and probiotics, all essential for supporting a healthy gut microbiome. Such gut dysbiosis may activate inflammatory pathways that not only worsen symptoms of anxiety and depression but also interfere with brain function and mental health (Firth et al. 2020).

Beyond its impact on physical health, the Western diet is implicated in mental health issues and cognitive decline through its effects on the brain. Notably, Western diet patterns have been linked to disrupted blood-brain barrier integrity, which may allow harmful substances to reach the brain more easily, promoting neuroinflammation and impairing cognitive function. Additionally, Western dietary patterns have been associated with hippocampal dysfunction, which is critical since the hippocampus plays a major role in memory and emotional regulation (Noble et al. 2017).

Studies have shown that the consumption of refined carbohydrates, a staple in the Western diet, can exacerbate depressive symptoms by increasing inflammation and altering brain function (Firth et al. 2020). Moreover, a diet high in fast foods has been linked to an elevated risk of anxiety and depressive symptoms, likely due to a combination of inflammatory processes and nutrient deficiencies (Tang and Lee 2024).

Research has shown that the Western diet can also exacerbate symptoms of ADHD, as foods high in refined sugars and unhealthy fats may impair blood sugar regulation and promote inflammation. Studies confirm that children and adolescents with ADHD are more likely to consume diets rich in refined sugars and saturated fats while lacking essential nutrients such as omega-3 fatty acids, zinc, iron, and magnesium—all of which are crucial for brain function and attention regulation (Del-Ponte et al. 2019; Talib et al. 2024; Ghoreishy et al. 2021; McWilliams et al. 2022; Effatpanah et al. 2019).

These findings underscore the need to understand how dietary patterns influence mental health and cognition as the prevalence of Western dietary habits continues to rise globally.

9.4.4 Caloric Restriction and Mitochondrial-Epigenetic Modulation

Caloric restriction has been extensively investigated for its possible effect on improving mitochondrial function and influencing epigenetic regulation in psychiatric disorders. Caloric restriction enhances mitochondrial biogenesis and reduces oxidative stress, promoting better cellular function and resilience. It also affects DNA methylation and histone acetylation in ways that enhance resilience to stress and improve cognitive function (Kautzky et al. 2021; Manchishi et al. 2018; Hahn et al. 2017).

Additionally, caloric restriction has been shown to improve depressive symptoms by enhancing mitochondrial function and promoting the expression of brain-derived neurotrophic factor (BDNF) through epigenetic mechanisms. These effects collectively contribute to improved mood and cognitive health (Igwe et al. 2021).

9.5 Special Considerations in Specific Mental Health Conditions

9.5.1 Depression

Recent studies have emphasized omega-3s' significant anti-inflammatory effects, which play an important role in managing depression. Inflammation is increasingly recognized as a contributing factor in mood disorders. Omega-3s are believed to modulate neurotransmission, especially in serotonin and dopamine systems, which are critical for mood regulation (Liu et al. 2015). Deficiencies in omega-3s have been linked to higher rates of depression, and supplementation has been shown to reduce depressive symptoms, particularly when combined with antidepressant therapies (Grosso et al. 2014; Liao et al. 2019).

Research indicates that zinc deficiency is linked to a heightened risk of depression, with supplementation showing promising antidepressant effects by modulating neurotransmitter systems and reducing inflammation. Magnesium is also noted for its potential to alleviate depressive symptoms, as it plays a key role in maintaining neurotransmitter balance and supporting neuroplasticity, which are crucial for mood regulation. Selenium, known for its powerful antioxidant properties, contributes to brain health by reducing oxidative stress, potentially lowering depression risk, and promoting cognitive resilience (Wang et al. 2018).

Low folic acid levels are associated with depression, causing low levels of serotonin, dopamine, and norepinephrine. It has been shown that the risk of major depression is lower in individuals with high folic acid levels, and the effects of antidepressant drugs are lower in patients who have low folic acid levels (Ekinci and Sanlier 2023).

9.5.2 Anxiety

Caffeine is the primary active compound in coffee and has a complex relationship with anxiety. While moderate coffee consumption is associated with enhanced alertness and cognitive performance, high doses of caffeine can increase anxiety levels by stimulating the central nervous system and triggering the release of stress hormones such as cortisol and adrenaline (Rogers et al. 2008). Excessive caffeine intake has been linked to heightened feelings of nervousness, restlessness, and even panic attacks in individuals sensitive to its effects (Klevebrant and Frick 2022).

Adenosine is a neurotransmitter that promotes relaxation and sleep. Caffeine acts as an adenosine receptor antagonist. This antagonism increases excitatory neurotransmitter release, leading to heightened arousal, which may intensify anxiety symptoms in susceptible individuals. For individuals with pre-existing anxiety disorders, high caffeine consumption can further aggravate symptoms, often impairing sleep quality and increasing the risk of insomnia, which can negatively affect mental health and emotional regulation (Temple et al. 2017).

Alcohol has a complex and bidirectional relationship with anxiety. While many individuals consume alcohol to alleviate stress or reduce social anxiety, its effects are often short-lived and can lead to increased anxiety in the long term. Initially, alcohol acts as a depressant, enhancing the activity of GABA, a neurotransmitter associated with relaxation and decreased neural excitability. This effect temporarily reduces feelings of anxiety; however, as the body metabolizes alcohol, these calming effects wane, and symptoms of anxiety can often rebound, resulting in heightened stress and nervousness (Hurlocker et al. 2023). Chronic alcohol use can also lead to changes in brain chemistry that heighten anxiety over time. Research suggests that repeated alcohol consumption disrupts the brain's natural balance of neurotransmitters, particularly GABA and glutamate, increasing neural excitability and contributing to anxiety during withdrawal periods. This cycle of drinking to alleviate anxiety, followed by withdrawal-induced anxiety, can lead to a harmful feedback loop known as "alcohol-induced anxiety," which exacerbates mental health issues and often contributes to the development of alcohol dependence (Gilpin and Koob 2008). Individuals with pre-existing anxiety disorders are particularly vulnerable to this cycle, as they may rely on alcohol as a coping mechanism for their symptoms, leading to both a worsening of anxiety and a higher risk of alcohol use disorder. Studies have shown that approximately 20% of people with social anxiety disorder, for example, meet the criteria for alcohol use disorder and that their alcohol consumption often exacerbates the severity of their anxiety symptoms in the long term (Smith and Randall 2012).

Additionally, alcohol consumption disrupts sleep quality, which is closely linked to anxiety. Poor sleep due to alcohol can impair emotional regulation, increase sensitivity to stress, and worsen overall mental health, contributing to a cyclical relationship between alcohol use and anxiety (Roehrs and Roth 2001). Given these factors, it is recommended that individuals prone to anxiety limit or avoid alcohol consumption, as it may ultimately contribute to increased symptoms and poorer mental health outcomes.

9.5.3 Bipolar Disorder

9.5.3.1 Nutritional Strategies for Mood Stabilization and Brain Health.

Vitamins and minerals are essential for neurotransmitter synthesis and overall brain health, suggesting that deficiencies in these nutrients may contribute to the mood dysregulation seen in bipolar disorder. Omega-3 fatty acids have anti-inflammatory properties and could reduce the frequency and severity of mood episodes in bipolar disorder. A nutrient-dense diet should be a complementary approach to traditional treatments for bipolar disorder. Individuals may improve their mood stability and well-being by prioritizing whole foods rich in essential nutrients. It has been shown that certain nutrients, such as omega-3 fatty acids, vitamins (including vitamin D and B vitamins), and minerals (like magnesium and zinc), play crucial roles in mood stabilization and cognitive function in patients with bipolar disorder (Gabriel et al. 2023).

9.5.4 Eating Disorders

Anorexia nervosa (AN) and bulimia nervosa (BN) are severe eating disorders that significantly impact physical and mental health. Both conditions are characterized by abnormal eating behaviors, body image disturbances, and a preoccupation with weight. While AN is primarily defined by restrictive eating and extreme weight loss, BN involves episodes of binge eating followed by compensatory behaviors such as vomiting, excessive exercise, or laxative use (Bühren 2022).

Nutritional rehabilitation is a crucial component of the treatment for both AN and BN, as it aims to restore healthy eating patterns and weight while addressing the underlying psychological factors. In individuals with AN, the primary goal is to achieve a safe and gradual weight restoration, which is essential for physical recovery and improved mental health outcomes. For those with BN, nutritional rehabilitation focuses on normalizing eating patterns and breaking the cycle of bingeing and purging, which can lead to nutritional deficiencies and other health complications (Bühren 2022; Marzola et al. 2013).

Despite its importance, nutritional rehabilitation presents several challenges. One significant hurdle is the patient's resistance to treatment, which is often rooted in their fear of weight gain and anxiety about food. This resistance can lead to noncompliance with nutritional guidelines and therapeutic interventions. Additionally, the psychological aspects of these disorders complicate the rehabilitation process. Many individuals struggle with distorted body image and self-esteem issues, making it difficult to accept the necessary dietary changes (Marzola et al. 2013).

Furthermore, the nutritional needs of individuals recovering from AN and BN are complex. Nutritional deficiencies are common in both disorders, often resulting from restrictive eating or disordered eating behaviors. Clinicians need to assess and address these deficiencies to prevent complications such as electrolyte imbalances, which can be life-threatening. A comprehensive approach that includes individualized meal plans, nutritional education, and ongoing support is crucial for successful rehabilitation (Marzola et al. 2013).

9.6 Nutritional Therapies in Mental Health

9.6.1 Omega-3 Fatty Acids

Omega-3s have significant anti-inflammatory effects, and they modulate neurotransmission, especially in serotonin and dopamine systems, which are critical for mood regulation (Liu et al. 2015). In addition to their anti-inflammatory properties, omega-3s support mitochondrial function by maintaining membrane integrity and reducing oxidative stress. For instance, omega-3 supplementation improves mitochondrial biogenesis and function, correlating with reduced depressive symptoms and enhanced overall brain health (Borja-Magno et al. 2023). Omega-3 fatty acids may also affect the regulation of microRNAs, small non-coding RNAs that modulate gene expression at the post-transcriptional level. Dysregulated microRNA expression has been observed in schizophrenia, affecting genes involved in synaptic plasticity and neurotransmitter function (Khavari and Cairns 2020; Beveridge and Cairns 2012).

9.6.2 Herbal Interventions

The increasing interest in herbal and nutraceutical interventions for mental health reflects a growing recognition of natural products' role in promoting psychological well-being. Herbal remedies and nutraceuticals are often considered complementary or alternative treatments for various mental health conditions, including anxiety, depression, and cognitive decline. Several herbs have been studied for their potential effects on mental health. For instance, St. John's Wort (Hypericum perforatum) is one of the most researched herbs for depression. Studies have demonstrated its efficacy in alleviating mild to moderate depressive symptoms, often with fewer side effects compared to conventional antidepressants (Linde et al. 2008). Another notable herb, Rhodiola rosea, is believed to have adaptogenic properties that may help the body cope with stress. Studies suggest that Rhodiola can reduce fatigue and improve mood, particularly in individuals facing stressful situations (Panossian and Wikman 2010). Chamomile (Matricaria chamomilla) has also gained attention for its anxiolytic properties. Research indicates that chamomile extract can significantly reduce the symptoms of generalized anxiety disorders, potentially due to its ability to modulate neurotransmitter activity (Srivastava et al. 2010).

Ginkgo biloba, one of the oldest living tree species, has been used in traditional medicine for centuries, particularly in Asian cultures. Its leaves contain various active compounds, including flavonoids and terpenoids, which are thought to contribute to its therapeutic effects. Ginkgo biloba is primarily recognized for its potential benefits in cognitive function and mental health, particularly in older adults. Numerous studies have investigated the effects of Ginkgo biloba on cognitive function, especially in the context of age-related cognitive decline and dementia. Research suggests that Ginkgo biloba may enhance cognitive performance, particularly in memory, attention, and information processing. A meta-analysis indicated that Ginkgo biloba extract may improve cognitive function in individuals with dementia and mild cognitive impairment, providing a degree of protection against further decline. The proposed mechanisms behind these cognitive benefits include enhancing cerebral blood flow, reducing oxidative stress, and modulating neurotransmitter systems. The flavonoids in Ginkgo biloba are believed to improve blood circulation to the brain, thereby promoting oxygen and nutrient delivery, which is vital for optimal brain function (Tan et al. 2015).

Beyond its cognitive benefits, Ginkgo biloba has also been studied for its potential effects on mood and anxiety disorders. Some studies suggest that Ginkgo biloba may have anxiolytic properties, helping to reduce symptoms of anxiety and improve overall mood. A randomized controlled trial found that Ginkgo biloba supplementation significantly reduced anxiety levels in participants compared to a placebo group, indicating its potential as an adjunctive treatment for anxiety disorders (Woelk et al. 2007). The anxiolytic effects may be attributed to Ginkgo biloba's ability to modulate neurotransmitter systems, including serotonin and dopamine, which are crucial for mood regulation. Additionally, its antioxidant properties may help reduce the impact of stress on mental health by combating oxidative stress and inflammation, both of which are implicated in mood disorders (Woelk et al. 2007).

9.6.3 Functional Foods for Mental Health

Cognitive health is crucial for overall well-being, and dietary interventions can significantly support brain function. Various foods, such as cocoa, dark chocolate, green tea, and turmeric, have garnered attention for their potential cognitive-enhancing properties. Each of these foods contains bioactive compounds that may positively influence cognitive function, memory, and mood.

Cocoa is rich in flavonoids, particularly epicatechin, and is associated with improved cognitive function due to its antioxidant and anti-inflammatory properties. Cocoa products can enhance cerebral blood flow and promote neuroplasticity, which may contribute to better cognitive outcomes. Moderate intake of cocoa and cocoa-rich products is linked to improved cognitive performance in tasks requiring memory and attention (Martín et al. 2020). Dark chocolate, particularly varieties containing 70% cocoa or higher, is rich in flavonoids, especially epicatechin. These flavonoids are known for their antioxidant and anti-inflammatory properties, which may protect the brain from oxidative stress and inflammation, both of which are associated with cognitive decline. Research has indicated that the consumption of dark chocolate can improve cognitive performance, particularly in tasks requiring memory and attention. A study showed that participants who consumed dark chocolate experienced enhanced cognitive function and mood compared to those who consumed milk chocolate (Sumiyoshi et al. 2019).

Green tea is another potent source of cognitive-enhancing compounds, primarily due to its high levels of polyphenols, particularly catechins, and L-theanine, an amino acid known for its calming effects. The combination of caffeine and

L-theanine in green tea has a synergistic effect, improving alertness and reducing anxiety, which can further support cognitive performance. Research suggests that the consumption of green tea can enhance cognitive function, particularly in areas related to working memory and attention, and decrease anxiety (Mancini et al. 2017).

Turmeric, a spice derived from the root of the Curcuma longa plant, contains curcumin, a polyphenolic compound known for its anti-inflammatory and antioxidant properties. Curcumin has been shown to cross the blood-brain barrier and may exert neuroprotective effects, making it a candidate for enhancing cognitive health (Sanmukhani et al. 2014). Research indicates that curcumin supplementation may improve cognitive function in older adults and individuals with mild cognitive impairment. A systematic review revealed that curcumin could enhance memory and attention, potentially due to its ability to reduce inflammation and oxidative stress in the brain. Additionally, curcumin may promote the growth of new neurons and improve synaptic function, which is essential for maintaining cognitive health (Francis et al. 2024).

9.7 Integrating Nutrition into Mental Health Care

Integrating nutrition into mental health care involves a collaborative, interdisciplinary approach. Dietitians, nutritionists, and mental health professionals collaborate to create comprehensive treatment plans. This holistic approach recognizes the interconnectedness of physical and mental health, with nutrition as a key factor influencing mental well-being.

9.7.1 Interdisciplinary Approaches, Patient Education, and Nutritional Counseling

Dietitians and nutritionists play a vital role in assessing and optimizing patients' diets to support mental health. These professionals evaluate dietary intake, identify nutrient deficiencies, and design individualized nutrition plans to address both physical and psychological needs. For instance, dietitians can guide patients on nutrient-dense diets rich in omega-3 fatty acids, B vitamins, zinc, and magnesium, all known to support cognitive function and mood stability. Additionally, they can educate patients on the impact of certain dietary patterns, like high sugar and processed food intake, which are associated with increased inflammation and mental health risks. In cases of specific mental health disorders like depression or anxiety, dietitians can recommend foods or supplements that may alleviate symptoms, such as probiotics for gut health, which are increasingly linked to mental health outcomes (Adan et al. 2019).

Mental health professionals, including psychologists, psychiatrists, and counselors, integrate nutritional insights into therapeutic practices to enhance the effectiveness of interventions. Psychiatrists, for example, can consider nutritional deficiencies when diagnosing or adjusting medications, as some micronutrient imbalances can mimic or exacerbate symptoms of mental health disorders (Adan et al. 2019). Meanwhile, psychologists and counselors can offer psychoeducation to help patients understand the relationship between nutrition and mental health, potentially motivating them to adopt healthier dietary habits as part of their self-care routines. In a holistic care model, interdisciplinary collaboration is essential. Dietitians, nutritionists, and mental health professionals collaborate to align their expertise, ensuring that patients receive well-rounded support. Regular team meetings, shared health records, and open communication channels between professionals enable comprehensive care that addresses both dietary and psychological factors. For example, a patient with depression who exhibits symptoms of fatigue and low motivation might benefit from both psychological counseling and nutritional interventions targeting iron and vitamin B12 levels. Further, integrative clinics that include dietitians, nutritionists, and mental health professionals facilitate a seamless experience for patients. These settings make implementing individualized treatment plans easier, tracking progress holistically, and adjusting as needed.

9.8 Challenges and Future Directions

9.8.1 Barriers to Implementing Nutritional Interventions

Incorporating nutritional interventions into mental health care is promising, yet several barriers can hinder their widespread adoption and effectiveness. Socioeconomic challenges, cultural considerations, and limited accessibility to nutrient-rich foods are prominent obstacles that affect individuals' ability to achieve and maintain a diet that supports mental well-being. Socioeconomic status significantly influences dietary choices and the ability to adopt nutrition-focused interventions. Individuals with limited financial resources often experience food insecurity, making it difficult to prioritize nutrient-dense foods that support mental health. Fresh produce, lean proteins, and whole grains can be more expensive than processed foods, making them less accessible for individuals from low-income backgrounds. Additionally, these financial limitations may lead to prioritizing cheaper, calorie-dense foods often high in sugar, unhealthy fats, and additives—ingredients that can contribute to inflammation and negatively affect mood and cognitive health (Desbouys et al. 2020).

Cultural beliefs and practices shape dietary habits and perceptions of mental health, and they can serve as both facilitators and barriers to implementing nutritional interventions. Some cultures emphasize specific foods or dietary practices that may not align with recommended nutritional guidelines for mental health, creating challenges for health professionals attempting to provide culturally sensitive dietary recommendations. For instance, certain dietary traditions might exclude particular food groups, like animal proteins or dairy, which could impact nutrient availability if not appropriately substituted (Desbouys et al. 2020).

9.8.2 Future Research in Nutritional Psychiatry

The field of nutritional psychiatry is rapidly evolving, with emerging research areas that promise to deepen our understanding of the connection between diet and mental health. Future studies aim to clarify how specific nutrients, dietary patterns, and gut-brain interactions influence mental well-being, opening avenues for more effective, personalized interventions. One of the most promising areas of future research in nutritional psychiatry is the exploration of the gut-brain axis and its impact on mental health. The gut microbiome, a complex community of microorganisms in the digestive tract, has been linked to mood regulation, stress response, and cognitive function. Studies are beginning to examine how dietary interventions—such as probiotics, prebiotics, and fermented foods—affect the gut microbiome and, in turn, mental health outcomes (Cryan et al. 2019). As research continues, scientists hope to identify specific bacterial strains or microbial metabolites that support mental health, potentially paving the way for microbiome-targeted therapies.

Another key area of future research is the study of nutraceuticals—nutrient-derived supplements—and plant-based compounds, such as polyphenols, flavonoids, and adaptogens, which show potential benefits for mental health. Compounds from sources like turmeric (curcumin), green tea (L-theanine), and saffron have shown preliminary efficacy in alleviating symptoms of depression, anxiety, and cognitive decline (Sanmukhani et al. 2014; Mancini et al. 2017). Future studies are expected to provide more robust evidence on the mechanisms of these compounds, as well as establish effective dosages and safety profiles, enabling them to be incorporated more widely into treatment protocols.

As nutritional psychiatry matures, there is growing interest in integrative approaches that combine dietary interventions with traditional mental health treatments, such as cognitive behavioral therapy and pharmacotherapy. Future research may provide insights into how these approaches can synergistically improve outcomes for patients with mood and anxiety disorders. For instance, combining dietary changes with therapy could enhance patients' resilience to stress and increase the effectiveness of treatments for conditions like depression and anxiety (Adan et al. 2019).

9.9 Conclusion

9.9.1 Summary of Key Points

Diet and nutrition have a profound impact on mental health and behavior. Specific nutrients, such as omega-3 fatty acids, B vitamins, antioxidants, polyphenols, and tryptophan, play critical roles in neurotransmitter synthesis, mitochondrial function, brain inflammation, and oxidative stress, all influencing the development and course of psychiatric disorders. Dietary patterns that promote brain health, such as the Mediterranean diet, can help reduce the risk of psychiatric disorders, while unhealthy diets rich in processed foods can exacerbate symptoms. Understanding the mechanisms of nutrition on mental health provides valuable insights for integrating nutritional interventions into psychiatric care.

9.9.2 Recommendations for Clinical Practice

A diet based on whole foods—such as fruits, vegetables, whole grains, lean proteins, nuts, and seeds—has been associated with improved mood, lower levels of stress, and reduced symptoms of depression and anxiety. Clinicians can recommend that patients prioritize minimally processed foods and increase their intake of fruits and vegetables, as they are rich in fiber, vitamins, and antioxidants, which are essential for brain health and reducing inflammation.

9.9.3 Call for Further Research

The future of nutritional psychiatry holds exciting potential to transform mental health care through targeted, evidence-based dietary interventions. As the field advances, emerging research areas—such as the gut-brain axis, precision nutrition, nutraceuticals, and integrative therapies—promise more personalized and effective treatments for mental health conditions, fostering a holistic approach to mental well-being.

References

Adan RAH, van der Beek EM, Buitelaar JK, Cryan JF, Hebebrand J, Higgs S, Schellekens H, Dickson SL (2019) Nutritional psychiatry: towards improving mental health by what you eat. Eur Neuropsychopharmacol 29(12):1321–1332. https://doi.org/10.1016/j.euroneuro.2019.10.011

American Psychiatric Association (2022) Diagnostic and statistical manual of mental disorders, fifth edn, text revision. https://doi.org/10.1176/appi.books.9780890425787

Basso M, Zorzan I, Johnstone N, Barberis M, Cohen KK (2024) Diet quality and anxiety: a critical overview with focus on the gut microbiome. Front Nutr 11:1346483. https://doi.org/10.3389/fnut.2024.1346483

Bender A, Hagan KE, Kingston N (2017) The association of folate and depression: a meta-analysis. J Psychiatr Res 95:9–18. https://doi.org/10.1016/j.jpsychires.2017.07.019

Berk M, Williams LJ, Jacka FN, O'Neil A, Pasco JA, Moylan S, Allen NB, Stuart AL, Hayley AC, Byrne ML, Maes M (2013) So depression is an inflammatory disease, but where does the inflammation come from? BMC Med 11:200. https://doi.org/10.1186/1741-7015-11-200

Beveridge NJ, Cairns MJ (2012) MicroRNA dysregulation in schizophrenia. Neurobiol Dis 46(2):263–271. https://doi.org/10.1016/j.nbd.2011.12.029

Borja-Magno AI, Furuzawa-Carballeda J, Guevara-Cruz M, Arias C, Granados J, Bourges H, Tovar AR, Sears B, Noriega LG, Gómez FE (2023) Supplementation with EPA and DHA omega-3 fatty acids improves peripheral immune cell mitochondrial dysfunction and inflammation in subjects with obesity. J Nutr Biochem 120:109415. https://doi.org/10.1016/j.jnutbio.2023.109415

Borkent J, Ioannou M, Laman JD, Haarman BCM, Sommer IEC (2022) Role of the gut microbiome in three major psychiatric disorders. Psychol Med 52(7):1222–1242. https://doi.org/10.1017/S0033291722000897

Bottero V, Potashkin JA (2020) A comparison of gene expression changes in the blood of individuals consuming diets supplemented with olives, nuts or long-chain omega-3 fatty acids. Nutrients 12(12):3765. https://doi.org/10.3390/nu12123765

Bühren K (2022) 3.14 nutritional rehabilitation in eating disorders. World Rev Nutr Diet 124:315–320. https://doi.org/10.1159/000516734

Cryan JF, O'Riordan KJ, Cowan CSM, Sandhu KV, Bastiaanssen TFS, Boehme M, Codagnone MG, Cussotto S, Fulling C, Golubeva AV, Guzzetta KE, Jaggar M, Long-Smith CM, Lyte JM, Martin JA, Molinero-Perez A, Moloney G, Morelli E, Morillas E, O'Connor R, Cruz-Pereira JS, Peterson VL, Rea K, Ritz NL, Sherwin E, Spichak S, Teichman EM, van de Wouw M, Ventura-Silva AP, Wallace-Fitzsimons SE, Hyland N, Clarke G, Dinan TG (2019) The microbiota-gut-brain axis. Physiol Rev 99(4):1877–2013. https://doi.org/10.1152/physrev.00018.2018

Dalile B, Van Oudenhove L, Vervliet B, Verbeke K (2019) The role of short-chain fatty acids in microbiota-gut-brain communication. Nat Rev Gastroenterol Hepatol 16(8):461–478. https://doi.org/10.1038/s41575-019-0157-3

Del-Ponte B, Quinte GC, Cruz S, Grellert M, Santos IS (2019) Dietary patterns and attention deficit/hyperactivity disorder (ADHD): a systematic review and meta-analysis. J Affect Disord 252:160–173. https://doi.org/10.1016/j.jad.2019.04.061

Desbouys L, Méjean C, De Henauw S, Castetbon K (2020) Socio-economic and cultural disparities in diet among adolescents and young adults: a systematic review. Public Health Nutr 23(5):843–860. https://doi.org/10.1017/S1368980019002362

Dinan TG, Cryan JF (2015) The impact of gut microbiota on brain and behaviour: implications for psychiatry. Curr Opin Clin Nutr Metab Care 18:552–558. https://doi.org/10.1097/MCO.0000000000000221

Du J, Zhu M, Bao H, Li B, Dong Y, Xiao C, Zhang GY, Henter I, Rudorfer M, Vitiello B (2016) The role of nutrients in protecting mitochondrial function and neurotransmitter signaling: implications for the treatment of depression, PTSD, and suicidal behaviors. Crit Rev Food Sci Nutr 56(15):2560–2578. https://doi.org/10.1080/10408398.2013.876960

Effatpanah M, Rezaei M, Effatpanah H, Effatpanah Z, Varkaneh HK, Mousavi SM, Fatahi S, Rinaldi G, Hashemi R (2019) Magnesium status and attention deficit hyperactivity disorder (ADHD): a meta-analysis. Psychiatry Res 274:228–234. https://doi.org/10.1016/j.psychres.2019.02.043

Ekinci GN, Sanlier N (2023) The relationship between nutrition and depression in the life process: a mini-review. Exp Gerontol 172:112072. https://doi.org/10.1016/j.exger.2022.112072

Eyles DW, Burne TH, McGrath JJ (2013) Vitamin D, eff ects on brain development, adult brain function and the links between low levels of vitamin D and neuropsychiatric disease. Front Neuroendocrinol 34:47–64. https://doi.org/10.1016/j.yfrne.2012.07.001

Field DT, Cracknell RO, Eastwood JR, Scarfe P, Williams CM, Zheng Y, Tavassoli T (2022) High-dose vitamin B6 supplementation reduces anxiety and strengthens visual surround suppression. Hum Psychopharmacol 37(6):e2852. https://doi.org/10.1002/hup.2852

Firth J, Gangwisch JE, Borisini A, Wootton RE, Mayer EA (2020) Food and mood: how do diet and nutrition affect mental wellbeing? BMJ 369:m2382. https://doi.org/10.1136/bmj.m2382. Erratum in: BMJ. 2020;371:m4269. 10.1136/bmj.m4269

Francis AJ, Sreenivasan C, Parikh A, AlQassab O, Kanthajan T, Pandey M, Nwosu M (2024) Curcumin and cognitive function: a systematic review of the effects of curcumin on adults with and without neurocognitive disorders. Cureus 16(8):e67706. https://doi.org/10.7759/cureus.67706

Gabriel FC, Oliveira M, Bruna De M, Martella BM, Brietzke E, Jacka FN, Lafer B (2023) Nutrition and bipolar disorder: a systematic review. Nutr Neurosci 26(7):637–651. https://doi.org/10.1080/1028415X.2022.2077031

Ghoreishy SM, Ebrahimi Mousavi S, Asoudeh F, Mohammadi H (2021) Zinc status in attention-deficit/hyperactivity disorder: a systematic review and meta-analysis of observational studies. Sci Rep 11(1):14612. https://doi.org/10.1038/s41598-021-94124-5

Gilpin NW, Koob GF (2008) Neurobiology of alcohol dependence: focus on motivational mechanisms. Alcohol Res Health 31(3):185–195

Glenn JM, Madero EN, Bott NT (2019) Dietary protein and amino acid intake: links to the maintenance of cognitive health. Nutrients 11(6):1315. https://doi.org/10.3390/nu11061315

Goh KK, Chen CY, Chen CH, Lu ML (2021) Effects of omega-3 polyunsaturated fatty acids supplements on psychopathology and metabolic parameters in schizophrenia: a meta-analysis of randomized controlled trials. J Psychopharmacol 35(3):221–235. https://doi.org/10.1177/0269881120981392

Grosso G, Pajak A, Marventano S, Castellano S, Galvano F, Bucolo C, Drago F, Caraci F (2014) Role of omega-3 fatty acids in the treatment of depressive disorders: a comprehensive meta-analysis of randomized clinical trials. PLoS One 9(5):e96905. https://doi.org/10.1371/journal.pone.0096905

Hahn O, Grönke S, Stubbs TM, Ficz G, Hendrich O, Krueger F, Andrews S, Zhang Q, Wakelam MJ, Beyer A, Reik W, Partridge L (2017) Dietary restriction protects from age-associated DNA methylation and induces epigenetic reprogramming of lipid metabolism. Genome Biol 18(1):56. https://doi.org/10.1186/s13059-017-1187-1

Han QQ, Shen TT, Wang F, Wu PF, Chen JG (2018) Preventive and therapeutic potential of vitamin C in mental disorders. Curr Med Sci 38(1):1–10. https://doi.org/10.1007/s11596-018-1840-2

Hedelin M, Lof M, Olsson M, Lewander T, Nilsson B, Hultman CM, Weiderpass E (2010) Dietary intake of fish, omega-3, omega-6 polyunsaturated fatty acids and vitamin D and the prevalence of psychotic-like symptoms in a cohort of 33000 women from the general population. BMC Psychiatry 10:38. https://doi.org/10.1186/1471-244X-10-38

Hemamy M, Pahlavani N, Amanollahi A, Islam SMS, McVicar J, Askari G, Malekahmadi M (2021) The effect of vitamin D and magnesium supplementation on the mental health status of attention-deficit hyperactive children: a randomized controlled trial. BMC Pediatr 21(1):178. https://doi.org/10.1186/s12887-021-02631-1

Hurlocker MC, Carlon H, Pearson MR, Hijaz D (2023) Trajectories of change in subclinical anxiety and alcohol use during alcohol treatment: a parallel process growth model. Drug Alcohol Depend 246:109838. https://doi.org/10.1016/j.drugalcdep.2023.109838

Igwe O, Sone M, Matveychuk D, Baker GB, Dursun SM (2021) A review of effects of calorie restriction and fasting with potential relevance to depression. Prog Neuro-Psychopharmacol Biol Psychiatry 111:110206. https://doi.org/10.1016/j.pnpbp.2020.110206

Jacka FN, O'Neil A, Opie R, Itsiopoulos C, Cotton S, Mohebbi M, Castle D, Dash S, Mihalopoulos C, Chatterton ML, Brazionis L, Dean OM, Hodge AM, Berk M (2017) A randomised controlled trial of dietary improvement for adults with major depression (the 'SMILES' trial). BMC Med 15:23. https://doi.org/10.1186/s12916-017-0791-y

Kautzky A, Heneis K, Stengg K, Fröhlich S, Kautzky-Willer A (2021) Short term caloric restriction and biofeedback enhance psychological wellbeing and reduce overweight in healthy women. J Pers Med 11(11):1096. https://doi.org/10.3390/jpm11111096

Khalid S, Barfoot KL, May G, Lamport DJ, Reynolds SA, Williams CM (2017) Effects of acute blueberry flavonoids on mood in children and young adults. Nutrients 9(2):158. https://doi.org/10.3390/nu9020158

Khavari B, Cairns MJ (2020) Epigenomic dysregulation in schizophrenia: in search of disease etiology and biomarkers. Cells 9(8):1837. https://doi.org/10.3390/cells9081837

Kikuchi AM, Tanabe A, Iwahori Y (2021) A systematic review of the effect of L-tryptophan supplementation on mood and emotional functioning. J Diet Suppl 18(3):316–333. https://doi.org/10.1080/19390211.2020.1746725

Kirkbride JB, Anglin DM, Colman I, Dykxhoorn J, Jones PB, Patalay P, Pitman A, Soneson E, Steare T, Wright T, Griffiths SL (2024) The social determinants of mental health and disorder: evidence, prevention and recommendations. World Psychiatry 23(1):58–90. https://doi.org/10.1002/wps.21160

Klevebrant L, Frick A (2022) Effects of caffeine on anxiety and panic attacks in patients with panic disorder: a systematic review and meta-analysis. Gen Hosp Psychiatry 74:22–31. https://doi.org/10.1016/j.genhosppsych.2021.11.005

Kotsi E, Kotsi E, Perrea DN (2019) Vitamin D levels in children and adolescents with attention-deficit hyperactivity disorder (ADHD): a meta-analysis. Atten Defic Hyperact Disord 11(3):221–232. https://doi.org/10.1007/s12402-018-0276-7

Lassale C, Batty GD, Baghdadli A, Jacka F, Sánchez-Villegas A, Kivimäki M, Akbaraly T (2019) Healthy dietary indices and risk of depressive outcomes: a systematic review and meta-analysis of observational studies. Mol Psychiatry 24(7):965–986. https://doi.org/10.1038/s41380-018-0237-8

Liao Y, Xie B, Zhang H, He Q, Guo L, Subramanieapillai M, Fan B, Lu C, McIntyre RS (2019) Efficacy of omega-3 PUFAs in depression: a meta-analysis. Transl Psychiatry 9(1):190. https://doi.org/10.1038/s41398-019-0515-5

Linde K, Berner MM, Kriston L (2008) St John's wort for major depression. Cochrane Database Syst Rev 2008(4):CD000448. https://doi.org/10.1002/14651858.CD000448.pub3

Liu JJ, Green P, John Mann J, Rapoport SI, Sublette ME (2015) Pathways of polyunsaturated fatty acid utilization: implications for brain function in neuropsychiatric health and disease. Brain Res 1597:220–246. https://doi.org/10.1016/j.brainres.2014.11.059

Liu Y, Tao H, Yang X, Huang K, Zhang X, Li C (2019) Decreased serum oxytocin and increased homocysteine in first-episode schizophrenia patients. Front Psych 10:217. https://doi.org/10.3389/fpsyt.2019.00217

Luchtman DW, Song C (2013) Cognitive enhancement by omega-3 fatty acids from child-hood to old age: findings from animal and clinical studies. Neuropharmacology 64:550–565. https://doi.org/10.1016/j.neuropharm.2012.07.019

Manchishi SM, Cui RJ, Zou XH, Cheng ZQ, Li BJ (2018) Effect of caloric restriction on depression. J Cell Mol Med 22(5):2528–2535. https://doi.org/10.1111/jcmm.13418

Mancini E, Beglinger C, Drewe J, Zanchi D, Lang UE, Borgwardt S (2017) Green tea effects on cognition, mood and human brain function: a systematic review. Phytomedicine 34:26–37. https://doi.org/10.1016/j.phymed.2017.07.008

Mantilla-Escalante DC, de Las L, Hazas MC, Crespo MC, Martín-Hernández R, Tomé-Carneiro J, Del Pozo-Acebo L, Salas-Salvadó J, Bulló M, Dávalos A (2021) Mediterranean diet enriched in extra-virgin olive oil or nuts modulates circulating exosomal non-coding RNAs. Eur J Nutr 60(8):4279–4293. https://doi.org/10.1007/s00394-021-02594-0

Martín MA, Goya L, de Pascual-Teresa S (2020) Effect of cocoa and cocoa products on cognitive performance in young adults. Nutrients 12(12):3691. https://doi.org/10.3390/nu12123691

Marzola E, Nasser JA, Hashim SA, Shih PA, Kaye WH (2013) Nutritional rehabilitation in anorexia nervosa: review of the literature and implications for treatment. BMC Psychiatry 13:290. https://doi.org/10.1186/1471-244X-13-290

Mayer EA, Knight R, Mazmanian SK, Cryan JF, Tillisch K (2014) Gut microbes and the brain: paradigm shift in neuroscience. J Neurosci 34:15490–15496. https://doi.org/10.1523/JNEUROSCI.3299-14.2014

McWilliams S, Singh I, Leung W, Stockler S, Ipsiroglu OS (2022) Iron deficiency and common neurodevelopmental disorders – a scoping review. PLoS One 17(9):e0273819. https://doi.org/10.1371/journal.pone.0273819

Melguizo-Ibáñez E, González-Valero G, Badicu G, Yagin FH, Alonso-Vargas JM, Ardigò LP, Puertas-Molero P (2023) Mediterranean diet adherence on self-concept and anxiety as a function of weekly physical activity: an explanatory model in higher education. Front Nutr 10:1215359. https://doi.org/10.3389/fnut.2023.1215359

Messaoudi M, Lalonde R, Violle N et al (2011) Assessment of psychotropic-like properties of a probiotic formulation (Lactobacillus helveticus R0052 and Bifidobacterium longum R0175) in rats and human subjects. Br J Nutr 105(5):755–764. https://doi.org/10.1017/S0007114510004319

Molendijk M, Molero P, Ortuño Sánchez-Pedreño F, Van der Does W, Angel Martínez-González M (2018) Diet quality and depression risk: a systematic review and dose-response meta-analysis of prospective studies. J Affect Disord 226:346–354. https://doi.org/10.1016/j.jad.2017.09.022

Moritz B, Schmitz AE, Rodrigues ALS, Dafre AL, Cunha MP (2020) The role of vitamin C in stress-related disorders. J Nutr Biochem 85:108459. https://doi.org/10.1016/j.jnutbio.2020.108459

Ng QX, Peters C, Ho CYX, Lim DY, Yeo WS (2018) A meta-analysis of the use of probiotics to alleviate depressive symptoms. J Affect Disord 228:13–19. https://doi.org/10.1016/j.jad.2017.11.063

Nigg JT, Lewis K, Edinger T, Falk M (2012) Meta-analysis of attention-deficit/hyperactivity disorder or attention-deficit/hyperactivity disorder symptoms, restriction diet, and synthetic food color additives. J Am Acad Child Adolesc Psychiatry 51(1):86–97.e8. https://doi.org/10.1016/j.jaac.2011.10.015

Noble EE, Hsu TM, Kanoski SE (2017) Gut to brain dysbiosis: mechanisms linking Western diet consumption, the microbiome, and cognitive impairment. Front Behav Neurosci 11:9. https://doi.org/10.3389/fnbeh.2017.00009

Ocean N, Howley P, Ensor J (2019) Lettuce be happy: a longitudinal UK study on the relationship between fruit and vegetable consumption and well-being. Soc Sci Med 222:335–345. https://doi.org/10.1016/j.socscimed.2018.12.017

Panossian A, Wikman G (2010) Effects of adaptogens on the central nervous system and the molecular mechanisms associated with their stress-protective activity. Pharmaceuticals (Basel) 3(1):188–224. https://doi.org/10.3390/ph3010188

Rea K, Dinan TG, Cryan JF (2016) The microbiome: a key regulator of stress and neuroinflammation. Neurobiol Stress 4:23–33. https://doi.org/10.1016/j.ynstr.2016.03.001

Ríos-Hernández A, Alda JA, Farran-Codina A, Ferreira-García E, Izquierdo-Pulido M (2017) The Mediterranean diet and ADHD in children and adolescents. Pediatrics 139(2):e20162027. https://doi.org/10.1542/peds.2016-2027

Roehrs T, Roth T (2001) Sleep, sleepiness, and alcohol use. Alcohol Res Health 25(2):101–109

Rogers PJ, Smith JE, Heatherley SV, Pleydell-Pearce CW (2008) Time for tea: mood, blood pressure and cognitive performance effects of caffeine and theanine administered alone and together. Psychopharmacology 195(4):569–577. https://doi.org/10.1007/s00213-007-0938-1

Saleem A, Qurat-Ul-Ain AMF (2022) Alternative therapy of psychosis: potential phytochemicals and drug targets in the management of schizophrenia. Front Pharmacol 13:895668. https://doi.org/10.3389/fphar.2022.895668

Sangle P, Sandhu O, Aftab Z, Anthony AT, Khan S (2020) Vitamin B12 supplementation: preventing onset and improving prognosis of depression. Cureus 12(10):e11169. https://doi.org/10.7759/cureus.11169

Sanmukhani J, Satodia V, Trivedi J, Patel T, Tiwari D, Panchal B, Goel A, Tripathi CB (2014) Efficacy and safety of curcumin in major depressive disorder: a randomized controlled trial. Phytother Res 28(4):579–585. https://doi.org/10.1002/ptr.5025

Schmidt K, Cowen PJ, Harmer CJ, Tzortzis G, Errington S, Burnet PW (2015) Prebiotic intake reduces the waking cortisol response and alters emotional bias in healthy volunteers. Psychopharmacology 232(10):1793–1801. https://doi.org/10.1007/s00213-014-3810-0

Smith JP, Randall CL (2012) Anxiety and alcohol use disorders: comorbidity and treatment considerations. Alcohol Res 34(4):414–431

Smith MA, Riby LM, Eekelen JA, Foster JK (2011) Glucose enhancement of human memory: a comprehensive research review of the glucose memory facilitation effect. Neurosci Biobehav Rev 35(3):770–783. https://doi.org/10.1016/j.neubiorev.2010.09.008

Srivastava JK, Shankar E, Gupta S (2010) Chamomile: a herbal medicine of the past with bright future. Mol Med Rep 3(6):895–901. https://doi.org/10.3892/mmr.2010.377

Steenbergen L, Sellaro R, van Hemert S, Bosch JA, Colzato LS (2015) A randomized controlled trial to test the effect of multispecies probiotics on cognitive reactivity to sad mood. Brain Behav Immun 48:258–264. https://doi.org/10.1016/j.bbi.2015.04.003

Sumiyoshi E, Matsuzaki K, Sugimoto N, Tanabe Y, Hara T, Katakura M, Miyamoto M, Mishima S, Shido O (2019) Sub-chronic consumption of dark chocolate enhances cognitive function and releases nerve growth factors: a parallel-group randomized trial. Nutrients 11(11):2800. https://doi.org/10.3390/nu11112800

Talib M, Rachdi M, Papazova A, Nicolis H (2024) The role of dietary patterns and nutritional supplements in the management of mental disorders in children and adolescents: an umbrella review of meta-analyses: Le rôle des habitudes alimentaires et des supplḿents nutritionnels dans la prise en charge des troubles mentaux chez les enfants et les adolescents: une méta-revue de méta-analyses. Can J Psychiatr 69(8):567–589. https://doi.org/10.1177/07067437241248070

Tan MS, Yu JT, Tan CC, Wang HF, Meng XF, Wang C, Jiang T, Zhu XC, Tan L (2015) Efficacy and adverse effects of ginkgo biloba for cognitive impairment and dementia: a systematic review and meta-analysis. J Alzheimer's Dis 43(2):589–603. https://doi.org/10.3233/JAD-140837

Tang WK, Lee JC (2024) Association of fast-food intake with depressive and anxiety symptoms among young adults: a pilot study. Nutrients 16(19):3317. https://doi.org/10.3390/nu16193317

Temple JL, Bernard C, Lipshultz SE, Czachor JD, Westphal JA, Mestre MA (2017) The safety of ingested caffeine: a comprehensive review. Front Psych 8:80. https://doi.org/10.3389/fpsyt.2017.00080

Toker L, Amar S, Bersudsky Y, Benjamin J, Klein E (2010) The biology of tryptophan depletion and mood disorders. Isr J Psychiatry Relat Sci 47(1):46–55

Trebatická J, Ďuračková Z (2015) Psychiatric disorders and polyphenols: can they be helpful in therapy? Oxidative Med Cell Longev 2015:248529. https://doi.org/10.1155/2015/248529

Vauzour D, Camprubi-Robles M, Miquel-Kergoat S, Andres-Lacueva C, Bánáti D, Barberger-Gateau P, Bowman GL, Caberlotto L, Clarke R, Hogervorst E, Kiliaan AJ, Lucca U, Manach C, Minihane AM, Mitchell ES, Perneczky R, Perry H, Roussel AM, Schuermans J, Sijben J, Spencer JP, Thuret S, van de Rest O, Vandewoude M, Wesnes K, Williams RJ, Williams RS, Ramirez M (2017) Nutrition for the ageing brain: towards evidence for an optimal diet. Ageing Res Rev 35:222–240. https://doi.org/10.1016/j.arr.2016.09.010

Wang J, Um P, Dickerman BA, Liu J (2018) Zinc, magnesium, selenium and depression: a review of the evidence, potential mechanisms and implications. Nutrients 10(5):584. https://doi.org/10.3390/nu10050584

Woelk H, Arnoldt KH, Kieser M, Hoerr R (2007) Ginkgo biloba special extract EGb 761 in generalized anxiety disorder and adjustment disorder with anxious mood: a randomized, double-blind, placebo-controlled trial. J Psychiatr Res 41(6):472–480. https://doi.org/10.1016/j.jpsychires.2006.05.004

Xu Y, Ku B, Tie L, Yao H, Jiang W, Ma X, Li X (2006) Curcumin reverses the effects of chronic stress on behavior, the HPA axis, BDNF expression and phosphorylation of CREB. Brain Res 1122(1):56–64. https://doi.org/10.1016/j.brainres.2006.09.009

Zhao M, Liu T, Qi S, Li W, Liu X, Li X, Xun G (2022) Individuals with bipolar disorder have a higher level of homocysteine than major depressive disorder: a retrospective chart review and observational study. Neuropsychiatr Dis Treat 18:2671–2682. https://doi.org/10.2147/NDT.S387063

Zhu LN, Mei X, Zhang ZG, Xie YP, Lang F (2019) Curcumin intervention for cognitive function in different types of people: a systematic review and meta-analysis. Phytother Res 33(3):524–533. https://doi.org/10.1002/ptr.6257

The Role of Omega-3 Fatty Acids in Mental Health

10

Abraham Olufemi Asuku, Adeyemi Fatai Odetayo, Mosunmola Busayo Oyeleke, and Asiat Na'Allah

Abstract

Omega-3 fatty acids (O3FAs) play vital functions in healthy cell membranes and mental well-being, and this is becoming more acknowledged. The role of omega-3 fatty acids, specifically eicosapentaenoic acid (EPA) and docosahexaenoic acid (DHA), in regulation of mental health issues is considered in this chapter. A plethora of studies reported that they may be useful in attenuating depression, anxiety, and symptoms of cognitive impairment. Based on clinical research, omega-3 supplement intake could be beneficial in cognitive performance, mood stabilization, and general mental health. These effects of omega-3 fatty acids are linked to their mechanistic influence on inflammation, neurotransmitter function, and neuroplasticity. However, despite the promising and encouraging results in this area, more investigation is necessary to completely comprehend the best dosages, impacts on long-term use, and the mechanisms

A. O. Asuku (✉)
Bioresources Development Centre, National Biotechnology Research and Development Agency, Ogbomoso, Nigeria

A. F. Odetayo
Department of Physiology, Faculty of Basic Medical Science, Federal University of Health Sciences, Ila Orangun, Nigeria

M. B. Oyeleke
Department of Physiology, Faculty of Basic Medical Sciences, Afe Babalola University, Ado-Ekiti, Nigeria

A. Na'Allah
Department of Biochemistry, Faculty of Pure and Applied Sciences, Kwara State University, Malete, Kwara State, Nigeria

W. Mohamed, A. Özge (eds.), *Feeding the Mind: The Connection Between Diet, Drugs, and Mental Health Volume 1*, Nutritional Neurosciences,
https://doi.org/10.1007/978-981-95-1414-4_10

underlying these advantages. Still, O3FAs as an adjuvant therapy is promising for managing mental health.

Keywords

Neuroplasticity · Cognitive impairment · Neurotransmitter · Neuroinflammation · Oxidative stress

10.1 Introduction

Omega-3 fatty acids (O3FAs) are polyunsaturated fatty acids (PUFAs) that are obtained directly from diets; hence, they are referred to as essential fatty acids. There are three major types of O3FAs which include alpha-linolenic acid (ALA), docosahexaenoic acid (DHA), and eicosapentaenoic acid (EPA). These are the most abundant components of the human brain total fatty acids (Dighriri et al. 2022). DHA, for example, occupies about 40% of the brain lipids, while EPA represents below 1% (Su et al. 2008; Lazzarin et al. 2009). To further break it down, about 55% of the brain weight is made up of lipids, out of which 35% of them consist of O3FAs.

Furthermore, DHA represents over 40% of the total O3FAs in the neuronal tissue, particularly the grey matter. Also, O3FAs rapidly accumulates in the brain during the late phase of intra-uterine life and one and half year after birth (Dighriri et al. 2022). Hence, O3FAs are abundantly present in the brain, indicating their dominant roles in normal brain functions. In fact, O3FAs are mostly added to newborn formula due to their important contributions to rapid brain growth and development (Conquer et al. 2000). Also, breast milk is rich in O3FAs where they help to maintain optimal mental health in the baby (Smith et al. 2011). Additionally, cell membrane signaling and structure are modified by O3FAs, demonstrating that O3FAs can improve brain functions by acting as an antidepressant.

According to the study of He et al. (2009), a transgenic mouse with the ability to convert the brain omega-6 fatty acids to O3FAs revealed a better hippocampal neurogenesis and improved learning and memory. Also, O3FAs have been shown to improve thyroid functions (Odetayo and Olayaki 2024), which are associated with mental health since thyroid dysfunction is a contributor to different neuropsychiatric manifestations such as depression (Lekurwale et al. 2023). Furthermore, inflammatory response has been shown to negatively impact most of the mechanisms associated with the maintenance of optimal mental health (Ouabbou et al. 2020; Asuku and Ogungbangbe 2024). On the other hand, O3FAs possess the ability to combat inflammation (Odetayo et al. 2023), indicating another possible relationship between O3FAs and mental health status. In addition, EPA and DHA which are the main types of O3FAs have also been recognized as precursors of various metabolites known as potent lipid mediators, identified to be efficient for preventing and treating diverse disorders (Swanson et al. 2012). Hence, this chapter is aimed at exploring the role of O3FAs on mental health and common disorders (e.g., depression, anxiety, and cognitive decline).

10.2 Biochemical and Physiological Role of O3FAs

10.2.1 Structural Characteristics of O3FAs

Similar to other PUFAs, O3FAs are made up of long carbon atom chains consisting of a carboxyl group and methyl group at each end. O3FAs consist of a carbon-to-carbon double bond present at the three carbons from end of the chain containing methyl.

Although, numerous types of O3FAs have been identified, but the three most important of them are the EPA, ALA, and DHA. ALA is made up of 18 carbon atoms, while EPA and DHA are referred to as long chain fatty acid because they have 20 and 22 carbon atoms, respectively. PUFAs are normally named according to their carbon atoms number and carbon–carbon double bonds.

10.2.2 Sources of O3FAs

The commonest O3FA sources are fish oil and fish (Asuku 2024). These two are specifically rich in DHA and EPA. Unlike DHA and EPA, ALA is present in chia seeds, perilla, flax seeds and oils, and green leafy vegetables. Additionally, ALA is present in nuts (e.g., walnuts) and beans (e.g., soybeans) (Covington 2004). Other sources are beef, flaxseed oil, salad oil, black nut, dried hemp seeds, and dried Brazil nuts (Akhigbe et al. 2021).

It is vital to know that although fish is rich in O3FAs, the content of O3FA in fish differs greatly. For instance, cold-water fish (including mackerel, tuna, salmon, herring, and sardines) are exceptionally rich in O3FAs compared to low fatty fish (bass, tilapia, cod, and shellfish). Also, the O3FAs also depend on the content of their feed. It was observed that a decline in the O3FAs content of farm-raised Atlantic salmon from Scotland between 2006 and 2015 resulted from the replacement of fish feed content with another ingredients (Cladis et al. 2014; Sprague et al. 2016).

10.3 Metabolism and Mechanisms

10.3.1 Conversion and Bioavailability

O3FAs can be converted from one form to another. For example, the Δ6-desaturate present in the hepatic cells is responsible for adding a double bond at the carbon six from the –COOH end, thereby producing stearidonic acid. This acid is then converted to eicosatetraenoic acid by undergoing two carbon elongations. After this, a double bond is added at the fifth carbon bond and another desaturation is performed to produce EPA. Also, EPA can also be produced from arachidonic acid (omega-6 fatty acids by-product) via the activities of Δ17-desaturase. This EPA generated can be converted via the Sprecher pathway to form DHA in mammals. This process involves EPA undergoing two successive cycles of elongation to form docosapentaenoic acid at first, and later tetracosanolpentaenoic. The produced tetracosanolpentaenoic is then converted by Δ6-desaturation to tetracosahexanoic acid. The length of this acid is then reduced by 2 carbons via β-oxidation to produce DHA which is the final product (Schmitz and Ecker 2008; Saini and Keum 2018). After ingestion, ALA can be rapidly converted to EPA but slowly transformed to produce DHA. These biosynthesized O3FAs are stored in the esterified form either in phospholipids or glycerides and are mobilized by the activities of phospholipase A2 (Saini and Keum 2018).

O3FAs bioavailability, the extent to which O3FAs are absorbed by the intestine into the portal circulation, depends on so many factors. Aside the chemical bond type, the type of food intake along O3FAs ingestion and the presence of other components can also influence O3FAs uptake. For example, the presence of calcium ions can form a complex with O3FAs and lessen their bioavailability (Schuchardt and Hahn 2013). Again, O3FAs availability can be influenced by the type of form they exist. For example, esters from O3FAs enriched diet have lower bioavailability compared to free fatty acids (Punia et al. 2019). Also, ALA has a low bioavailability because of high rate of β-oxidation during its utilization by the body (Libinaki and Gavin 2017). As a result of this, about 85% of ALA will disappear after consumption. On the other hand, DHA is a poor substrate for β-oxidation, and it is comparably more available than ALA. Additionally, bioavailability of O3FAs can be influenced by the glycerol molecule position of the triglycerol. For example, the best position is sn-2 and has easy absorption than others, which consequently influenced its bioavailability (Punia et al. 2019). According to the study of Poumès-Ballihaut et al. (2001), the bioaccumulation of DHA is about 86%, and it is about two-fold higher than ALA that is about 30% of the total intake. Another factor that can influence O3FAs bioavailability is the dietary source. Fatty acids obtained from milk tend to have higher bioavailability because fat obtained from milk is greatly dispersed in small micelles which tend to increase bioavailability and absorption of compound which is soluble in lipid such as O3FAs (Visioli et al. 2000). In addition, fish O3FAs are more adequately adapted to the plasma than the capsules-administered ones (Visioli et al. 2003).

O3FAs, particularly DHA, perform a remarkable function in the fluidity of cell membranes (Hathaway III et al. 2020). Certain structural features of DHA, including several double bonds, establish itself within the phospholipid bilayer and enhance membrane flexibility and permeability (Ali and Szabó 2023). This fluidity is required for optimal cell signaling, enabling neurotransmitter release to occur and receptor functions to take place efficiently. More importantly, DHA contributes to cell membrane fluidity and, as such, allows the adaptability and resilience of neurons so crucial for cognitive and affective stability (Scuto et al. 2024). Enhanced fluidity further enables the formation of lipid rafts, key modulators in synaptic transmission, and neuroplasticity, and this offers wide health benefits to the brain.

10.4 Influence on Neurotransmitter Systems

A sufficient amount of omega-3 polyunsaturated fatty acids (PUFAs) in the brain is necessary for the neurotransmitter systems to operate normally, according to increasing evidence (Zhou et al. 2022; DiNicolantonio and O'Keefe 2020). The synthesis, synaptic release, and absorption of a number of neurotransmitters, such as norepinephrine (NE), serotonin (5-HT), dopamine (DA), γ-amino-butyric acid (GABA), and glutamate (Glu), are all impacted (Yang et al. 2019; DiNicolantonio and O'Keefe 2020). The underlying mechanism of omega-3 PUFAs' positive effects on depression has been hypothesized to be improved transmission of DA and 5-HT, as well as increased D2 receptor and decreased 5-HT2 receptor in the frontal brain (DiNicolantonio and O'Keefe 2020).

In the LPS-induced model of depression, fish oil supplementation has been shown to have an antidepressant-like effect. This effect is linked to increased levels of 5-HT in the hippocampus and decreased expression of indoleamine-2, 3-Dioxygenase (Carabelli et al. 2020). The expression levels of different neurotransmitter receptors may also be impacted by omega-3 PUFAs. Supplementing with DHA can stop the rise in binding density of CB1, 5-HT, and GABA-A receptors that is brought on by a high fat diet. This has been connected to the brain's cognitive function (Zhou et al. 2022).

O3FAs, especially EPA and DHA, perform a determining role in neurotransmitter systems including serotonin and dopamine (Dighriri et al. 2022; Odetayo and Olayaki 2023). Fatty acids are integral parts of synaptic membrane composition and hence efficient neurotransmitter release and receptor binding are ensured. DHA has been described to increase serotonin transmission by supporting receptor sensitivity and synaptic plasticity, which underlies mood stabilization. Similarly, EPA regulates dopamine pathways, which creates the basis for improved reward processing and emotional regulation (Fernandez et al. 2022). O3FAs confer fluidity that allows the assembly and functionality of proteins involved in neurotransmission (Fig. 10.1). Similarly, clinical trials, such as Sikka et al. (2021) and Zhou et al. (2022), indicate the possible use of O3FAs supplementation to reduce mood disorders by normalizing neurotransmitter activities. Moreover, their anti-inflammatory action indirectly favors neurotransmitter function by depressing neuroinflammation, which impairs synaptic activity. This dual mechanism underlines the therapeutic role of O3Fas in managing mental health disorders by their modulating action on key neurotransmitter systems.

10.5 Effects on Neuroinflammation and Oxidative Stress

Neuroinflammatory process takes place in the nervous system and disturbs homeostasis (Asuku and Ogungbangbe 2024). According to Asuku et al. (2024), chronic inflammation has been associated with mental health problems. Omega-3 PUFA,

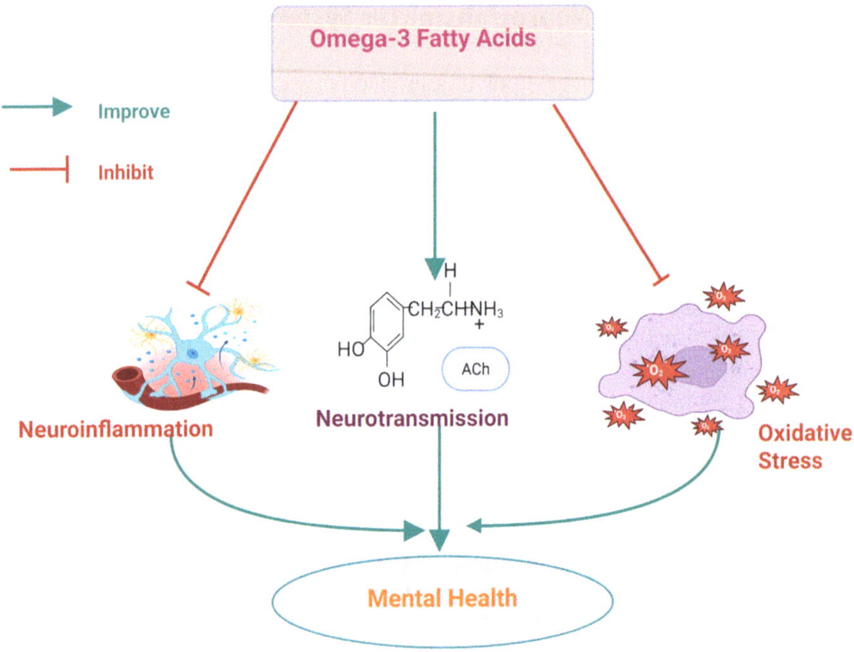

Fig. 10.1 Mechanisms of action of O3FA on mental health

particularly eicosapentaenoic acid (EPA, C20:5n-3) and docosahexaenoic acid (DHA, C22:6n-3), are essential fatty acids that have been reported to play important roles in modulating inflammation and neurobiological mechanisms of depression (Hallahan et al. 2016). Additionally, omega-3 PUFA have effects on lipids synthesis, regulating both the inflammatory response and releasing inflammatory factors; it also has antioxidative and anti-inflammation effects; all of which influence the onset of many diseases, including stroke, Parkinson's disease, multiple sclerosis, and Alzheimer's disease (AD). Omega-3 PUFAs possess an anti-inflammatory effect by replacing arachidonic acid (AA) in membrane phospholipids, decreases the synthesis of pro-inflammatory eicosanoids, and supports inflammation regulation, because they represent the precursors of a group of lipid mediators called "specialized pro-resolving mediators" (SPMs).

Likewise, EPA and DHA have inhibitory impacts on immune cells and reduce the release of pro-inflammatory cytokines. The omega-3 PUFAs, such as DHA and EPA, and the omega-6 PUFAs, such as arachidonic acid, are present at the sites of acute inflammation where they are converted into bioactive SPMs. Numerous anti-inflammatory properties are shared by EPA and DHA (Calder 2015, 2020; Giacobbe et al. 2020). The synthesis of AA-derived eicosanoids is significantly decreased by EPA and DHA. They accomplish this in part through competition with AA for integration into phospholipids found in cell membranes (implying that EPA and DHA reduce the amount cell membranes AA), decreasing AA release from membranes, and blocking the activity of cyclo-oxygenase-2 (COX-2) and 5-lipoxygenase (5-LOX) enzymes (Calder 2015). Research has shown that consuming more DHA and EPA raises the amounts of these two nutrients in the membranes of inflammatory cells (and probably many other cell types) (Calder 2015).

According to Farag and Gad (2022), O3FAs are actively anti-inflammatory and have potent antioxidative qualities that perform an important function in fighting neuroinflammation and oxidative process. According to Tułowiecka et al. (2020), EPA and DHA are precursors for resolving and protecting active lipid mediators that alleviate inflammation by lowering the production of pro-inflammatory cytokines. Omega-3 modulates neuroinflammatory responses by inhibiting the inflammatory genes expression via the NF-κB pathway, thereby protecting against the disorders related to mental health. Their scavenging action toward free radicals further reduces the possibility of oxidative neuronal damage, thus preserving cellular integrity. These effects are specifically helpful in protecting against neuronal apoptosis and preserving synaptic function. According to a study by Zhou et al. (2022), the ability of O3FA to combat inflammation and oxidative process may lower one's risk for depression, anxiety disorders, and cognitive decline. O3FAs protect the health of the brain by mitigating oxidative and inflammatory processes, thereby being regarded as neuroprotective in the management of mental health.

10.6 Role of O3FAs in Neuroplasticity and Brain Structure

Impaired neuronal plasticity could primarily result in reduced neurogenesis, decreased density of dendritic spine, reduced number and strength of synapses, decreased synaptic remodeling, dendritic atrophy, and even dysregulation of the reward circuit (Wohleb et al. 2018; Asuku et al. 2024; Asuku et al. 2025). Aberrant neuronal plasticity which are frequently seen in the hippocampus, prefrontal cortex, and amygdala are closely linked to depression (Yang et al. 2019). Several treatments for depression, including pharmacotherapy (ketamine, fluoxetine, and TJZL184) and physiotherapy (transcranial stimulation, direct and alternating currents, and electroconvulsive therapy), have been found to primarily work by controlling neuronal plasticity (Micheli et al. 2018; Li et al. 2021).

It has been demonstrated that omega-3 PUFAs support neuroplasticity in a number of ways (Fig. 10.2). The production of new synapses and dendritic spines can be accelerated by elevated brain omega-3 PUFA levels. The stimulation of pathways associated with proliferation involving molecular signaling, such as MAPK, cAMP-response element binding protein (CREB), p38, GPR40, and brain-derived neurotrophic factor (BDNF), may be one of the mechanisms behind omega-3 PUFAs' pro-neurogenesis effect (Zhao et al. 2019; Akerele and Cheema 2020). Through experiments with transgenic mice endogenously enriched with omega-3 PUFAs, a group of researchers observed that a considerable elevation of DHA the brain could dramatically accelerate neurogenesis in the hippocampus and improve dendritic spine generation of CA1 pyramidal neurons (Zhou et al. 2022).

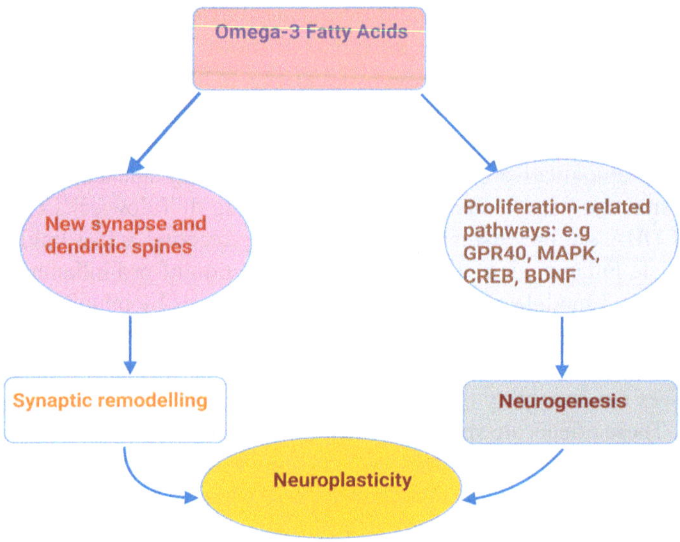

Fig. 10.2 Ways through which O3FAs support neuroplasticity

Supplementing with omega-3 PUFAs has the same effect on synaptic plasticity, transmission, and formation. Early growth response 1, postsynaptic density protein 95, E-cadherin, and signaling factors involved in synaptic plasticity [like Fyn and the N-methyl-D-aspartic acid (NMDA) receptor] are among the significant neural and glial proteins whose expression may be significantly regulated by omega-3 PUFAs (Zhou et al. 2022). Deficits in omega-3 PUFAs have been reported to reduce glutamate receptor subunit concentrations, synaptic vesicle proteins, and long-term potentiation (LTP) in hippocampus glutamatergic synapses (Aryal et al. 2019).

Docosahexaenoyl ethanolamide (DHEA) and eicosapentaenoyl ethanolamide (EPEA) are two endocannabinoids (eCBs) that activate eCB receptors to provide physiological effects, and this can also be produced from DHA and EPA. DHEA and EPEA have been shown to have a variety of immunomodulatory and anti-inflammatory properties, as well as impacts on mood and food consumption (McDougle et al. 2017). The eCBs produced in postsynaptic neurons could inversely operate to control the presynaptic input; hence, they can encourage both long- and short-term alterations and synaptic plasticity throughout the nervous system (Winters and Vaughan 2021). Weaned pups' hippocampal eCB gating of LTP may be hampered by maternal omega-3 PUFA deficit (Thomazeau et al. 2017).

Additionally, it has been discovered that a lifelong reduction in omega-3 PUFAs can specifically prevent eCB-mediated long-term synaptic depression in adult mice's accumbens and prelimbic prefrontal cortex. Reduced CB1 receptor function, which has been connected to emotional behavior impairment, is the cause of this effect (Zhou et al. 2022).

O3FAs have great importance in neuroplasticity and maintaining brain structure. Zhou et al. (2022) observed that incorporating DHA into neuronal membranes enhances synaptic connectivity and increases dendrite growth, processes supportive of learning and memory. O3FAs upregulate the expressions of BDNF, an important protein supporting neurogenesis and synaptic plasticity. This level of regulation of BDNF supports adaptive brain changes that help in recovery from both stress and injury. However, O3FAs are implicated in the maintenance of white matter integrity and grey matter volume, important for the sustenance of emotional control and cognitive functions. The review done by Dighriri et al. (2022) has now shown that those with higher intake of O3FAs represent improved connectivity in the brain and less atrophy in parts connected with memory and regulation of mood. Through the promotion of neuronal resilience and adaptability, O3FAs indicate their significance in maintaining brain health for mental health.

10.7 Impact of O3FAs on Depression

Numerous studies have linked O3FAs, particularly EPA and DHA, to alleviation of depressive symptoms (Lekurwale et al. 2023). Anti-inflammatory roles of EPA form a critical component in mitigating neuroinflammation, now established as part of the etiology of depression. Clinical studies by Serefko et al. (2024) have demonstrated that omega-3 supplementation, in particular preparations containing high amounts of EPA, leads to significant reductions in depressive symptoms among patients, especially those with major depressive disorder. Lipid mediators from EPA and DHA further promote neuronal survival and synaptic function, which are often compromised in depression.

Moreover, O3FAs have one more influence on the control of the hypothalamic-pituitary-adrenal (HPA) axis, which causes a depleted response to stress, hence alleviating depressive symptoms. Mehdi et al. (2023) and Puzantian (2024) also point out that O3FAs are good as an augmentation strategy alongside conventional antidepressants and provide a complementary mode of treatment. Despite these merits, variations in individual responses and optimal dosages stand as points for further research. Nevertheless, these evidences underline the therapeutic role of O3FAs in depression management, in alleviating symptom severity and improving general mental health.

10.8 Effects of O3FAs on Anxiety Disorders

Omega-3 fatty acids, in some ways, have shown the reduction of symptoms in anxiety disorders through anti-inflammation and neuroprotection. Anxiety in chronic states is more related to increased neuroinflammation and oxidative stress events that O3FAs counteract by mechanisms involving a reduction in pro-inflammatory cytokines and ROS. According to Suryavanshi et al. (2024), EPA and DHA modulate neurotransmitter systems, particularly serotonin and GABA. Anxiety is further put under control by promoting calmness and emotional stability.

The basic levels of O3FAs in the body, as identified by the findings of Norwitz and Naidoo (2021) and Zielińska et al. (2023), indicated deficiency of this nutrient in diseased populations. As Kelaiditis et al. (2022) have identified, supplementation with formulations rich in EPA results in significant decreases in symptoms of anxiety, including generalized anxiety disorder and social anxiety. Aside from the effects on the HPA axis, moderation of the stress response via O3FAs would tend to reduce the physiological symptoms of anxiety. Although the exact mechanisms are under investigation, the cumulative evidence points toward O3FAs as part of the safe and complementary options for interventions against anxiety disorders. However, doses should be standardized, and the populations most likely to benefit from them must be identified. Nevertheless, so far, the findings underline their efficacy by enhancing emotional resilience and reducing symptoms of anxiety.

10.9 Impact of O3FAs on Cognitive Decline and Neurodegenerative Disorders

According to Lange (2020) O3FAs, especially DHA, play a significant role in limiting cognitive impairment and neurodegenerative conditions, like AD. In such cases, it is known to maintain neuronal membrane integrity and promote synaptic function in order to sustain cognitive performance. O3FAs support the reduction of neuroinflammation and oxidative stress, which are characteristic processes in neuronal degeneration.

Epidemiological studies, like Alex et al. (2020), have shown that a large consumption of O3FAs is connected to a lesser cognitive decline. Clinical trials, like Chu et al. (2022), have demonstrated that DHA reduces cognitive impairment in persons with AD and mild cognitive impairment. The mechanistic basis involves the induction of neurogenesis and enhancement of connectivity within the brain, thereby maintaining hippocampal function, which is important for memory. They also affect the aggregation of amyloid-β proteins, a biomarker of Alzheimer's pathology, thus protecting against further neuronal damage.

Furthermore, the effect of O3FAs on vascular health underlines flow in cerebral blood to the brain for oxygen and nutrient delivery. This multi-modal mechanism underlines their potential to delay the onset and slowing of neurodegenerative diseases. While further more longitudinal studies are needed to confirm this hypothesis, the current evidence suggests that O3FAs is a very important dietary adjunct to the strategies aimed at preserving cognitive health and neurodegenerative disorders.

10.10 Conclusion and Future Perspectives

In conclusion, O3FAs are very promising panacea for mental problems, thanks to its antioxidative, anti-inflammatory, and neurotransmitters modulatory qualities. However, there is need for large-scale, long-term studies before it can be adopted for the management of anxiety, depression, and cognitive impairment. Furthermore, the issues of standardization of O3FAs supplementation, measurement techniques, genetic responses, and variability in individual responses to O3FAs need to be carefully considered for effective and efficient results.

References

Akerele OA, Cheema SK (2020) Maternal diet high in Omega-3 fatty acids upregulate genes involved in neurotrophin signalling in fetal brain during pregnancy in C57BL/6 mice. Neurochem Int 138:104778

Akhigbe RE, Hamed MA, Odetayo AF, Akhigbe TM, Ajayi AF, Ajibogun F (2021) Omega-3 fatty acid rescues ischaemia/perfusion-induced testicular and sperm damage via modulation of lactate transport and xanthine oxidase/uric acid signaling. Biomed Pharmacother 142:111975

Alex A, Abbott KA, McEvoy M, Schofield PW, Garg ML (2020) Long-chain omega-3 polyunsaturated fatty acids and cognitive decline in non-demented adults: a systematic review and meta-analysis. Nutr Rev 78(7):563–578

Ali O, Szabó A (2023) Review of eukaryote cellular membrane lipid composition, with special attention to the fatty acids. Int J Mol Sci 24(21):15693

Aryal S, Hussain S, Drevon CA, Nagelhus E, Hvalby Ø, Jensen V, Walaas SI, Davanger S (2019) Omega-3 fatty acids regulate plasticity in distinct hippocampal glutamatergic synapses. Eur J Neurosci 49(1):40–50

Asuku AO (2024) Pharmacological applications of marine metabolites: anticancer, antimicrobial, anti-inflammatory, antiviral, and other therapeutic properties. In: Marine metabolites in disease management. IGI Publisher. https://doi.org/10.4018/979-8-3693-5878-8.ch003

Asuku AO, Ogungbangbe GO (2024) Neuroinflammation and binge eating. In: Neuroscientific insights and therapeutic approaches to eating disorders, vol 10. IGI Publisher, pp 147–170. https://doi.org/10.4018/979-8-3693-3230-56

Asuku AO, Ayinla MT, Ajibare AJ (2024) Measuring mood in nutritional research. In: Mohamed W, Kobeissy F (eds) Nutrition and psychiatric disorders. Nutritional neurosciences. Springer, Singapore. https://doi.org/10.1007/978-981-97-2681-3_10

Asuku AO, Arinde E, Saliu SB, Oyerinde T, Olajide TS, Afolabi H, Kehinde A (2025) Neuroplasticity: harnessing the brain ability to heal itself. In: Advancing medical research through neuroscience. IGI Global Publisher. https://doi.org/10.4018/979-8-3693-5464-3.ch005

Calder PC (2015) Marine omega-3 fatty acids and inflammatory processes: effects, mechanisms and clinical relevance. Biochimica et Biophysica Acta (BBA)-Molecular and Cell Biology of Lipids 1851(4):469–484

Calder PC (2020) n-3 PUFA and inflammation: from membrane to nucleus and from bench to bedside. Proc Nutr Soc 79(4):404–416

Carabelli B, Delattre AM, Waltrick APF, Araujo G, Suchecki D, Machado RB, de Souza LER, Zanata SM, Zanoveli JM, Ferraz AC (2020) Fish-oil supplementation decreases Indoleamine-2, 3-dioxygenase expression and increases hippocampal serotonin levels in the LPS depression model. Behav Brain Res 390:112675

Chu CS, Hung CF, Ponnusamy VK, Chen KC, Chen NC (2022) Higher serum DHA and slower cognitive decline in patients with Alzheimer's disease: two-year follow-up. Nutrients 14(6):1159

Cladis DP, Kleiner AC, Freiser HH, Santerre CR (2014) Fatty acid profiles of commercially available finfish fillets in the United States. Lipids 49(10):1005–1018

Conquer JA, Tierney MC, Zecevic J, Bettger WJ, Fisher RH (2000) Fatty acid analysis of blood plasma of patients with Alzheimer's disease, other types of dementia, and cognitive impairment. Lipids 35(12):1305–1312

Covington MB (2004) Omega-3 fatty acids. Am Fam Physician 70(1):133–140

Dighriri IM, Alsubaie AM, Hakami FM, Hamithi DM, Alshekh MM, Khobrani FA, Dalak FE, Hakami AA, Alsueaadi EH, Alsaawi LS, Alshammari SF, Alqahtani AS, Alawi IA, Aljuaid

AA, Tawhari MQ (2022) Effects of omega-3 polyunsaturated fatty acids on brain functions: a systematic review. Cureus 14(10):e30091

DiNicolantonio JJ, O'Keefe JH (2020) The importance of marine omega-3s for brain development and the prevention and treatment of behavior, mood, and other brain disorders. Nutrients 12(8):2333.180

Farag MA, Gad MZ (2022) Omega-9 fatty acids: potential roles in inflammation and cancer management. J Genet Eng Biotechnol 20(1):48

Fernandez F, Byrne MK, Batterham M, Grant L, Meyer BJ (2022) Associations between Omega-3 index, dopaminergic genetic variants and aggressive and metacognitive traits: a study in adult male prisoners. Nutrients 14(7):1379

Giacobbe J, Benoiton B, Zunszain P, Pariante CM, Borsini A (2020) The anti-inflammatory role of omega-3 polyunsaturated fatty acids metabolites in pre-clinical models of psychiatric, neurodegenerative, and neurological disorders. Front Psych 11:122

Hallahan B, Ryan T, Hibbeln JR, Murray IT, Glynn S, Ramsden CE, SanGiovanni JP, Davis JM (2016) Efficacy of omega-3 highly unsaturated fatty acids in the treatment of depression. Br J Psychiatry 209(3):192–201

Hathaway D III, Pandav K, Patel M, Riva-Moscoso A, Singh BM, Patel A, Min ZC, Singh-Makkar S, Sana MK, Sanchez-Dopazo R, Desir R (2020) Omega 3 fatty acids and COVID-19: a comprehensive review. Infect Chemother 52(4):478

He C, Qu X, Cui L, Wang J, Kang JX (2009) Improved spatial learning performance of fat-1 mice is associated with enhanced neurogenesis and neuritogenesis by docosahexaenoic acid. Proc Natl Acad Sci USA 106(27):11370–11375

Kelaiditis C, Gibson E, Dyall S (2022) Effects of long-chain omega-3 polyunsaturated fatty acids on reducing anxiety and/or depression in adults; a systematic review and meta-analysis of randomised controlled trials. Prostaglandins Leukot Essent Fatty Acids 192:102572. https://doi.org/10.1101/2022.10.17.22281092

Lange KW (2020) Omega-3 fatty acids and mental health. Glob Health J 4(1):18–30

Lazzarin N, Vaquero E, Exacoustos C, Bertonotti E, Romanini ME, Arduini D (2009) Low-dose aspirin and omega-3 fatty acids improve uterine artery blood flow velocity in women with recurrent miscarriage due to impaired uterine perfusion. Fertil Steril 92(1):296–300

Lekurwale V, Acharya S, Shukla S, Kumar S (2023) Neuropsychiatric manifestations of thyroid diseases. Cureus 15(1):e33987

Li Z, Ruan M, Chen J, Fang Y (2021) Major depressive disorder: advances in neuroscience research and translational applications. Neurosci Bull 37:863–880

Libinaki R, Gavin PD (2017) Changes in bioavailability of Omega-3 (DHA) through alpha-Tocopheryl phosphate mixture (TPM) after oral administration in rats. Nutrients 9(9):1042

McDougle DR, Watson JE, Abdeen AA, Adili R, Caputo MP, Krapf JE, Johnson RW, Kilian KA, Holinstat M, Das A (2017) Anti-inflammatory ω-3 endocannabinoid epoxides. Proc Natl Acad Sci 114(30):E6034–E6043

Mehdi S, Manohar K, Shariff A, Kinattingal N, Wani S, Alshehri S, Imam M, Shakeel F, Krishna K (2023) Omega-3 fatty acids supplementation in the treatment of depression: an observational study. J Pers Med 13. https://doi.org/10.3390/jpm13020224

Micheli L, Ceccarelli M, D'Andrea G, Tirone F (2018) Depression and adult neurogenesis: positive effects of the antidepressant fluoxetine and of physical exercise. Brain Res Bull 143:181–193

Norwitz NG, Naidoo U (2021) Nutrition as metabolic treatment for anxiety. Front Psych 12:598119

Odetayo AF, Olayaki LA (2023) Omega 3 fatty acid improves sexual and erectile function in BPF-treated rats by upregulating NO/cGMP signaling and steroidogenic enzymes activities. Sci Rep 13:18060

Odetayo AF, Olayaki LA (2024) Omega 3 fatty acids preserve testicular function by ameliorating BPF-induced dysthyroidism: role of p53/Bcl-2 signaling and proton pump activities. JBRA Assist Reprod 28(3):471–482

Odetayo AF, Adeyemi WJ, Olayaki LA (2023) Omega-3 fatty acid ameliorates bisphenol F-induced testicular toxicity by modulating Nrf2/NFkB pathway and apoptotic signaling. Front Endocrinol 14:1256154

Ouabbou S, He Y, Butler K, Tsuang M (2020) Inflammation in mental disorders: is the microbiota the missing link? Neurosci Bull 36(9):1071–1084
Poumès-Ballihaut C, Langelier B, Houlier F, Alessandri JM, Durand G, Latge C, Guesnet P (2001) Comparative bioavailability of dietary alpha-linolenic and docosahexaenoic acids in the growing rat. Lipids 36(8):793–800
Punia S, Sandhu KS, Siroha AK, Dhull SB (2019) Omega 3-Metabolism, Absorption, Bioavailability and health benefits – a review. PharmaNutrition 10:100162
Puzantian T (2024) Nutraceuticals, dietary, and herbal supplements. In: Age-adjusted psychiatric treatment for the older patient. Springer Nature Switzerland, Cham, pp 321–331
Saini RK, Keum YS (2018) Omega-3 and omega-6 polyunsaturated fatty acids: dietary sources, metabolism, and significance–a review. Life Sci 203:255–267
Schmitz G, Ecker J (2008) The opposing effects of n-3 and n-6 fatty acids. Prog Lipid Res 47(2):147–155
Schuchardt JP, Hahn A (2013) Bioavailability of long-chain omega-3 fatty acids. Prostaglandins Leukot Essent Fatty Acids 89(1):1–8
Scuto M, Majzúnová M, Torcitto G, Antonuzzo S, Rampulla F, Di Fatta E, Trovato Salinaro A (2024) Functional food nutrients, redox resilience Signaling and Neurosteroids for brain health. Int J Mol Sci 25(22):12155
Serefko A, Jach ME, Pietraszuk M, Świąder M, Świąder K, Szopa A (2024) Omega-3 polyunsaturated fatty acids in depression. Int J Mol Sci 25(16):8675
Sikka P, Behl T, Sharma S, Sehgal A, Bhatia S, Al-Harrasi A, Singh S, Sharma N, Aleya L (2021) Exploring the therapeutic potential of omega-3 fatty acids in depression. Environ Sci Pollut Res 28:43021–43034
Smithers LG, Gibson RA, Makrides M (2011) Maternal supplementation with docosahexaenoic acid during pregnancy does not affect early visual development in the infant: A randomized controlled trial. Am J Clin Nutr 93(6):1293–9
Sprague M, Dick JR, Tocher DR (2016) Impact of sustainable feeds on omega-3 long-chain fatty acid levels in farmed Atlantic salmon, 2006-2015. Sci Rep 6:21892
Su KP, Huang SY, Chiu TH, Huang KC, Huang CL, Chang HC, Pariante CM (2008) Omega-3 fatty acids for major depressive disorder during pregnancy: results from a randomized, double-blind, placebo-controlled trial. J Clin Psychiatry 69(4):644–651
Suryavanshi A, Kumar V, Shukla D, Asati V, Mahapatra DK, Bharti SK, Ahire ED (2024) Nutraceuticals and anxiety. In: Nutraceutical fruits and foods for neurodegenerative disorders. Academic Press, pp 537–559
Swanson D, Block R, Mousa SA (2012) Omega-3 fatty acids EPA and DHA: health benefits throughout life. Adv Nutr (Bethesda, Md) 3(1):1–7
Thomazeau A, Bosch-Bouju C, Manzoni O, Layé S (2017) Nutritional Omega-3 PUFA deficiency abolishes endocannabinoid gating of hippocampal long-term potentiation. Cereb Cortex 27(4):2571–2579
Tułowiecka N, Kotlęga D, Prowans P, Szczuko M (2020) The role of resolvins: EPA and DHA derivatives can be useful in the prevention and treatment of ischemic stroke. Int J Mol Sci 21(20):7628
Visioli F, Risè P, Plasmati E, Pazzucconi F, Sirtori CR, Galli C (2000) Very low intakes of N-3 fatty acids incorporated into bovine milk reduce plasma triacylglycerol and increase HDL-cholesterol concentrations in healthy subjects. Pharmacol Res 41(5):571–576
Visioli F, Risé P, Barassi MC, Marangoni F, Galli C (2003) Dietary intake of fish vs. formulations leads to higher plasma concentrations of n-3 fatty acids. Lipids 38(4):415–418
Winters BL, Vaughan CW (2021) Mechanisms of endocannabinoid control of synaptic plasticity,Neuropharmacology, 197:108736
Wohleb ES, Terwilliger R, Duman CH, Duman RS (2018) Stress-induced neuronal colony stimulating factor 1 provokes microglia-mediated neuronal remodeling and depressive-like behavior. Biol Psychiatry 83(1):38–49

Yang R, Zhang MQ, Xue Y, Yang R, Tang MM (2019) Dietary of n-3 polyunsaturated fatty acids influence neurotransmitter systems of rats exposed to unpredictable chronic mild stress. Behav Brain Res 376:112172

Zhao WN, Hylton NK, Wang J, Chindavong PS, Alural B, Kurtser I, Subramanian A, Mazitschek R, Perlis RH, Haggarty SJ (2019) Activation of WNT and CREB signaling pathways in human neuronal cells in response to the Omega-3 fatty acid docosahexaenoic acid (DHA). Mol Cell Neurosci 99:103386

Zhou L, Xiong JY, Chai YQ, Huang L, Tang ZY, Zhang XF, Liu B, Zhang JT (2022) Possible anti-depressant mechanisms of omega-3 polyunsaturated fatty acids acting on the central nervous system. Front Psych 13:933704

Zielińska M, Łuszczki E, Dereń K (2023) Dietary nutrient deficiencies and risk of depression (review article 2018–2023). Nutrients 15(11):2433

The Impact of Micronutrients on Cognitive Functions

11

Mohab Ali Hassan, Al-Hassan Soliman Wadan, Jana S. Elhelw, Nada E. Atia, and Mohamed M. Sarhan

Abstract

Micronutrients are essential for preserving brain health and adequate mental function. Their importance comes from their crucial effect on signal transduction, neural plasticity, neurogenesis, neurotransmitter synthesis, and brain development and evolution. Brain functions including memory, perception, and decision-making have all been found to be impacted by essential micronutrients deficiency, such as iron, zinc, iodine, and vitamins. On the other hand, sufficient consumption of these nutrients has been associated with enhanced cognitive performance and a decreased risk of neurodegenerative diseases. Recent studies demonstrate how micronutrient supplements may improve cognitive performance. It was found that effective nutritional supplementation can promote brain

M. A. Hassan (✉)
Faculty of Pharmacy, King Salman International University (KSIU), Ras Sudr, South Sinai, Egypt

Faculty of Pharmacy, Zagazig University, Zagazig, Egypt
e-mail: mohab.ali@ksiu.edu.eg; mohabali1@hotmail.com

A.-H. S. Wadan
Oral Biology Department, Faculty of Dentistry, Galala University (15888), Galala Plateau, Attaka, Suez Governorate, Egypt
e-mail: Al-Hassan.Mohamed@Gu.edu.eg

J. S. Elhelw
Faculty of Pharmacy, New Giza University, Giza, Egypt
e-mail: jana.sherif@ngu.edu.eg

N. E. Atia · M. M. Sarhan
Faculty of Pharmacy, King Salman International University (KSIU), Ras Sudr, South Sinai, Egypt
e-mail: nada.emad@ksiu.edu.eg; nada163563@bue.edu.eg; mohamed.mohe@ksiu.edu.eg

W. Mohamed, A. Özge (eds.), *Feeding the Mind: The Connection Between Diet, Drugs, and Mental Health Volume 1*, Nutritional Neurosciences,
https://doi.org/10.1007/978-981-95-1414-4_11

health and slow cognitive decline, and this requires an understanding of the interactions between micronutrients and cognitive functions. In this chapter, we shed light on the importance of micronutrients on brain health and how micronutrients deficiency affects brain function on all levels including cognition, memory, and perception.

Keywords

Micronutrients · Cognition · Neurodegenerative diseases · Micronutrient supplements

11.1 Background

A balanced diet containing the minimum requirement of nutrients is essential for body's physical and mental performance. Nutrients are classified into two main categories, namely macronutrients that are consumed in large amounts and micronutrients which are consumed in relatively smaller amounts. When it comes to the main functions of macronutrients, it was found that they are essential for energy production and cellular homeostasis while micronutrients are involved in maintaining the body performance. Micronutrients consist of vitamins and minerals. Vitamins are classified into fat-soluble vitamins including A, D, E, and K and water-soluble vitamins such as B and C. Minerals that are essential inorganic elements include potassium, sodium, calcium, phosphorus, chloride, manganese, selenium, molybdenum, iodine, and others (Lukaski 2004; Morris and Mohiuddin 2024; Shenkin 2006).

Micronutrients have a crucial effect on health and disease and their effect reaches all parts of the body. Further, vitamins and trace elements are important for brain structure and function. It was found that vitamin B1 is central to energy production and vitamin B6 is suggested to reduce premenstrual depression; in the presence of vitamin B12, both contribute to neurotransmitter synthesis. Vitamin B9 was found to be involved in memory development; vitamin D plays a significant role in inhibiting neurodegenerative and neuroimmune diseases. Other vitamins such as C, K, and E are of interest due to their influence on central nervous system (CNS) development. Trace elements also show importance in brain development. For example, iron plays an important role in cognitive development and its deficiency can cause neurological disorders. Zinc, which has a role in DNA and protein synthesis, was found to contribute to growth, development, and perception while magnesium is involved in ionic regulation and metabolism. Choline contributes mainly to stem cell proliferation and neurogenesis in addition to neurotransmitter synthesis such as acetylcholine (Bourre 2006; Prado and Dewey 2014; Wasantwisut 1997). We shed light on the importance of micronutrients on cognitive function and how lack of micronutrient intake affects brain function significantly.

11.2 Micronutrients and Brain Function

Brain development is a crucial stage in human life, especially during the first two years after birth. Brain development is genetic dependent; however, it also depends on the type of nutrition provided for the baby during the first two years of life. Nutrition for babies should be reinforced by micronutrients which play a significant role in building brain structure (see Fig. 11.1). Three weeks after conception, the progenitor cells (descendants of stem cells) start to differentiate and increase their size and dimensions until they reach 1000 days of life which marks the fully developed brain structure (Fox et al. 2010; Stiles and Jernigan 2010). During this period, micronutrients support the growth of the progenitor cells and their neuro-differentiation all over the body, especially the brain. In addition, micronutrients guarantee different functions such as cognition to be healthier if the baby receives the recommended dietary allowance (RDA) from each micronutrient. For example, the iron RDA for babies in their first 6 months is 0.27 mg/d. It is important to note that deficiency in micronutrients during early life and pregnancy could lead to epigenetic neural alteration which may cause structural changes in the brain besides behavioral and cognitive disorders (Mattei and Pietrobelli 2019). Babies who were breastfed were found to have better cognition than the non-breastfeeding babies due to better gamma and beta function in the cortex (Gilbreath et al. 2023).

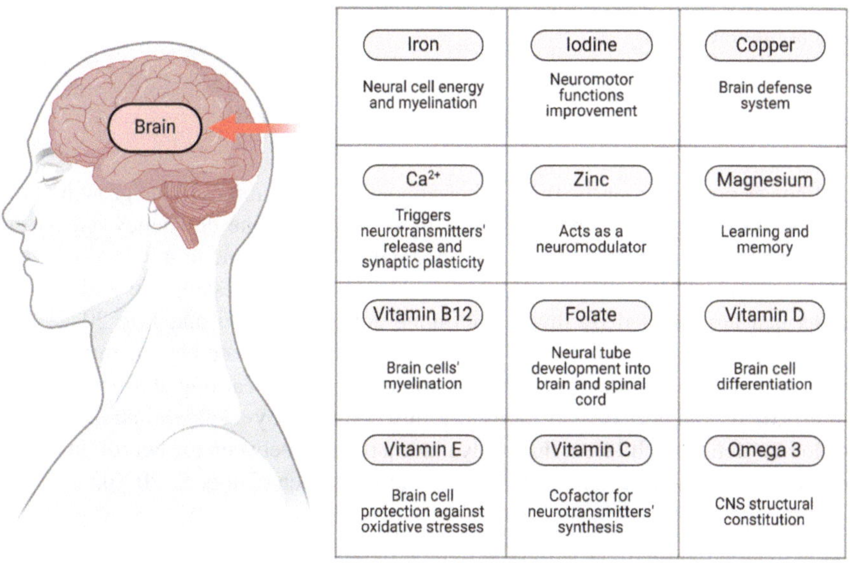

Fig. 11.1 Illustration of the significant role of micronutrients in the development of cognitive function. (All illustrations have been generated on BioRender.com)

11.2.1 Iron

Iron is important for neurodevelopment during early pregnancy and the first 6 months (McCann et al. 2020). Around 20% of oxygen in the body is utilized by the brain which in turn increases the iron in the mitochondria to be able to produce more adenosine triphosphate (ATP) (W. Mohamed and Yamashita 2024). Iron homeostasis plays a key role in building the central nervous system (CNS) and in its pathogenesis. First, iron enters the synthesis of the monoamine neurotransmitter through the hydroxylase enzyme which controls the socio-behavioral functions such as movement, emotion, and attention in humans (Georgieff 2023). In oligodendrocytes, iron takes part in the myelination process and oxidative metabolism (Y. Wang et al. 2019). Thus, neurodegeneration is a common structural defect that happens because of iron deficiency due to disruption of neurotransmitter, homeostasis, synaptogenesis, and myelin production. Iron deficiency could be the reason for the development of psychiatric disorders due to disturbance of the dopaminergic receptors (Levi et al. 2024). On the other hand, excessive iron accumulation could be fatal, especially in older people. The increment of iron levels could be a reason for developing Parkinson's disease, Alzheimer's disease, and Friedreich's ataxia due to iron deposition. Iron overload could counteract the effect of reactive oxygen species/reactive nitrogen species (ROS/RNS) balance and increase oxidative stress leading to unspecified brain degeneration and demyelination (W. Mohamed and Yamashita 2024). Thus, maintaining iron homeostasis in the brain is critical in different human stages.

11.2.2 Iodine

Iodine is a precursor for the thyroid hormone which contributes to the growth of the human body, especially the brain. Therefore, severe iodine deficiency during the first and second trimesters could cause mental disabilities due to low levels of thyroid hormones. Motor functions, memory, and visual processing are major functions that are affected by thyroid hormone deficiency (Fuzi and Loh 2022). For example, the plasticity of the hippocampus is partially regulated by the thyroid hormones. This plasticity has a direct effect on memory and learning abilities (Chaalal et al. 2019). The necessity of iodine was also revealed by a study done on a group of iodine-deficient children; this study made a relation between the neurotransmitter dopamine and the level of intelligence of those children (Cui et al. 2020).

11.2.3 Copper

Copper maintains brain functions as it is considered as the cofactor of various enzymes such as dopamine-β-hydroxylase (Rodan et al. 2015). This enzyme is used to produce norepinephrine from dopamine and norepinephrine will be used for balancing catecholamines (Vendelboe et al. 2016). Catecholamines are important for general homeostasis as they act as neurotransmitters in CNS. Those neurotransmitters influence emotion, motor, and cognitive control. Hence, copper deficiency is associated with significant symptoms such as seizures, neuronal hypomyelination, and even—if the deficiency is severe in Menkes disease—death (Lutsenko et al. 2019).

11.2.4 Calcium, Zinc, and Magnesium

Other micronutrients like calcium, zinc, and magnesium play crucial roles in brain development. For example, calcium acts as a secondary messenger in the CNS. This role by calcium is essential for maintaining homeostasis among the neural progenitor cells resulting in perfect proliferation and differentiation for those cells in case of good supplementation of this micronutrient. Disturbance of calcium levels during the embryonic stage could lead to the impairment of the migration of neurons which in turn could lead to neurodevelopmental disorders (Paudel et al. 2018). Those disorders could be as serious as autism and schizophrenia after long-term bad nutrition (Berridge 2014). Besides calcium, zinc is also critical in maintaining the neural progenitor cells differentiation via regulation of mitogen-activated protein kinases (MAPKs) (Chen et al. 2024). In addition, some pathways like Wnt signaling pathway are considered important for cells proliferation and are regulated by zinc-rich areas in the brain (Azbazdar et al. 2021). Magnesium has similar functions to zinc and plays the same role in brain development. It supports myelination and neurogenesis in addition to being a precursor for secondary neurotransmitters (de Baaij et al. 2015). Magnesium deficiency is highly associated with autism, attention-deficit hyperactivity disorder (ADHD), and cognitive impairment (Aschner et al. 2024).

11.2.5 Vitamin B12 and Folate

Vitamin B12 and folate play an essential role in promoting cognitive and brain functions via homocysteine re-methylation and the maintenance of tetrahydrofolate, the active form of folate. In homocysteine re-methylation, vitamin B12 is a cofactor for methionine synthase which employs the methyl group from 5-methlytetrahydrofolate, the dietary folate; this, in turn, deprives the neurotoxic homocysteine, provides tetrahydrofolate which plays a crucial role in hippocampus neurogenesis, and supplies S-adenosyl methionine which has a major role in methylation and neurotransmitter synthesis (Doshmanziari et al. 2021).

11.2.6 Vitamin D

Vitamin D has a significant function in neuroprotection and cognition preservation achieved through vitamin D receptor (VDR). VDRs are present in several areas in the CNS such as CA1, CA2, CA3, and CA4 regions in the hippocampus pyramidal cells, cerebral cortex, and limbic system. Therefore, they are very critical in cognitive development, learning, and memory. Vitamin D also affects the brain genetically through VDR. It is suggested that vitamin D deficiency can result in a dysregulation in the expression of about 74 genes and 36 proteins that are associated with the disruption of certain functions in the prefrontal cortex and hippocampus such as neurotransmission, synaptic plasticity, calcium homeostasis, and redox balance. Furthermore, calcitriol stimulates dopaminergic neuron proliferation and differentiation processes, and vitamin D deficiency can also dysregulate the dopaminergic system development by depriving certain vital neuron-maturation-related components for this process including neurotrophins, brain-derived neurotrophic factor (BDNF), transforming growth factor-beta 1 (TGF-β1), and nuclear receptor related 1 protein (Nurr1), and declining the levels of catechol-O-methyltransferase (COMT) and tyrosine hydroxylase expression (Gáll and Székely 2021).

11.2.7 Vitamin E

Vitamin E (alpha-tocopherol) acts as an antioxidant by limiting oxidation-reduction reactions and lipid oxidation in the membranes including those of the neuronal structures. Different signal transduction enzymes are regulated by vitamin E, and their actions subsequently impact gene expression. It also alters other enzymatic processes, which has an impact on inflammation, cellular adhesion, and cell proliferation (Fata et al. 2014). Vitamin E intake substantially enhanced memory performance, cognitive function, learning abilities, motor function, and brain signals associated with neuroregeneration and brain protection, according to experiments conducted with in vivo models of neurodegenerative disorders. In Alzheimer's disease experimental models, beta-amyloid (Aβ) neurotoxicity and deposition were found to decrease on vitamin E administration. Furthermore, it raised levels of superoxide dismutase and brain-derived neurotrophic factor (BDNF) and declined tau-protein hyperphosphorylation in rodents, suggesting that vitamin E is a good candidate to prevent and slow degeneration in the CNS (da Cunha Germano et al. 2023).

11.2.8 Vitamin C

Vitamin C contributes to cognitive function by influencing various neurological processes (Sim et al. 2022). It is essential for neuronal differentiation, functionalization, and maturation, facilitating formation of myelin and modulating neurotransmitters, particularly cholinergic and catecholaminergic pathways, which are crucial for maintaining synaptic integrity (Heo and Lee 2013; Moritz et al. 2020). As a potent antioxidant, it protects neurons from oxidative stress, with its high concentration in the brain underscoring its importance in neuronal health (Heo and Lee 2013).

It also helps regenerate the antioxidant function of vitamin D after free radical neutralization, and neurodegeneration and oxidative stress can occur drastically due to its insufficiency (Tardy et al. 2020).

11.2.9 Omega-3 Polyunsaturated Fatty Acids (Omega-3 PUFA)

Omega-3 PUFA constitutes about 35% of brain lipids which are about 50–60% of brain weight. Docosahexaenoic acid (DHA) alone contributes to more than 40% of all other omega-3 PUFAs in neuronal tissue, particularly the grey matter. Omega-3 has a crucial role in enhancing neurotransmission via altering membrane fluidity and by neurotransmitter release stimulation. Besides, it upregulates anti-apoptotic proteins gene expression, downregulates the expression of apoptotic ones, and diminishes the action of free radical-mediated oxygen species, so mitigating the rates and extent of the apoptotic processes in the brain (Dighriri et al. 2022).

11.2.10 Vitamin K

Besides its notorious importance in the blood coagulation processes, vitamin K is also necessary in cerebral homeostasis and maintaining brain cell membrane integrity and brain metabolism. Vitamin K, with its related compounds, contributes to protection against apoptosis and cell proliferation and myelin sheath production enhancement through growth-arrest specific 6 protein (Gas-6). The latter can offer these actions through several pathways. For example, it can stimulate a family of tyrosine kinase receptors, namely Tyro3, Axl, and Mer (TAM). Stimulation of Axl with PI3K signaling pathways harmonizes oligodendrocyte survival and microglial phenotype and thus guarantees the myelinating activity of Gas-6 and inhibiting TNF-α induced apoptosis. In addition, Gas-6 can decrease the proinflammatory response via interleukin-1beta induced nitric oxide synthase. It also inhibits β-amyloid induced apoptosis, which is one of the most significant pathophysiological features of Alzheimer's disease, by hindering low-voltage Ca^{2+} influx channels. Moreover, Axl activation is also required for gonadotrophin-releasing hormone, neuron cell boost, and survival and transfer from the olfactory bulb to the hypothalamus. Tyro3 may also have a similar activity, and Mer stimulation defends against reactive oxygen species. It was also reported that vitamin K can enhance neuroprotection and maintain the blood–brain barrier intact through ischemic/hypoxic injury both in vivo and in vitro through its stimulation of protein S. The latter does so via triggering Tyro3/Akt pathway, which has a defensive role against N-Methyl-D-Aspartate (NMDA)-induced toxicity and apoptosis as well, and through its action with tissue plasminogen activator. Protein S can exert some actions via TAM receptors as it has a similar structure to that of Gas-6. Vitamin K stimulates the production of sphingolipids, the most crucial constituents of cell membrane, modulates cell proliferation, differentiation, and survival and serves as a mentor for myelination in CNS. It is involved in stimulating Ceramides, sphingomyelin, cerebrosides, sulfatides, and gangliosides which are the most abundant sphingolipids in neural cell membranes and oligodendrocytes (Alisi et al. 2019).

11.2.11 Vitamin A

Vitamin A and its active metabolite, retinoic acid, play a crucial role in maintaining neuroplasticity and synaptic plasticity in the hippocampus which certainly maintains memory and learning abilities. Retinoic acid exerts its action inside the nucleus through its binding with receptors that exist in the CNS called α, β, and γ retinoic acid receptors (RARs) which form a dimer with other nuclear receptors, α, β, and γ retinoid X receptor (RXR) subtypes. The resulting heterodimers affects gene expression through its binding with retinoic acid response element (RARE) of the promoter of target genes. In addition, it promotes epidermal growth factor responsive stem cell generation, so enhancing the process of neurogenesis (Ravikumar et al. 2022; Wołoszynowska-Fraser et al. 2020).

11.3 Micronutrients and Cognition from a Physiological Perspective

Micronutrients affect multiple physiological processes involved in cognition, including the biochemical functions needed for neurotransmitter synthesis and brain energy regulation. The brain has a very high energy demand, which is greatly impacted by vitamins and minerals, and a reduction in energy severely affects cognitive function (Tardy et al. 2020). Moreover, cognition involves processes such as synaptic plasticity, neurotransmission (Speranza et al. 2021), and neurogenesis (Rajaram et al. 2019), all of which are regulated by micronutrients, such as vitamins D and B, iron, and zinc. Vitamins, such as E and C, are also crucial in the protection of neurons from free radicals by acting as antioxidants; hence, they are neuroprotective and reduce the chance of neuronal death (Tardy et al. 2020).

11.3.1 Neurotransmission and Synaptic Plasticity

Synaptic plasticity is a core process in cognitive function. It is the capacity of the nervous system to accommodate environmental changes by forming new synapses and modifying existing ones. This leads to learning and development through new experience. Synaptic changes occur through synaptic pruning, changes in spine density, and the formation of new synapses, known as synaptogenesis (Speranza et al. 2021).

Neurotransmitter synthesis is tightly regulated by the Vitamin B complex. Vitamin B1 or thiamine regulates the synthesis of acetylcholine and glutamate precursors. Moreover, it plays an important role in neuroplasticity by aiding synaptic formation and the development of neuronal axons. Vitamin B5 is essential for acetylcholine synthesis. Vitamin B6 acts as a cofactor for the biosynthesis of dopamine serotonin. Vitamin B9 is involved in serotonin and dopamine metabolism. Vitamin C was also found to affect neurotransmission, where catecholamines, such as noradrenaline and adrenaline, along with neuropeptides, require vitamin C as a cofactor during their synthesis (Tardy et al. 2020).

Vitamin D3 was shown to upregulate the gene expression of BDNF, which is important for synaptic plasticity. It also upregulates neurotrophin-3 (NFT3) and ciliary neurotrophic factor (CNTF). The dopaminergic system is highly dependent on vitamin D3 for proper functioning and development. Enzymes, such as tyrosine hydroxylase, which are involved in the dopaminergic pathway, are affected by vitamin D3. The distribution of synaptic vesicle proteins responsible for dopamine release is regulated by vitamin D3 and neurite growth of dopamine neurons (Pertile et al. 2023). Neuroplasticity is also influenced by vitamin D3, where it was found to improve the recycling of synaptic vesicles and aid in stabilizing myelin structures. Moreover, vitamin D3 enhances the expression of transcription factors involved in cognition, such as cAMP-responsive element binding protein 3 (CREB3) (Sailike et al. 2024).

Iron is involved in neuronal differentiation by affecting neural plasticity and myelination. Moreover, calcium release in hippocampal neurons is stimulated by ROS generated from the Fenton reaction of iron, which then leads to the overexpression of genes contributing to synaptic plasticity (Tardy et al. 2020).

Magnesium is essential for optimal nerve transmission and is involved in the regulation of NMDA receptors. Low magnesium intake or excessive magnesium loss can lead to hypomagnesemia, affecting synaptic transmission. Magnesium contributes to controlling neuromuscular processes by regulating calcium and potassium transport across cells. Moreover, it protects against excitotoxicity by blocking the calcium channels involved in glutamatergic neurotransmission (Patel et al. 2024). In contrast, zinc influences depression and neural transmission by interacting with NMDA receptors and voltage-gated Ca^{2+} channels. Zinc-containing neurons are types of neurons with zinc in their synaptic vesicles, which modulate excitability during glutamate and gamma-aminobutyric acid (GABA) neurotransmission. Both minerals are crucial for the regulation of synaptic plasticity, neurotransmitter release, and overall brain function (Tardy et al. 2020).

11.3.2 Neurogenesis and Brain Energy Metabolism

Neurogenesis involves the maturation of neural stem cells into neurons. In the brain of adult population, it was thought that neurogenesis only happens to glial cells, known as "gliogenesis." Nevertheless, it was discovered that neurogenesis in adult brain occurs in the subgranular zone in the dentate gyrus of the hippocampus and the subventricular zone of the lateral ventricles. However, this is restricted to specific neurons; these neurons are fewer in number than neurogenesis that occurs during the prenatal period (W. Mohamed and Yamashita 2024).

Folic acid (vitamin B9) aids in DNA synthesis during neurogenesis. Folate is essential for the biosynthesis of RNA and DNA, as well as for DNA repair. Deficiencies in folate result in a decrease in neural progenitor cells in the hippocampus, and hence, a reduction in neurogenesis needed for memory and cognition (Virdi and Jadavji 2022).

Several studies have reported that supplementation with 30 mg of zinc upregulates the activity of BDNF through an increase in serum levels. Zinc has been suggested to elevate BDNF expression by upregulating the activities of matrix metalloproteinases (MMP 2 and 9, which are enzymes that convert BDNF to its active form. BDNF is a growth factor involved in the regulation of multiple pathways related to neurogenesis, neuronal survival, synaptic plasticity, and neuroinflammation reduction, specifically in the hippocampus. The importance of BDNF in cognitive function has been elucidated through its reduction in patients with schizophrenia with impaired cognitive function. Moreover, its reduction is associated with neurodegenerative diseases and first-episode psychosis (FEP) (Agh et al. 2022).

Micronutrients, such as iron, are vital for maintaining proper neuronal function in neurons by supporting key cellular processes and metabolic pathways. They also play an important role in ATP production in neurons by influencing energy-yielding metabolism, oxygen transport, DNA synthesis, and neuronal functionality. For example, iron is essential for mitochondrial function and glucose metabolism, affecting ATP synthesis and cellular energy regulation. It is part of the cytochromes involved in oxidative phosphorylation, an important pathway for ATP production in the mitochondria. In addition, iron is reported to integrate in the synthesis and production of neurotransmitters and is essential for glucose metabolism (Tardy et al. 2020). Iron is a component of hemoglobin that acts as a carrier of oxygen. Hence, iron is crucial for oxygen transport and energy metabolism, which are essential for cognitive functions (Li and Sheng 2022).

Deficiencies in iron and glucose regulation can significantly impact cognitive functions such as memory and attention. Iron deficiency can lead to anemia, which reduces oxygen delivery to the brain. This can impair cognitive functions, including attention and memory, because neurons require adequate oxygen for optimal functioning. Mitochondrial dysfunction due to iron deficiency can also lead to elevated oxidative stress, which can cause injuries to neurons and impair cognitive function. Acute hypoglycemia (low blood sugar) can result in confusion, irritability, and difficulty in concentrating. These symptoms can severely impair attention and memory, making it difficult to perform tasks requiring cognitive effort. Chronic issues with glucose regulation, such as in diabetes, can lead to cognitive decline over time, affecting both memory and attention (Li and Sheng 2022).

11.3.3 The Integrity of the Blood–Brain Barrier

The blood–brain barrier (BBB) is a selective semipermeable membrane that prevents foreign substances, such as pathogens in circulation, from reaching the brain. The BBB actively transports metabolic products, such as glucose, using specific transport proteins, which are required for proper brain function and development (Segarra et al. 2021). Traumatic brain injury (TBI) can disrupt the BBB, leading to brain damage and cognitive impairments such as memory loss and attention deficits (Thapa et al. 2021).

Micronutrients play a crucial role in maintaining the integrity of the BBB, which is essential for cognitive health. For instance, vitamin C, when delivered in a complex form such as NXP031, has been shown to improve cognitive impairment by enhancing antioxidant efficacy and promoting BBB integrity (Lee et al. 2021).

Omega-3 fatty acids, particularly eicosapentaenoic acid (EPA) and DHA, have an important role in keeping BBB healthy. EPA and DHA contribute to a decrease in excitotoxicity, which is an important neurological protection mechanism provided by these fatty acids (Martami and Holton 2023). Furthermore, omega-3 fatty acids are an integral part of cell membranes, including those in the brain, and promote amino acid transport (Segarra et al. 2021). Alpha-linolenic acid (ALA), a precursor to DHA, has emerged as a potential brain savior that improves BBB functionality (Leikin-Frenkel et al. 2022).

Disruption of the BBB, related to micronutrient deficiencies, has thus been linked to cognitive decline, especially in aging and neurodegenerative diseases. The breakdown of the BBB has been reported to have a role in the diagnosis of age-related cognitive impairment and dementia. Moreover, disruption of the neurovascular unit (NVU) has been linked to cognitive impairment in neurodegenerative disorders. The integrity of the BBB and sufficiency of various nutrient levels have been brought into the spotlight and are critical for maintaining cognitive function, particularly in aging people and patients with neurodegenerative diseases (Che et al. 2024).

11.4 Therapeutic Perspective of Micronutrients in Cognitive Disorders

Recent studies indicate that early-life nutrition profoundly affects neurodevelopment, subsequently altering health outcomes, cognitive function, and illness susceptibility in later life (Mattei and Pietrobelli 2019). Certain mental disorders have been associated with insufficient nutrition during early development (Puri et al. 2023). Epigenetic mechanisms may significantly influence gene imprinting during early development, increasing the individual's vulnerability to diseases in later life (Jirtle et al. 2018; Vaiserman 2015). Well-nourished children are more likely to realize their developmental potential in cognitive, motor, and socio-emotional skills, positively influencing society. It is advisable to recommend specific dietary practices to our patients to enhance their early-life nutritional status and prevent long-term adverse consequences. Prevention measures must focus on the provision of superior quality meals to preconception, pregnant, and breastfeeding women, as well as young children, in both malnutrition-prone regions and industrialized nations.

11.4.1 Importance of Nutrition in Brain Health

Key nutritional components for brain health include antioxidants, B vitamins, omega-3 fatty acids, healthy fats, and flavonoids. Omega-3 fatty acids, found in fatty fish-like salmon and plant sources such as walnuts and flaxseeds, are essential for maintaining brain cell structure and function while reducing inflammation. Antioxidants, present in brightly colored fruits and vegetables such as berries and leafy greens, protect the brain from oxidative stress caused by free radicals. B vitamins, especially B6, B9, and B12, are crucial for neurotransmitter production (Puri et al. 2023), influencing mood regulation and cognitive function; they can be sourced from whole grains, legumes, and leafy greens (J. Li et al. 2023). Healthy fats, including monounsaturated and polyunsaturated fats from olive oil, avocados, and nuts, promote healthy blood perfusion to the brain and decrease inflammation significantly. Lastly, flavonoids found in foods like dark chocolate and tea have been reported to improve memory and cognitive functionality. A balanced diet incorporating these nutrients is central for brain health and its normal function (Steindler and Reynolds 2017; Zsolt 2021).

11.4.2 Therapeutic Interventions Associated with Vitamins for Brain Health

11.4.2.1 Vitamin B

Vitamin B contributes to keeping the brain healthy through its various components, each contributing to essential functions that support cognitive abilities and emotional well-being. The B vitamins, including B6, B9, and B12, are particularly important for synthesizing neurotransmitters, which are crucial for communication between nerve cells (Li et al. 2023). For example, vitamin B6, known as pyridoxine, is involved in the synthesis of serotonin and dopamine, which regulate mood and cognitive function (Alberto et al. 2020; Kennedy 2016). Even mild deficiencies in B6 can lead to impaired neurotransmitter synthesis, affecting brain health and function. In addition, B9 folates are vital for DNA synthesis and repair, where low levels of folate have been associated with cognitive impairment and an increased risk of neurological disorders (Kennedy 2016). Vitamin B12, known as cobalamin, plays a critical role in myelin formation, which protects nerve fibers and ensures efficient signal transmission (Kennedy 2016).

Adequate levels of these vitamins help regulate mood and cognitive function; deficiencies can lead to symptoms such as memory impairment, depression, and cognitive decline (Puri et al. 2023). Research indicates that vitamin B12 is important for neural myelination and brain development, especially during pregnancy and early childhood (Mattei and Pietrobelli 2019; Mccleery et al. 2018). In addition, studies have reported that low folate intake is associated with a high risk of mild cognitive impairment and dementia in older adults (Visioli and Burgos-ramos 2016). Furthermore, B vitamins help decrease homocysteine levels, an amino acid associated with an elevated risk of Alzheimer's disease (Haan et al. 2007; Wang et al. 2021).

11.4.2.2 Vitamin D

Vitamin D deficiency (VDD) impacts around one billion individuals (Pfotenhauer and Shubrook 2017). VDD is now acknowledged as a contributor to various extra-skeletal pathologies beyond osteomalacia and rickets, including multiple sclerosis, autoimmune disorders, increased infection rates, respiratory disorders, cardiovascular disorders, and different cancers (Pfotenhauer and Shubrook 2017). The emergence of VDD presents a specific risk to people with restricted intake, low sunshine exposure, or poor intestinal absorption (Kennel et al. 2010). Vitamin D3 is produced in the skin through ultraviolet (UV) radiation (Fig. 11.2). Air pollution, prevalent in developing nations, may also facilitate the onset of VDD, as particulate matter either absorbs UVB radiation or reflects sunshine (Mousavi et al. 2019). The practice of dressing may contribute to the development of vitamin D deficiency, as the production of vitamin D in the skin is inhibited (Jarrett and Scragg 2020). Reduced levels of 7-dehydrocholesterol (7-DHC) in the epidermis, insufficient intake of specific foods such as dairy products, and compromised vitamin D metabolism due to renal and hepatic insufficiency may also result in deficiency. Certain pharmaceuticals, such as immunosuppressants, glucocorticoids, antiepileptics, antifungals, and antiretrovirals, may induce heightened catabolism of vitamin D, leading to vitamin D deficiency (Kennel et al. 2010; Pfotenhauer and Shubrook 2017; Vaiserman 2015).

11.4.2.3 Vitamin C

Vitamin C acts as a cofactor in the biosynthesis and production of neurotransmitters such as dopamine and serotonin, regulating their release and reuptake, thereby impacting mood and cognitive functions like attention and memory (Travica et al. 2017). Recent studies have shown that vitamin C influences neuronal metabolism by preferring lactate over glucose as a source of energy. Clinical evidence indicates a positive relationship between plasma vitamin C levels and cognitive performance in healthy adults, with higher concentrations associated with better outcomes on processes involving attention and memory. In cognitively impaired individuals, vitamin C status appears lower compared to cognitively intact groups, although the relationship between vitamin C levels and specific cognitive assessments remains less clear. Supplementation studies suggest that vitamin C can enhance cognitive performance such as attention and motivation in those with inadequate status (Hansen et al. 2014; Travica et al. 2019).

11.4.3 Therapeutic Interventions of Minerals Associated with Brain Health

Research indicates that specific minerals, including iron, magnesium, copper, and zinc, may play a role in systems associated with cognitive functions of the brain. These micronutrients are implicated in DNA repair, protection against oxidative

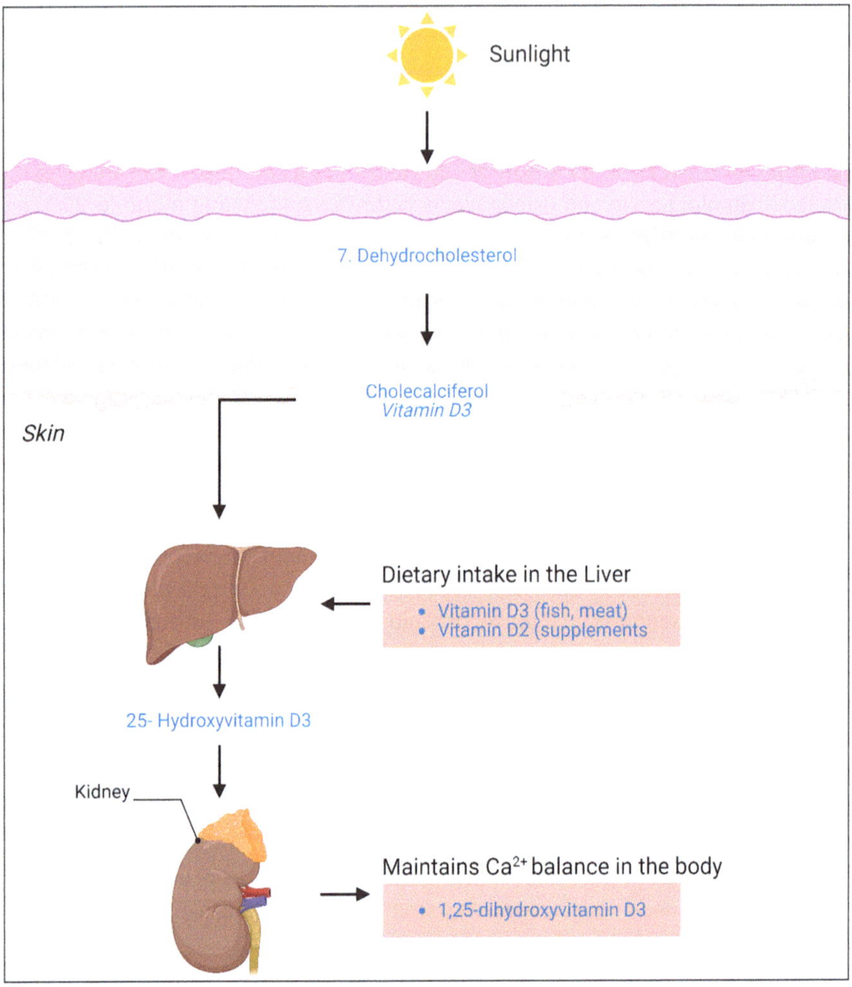

Fig. 11.2 Under sunlight, the skin synthesizes cholecalciferol, a form of vitamin D3 generated from steroid cholesterol. The liver converts cholecalciferol into calcidiol, which is subsequently turned into calcitriol by renal processes (the active form of the vitamin). (All illustrations have been generated on BioRender.com)

stress, and proper DNA methylation, among other functions. Consequently, they seem to be significant in the control of cellular function and neuromodulation, potentially playing a vital role in antioxidant protection. Moreover, their antioxidative characteristics may alleviate damage caused by free radicals, therefore preventing or slowing the cognitive deterioration associated with the neurotoxic effects of oxidative stress (Lorenzo-Mora et al. 2023; Meramat et al. 2015; Nooyens et al. 2015).

11.4.3.1 Iron

Iron plays a therapeutic role in cognitive performance, particularly in children and adolescents, where deficiency can lead to significant cognitive impairments. It is essential for various brain processes, including the production of neuronal myelin sheaths and neurotransmitter synthesis, which are vital for cognitive development (Travica et al. 2019). Evidence shows that iron deficiency is closely associated with retarded brain development, poorer school performance, and behavioral irregularities (Hansen et al. 2014). In addition, systematic reviews indicate that iron supplementation can significantly improve cognitive outcomes such as intelligence, attention, and memory in school-age children. For instance, one study found that iron supplementation led to a standardized mean difference (SMD) of 0.46 in intelligence and 0.44 in memory performance among children receiving supplementation compared to those on placebo (Sim et al. 2022). In low- and middle-income countries (LMICs), iron supplementation has been reported to positively impact intelligence test scores among children and adolescents, with higher doses correlating with better outcomes (Canals et al. 2023). However, while improvements in intelligence have been consistently reported, effects on attention and long-term memory are less clear, indicating variability in the effectiveness of supplementation across different contexts. Further, the benefits appear more pronounced in anemic individuals, suggesting that targeted interventions for those with iron deficiency may yield the most significant cognitive improvements (Canals et al. 2023; Meramat et al. 2015; Nooyens et al. 2015; Tariku et al. 2023).

11.4.3.2 Zinc

Zinc has a therapeutic role in cognitive function, particularly in children, where its deficiency can result in significant deterioration in memory, attention, and overall intellectual development. Research indicates that zinc is essential for neurogenesis, synaptogenesis, and neurotransmitter release, including GABA and glutamate, which are crucial for cognitive processes. A study involving school-age boys demonstrated that zinc supplementation significantly improved cognitive function, as assessed by the Raven IQ test, with the experimental group showing notable enhancements compared to the placebo group ($P < 0.05$) (Bhatnagar and Taneja 2001; Khodashenas et al. 2015). In addition, other studies have shown that zinc supplementation can positively affect neuropsychological functions and motor development in infants and toddlers, supporting its importance during critical periods of brain growth. In older children, while data is more controversial, some evidence suggests improved cognitive outcomes with zinc supplementation. For adults, particularly those aged 55–87 years, zinc supplementation has shown mixed results. While some measures of working memory and attention improved after supplementation, the effects were not uniformly beneficial across all cognitive domains (Bhatnagar and Taneja 2001; Khodashenas et al. 2015).

11.4.3.3 Magnesium

Magnesium is a crucial mineral that significantly impacts cognitive function, playing a vital role in various neurological processes. It participates in over 300 enzymatic reactions central to brain health, including DNA synthesis, energy production, and neurotransmitter regulation. Specifically, magnesium is critical for the proper functioning of neurons and helps regulate neurotransmitters like glutamate and GABA, which are pivotal for learning and memory (Kirkland et al. 2018; Mohamed et al. 2022; Xu et al. 2024). Research indicates that increased magnesium levels can enhance synaptic plasticity, which is vital for learning and memory retention (Maier et al. 2023). Studies on animals have shown that magnesium supplementation improves working memory and long-term memory, while human studies suggest that higher magnesium intake correlates with better cognitive performance, particularly in older adults (Kirkland et al. 2018; Xu et al. 2024). For example, a study found that individuals consuming over 550 mg of magnesium daily exhibited less age-related brain shrinkage, equating to a brain age approximately one year younger than those consuming lower amounts (Alateeq et al. 2023). Furthermore, magnesium has neuroprotective properties that may help mitigate the risk of neurodegenerative diseases such as Alzheimer's disease (Xu et al. 2024). However, deficiencies in magnesium can result in cognitive dysfunction and an elevated risk of neurological disorders (Maier et al. 2023).

11.4.3.4 Copper

Copper is essential for maintaining brain health, influencing various neurological processes essential for cognitive function (An et al. 2022). They are integral to neurotransmission, particularly at synapses, where copper is released to regulate the activity of neurotransmitter receptors, such as NMDA receptors. This modulation enhances synaptic plasticity (Sanaeifar et al. 2024; Wadan and Liaquat 2024), a fundamental mechanism underlying learning and memory. In addition, copper contributes to the regulation of oxidative stress within the brain; it participates in enzymatic reactions that produce ROS, which can have both protective and damaging effects depending on their concentration. An imbalance in copper levels can lead to neurodegenerative diseases; for instance, copper deficiency has been linked to cognitive impairment and conditions like Alzheimer's disease, while excessive copper accumulation can be toxic and contribute to neuronal damage (Bagheri et al. 2018).

11.4.4 Therapeutic Interventions of Omega-3 Fatty Acids

Omega-3 fatty acids, particularly EPA and DHA, contribute to maintaining and enhancing cognitive functions throughout the human lifespan. DHA comprises about 40% of the total fatty acids in the brain and is vital for neuronal structure and function, supporting neurotransmission by enhancing membrane fluidity and promoting the release of neurotransmitters. EPA, known for its anti-inflammatory effect, contributes to mental health by potentially alleviating symptoms of depression and anxiety. Research indicates that higher DHA levels correlate with improved memory performance and cognitive flexibility in adults, while omega-3 intake is associated with larger hippocampal volumes in midlife, which is crucial for memory processing. These fatty acids produce their neuroprotective effects by reducing neuroinflammation and oxidative stress, which are critical for preventing cognitive decline associated with aging, and they also support neurogenesis. Omega-3 influences brain performance through mechanisms such as enhancing membrane fluidity, promoting neurotransmitter release, and mitigating chronic inflammation. Health organizations advise a daily dose of 250–500 mg of mixed EPA and DHA for adults, with higher amounts suggested for pregnant women and individuals with certain health conditions (Cutuli et al. 2016; Dighriri et al. 2022).

11.5 Conclusion

Micronutrients are essential for brain health and normal functionality. In this chapter, we highlighted the importance of micronutrients that are involved in brain functions including neurogenesis, neural plasticity, signal processing, and brain development. A balanced diet that contains vitamins, minerals, and trace elements is essential for brain vitality and normal cognitive performance. We reviewed the literature which revealed that minerals regulate cognitive development and maintain their performance. In addition, vitamins such as the B group, C, and D are crucial for neurogenesis and neuroprotection. They also protect against neurodegeneration. Micronutrients also support neuronal structures, functions, and support signaling between neurons. Furthermore, they showed therapeutic potential due to their anti-inflammatory and antioxidant properties. They also aid in restoring cognitive function in patients with cognitive dysfunction and thereby play an important role in brain homeostasis.

References

Agh F, Hasani M, Khazdouz M, Amiri F, Heshmati J, Aryaeian N (2022) The effect of zinc supplementation on circulating levels of brain-derived neurotrophic factor (BDNF): A systematic review and meta-analysis of randomized controlled trials. Int J Prev Med 13(1):117). Wolters Kluwer Medknow Publications. https://doi.org/10.4103/ijpvm.IJPVM_478_20

Alateeq K, Walsh EI, Cherbuin N (2023) Dietary magnesium intake is related to larger brain volumes and lower white matter lesions with notable sex differences. Eur J Nutr 62(5):2039–2051. https://doi.org/10.1007/s00394-023-03123-x

Alberto C, Ospina C, Orlando M, Mesa N (2020) B vitamins in the nervous system: current knowledge of the biochemical modes of action and synergies of thiamine, pyridoxine, and cobalamin. CNS Neurosci Ther 2019:5–13. https://doi.org/10.1111/cns.13207

Alisi L, Cao R, De Angelis C, Cafolla A, Caramia F, Cartocci G, Librando A, Fiorelli M (2019) The relationships between vitamin k and cognition: a review of current evidence. Front Neurol 10. https://doi.org/10.3389/fneur.2019.00239

An Y, Li S, Huang X, Chen X, Shan H, Zhang M (2022) The role of copper homeostasis in brain disease. Int J Mol Sci 23(22):13850

Aschner M, Martins AC, Oliveira-Paula GH, Skalny AV, Zaitseva IP, Bowman AB, Kirichuk AA, Santamaria A, Tizabi Y, Tinkov AA (2024) Manganese in autism spectrum disorder and attention deficit hyperactivity disorder: the state of the art. Curr Res Toxicol 6:100170. https://doi.org/10.1016/j.crtox.2024.100170

Azbazdar Y, Karabicici M, Erdal E, Ozhan G (2021) Regulation of Wnt signaling pathways at the plasma membrane and their misregulation in cancer. Front Cell Dev Biol 9. https://doi.org/10.3389/fcell.2021.631623

Bagheri S, Squitti R, Haertlé T, Siotto M, Saboury AA, John D, Lane R, Gill AC (2018) Role of copper in the onset of Alzheimer's disease compared to other metals. Front Aging Neurosci 9(January):1–15. https://doi.org/10.3389/fnagi.2017.00446

Berridge MJ (2014) Calcium signalling and psychiatric disease: bipolar disorder and schizophrenia. Cell Tissue Res 357(2):477–492. https://doi.org/10.1007/s00441-014-1806-z

Bhatnagar S, Taneja S (2001) Zinc and cognitive development. Br J Nutr 85(S2):S139–S145. https://doi.org/10.1079/BJN2000306

Bourre JM (2006) Effects of nutrients (in food) on the structure and function of the nervous system: update on dietary requirements for brain. Part 1: micronutrients. J Nutr Health Aging 10(5):377–385

Canals J, Coronel P, Gimeno M, Basora J, Arija V, Edh TAG, Tag M (2023) Importance of maternal iron status on the improvement of cognitive function in children after prenatal iron supplementation. Am J Prev Med 65(3):395–405. https://doi.org/10.1016/j.amepre.2023.02.006

Chaalal A, Poirier R, Blum D, Laroche S, Enderlin V (2019) Thyroid hormone supplementation restores spatial memory, hippocampal markers of neuroinflammation, plasticity-related signaling molecules, and β-amyloid peptide load in hypothyroid rats. Mol Neurobiol 56(1):722–735

Che J, Sun Y, Deng Y, Zhang J (2024) Blood-brain barrier disruption: a culprit of cognitive decline? Fluids Barriers CNS 21(1):63. https://doi.org/10.1186/s12987-024-00563-3

Chen B, Yu P, Chan WN, Xie F, Zhang Y, Liang L, Leung KT, Lo KW, Yu J, Tse GMK, Kang W, To, K. F (2024) Cellular zinc metabolism and zinc signaling: from biological functions to diseases and therapeutic targets. Signal Transduct Target Ther 9(1):6. https://doi.org/10.1038/s41392-023-01679-y

Cui Y, Yu J, Zhang B, Guo B, Gao T, Liu H (2020) The relationships between thyroid-stimulating hormone and/or dopamine levels in peripheral blood and IQ in children with different urinary iodine concentrations. Neurosci Lett 729:134981. https://doi.org/10.1016/j.neulet.2020.134981

Cutuli D, Pagani M, Caporali P, Galbusera A, Laricchiuta D, Foti F, Neri C, Spalletta G, Caltagirone C, Petrosini L, Gozzi A (2016) Effects of Omega-3 fatty acid supplementation on cognitive functions and neural substrates: A voxel-based morphometry study in aged mice. Front Aging Neurosci 8(March):1–14. https://doi.org/10.3389/fnagi.2016.00038

da Cunha Germano BC, de Morais LCC, Idalina Neta F, Fernandes ACL, Pinheiro FI, do Rego ACM, Araújo Filho I, de Azevedo EP, de Paiva Cavalcanti JRL, Guzen FP, Cobucci RN (2023) Vitamin E and its molecular effects in experimental models of neurodegenerative diseases. Int J Mol Sci 24(13):11191. https://doi.org/10.3390/ijms241311191

de Baaij JHF, Hoenderop JGJ, Bindels RJM (2015) Magnesium in man: implications for health and disease. Physiol Rev 95(1):1–46. https://doi.org/10.1152/physrev.00012.2014

Dighriri IM, Alsubaie AM, Hakami FM, Hamithi DM, Alshekh MM, Khobrani FA, Dalak FE, Hakami AA, Alsueaadi EH, Alsaawi LS, Alshammari SF, Alqahtani AS, Alawi IA, Aljuaid AA, Tawhari MQ (2022) Effects of omega-3 polyunsaturated fatty acids on brain functions: a systematic review. Cureus 14(10). https://doi.org/10.7759/CUREUS.30091

Doshmanziari M, Shirian S, Kouchakian M-R, Moniri SF, Jangnoo S, Mohammadi N, Zafari F (2021) Mesenchymal stem cells act as stimulators of neurogenesis and synaptic function in a rat model of Alzheimer's disease. Heliyon:e07996. https://doi.org/10.1016/J.HELIYON.2021.E07996

Fata G, Weber P, Mohajeri M (2014) Effects of vitamin E on cognitive performance during ageing and in Alzheimer's disease. Nutrients 6(12):5453–5472. https://doi.org/10.3390/nu6125453

Fox SE, Levitt P, III, C. A. N. (2010) How the timing and quality of early experiences influence the development of brain architecture. Child Dev 81(1):28–40. https://doi.org/10.1111/j.1467-8624.2009.01380.x

Fuzi SFA, Loh SP (2022) Iodine: A critical micronutrient in brain development. In: Role of micronutrients in brain health. Springer Singapore, Singapore, pp 49–67. https://doi.org/10.1007/978-981-16-6467-0_4

Gáll Z, Székely O (2021) Role of vitamin d in cognitive dysfunction: new molecular concepts and discrepancies between animal and human findings. Nutrients 13(11) MDPI. https://doi.org/10.3390/nu13113672

Georgieff MK (2023) The importance of iron deficiency in pregnancy on fetal, neonatal, and infant neurodevelopmental outcomes. Int J Gynecol Obstet 162(S2):83–88. https://doi.org/10.1002/ijgo.14951

Gilbreath D, Hagood D, Alatorre-Cruz GC, Andres A, Downs H, Larson-Prior LJ (2023) Effects of early nutrition factors on baseline neurodevelopment during the first 6 months of life: an EEG study. Nutrients 15(6):1535. https://doi.org/10.3390/nu15061535

Haan MN, Miller JW, Aiello AE, Whitmer RA, Jagust WJ, Mungas DM, Allen LH, Green R (2007) Homocysteine, B vitamins, and the incidence of dementia and cognitive impairment: results from the Sacramento Area Latino. Am J Clin Nutr 85(2):511–517. https://doi.org/10.1093/ajcn/85.2.511

Hansen SN, Tveden-nyborg P, Lykkesfeldt J (2014) Does vitamin C deficiency affect cognitive development and function. Nutrients 6(9):3818–3846. https://doi.org/10.3390/nu6093818

Heo J, Lee K (2013) The possible role of antioxidant vitamin C in Alzheimer's disease treatment and prevention experimental studies supporting the efficacy of vitamin. Am J Alzheimers Dis Other Dement 28(2):120–125. https://doi.org/10.1177/1533317512473193

Jarrett P, Scragg R (2020) Evolution, prehistory and vitamin D. Int J Environ Res Public Health 17(2):646. https://doi.org/10.3390/ijerph17020646

Jirtle RL, Skinner MK, Carolina N (2018) HHS public access. Nat Rev Genet 8(4):253–262. https://doi.org/10.1038/nrg2045.Environmental

Kennedy DO (2016) B vitamins and the brain: mechanisms, dose and efficacy – a review. Nutrients 8(2):68. https://doi.org/10.3390/nu8020068

Kennel KA, Drake MT, Hurley DL (2010) Vitamin D deficiency in adults: when to test and how to treat. Mayo Clin Proc 85(8):752–758. https://doi.org/10.4065/mcp.2010.0138

Khodashenas E, Mohammadzadeh A, Sohrabi M, Izanloo A (2015) The effect of zinc supplementation on cognitive performance in schoolchildren. J Pediatr Perspect 3(23):1033–1038

Kirkland AE, Sarlo GL, Holton KF (2018) The role of magnesium in neurological disorders. Nutrients 10(6):730. https://doi.org/10.3390/nu10060730

Lee J-M, Lee J-H, Song M-K, Kim Y-J (2021) NXP031 improves cognitive impairment in a chronic cerebral Hypoperfusion-induced vascular dementia rat model through Nrf2 Signaling. Int J Mol Sci 22(12):6285. https://doi.org/10.3390/ijms22126285

Leikin-Frenkel A, Schnaider Beeri M, Cooper I (2022) How alpha linolenic acid may sustain blood–brain barrier integrity and boost brain resilience against alzheimer's disease. Nutrients 14(23):5091. https://doi.org/10.3390/nu14235091

Levi S, Ripamonti M, Moro AS, Cozzi A (2024) Iron imbalance in neurodegeneration. Mol Psychiatry 29(4):1139–1152. https://doi.org/10.1038/s41380-023-02399-z

Li S, Sheng ZH (2022) Energy matters: presynaptic metabolism and the maintenance of synaptic transmission. Nat Rev Neurosci 23(1):4–22). Nature Research. https://doi.org/10.1038/s41583-021-00535-8

Li J, Cao Y, Xu J, Li J, Lv C, Gao Q, Zhang C (2023) Vitamin D improves cognitive impairment and alleviates Ferroptosis via the Nrf2 Signaling pathway in aging mice. Int J Mol Sci 24(20):15315

Lorenzo-Mora AM, López-Sobaler AM, Bermejo LM, González-Rodríguez LG, Cuadrado-Soto E, Peral-Suárez Á, Salas-González MD, Delgado-Losada ML, Rodríguez-Rojo IC, Barabash A, Maestú-Unturbe F, Aparicio A (2023) Association between mineral intake and cognition evaluated by Montreal Cognitive Assessment (MoCA): a cross-sectional study. Nutrients 15(21):4505. https://doi.org/10.3390/nu15214505

Lukaski HC (2004) Vitamin and mineral status: effects on physical performance. Nutrition 20(7–8):632–644. https://doi.org/10.1016/j.nut.2004.04.001

Lutsenko S, Washington-Hughes C, Ralle M, Schmidt K (2019) Copper and the brain noradrenergic system. JBIC J Biol Inorg Chem 24(8):1179–1188. https://doi.org/10.1007/s00775-019-01737-3

Maier JAM, Locatelli L, Fedele G, Cazzaniga A (2023) Magnesium and the brain: a focus on Neuroinflammation and neurodegeneration. Int J Mol Sci

Martami F, Holton KF (2023) Targeting glutamate neurotoxicity through dietary manipulation: potential treatment for migraine. Nutrients 15(18):3952. https://doi.org/10.3390/nu15183952

Mattei D, Pietrobelli A (2019) Micronutrients and brain development. Curr Nutr Rep 8(2):99–107. https://doi.org/10.1007/s13668-019-0268-z

McCann S, Amadó MP, Moore SE (2020) The role of iron in brain development: a systematic review. Nutrients 12(7):2001. https://doi.org/10.3390/nu12072001

Mccleery J, Abraham RP, Denton DA, Rutjes AW, Chong L-Y, Al-Assaf AS, Griffith DJ, Rafeeq S, Yaman H, Malik MA, Nisio MD, Martínez G, Vernooij RW, Tabet N (2018) Vitamin and mineral supplementation for preventing dementia or delaying cognitive decline in people with mild cognitive impairment (review). Cochrane Database Syst Rev 11(11):CD011905. https://doi.org/10.1002/14651858.CD011905.pub2. www.cochranelibrary.com

Meramat A, Rajab NF, Shahar S, Sharif R (2015) Cognitive impairment, genomic instability and trace elements. J Nutr Health Aging 19(1):48–57. https://doi.org/10.1007/s12603-014-0489-1

Mohamed W, Yamashita T (2024) Role of micronutrients in brain health. https://link.springer.com/bookseries/16639

Mohamed AA, Hassaan MMM, Hussien AG, Allam AS (2022) Correlation between cognitive dysfunction and serum magnesium, calcium and phosphorus level in the elderly Egyptian patients in Zagazig University. Hospitals 87(April):1097–1105

Moritz B, Schmitz AE, Rodrigues ALS, Dafre AL, Cunha MP (2020) The role of vitamin C in stress-related disorders. J Nutr Biochem 85:108459. https://doi.org/10.1016/j.jnutbio.2020.108459

Morris, A. L., & Mohiuddin, S. S. (2024). Biochemistry, nutrients

Mousavi SE, Amini H, Heydarpour P, Amini Chermahini F, Godderis L (2019) Air pollution, environmental chemicals, and smoking may trigger vitamin D deficiency: evidence and potential mechanisms. Environ Int 122:67–90. https://doi.org/10.1016/j.envint.2018.11.052

Nooyens ACJ, Milder IEJ, van Gelder BM, Bueno-de-Mesquita HB, van Boxtel MPJ, Verschuren WMM (2015) Diet and cognitive decline at middle age: the role of antioxidants. Br J Nutr 113(9):1410–1417. https://doi.org/10.1017/S0007114515000720

Patel V, Akimbekov NS, Grant WB, Dean C, Fang X, Razzaque MS (2024) Neuroprotective effects of magnesium: implications for neuroinflammation and cognitive decline. Front Endocrinol 15. https://doi.org/10.3389/fendo.2024.1406455

Paudel S, Sindelar R, Saha M (2018) Calcium signaling in vertebrate development and its role in disease. Int J Mol Sci 19(11):3390. https://doi.org/10.3390/ijms19113390

Pertile RAN, Brigden R, Raman V, Cui X, Du Z, Eyles D (2023) Vitamin D: A potent regulator of dopaminergic neuron differentiation and function. J Neurochem 166(5):779–789. https://doi.org/10.1111/jnc.15829

Pfotenhauer KM, Shubrook JH (2017) Vitamin D deficiency, its role in health and disease, and current supplementation recommendations. JOM 117(5):301–305. https://doi.org/10.7556/jaoa.2017.055

Prado EL, Dewey KG (2014) Nutrition and brain development in early life. Nutr Rev 72(4):267–284. https://doi.org/10.1111/nure.12102

Puri S, Shaheen M, Grover B (2023) Nutrition and cognitive health: a life course approach. Front Public Health 11:1023907. https://doi.org/10.3389/fpubh.2023.1023907

Rajaram S, Jones J, Lee GJ (2019) Plant-based dietary patterns, plant foods, and age-related cognitive decline. Adv Nutr 10(4):422–436. https://doi.org/10.1093/advances/nmz081

Ravikumar N, Chegukrishnamurthi M, Gadde Venkata S (2022) Role of micronutrients in neurological development. In: Role of nutrients in neurological disorders. Springer Singapore, Singapore, pp 177–199. https://doi.org/10.1007/978-981-16-8158-5_9

Rodan LH, Gibson KM, Pearl PL (2015) Clinical use of CSF neurotransmitters. Pediatr Neurol 53(4):277–286. https://doi.org/10.1016/j.pediatrneurol.2015.04.016

Sailike B, Onzhanova Z, Akbay B, Tokay T, Molnár F (2024) Vitamin D in central nervous system: implications for neurological disorders. Int J Mol Sci 25(14) Multidisciplinary Digital Publishing Institute (MDPI). https://doi.org/10.3390/ijms25147809

Sanaeifar F, Pourranjbar S, Pourranjbar M, Ramezani S (2024) Beneficial effects of physical exercise on cognitive-behavioral impairments and brain-derived neurotrophic factor alteration in the limbic system induced by neurodegeneration. Exp Gerontol 195(August):112539. https://doi.org/10.1016/j.exger.2024.112539

Segarra M, Aburto MR, Acker-Palmer A (2021) Blood–brain barrier dynamics to maintain brain homeostasis. Trends Neurosci 44(5):393–405). Elsevier Ltd. https://doi.org/10.1016/j.tins.2020.12.002

Shenkin A (2006) The key role of micronutrients☆. Clin Nutr 25(1):1–13. https://doi.org/10.1016/j.clnu.2005.11.006

Sim M, Hong S, Jung S, Soo J, Young K, Goo T, Young W, Dong C, Shin M (2022) Vitamin C supplementation promotes mental vitality in healthy young adults: results from a cross sectional analysis and a randomized, double blind, placebo controlled trial. Eur J Nutr 61(1):447–459. https://doi.org/10.1007/s00394-021-02656-3

Speranza L, Di Porzio U, Viggiano D, De Donato A, Volpicelli F (2021) Dopamine: the neuromodulator of long-term synaptic plasticity, reward and movement control. Cells 10(4):735. https://doi.org/10.3390/cells

Steindler DA, Reynolds BA (2017) Perspective: Neuroregenerative nutrition. Adv Nutr 8(4):546–557

Stiles J, Jernigan TL (2010) The basics of brain development. Neuropsychol Rev 20(4):327–348. https://doi.org/10.1007/s11065-010-9148-4

Tardy AL, Pouteau E, Marquez D, Yilmaz C, Scholey A (2020) Vitamins and minerals for energy, fatigue and cognition: a narrative review of the biochemical and clinical evidence. Nutrients 12(1). https://doi.org/10.3390/nu12010228

Tariku B, Id G, Sorrie MB, Megersa ND, Yesera E, Yeshitila YG, Pauwels NS, De S, Id SA (2023) Effects of iron supplementation on cognitive development in school-age children: systematic review and meta-analysis. PLoS One 18(6):e0287703. https://doi.org/10.1371/journal.pone.0287703

Thapa K, Khan H, Singh TG, Kaur A (2021) Traumatic brain injury: mechanistic insight on pathophysiology and potential therapeutic targets. J Mol Neurosci 71(9):1725–1742. https://doi.org/10.1007/s12031-021-01841-7

Travica N, Ried K, Sali A, Scholey A, Pipingas A (2017) Vitamin C status and cognitive function: a systematic review. Nutrients:1–21. https://doi.org/10.3390/nu9090960

Travica N, Ried K, Sali A, Hudson I, Scholey A, Pipingas A (2019) Plasma vitamin C concentrations and cognitive function: a cross-sectional study. Front Aging Neurosci 11(April):1–21. https://doi.org/10.3389/fnagi.2019.00072

Vaiserman A (2015) Epidemiologic evidence for association between adverse environmental exposures in early life and epigenetic variation: a potential link to disease susceptibility? Clin Epigenetics 7(1):96. https://doi.org/10.1186/s13148-015-0130-0

Vendelboe TV, Harris P, Zhao Y, Walter TS, Harlos K, El Omari K, Christensen HEM (2016) The crystal structure of human dopamine β-hydroxylase at 2.9 Å resolution. Sci Adv 2(4). https://doi.org/10.1126/sciadv.1500980

Virdi S, Jadavji NM (2022) The impact of maternal folates on brain development and function after birth. Meta 12(9):876. https://doi.org/10.3390/metabo12090876

Visioli F, Burgos-ramos E (2016) Selected micronutrients in cognitive decline prevention and therapy. Mol Neurobiol:4083–4093. https://doi.org/10.1007/s12035-015-9349-1

Wadan AS, Liaquat M (2024) Synaptic modulation by coffee compounds: insights into neural plasticity. In: Neuroscience of coffee part B, 1st edn, Elsevier B.V. https://doi.org/10.1016/bs.pbr.2024.06.008

Wang Y, Wu Y, Li T, Wang X, Zhu C (2019) Iron metabolism and brain development in premature infants. Front Physiol 10. https://doi.org/10.3389/fphys.2019.00463

Wang Q, Zhao J, Chang H, Liu X, Zhu R (2021) Homocysteine and folic acid: risk factors for Alzheimer' s disease—an updated meta-analysis. Front Aging Neurosci 13(May):1–15. https://doi.org/10.3389/fnagi.2021.665114

Wasantwisut E (1997) Nutrition and development: other micronutrients' effect on growth and cognition. Southeast Asian J Trop Med Public Health 28(Suppl 2):78–82

Wołoszynowska-Fraser MU, Kouchmeshky A, McCaffery P (2020) Vitamin A and retinoic acid in cognition and cognitive disease. In. Annu Rev Nutr 40. https://doi.org/10.1146/annurev-nutr-122319-034227

Xu P, Cui D, Jin M, Sun L (2024) Magnesium ions and dementia. J Neurorestoratology 12(1):100094. https://doi.org/10.1016/j.jnrt.2024.100094

Zsolt G (2021) Role of vitamin D in cognitive dysfunction: new molecular concepts and discrepancies between animal and human findings. Nutrients:1–24

The Effect of Probiotics on Mental Health 12

Gülcan Göçmez Yılmaz

Abstract

Good mental health represents a status of mental and psychological well-being. The necessity of a healthy functioning digestive system for good mental health has been demonstrated in many studies. The mutual communication and interaction between the brain and the intestines is called the brain-gut axis. All living microorganisms found in the intestines are defined as microbiota. There are neuronal, hormonal, and metabolic pathways between the gut microbiota and the brain that form the brain-gut axis. In addition, the gut microbiota also has regulatory functions on the immune system. It has been thought that the interaction of the gut microenvironment with these pathways and the immune system may play a role in many diseases related to brain health. The role of the microbiota has been shown in studies in psychiatric diseases such as depression and anxiety, autism spectrum disorder, sleep disorders, neurodegenerative diseases such as Parkinson's and Alzheimer's diseases, neuroinflammatory pathogenesis, multiple sclerosis and other demyelinating diseases, and even neuropathic pain and migraine headaches.

Probiotics are living microorganisms that have beneficial effects on the host. It encourages the proliferation of beneficial microorganisms and prevents pathogenic microorganisms from adhering to the intestines. It is thought that probiotics can be an alternative treatment option in diseases that develop with the role of microbiota due to their potential benefits on intestinal microbiota. There are positive results in this regard, especially in animal experiments. However, studies are still ongoing on which patients, in which doses, and for which durations pro-

G. G. Yılmaz (✉)
Mersin City Training and Research Hospital, Mersin, Turkey
e-mail: gocmezgulcan@gmail.com

W. Mohamed, A. Özge (eds.), *Feeding the Mind: The Connection Between Diet, Drugs, and Mental Health Volume 1*, Nutritional Neurosciences,
https://doi.org/10.1007/978-981-95-1414-4_12

biotics will be used in clinical studies. There are also contrary studies; some studies show that probiotics do not have therapeutic effects.

The gut microbiome of the brain, which is in direct and indirect interaction with the brain-gut axis, has beneficial effects on mental health. The use of probiotics that will externally modify the gut microbiome, which is accused in the pathophysiology of many diseases, seems to be a promising option in diseases. However, multi-center, longitudinal clinical studies are needed on the use of probiotics in appropriate patients, with appropriate doses and for appropriate durations in the protection of mental health and in brain diseases.

In this section, the brain-gut axis, microbiota and related mental diseases, probiotics, and their use for mental health will be discussed.

Keywords

Probiotics · Gut-brain axis · Microbiota · Gut microbiota · Mental health

12.1 Introduction

12.1.1 Overview of Mental Health Disorders

According to the WHO, mental health is defined as a state of emotional, psychological, and social well-being (Galderisi et al. 2017).

Mental health disorders are a significant cause of morbidity and economic burden today. In developed societies, depression and anxiety are the most common mental health disorders. Mental illnesses can be grouped as neurodevelopmental diseases, primarily attention-deficit hyperactivity disorder and autism spectrum disorder (ASD); neurodegenerative diseases, such as Parkinson's disease and Alzheimer's disease; and neuropsychiatric diseases, including depression, anxiety, psychosis, and sleep disorders.

Mental health disorders significantly reduce the quality of life by causing decreased ability, altered body condition, social and occupational impairment, and a heavy economic burden. Therefore, recognizing these disorders and taking preventive measures for their development is essential.

12.1.2 The Gut and Brain: A Bidirectional Communication

A close relationship between the gut and mental health has been acknowledged. Therefore, nutrition plays a key role in maintaining good mental health. The relationship between mental health and diet quality has been highlighted in many studies (O'Neil et al. 2014). A healthy digestive system is essential for good mental health.

There are many bacterial colonizations in our digestive system, the intestines. Living microorganisms consisting of bacteria and archaea in the digestive system are defined as microbiota. These bacteria and the metabolic wastes they produce can cause changes in brain functions and behaviors through various pathways. Changes in the composition or amount of microflora in the intestinal microbiome can directly or indirectly affect physical and mental health (Liu et al. 2022a). The importance of the relationship between the gut microbiome and mental illnesses has been demonstrated in many studies recently (Clapp et al. 2017). These studies have evaluated that the microbiome can be a shaper in neural development, regulation of neurotransmission, and behavioral effects and contribute to the pathogenesis and progression of neurodevelopmental, neuropsychiatric, and neurological illnesses (Galderisi et al. 2017; Allen et al. 2017).

There is a complex communication system between the gastrointestinal tract, the microorganisms that inhabit it, and the peripheral and central nervous systems. Within the digestive system lies the enteric nervous system (ENS), which is a crucial part of the nervous system. The ENS contains approximately 100 million autonomously functioning neurons. Major brain neurotransmitters within the ENS, such as acetylcholine, serotonin, and norepinephrine, can be triggered by various factors, influencing the gut system and microbiota. In the gut microbiota, neurotransmitter metabolism, immune regulation, short-chain fatty acids (SCFAs), and vagal nerve efferent pathways affect brain functions. This reciprocal bidirectional interaction system is known as the gut-brain axis (Liu et al. 2022b).

The composition of the gut microbiome is influenced by both genetic and environmental factors. Factors such as an individual's dietary habits, medication use, and infection status can alter the microbiota in the gut system, impacting mental health through the gut-brain axis (Cryan and Dinan 2012).

The gut microbiota is a modifiable target with the potential for epigenetic modification. The gut-brain axis appears to be modifiable through various interventions such as diverse diets, probiotics, and fecal microbiota transplantation (Mörkl et al. 2020). Probiotics are live microorganisms that can cause changes in the gut microbiota balance.

12.1.3 Probiotics: Definition and Mechanism of Action

Probiotics are defined as live microorganisms that confer beneficial health effects on the host (Bistas and Tabet 2023; Kim et al. 2019). Probiotics are thought to interact with host microorganisms by suppressing pathogenic microorganisms and promoting the establishment of beneficial ones. They balance the gut microbiota composition by improving the gut environment and preventing harmful microorganisms from adhering to epithelial membranes (Sanders 2016).

The effect of the intestines on brain health has recently attracted attention, and research on this subject is increasing (Madabushi et al. 2023). Regulation of the intestinal microenvironment, which is an important factor in brain health, with probiotics, which are digestible organisms taken from outside, has led to treatment

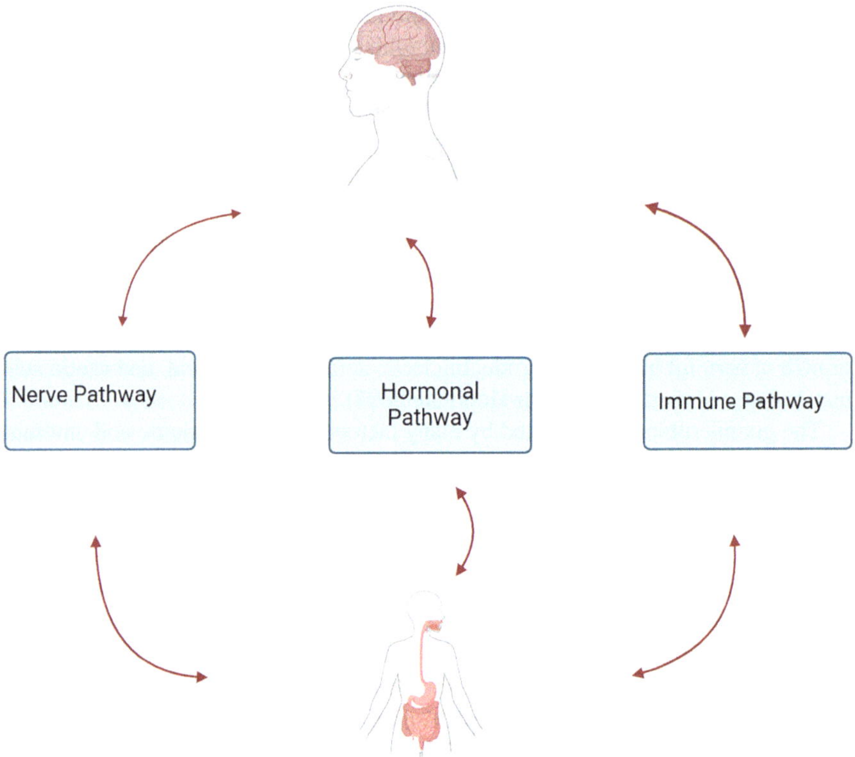

Fig. 12.1 Gut-brain axis and bidirectional communication pathways

The gut is connected to the brain through the ENS, which involves parallel outflow from both the parasympathetic (vagus nerve) and sympathetic (prevertebral ganglia) nervous systems, as well as the hypothalamic-pituitary-adrenal axis (Carabotti et al. 2015; Rhee et al. 2009). The ENS is divided into two types of ganglia: the myenteric and submucosal plexuses.

The vagus nerve is the tenth cranial nerve, with its hepatic and celiac branches innervating the gut. Sensory (afferent) and motor (efferent) fibers are intermingled within the vagus nerve. The sensory fibers originate from neurons in the nodose ganglion, while the motor fibers arise from neurons in the dorsal motor nucleus of the vagus and the nucleus ambiguous (Wang et al. 2023).

Vagal chemoreceptors appear to modulate the relationship between the microbiota and the brain. The vagus nerve modulates up to 90% of impulses going from the gut to the brain and just 10% going the other way.

The activation of vagal afferent nerves by microorganisms has been demonstrated through the injection of *Lactobacillus johnsonii* colonies into the duodenum, leading to an improvement in gastric vagal activity.

The brain regulates intestinal functions via the hypothalamic-pituitary-adrenal pathway and the autonomic nervous system. Conversely, intestinal CNS functions are modulated by various microbiota metabolites and products, neuroactive substances, intestinal hormones, the enteric nervous system, the vagal nerve, the circulatory system, and immune system pathways. All of these pathways are collectively referred to as the gut-brain axis (Ja¨rbrink-Sehgal and Andreasson 2020). The interaction between the brain and gut axis involves three major pathways: the immune pathway, the neuronal pathway, and the endocrine/systemic pathway, with interactions and crosstalk occurring between these three (Wang et al. 2023).

Various neurotransmitter syntheses are provided by the gastrointestinal microbiome and its metabolites. These neurotransmitters involved in brain-gut communication are free fatty acids, tryptophan metabolites, serotonin, GABA, dopamine, and noradrenaline (Ustianowsk et al. 2022). The main transmitter at both terminal ends of the gut-brain axis is serotonin. Almost 90.0% of serotonin is produced by tryptophan, which is synthesized by the gut microbiota. Serotonin is produced by some strains of Lactobacillus, Streptococcus, and Lactococcus.

Some behavioral changes are regulated by serotonergic transmission under the influence of the microbiome environment. Changes in the gut microbiome trigger systemic inflammation and affect various systems in the body (Clemente et al. 2018; Tilg et al. 2019). It has also been determined that changes in the gut microbiome cause an increase in the permeability of the blood-brain barrier (Braniste et al. 2014).

Inflammatory changes in the intestines have been shown to cause increased permeability of the blood-brain barrier in fetal and adult mice. With an increase in BBB permeability, the inflammatory process that begins in the intestines results in inflammatory changes in the brain (Micheal and Prat 2016).

The gut microbiota represents a promising reservoir for novel therapeutic avenues, offering the intriguing prospect of targeting and influencing the microbiota-gut-brain axis to address numerous complex diseases of the CNS.

12.2.4 The Relationship Between Gut Microbiota and Mental Health

Dysbiosis refers to the damage or disruption of the gut microbiome, resulting in an imbalance in the microbial environment. This common issue is caused by bacterial infections, changes in diet, or the use of antibiotics. A dysbiotic gut microbiome is characterized by altered composition and function, with lower microbial diversity, an increase in pathogenic microbiota, or a loss of beneficial microbes (Matenchuk et al. 2020).

The dysbiosis of the intestinal microbiota contributes to the development of metabolic diseases, including obesity and diabetes, as well as various intestinal diseases such as antibiotic-associated diarrhea (AAD), inflammatory bowel disease (IBD), Crohn's disease, and colorectal cancer (Kim et al. 2019).

Dysbiosis of the gut microbiota may lead to alterations in the gut barrier, resulting in a "leaky gut" and making for the translocation of pathogens into the portal and systemic circulation, which contributes to neuroinflammation in CNS disorders (Fasano 2020). The gut microbiome regulates neuroinflammation, and

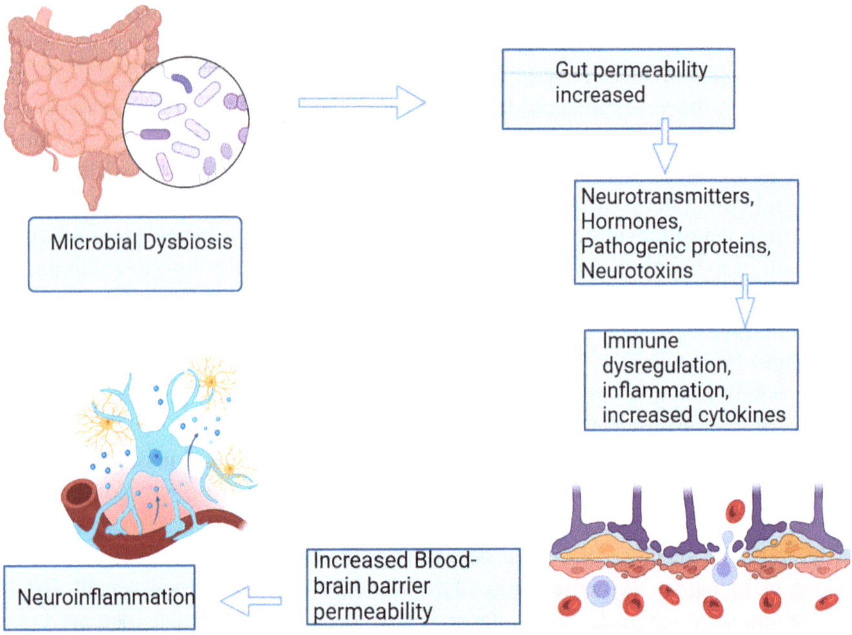

Fig. 12.2 Dysbiosis and neuroinflammation

proinflammatory factors increase with dysbiosis. The gut microbiome plays a role in the development of neurological diseases through neuroinflammation.

Inflammatory changes in the intestines have been shown to cause increased permeability of the blood-brain barrier in fetal and adult mice. With an increase in BBB permeability, the inflammatory process that begins in the intestines results in inflammatory changes in the brain (Kazemi et al. 2019). Neuroinflammation due to dysbiosis is shown in the flowchart in Fig. 12.2.

The gut microbiome affects appetite, memory, and mood by activating various neuroreceptors, neurotransmitters, and endocrine pathways. Many studies have shown different microbiota compositions in patients with major depression compared to healthy controls (Kazemi et al. 2019; Nikolova et al. 2021).

Alterations in the composition of the gut microbiota have been associated with various psychiatric diseases, such as ASD, major depression, and anxiety. In particular, it has been suggested that in the development of depression, a decrease in the diversity and richness of the gut microbiota increases the proinflammatory state, increases cortisol levels, and causes changes in tryptophan metabolism (Kelly et al. 2016).

Bacteria such as *Escherichia*, *Fusobacterium*, *Prevotella*, *Enterococcus casseliflavus*, and *Bacteroides* synthesize tryptophan. Tryptophan crosses the blood-brain barrier and affects serotonergic transmission in the CNS. Approximately 90% of serotonin produced in the CNS originates from the gut. In gut inflammation, the conversion of tryptophan to kynurenine is increased, and it is believed that this conversion may be responsible for anxiety and mood disorders. Kynurenine has been described as a neurotoxic agent (Agus et al. 2018).

Numerous researchers have shown that neuroinflammation plays an important role in the progression of cognitive deficits associated with aging. Various lines of evidence suggest that aging-induced changes in gut microbiota composition might lead to inflammaging (a chronic low-grade inflammation), a risk factor linked to cognitive decline in the elderly (Yanga et al. 2020; Atienza et al. 2018).

Histological and behavioral manifestations of Alzheimer's Disease (AD) are correlated with gut microbiome dysbiosis, and the potential use of probiotics to prevent AD progression is one of the recent innovations in the field of neurodegenerative diseases (Mancuso and Santangel 2018; Shen et al. 2017).

The aggregation of alpha-synuclein, responsible for the pathology of Parkinson's disease, a neurodegenerative disorder, has been associated with immune system-mediated inflammation and changes in the gut microbiome. According to Braak's theory, in Parkinson's disease patients, alpha-synuclein accumulation has been observed in the appendix and colonic tissues, starting from the enteric nervous system (Kaur et al. 2021). In their studies, Hauser and Tansey suggested that changes in the gut microbiome lead to increased intestinal permeability, allowing inflammatory metabolites and bacteria to leak into the system, which contributes to the pathogenesis of Parkinson's disease (Houser and Tansey 2017; Tansey and Romero-Ramos 2019). This leakage results in increased permeability of the blood-brain barrier and alpha-synuclein aggregation, leading to neuroinflammation and neurodegeneration.

It has been suggested in studies that the gut microbiome may also play a role in the pathophysiology of multiple sclerosis (MS), a chronic, neurodegenerative, and neuroinflammatory disease. Microorganisms associated with MS pathology include species from the *Pseudomonas*, *Mycoplasma*, *Haemophilus*, *Blautia*, *Dorea*, *Faecalibacterium*, *Methanobrevibacter*, *Akkermansia*, and *Desulfovibrionaceae* genera (Dunalska et al. 2023). The proposed mechanism in this pathophysiology is that a virus, bacterium, or fungal pathogen induces changes in the microbiota composition, leading to immune system activation (Perlejewski et al. 2016; Yuan et al. 2020). Another proposed mechanism is that gut bacterial components become molecular mimics of CNS autoantigens, thereby stimulating Th17 cells and activating the immune system (Miyauchi et al. 2020).

Additionally, the gut microbiome is thought to be a factor influencing central sensitization in pain pathways. Any changes in the gut-brain axis are suggested to affect cytokine expression, potentially triggering migraine attacks (Ustianowsk et al. 2022; Santoni et al. 2021).

12.3 Probiotics as a Therapeutic Approach

There are various strategies to modulate the gut microbiota, which is related to many systems and has direct and indirect bidirectional interactions with the brain. These are dietary modification, fecal transplantation, antimicrobial therapies, and probiotic use. In particular, the effects of probiotics on the gut microenvironment are a topic that has been investigated recently in the treatment of many neurological and psychiatric diseases related to brain health.

Probiotics emerged in the early nineteenth century when it was determined that bifidobacteria decreased in children with diarrhea, and oral administration of bifidobacteria was recommended (Wieërs et al. 2020). Although studies on regulating intestinal microbiota with fecal material have been conducted for many years, specific clinical studies have been conducted for the last 50 years (Wieërs et al. 2020).

The first available probiotic was a preparation made from *Saccharomyces* or *Lactobacillus* species, used for diarrhea caused by *Clostridium difficile* or AAD (Goldenberg et al. 2017).

Following antibiotic treatment, probiotics are one of the key solutions to restore gut microbial balance and prevent infections in patients, which can induce gut microbiota dysbiosis. Probiotics are preparations consisting of microorganisms that improve intestinal balance (Alli et al. 2022). Subsequent forms of probiotics contained a larger variety and number of microorganisms, ranging from 10^8 to more than 10^{10} organisms.

The microbiota organizes the intestines, which are a focal ecosystem. The composition of the microbiota is a dynamic structure that is customized to the individual by being affected by the person's diet, probiotic use, and environmental factors (2015). The parietal microbiota living in the mucus or on the wall of the intestines and the luminal microbiota reaching through food and excreted with feces constitute the intestinal system microbiota, where there is a transition from one group to another. The luminal microbiota changes with probiotic treatment. It is suggested that probiotics affect human health by affecting the resident microbiota and regulating the intestinal epithelial cells and the immune system.

Prebiotics, first defined in 1995, stimulate the growth and/or activity of a limited number of beneficial bacteria in the gut to improve host health (Gibson and Roberfroid 1995). In subsequent years, this definition has expanded to include certain food components that act through the microbiota. Prebiotics contain components that are metabolized by gut microorganisms and promote the regulation of the gut microbiome in ways that benefit the host.

12.3.1 Common Probiotic Strains with Mental Health Benefits (e.g., Lactobacillus, Bifidobacterium)

Probiotics most often consist of combinations of *Lactobacillus* and *Bifidobacterium* genera, with some strains of Enterococcus and Saccharomyces species being among the exceptions. It has been demonstrated in numerous studies that these microorganisms suppress inflammation and modulate the immune system (Bravo et al. 2011).

The beneficial effects of probiotics on human health are summarized in one source into eight main categories: (1) Preventing harmful microbes from adhering to the intestinal epithelium, (2) Inhibiting the growth of pathogenic bacteria by binding to gram-negative bacteria, (3) Maintaining SCFAs at normal levels, (4)

Repairing intestinal permeability, (5) Increasing electrolyte absorption in the intestines, (6) Improving the immunological response in the gut, (7) Regulating lipid metabolism, (8) Suppressing proinflammatory cytokines in the gut (Kopp-Hoolihan 2001; Ghasemian et al. 2018; Khan et al. 2013).

12.3.2 Delivery Mechanisms: Supplements vs. Functional Foods

Probiotics are available in liquid, paste, and powdered forms and come in sachets as well as capsules. They have been widely used in the food industry for the development of novel functional foods, medicinal foods, and dietary supplements. WHO and FAO suggest that, to provide satisfactory health benefits to the host, the viability of probiotic cells in food should be greater than 10^6 log CFU/ml or g (Sinhg et al. 2022).

Probiotics containing live organisms must be preserved and delivered to the intestines in a viable and effective manner. Probiotic delivery systems are critical for ensuring that a sufficient amount of probiotics reaches the digestive system. In the food industry, this has been a challenging limiting factor. One of the methods used to successfully deliver probiotics to the target site is encapsulation techniques. In the encapsulation process, the selection of the encapsulating material, the type of encapsulation method—such as spray drying, emulsion, and extrusion—and the probiotics' resistance to pH, heat, oxygen exposure, and other stress factors are crucial (Sinhg et al. 2022).

Various encapsulation techniques are also used using biopolymers produced from living sources, plants, animals, or microbes. Biopolymers make probiotics ideal for factors such as biocompatibility, biodegradability, and sustainability (Vijayaram et al. 2024).

Many methods are being studied for daily use in the use of probiotics, either in capsule form or in easy-to-use tablets or integrated into various food products (Vijayaram et al. 2024). However, the method that will successfully pass the processes in the digestive system and continue to function most effectively in the target organ is not yet clear. More effective and targeted delivery methods for probiotics need to be developed.

12.4 Mechanisms Linking Probiotics to Mental Health

12.4.1 Serotonin Production and Gut-Derived Neurotransmitters

Gut microbiota has an important role in neurotransmitter modulation in the brain. Metabolites produced in the intestinal flora are used as precursors in neurotransmitter synthesis in the CNS. Gut microbiota metabolites enter the brain largely through vagal nerve afferent input (Bienenstock 2014).

Serotonin is a neurotransmitter that plays a key role in human body functions such as behavior and mood, gastrointestinal secretion, and peristalsis (Berger et al.

2009). Serotonin is distributed throughout the body, but 90–95% is formed in the gastrointestinal tract (Gershon and Tack 2007). Therefore, the relationship between gut microbiota and serotonin levels can be predicted. For example, Bifidobacterium infantis has been shown to increase plasma tryptophan levels, thus increasing the transmission of 5-hydroxytryptophan to the brain (Desbonet et al. 2010). Microbiota produces various neurotransmitters. Lactobacillus brevis and Bifidobacterium dentium are able to produce γ-aminobutyric acid (GABA).

12.4.2 HPA Axis Regulation

In response to stress, the paraventricular nucleus of the hypothalamus releases corticotropin-releasing factor (CRF). CRF triggers the release of adrenocorticotropic hormone (ACTH) from the pituitary gland. ACTH, in turn, induces the release of the hormone cortisol from the adrenal glands. Increased activation of the hypothalamic-pituitary-adrenal axis (HPA axis) leads to elevated levels of the stress hormone cortisol (Farzi et al. 2018).

HPA axis activation increases gut microbiota composition and gastrointestinal permeability (Farzi et al. 2018). Inflammatory processes are also known to activate the HPA axis. The intestines are exposed to stress hormone-induced interactions to increase barrier permeability, and the microbiota is involved in many ways, including epigenetic programming of the intestinal epithelium. The increased intestinal barrier permeability due to stress exposure in the HPA axis leads to the release of inflammatory cytokines (IL-1, IL-6, and TNF-a) and microbial antigens into the systemic circulation and to subclinical neuroinflammation by crossing the blood-brain barrier. In addition, this pathway indirectly activates the negative feedback mechanism of the HPA axis by disrupting the function of nerve regions (Freimer et al. 2022). HPA is a component of our limbic network, engaging in emotional and memory actions, and acts as central tension affecting nerve-managing reactions to stressors of some types.

The gut microbiota determines stress responsivity by influencing the HPA axis, and stress-induced cortisol responses can be altered by certain probiotics. In particular, numerous animal studies have demonstrated that various species of *Bifidobacterium* and *Lactobacillus* regulate the endocrine system and, consequently, affect systemic inflammation (Farzi et al. 2018).

12.4.3 Inflammatory Pathways and Immune Modulation

Probiotics regulate T cell differentiation and function. They reduce inflammatory responses and modulate the gut environment. They increase SCFAs and microorganisms that produce beneficial metabolites (Tegegne and Kebede 2022). SCFAs play a very important role in maintaining intestinal and immune homeostasis in the human body. SCFAs are speculated to have a mediational role in the microbiota gut-brain axis crosstalk. They show anti-inflammatory activity, which involves inhibiting the activity of inflammatory mediators in the intestinal epithelium and thus inhibiting the activation of NFκB macrophages, which are the main source of cytokines in the course of the inflammatory process (Van der Beek et al. 2015).

SCFAs also play an important role in the regulation of several physiological pathways within the nervous system. First, SCFAs modulate brain-induced intestinal gluconeogenesis. SCFAs control systemic and neuroinflammation through the modulation of functions and structures of microglia cells, resulting in the modulation of emotion, cognition, and mental disorders. Additionally, high concentrations of SCFAs seem to be related to the major expression of neurotrophic factors (Yang et al. 2020).

12.5 Evidence from Preclinical Studies

12.5.1 Animal Models of Stress, Anxiety, and Depression

Several studies suggest that probiotic use provides physical and mental health benefits to the host (Pelto et al. 1996; DeVrese et al. 2001). In animal models, probiotic supplementation has been shown to at least partially reverse behavioral alterations observed in germ-free models, as well as in models of maternal separation-induced depression or those induced by exposure to acute stressors.

In human studies, probiotics were shown to improve the inflammatory status in irritable bowel syndrome, decrease anxiety in chronic fatigue syndrome, reduce psychological distress in healthy volunteers, and modify the brain's processing of emotional information, as measured by fMRI scans. The mechanism of action of probiotic supplementation has been shown not to radically alter the gut microbiota composition.

A large number of human and animal studies support the fact that the gut microbiota plays an important role in cognitive development and function, mood and emotion regulation, and interpersonal interactions and communication. This is partly through the role of the immune system in neuronal differentiation, axonal development, and synaptic plasticity, serving as a major factor (Wieërs et al. 2020).

12.5.2 Probiotic-Induced Behavioral Changes in Rodents

It has been shown that the use of probiotics containing Bifidobacterium or Lactobacillus in rats reduces depressive symptoms (Desbonnet et al. 2010; Bravo et al. 2011). Bifidobacterium longum, Bifidobacterium breve, Bifidobacterium infantis, and Bifidobacterium bifidum from Bifidobacterium species; and L. acidophilus, L. helveticus, L. rhamnosus, and L. plantarum from Lactobacillus species may impact neurotransmission and emotional states via the vagus nerve-mediated gut-brain axis (El Khatib et al. 2019). Many studies have been conducted on probiotics and mood disorders such as depression and anxiety (Ja¨rbrink-Sehgal and Andreasson 2020). The use of probiotics has been accepted as an application that has positive effects on mood in the treatment of depression and is also considered a method to prevent depression in the healthy population. Probiotics that have positive effects on mental health are also called psychobiotics (Sharma et al. 2021).

12.6 Clinical Trials and Human Studies

12.6.1 Summary of Key Studies on Probiotics and Depression

In a study in which 38 healthy volunteers received a probiotic mixture of Lactobacillus fermentum LF16, L. rhamnosus LR06, L. plantarum LP01, and Bifidobacterium longum BL04 for 6 weeks, mood, personal dimensions, and sleep quality were evaluated with various tests before and after probiotic use. In the probiotic group, a decrease in depressive mood, irritability, and fatigue, and an improvement in sleep quality were found between the control weeks. However, no significant difference was found in the tests between the groups using and not using probiotics (Marotta et al. 2019). In a study conducted in 2022 with 45 patients, a positive effect was observed on depression findings in patients given probiotics containing Bifidobacterium breve (Tian et al. 2022). There is an ongoing debate on whether gut microbiome dysregulation in bidirectional communication between the gut and brain triggers depression or whether depression involves this dysregulation (Macrea et al. 2023). As a result, the use of probiotics in the treatment of depression is not seen as a treatment that will replace standard treatment, but studies have emphasized that it is safe when used together with standard treatment (Macrea et al. 2023).

12.6.2 Impact on Anxiety Disorders

Changes in neurotransmitters are considered the essential neurobiochemical mechanism for the occurrence of anxiety disorders.

Animal studies have shown that gut microbiota regulation can alter anxiety-related behaviors (Neufeld et al. 2011; Collins et al. 2013). In a study conducted on zebrafish, a probiotic mixture containing *Lactobacillus rhamnosus* and *Bifidobacterium longum* was administered for 4 months, and their swimming behavior in a new environment was observed. Probiotic-treated zebrafish exhibited a reduction in bottom-dwelling geotactic behavior. This behavioral response was thought to be correlated with reduced anxiety. It has been suggested that this probiotic mixture may be beneficial in the treatment of anxiety.

In a review of human clinical studies on anxiety, 21 studies were included, 14 of which investigated the effect of probiotic treatment and 7 non-probiotic interventions on the improvement of anxiety symptoms by regulating intestinal flora. In 14 studies in which probiotic treatment was given, it was shown that regulating intestinal flora may affect the improvement of anxiety (Yang et al. 2019).

According to a review investigating the effects of probiotics on physiological stress and anxiety, L. casei and Bifidobacterium longum are the most studied strains on this subject. In this review, out of four clinical studies on anxiety, all but one showed improvement in anxiety. In the study conducted by Pinto-Sanchez et al., no improvement was seen in anxiety scores. In the study conducted by Colica et al., where multiple probiotic strains were administered, improvement in anxiety was found in 3 weeks (Vitellio et al. 2020).

In addition to studies showing that probiotics have beneficial effects on anxiety, there are also studies that show no difference. In a very comprehensive meta-analysis examining psychiatric symptoms, for qualitative group comparisons (probiotic and placebo treatment), 32 samples (78%) showed no difference between the groups, and nine studies (22%) were found in favor of the probiotic group (Le Morvan et al. 2022).

12.6.3 Impact on Other Mental Diseases

Some studies have shown that ASD is associated with some bacterial species (Marotta et al. 2019). Reduced bacterial diversity and increased microbial biomass have been reported in individuals with ASD. These alterations are typically characterized by a shift from beneficial microorganisms to those that produce spores, exhibit antibiotic resistance, and/or generate neurotoxins. Notably, there is often a significant reduction in the abundance of *Prevotella*, *Corprococcus*, and *Veillonellaceae*, which play crucial roles in carbohydrate digestion and fermentation, as well as other bacterial groups such as *Bacteroidetes*, *Actinobacteria*, and *Proteobacteria* (Kang et al. 2013).

The positive effects of using probiotics in MS have been shown in studies (Tankou et al. 2018a; Kouchaki et al. 2016). The use of probiotics in MS increases the relatively decreased number of Lactobacillus and decreases the Akkermansia and Blautia species that cause dysbiosis (Tankou et al. 2018b). Administration of Lactobacillus acidophilus, Lactobacillus casei, Bifidobacterium bifidum, and Lactobacillus fermentum has favorable effects on disability levels assessed with EDSS, parameters of mental health, inflammatory factors, markers of insulin resistance, and HDL- and total-/HDL-cholesterol levels in MS patients (Kouchaki et al. 2016).

It has been observed that patients with migraines have a higher prevalence of gastrointestinal disorders compared to healthy controls and that individuals with gastrointestinal disorders experience more frequent headaches than normal controls. This association between migraine headaches and gastrointestinal disorders may be explained by increased intestinal permeability and inflammatory responses. With increased intestinal permeability, undigested food particles and bacterial metabolites enter the circulation, potentially activating the trigeminovascular system.

In a study where patients with migraine headaches were given a multispecies probiotic for 12 weeks, their condition was assessed using the Migraine Disability Assessment Scale (MIDAS) and the Henry Ford Hospital Headache Disability

Butyric acid, which is important for human health in cognitive functions, is produced in the presence of Firmicutes. SCFAs such as acetic acid and propionic acid are formed either by fermentation or by bacteroids such as lactobacilli and bifidobacteria. SCFAs are important for cognitive neuroeducation and autistic behavior (Verbeke et al. 2015). SCFAs can have neurotoxic effects when they reach the brain, and it is stated that propionate, in particular, can cause autism-like behavior in animal models (MacFabe 2012). Butyrate is effective in intestinal epithelial permeability and the formation of T regulatory cells.

In the pathophysiology of autism, increased intestinal permeability is assumed to be the link between the gut and the brain. Low-grade systemic inflammation, increased intestinal permeability, and neuroinflammation are all symptoms of dysbiosis (Ağagündüz et al. 2022). It has also been noted that gastrointestinal symptoms such as constipation, diarrhea, vomiting, flatulence, reflux, abdominal pain, and foul-smelling stools are frequently seen in children with ASD (Ağagündüz et al. 2022).

Another study found that mothers of children with ASD had higher levels of Alphaproteobacteria, Proteobacteria, Acinetobacter, and Moraxellaceae compared to mothers of healthy children (Li et al. 2019).

The microbiome of children with autism showed a higher abundance of Bacteroidetes and a lower amount of Firmicutes compared to controls. In children with autism exhibiting this dysbiosis, cytokine levels and tryptophan homeostasis were found to be correlated with the microbial imbalance. In another study, it was observed that Bacteroidetes and Proteobacteria were more abundant, while Actinobacteria and Firmicutes were reduced in a group of children with autism (Al-Beltagi 2012). Studies have shown that Firmicutes, which are important in gastrointestinal regulation, are low in children with ADHD (Verbeke et al. 2015).

For 3 months, a probiotic mixture containing *Lactobacillus acidophilus, Lacticaseibacillus rhamnosus,* and *Bifidobacterium longum* (each gram containing 100×10^6 CFU) was administered once daily at a dose of 5 g/day, diluted in water, to 30 children with ASD aged 5–9 years and a matched control group of 30 healthy children. Before the intervention, *Bifidobacteria* levels in the feces of children with ASD were lower than those in the control group. After probiotic treatment, *Bifidobacteria* and *Lactobacillus* levels increased in ASD children. The Autism Treatment Evaluation Checklist (ATEC) overall scores decreased, indicating a reduction in ASD symptom severity. Additionally, overall gastrointestinal (GI) symptom severity, including constipation, gas, stool consistency, and abdominal pain, was also reduced (Shaaban et al. 2018).

Epilepsy is a common neurological disorder in childhood. Studies have shown a close relationship between gut microbiota and epilepsy. In one study comparing the microbiota composition of individuals with idiopathic focal epilepsy and healthy volunteers, an increased prevalence of *Fusobacteria* species (10.6%) was found in patients with epilepsy but not in healthy volunteers (Şafak et al. 2020).

12.7.2 Probiotics in Aging Populations

Nowadays, societies are increasingly aging, and as aging progresses, age-related chronic diseases become a significant issue. Along with aging, various metabolic diseases, as well as declines in cognitive functions, dementia, and psychiatric conditions such as depression, anxiety, and agitation, are commonly observed. Global solutions are being sought for the prevention and management of these conditions. In this regard, as the relationship between mental health and the gut microbiota is increasingly understood, the influence of the gut on the development and treatment of diseases associated with aging has become an important area of research. There are studies on the use of probiotics that affect the gut microbiota in the fight against age-related diseases (Hamilton-Miller 2004).

Compared to the adult group, a decrease in microbiota diversity was shown in older ages, with interindividual variability. Bifidobacteria and Enterobacteriaceae were found to decrease with age (Claesson et al. 2011; Biagi et al. 2010). Another study showed a decrease in Firmicutes and an increase in Bacteroides groups in the elderly (Woodmansey 2007). These changes in microbiota are not primarily caused by old age, but nutritional deficiencies, other diseases, and increased drug consumption lead to this change.

It is known that neuroinflammation is an important risk factor for the progression of neurodegenerative diseases. It is suggested that age-related neuroinflammation triggers intestinal-related changes with aging and develops as a part of systemic inflammation. In this context, it is thought that a healthy intestinal function can reduce inflammation and be effective against the development of age-related neurodegenerative symptoms (Anderson 2022). Probiotics are predicted to reduce systemic inflammation by changing the composition of the microbiota and strengthening the gut barrier function.

In a study involving healthy elderly individuals over the age of 65, 30 participants were given a placebo, and 31 participants were given probiotic capsules containing *Bifidobacterium bifidum* BGN4 and *Bifidobacterium longum* BORI for 12 weeks. The results, evaluated using the Korean version of the Consortium to Establish a Registry for AD (CERAD-K) test, showed an increase in attention and executive function scores in the probiotic group compared to the placebo group after 12 weeks. Additionally, at the end of the 12 weeks, an increase in serum brain-derived neurotrophic factor (BDNF) levels was observed in the probiotic group. BDNF is a neuroprotective protein that combats gut-originating neuroinflammation, and its levels are known to correlate with *Eubacterium* and *Clostridiales* subpopulations (Kim et al. 2021).

In another randomized, double-blind, placebo-controlled crossover trial involving healthy elderly adults, participants were given a probiotic containing *Lactobacillus rhamnosus* and *Bifidobacterium lactis* multispecies for 10 weeks. The participants' cognitive functions (assessed using the MMSE and the Psychological Experiments Construction Language Test Battery), the Beck Depression Inventory (BDI), the State-Trait Anxiety Inventory, and emotional states (using the BDI and the State-Trait Anxiety Inventory) were evaluated using various inventories.

Statistically significant results showed noticeable improvements in cognitive function, memory, and depressive symptoms. This study is important in terms of evaluating probiotic use, specifically in the >55 age group, and assessing age-related cognitive and emotional decline (Ruiz-Gonzalez et al. 2024).

In another study, a patient population with mild cognitive impairment (MCI) aged 65–88 and a placebo group were included, and they were given a probiotic containing *Bifidobacterium breve* for 24 weeks. For cognitive assessments, the ADAS-Jcog (Japanese version of the Alzheimer's Disease Assessment Scale) and MMSE tests were used, while the Voxel-based Specific Regional Analysis System for Alzheimer's Disease (VSRAD) test was used to evaluate changes in brain atrophy. Significant improvements were observed in the orientation scale of the ADAS-Jcog test and the writing scale of the MMSE in the probiotic group compared to the placebo group. VSRAD scores were found to be higher in the placebo group. This study highlighted the potential effectiveness of probiotics in preventing the progression of dementia (Asaoka et al. 2022).

12.7.3 Role in Postpartum Depression

Postpartum depression (PPD) is a condition that can frequently develop in mothers after birth and negatively affects the mother's ability to continue her daily functions and care for the newborn. The quality of life of the mother with PPD deteriorates and also affects the baby's cognitive and social-emotional functions (Slykerman et al. 2017). Postnatal depression is often accompanied by an anxiety disorder. During pregnancy and breastfeeding, mothers are reluctant to receive antidepressant treatment for these diseases and to continue the treatment. In addition, even if medication is used, it takes several weeks for the treatment to be effective (Gartlehner et al. 2005). For these reasons, effective and reliable treatment options are needed for PPD.

The gut microbiota profile during the postpartum period is influenced by various factors, including the mode of delivery, type of nutrition, breastfeeding status, environmental exposures, and lifestyle. Along with these changes, postpartum ovarian hormone levels, the mother's perinatal physical health, social support, and communication are factors that contribute to PPD. Additionally, chronic gestational stress, modulation of 5-hydroxytryptamine (5-HT) and its receptors, and low levels of serum polyunsaturated fatty acids can lead to the development of PPD. The effects of probiotics on mood disorders and mental health have been demonstrated in both animal and human studies. It has been suggested that regulating the maternal microbiota with probiotics could improve maternal mental health and mood and indirectly improve infant health by modulating the infant microbiome.

Positive effects of probiotics on mood disorders have been demonstrated in animal studies, particularly those showing depressive-like and anxiety-like behavior (Bravo et al. 2011). In a study conducted on rats with a postpartum depression (PPD) model, the rats were administered *Lactobacillus casei*. Rat behaviors were measured using the sucrose preference test, the elevated plus maze, the forced swim test, and the tail suspension test. Administration of *Lactobacillus casei* improved depressive-like behaviors, intestinal microflora, and oxidative stress in the PPD model rats. The *Lactobacillus casei* intervention reversed changes in BDNF, N-methyl-D-aspartic acid receptor 1 (NR1), ERK1/2, and monoamines in the brains of the PPD rats.

In a study involving 193 patients diagnosed with PPD, *Lactobacillus rhamnosus* HN001 was administered to the treatment group, while 187 patients were assigned to the placebo group. In the group receiving probiotic treatment, significant reductions were observed in depression and anxiety scores. The results showed depression scores of *HN001* mean = 7.7 (SD = 5.4) compared to placebo 9.0 (6.0), with an effect size of −1.2 (95% CI −2.4, −0.1), $p = 0.035$, and anxiety scores of *HN001* 12.0 (4.0) compared to placebo 13.0 (4.3), with an effect size of −1.1 (−1.9, −0.2), $p = 0.014$ (Slykerman et al. 2017).

In a large meta-analysis investigating the effects of probiotic use during pregnancy and the postpartum period on postnatal depression and anxiety, no clear evidence was found to show the exact beneficial effects of the investigated probiotics, but some of the included strains were found to have maternal mental health-enhancing effects. In addition, this review suggested that probiotic supplementation in infants reduced colic symptoms, thus indirectly reducing postnatal depression symptoms (Trifkoviˇ et al. 2022).

There are still question marks about whether probiotics can be used as primary treatment for maternal mental health problems in time and as adjuvant therapy in addition to standard treatment. There seems to be a need for clinical studies on the effects of probiotic use on maternal and infant health in postnatal depression.

12.8 Challenges and Controversies

12.8.1 Variability in Probiotic Strains and Dosage

In many studies involving probiotics, the administered probiotic has been found to be effective in correlation with clinical tests within the specified duration and dosage. However, the same effect may not be achieved if the dosage or duration differs. In addition to this uncertainty regarding probiotics, due to the fact that each host has a unique gut microenvironment, the interaction between the given probiotic strain and the microbiome may not occur as anticipated, and the expected results may not be obtained (Slykerman et al. 2017).

Beneficial effects on maternal mental health were observed in a clinical study conducted by Slykerman et al. with the Lacticaseibacillus rhamnosus HN001 strain 116 (ö). No beneficial effects on mental health were observed in a study conducted by Dawe et al. with another strain of the same species, namely Lacticaseibacillus rhamnosus GG, compared with a placebo group (Dawe et al. 2020).

Bacteria of the Lacticaseibacillus genus, in particular the species Lacticaseibacillus casei, paracasei, rhamnosus, and zeae, colonize the gastrointestinal tract and exhibit probiotic effects. These are strains that are taken with a daily diet and plant-based fermented foods. The probiotic effects of these species have been studied in many studies for different medical conditions, the most studied of which is L. rhamnosus GG (Colombo Pimentel et al. 2021). The capacity of some strains to exhibit beneficial effects may be variable; therefore, when developing a probiotic, it may be useful to select those with potential in terms of probiotic properties. Based on the efficacy of high-potency strains, a study on probiotic strains focused on determining genetic characteristics, if any, that could increase the potential of a bacterial strain belonging to the Lacticaseibacillus genus to act as a probiotic (Rossi et al. 2022). In this study, the genome of the Lacticaseibacillus probiotic, L. rhamnosus GG, was used as a reference to identify gene loci encoding traits such as survival and colonization capacity in the GIT and adhesion to mucus or other host molecules. Eighty-two gene loci were retrieved, and their presence and variability in other Lacticaseibacillus spp. genomes were assessed by alignment with the publicly available fully annotated genome sequences of L. casei, L. paracasei, L. rhamnosus, and L. zeae. This study resulted in the identification of variable genetic traits to be analyzed in the preliminary selection of natural Lacticaseibacillus strains to be used to enhance the beneficial properties of fermented products and probiotic candidates (Rossi et al. 2022).

With the use of probiotics, the targeted effects can be provided by a single strain, a combination of a group of strains, or a group of micronutrients with similar abilities. These effects can be provided by immunomodulatory interactions or by the production of beneficial metabolites (Hill and Sanders 2013). In addition, the possible synergistic effects of probiotic mixtures are also important. When developing a probiotic strain or group for a specific disease, studies should be conducted considering these multifactorial factors. In addition, the possible synergistic effects of probiotic mixtures are also important.

12.8.2 Regulatory and Safety Concerns

The global probiotic market is rapidly growing, and strict quality control measures are required to ensure the effectiveness and safety of probiotic products.

The tests used for quality control of probiotics launched in the market should be reliable, accurate, sensitive, and preferably fast and simple. In a study on the safety of probiotics, several DNA-based methods are used for probiotic strain or species identification, including targeted methods and non-targeted methods (Shehata and Newmaster 2023). Targeted methods and non-targeted methods for testing the reliability of probiotic strains have been compared. According to the results of this study, targeted methods are able to quickly and accurately confirm the presence or absence of target taxa in a product with high taxonomic resolution, high sensitivity, and low cost. In contrast, non-targeted methods can identify the species present in a product, including unlisted species and contaminating species, but this technique has been found to be more expensive, more complex, and more time-consuming and to have lower taxonomic resolution (Shehata and Newmaster 2023).

The safety of a probiotic strain is determined by its origin, resistance to antibiotics, and lack of pathogenicity (Markowiak and Katarzyna Sli'zewska. 2017). Although probiotics are generally known to be safe, some side effects have been reported. Probiotics may cause systemic infections, stimulate the immune system, disturb metabolism, and potentially transfer genes. Lactobacillus and Bifidobacterium are the most widely studied and most frequently used probiotics. The Lactobacillus and Bifidobacterium strains, as commensals of the human microbiome, are safe and nonpathogenic strains. There is no evidence that probiotic lactobacilli or bifidobacterial strains cause infection when taken more than commensal strains. However, infection with these strains has been shown to be very rare in both adults and children (Zawistowska-Rojek and Tyskia 2018).

It is thought that intraspecies genomic variation for the same probiotic strains in any disease may lead to a large effect difference in cellular physiology and therefore in probiotic efficacy. As a result of this study, it was noted that microbial behavior and physiology are related to genomic data (Arnold et al. 2018).

In studies on the use of probiotics, factors such as small sample sizes, short and limited durations of use, and the use of different probiotic species and dosages create limitations in determining the safety of a probiotic that shows beneficial effects.

12.8.3 Placebo Effect in Mental Health Studies

New scientific evidence shows that placebo effects arise from highly active neurobiological processes that are influenced by psychological mechanisms such as expectations and conditions. These neurophysiological processes can be measured and modified, which represents the basis for their application in clinical medicine.

Whether the doctor or patient is aware of it or not, placebo effects are extremely powerful and represent an important part of the treatment process, both in ancient cultural practices and modern medicine.

While the importance of a drug's efficacy is significant, the role of the placebo effect should not be overlooked (Klarić et al. 2017).

12.9 Future Directions

12.9.1 Personalized Probiotic Therapy

The potential effects of probiotics on human health through gut-brain axis pathways are now widely accepted. However, uncertainties remain regarding which probiotic strains are effective for specific diseases, which age groups should use them, the duration of use, and the appropriate dosage. Additionally, after being ingested, probiotics interact with various factors in the host body, including lifestyle, diet, medication use, antibiotic consumption, and the geographical region in which the individual resides. Variables affecting probiotic use, including probiotic factors and host factors, are shown in Fig. 12.3.

12.10.2 Implications for Mental Health Treatment

Mental health is an important complementary part of the body for an individual's productivity in daily life. Once mental health disorders develop, they cause significant morbidity and societal burden. For many mental health disorders, only symptomatic treatments are available. There is a need for mechanistic studies and treatments related to the development of these disorders.

As the relationship between the gut-brain axis is revealed, the effects of the gut microbiome environment on mental health are becoming better understood. In this context, studies are being conducted on the interaction of probiotics, which have a microbiome-regulating function, with mental health. Positive effects of probiotics have been observed on several mental health disorders mentioned in the previous section. However, there is still not enough evidence to include probiotics in treatment protocols for mental health disorders. More rigorous research on specific genera, species, and strains of probiotics is needed to observe their benefits for mental health and human health.

References

Derrien M, van Hylckama Vlieg JET (2015 Jun) Fate, activity, and impact of ingested bacteria within the human gut microbiota. Trends Microbiol 23(6):354–366. https://doi.org/10.1016/j.tim.2015.03.002

Adak A, Khan MR (2019 Feb) An insight into gut microbiota and its functionalities. Cell Mol Life Sci 76(3):473–493. https://doi.org/10.1007/s00018-018-2943-4

Ağagündüz D et al (2022) Recent developments in the probiotics as live biotherapeutic products (LBPs) as modulators of gut brain axis related neurological conditions. J Transl Med. https://doi.org/10.1186/s12967-022-03609-y

Agus A, Planchais J, Sokol H (2018) Gut microbiota regulation of tryptophan metabolism in health and disease. Cell Host Microbe 23:716–724

Ahmad A, Khalid S (2018) Therapeutic aspects of probiotics and prebiotics. In: Diet, microbiome and health. Elsevier, pp 53–91. https://doi.org/10.1016/B978-0-12-811440-7.00003-X

Akbari E, Asemi Z, Daneshvar Kakhaki R, Bahmani F, Kouchaki E, Tamtaji OR et al (2016) Effect of probiotic supplementation on cognitive function and metabolic status in Alzheimer's disease: a randomized, double-blind and controlled trial. Front Aging Neurosci 8:256

Al-Beltagi M (2012) Autism medical comorbidities. World J Clin Pediatr 10(3):15–28. https://doi.org/10.5409/wjcp.v10.i3.15

Allen AP, Dinan TG, Clarke G, Cryan JF (2017) A psychology of the human brain–gut–microbiome axis. Soc Personal Psychol Compass 11:e12309. https://doi.org/10.1111/spc3.12309

Alli SR, Gorbovskaya I, Liu JCW, Kolla NJ, Brown L, Müller DJ (2022) The GutMicrobiome in depression and potential benefit of prebiotics, probiotics and synbiotics: a systematic review of clinical trials and observational studies. Int J Mol Sci 23:4494. https://doi.org/10.3390/ijms23094494

Anderson RC (2022) Can probiotics mitigate age-related neuroinflammation leading to improved cognitive outcomes? Front Nutr 9. Sec. Nutrition, Psychology and Brain Health. https://doi.org/10.3389/fnut.2022.1012076

Arnold JW, Simpson JB, Roach J, Kwintkiewicz J, Andrea Azcarate-Peril M (2018) Intra-species genomic and physiological variability impact stress resistance in strains of probiotic potential. Front Microbiol 9:242. https://doi.org/10.3389/fmicb.2018.00242

Asaoka D et al (2022) Effect of probiotic Bifidobacterium breve in improving cognitive function and preventing brain atrophy in older patients with suspected mild cognitive impairment: results of a 24-week randomized, double-blind, placebo-controlled trial. J Alzheimers Dis 88:75–95. https://doi.org/10.3233/JAD-220148

Atienza M, Ziontz J, Cantero JL (2018) Low-grade inflammation in the relationship between sleep disruption, dysfunctional adiposity, and cognitive decline in aging. Sleep Med Rev 42:171–183. https://doi.org/10.1016/j.smrv.2018.08.002

Azm SAN, Djazayeri A, Safa M, Azami K, Ahmadvand B, Sabbaghziarani F et al (2018) Lactobacilli and Bifidobacteria ameliorate memory and learning deficits and oxidative stress in beta-amyloid1e42 injected rats. Appl Physiol Nutr Metabol 43:718e26

Backhed F, Ley RE, Sonnenburg JL, Peterson DA, Gordon JI (2005) Host-bacterial mutualism in the human intestine. Science 307(5717):1915–1920. https://doi.org/10.1126/science.1104816

Berger M, Gray JA, Roth BL (2009) The expanded biology of serotonin. Annu Rev Med 60:355–366. https://doi.org/10.1146/annurev.med.60.042307.110802

Biagi E, Nylund L, Candela M, Ostan R, Bucci L, Pini E, Nikkïla J, Monti D, Satokari R, Franceschi C, Brigidi P, De Vos W (2010) Through ageing, and beyond: gut microbiota and inflammatory status in seniors and centenarians. PLoS One 5:e10667. https://doi.org/10.1371/journal.pone.0010667

Bistas KG, Tabet J (2023) The benefits of prebiotics and probiotics on mental health. Cureus 15(8):e43217. https://doi.org/10.7759/cureus.43217

Boyle RJ, Robins-Browne RM, Tang MLK (2006) Probiotic use in clinical practice: what are the risks? Am J Clin Nutr 83:1256–1264

Braniste V, Al-Asmakh M, Kowal C, Anuar F, Abbaspour A, Tóth M, Korecka A, Bakocevic N, Ng LG, Kundu P et al (2014) The gut microbiota influences blood-brain barrier permeability in mice. Sci Transl Med 6:263ra158

Bravo JA, Forsythe P, Chew MV, Escaravage E, Savignac HM, Dinan TG, Bienenstock J, Cryan JF (2011) Ingestion of Lactobacillus strain regulates emotional behavior and central GABA receptor expression in a mouse via the vagus nerve. Proc Natl Acad Sci USA 108:16050–16055. https://doi.org/10.1073/pnas.1102999108

Carabotti M, Scirocco A, Maselli MA, Severi C (2015) The gut brain axis: interactions between enteric microbiota, central and enteric nervous systems. Ann Gastroenterol 28(2):203

Chuang Y-F et al (2024) Precision probiotics supplement strategy in aging population based on gut microbiome composition. Brief Bioinform 25(4):bbae351. https://doi.org/10.1093/bib/bbae351

Claesson MJ, Cusack S, O'Sullivan O, Greene-Diniz R, de Weerd H, Flannery E, Marchesi JR, Falush D, Dinan T, Fitzgerald G, Stanton C, van Sinderen D, O'Connor M, Harnedy N, O'Connor K, Henry C, O'Mahony D, Fitzgerald AP, Shanahan F, Twomey C, Hill C, Ross RP, O'Toole PW (2011) Composition, variability, and temporal stability of the intestinal microbiota of the elderly. Proc Natl Acad Sci USA 108(Suppl 1):4586–4591. https://doi.org/10.1073/pnas.1000097107

Clapp M et al (2017) Gut microbiota's effect on mental health. The gut-brain axis. Clin Pract 7(4):987. https://doi.org/10.4081/cp.2017.987

Clemente JC, Manasson J, Scher JU (2018) The role of the gut microbiome in systemic inflammatory disease. BMJ 360:j5145

Collins SM, Kassam Z, Bercik P (2013) The adoptive transfer of behavioral phenotype via the intestinal microbiota: experimental evidence and clinical implications. Curr Opin Microbiol 16:240–245

Colombo Pimentel T, Ramalho Brandão L, Pereira de Oliveira M, Almeida da Costa WK, Magnani M (2021) Health benefits and technological effects of Lacticaseibacillus casei-01: an overview of the scientific literature. Trends Food Sci Technol 114:722–737

Cryan JF, Dinan TG (2012) Mind-altering microorganisms: the impact of the gut microbiota on brain and behaviour. Nat Rev Neurosci 13(10):701–712. https://doi.org/10.1038/nrn3346

Cryan JF, Dinan TG (2015) More than a gut feeling: the microbiota regulates neurodevelopment and behavior. Neuropsychopharmacology 40:241–242. [PMID: 25482171 DOI: 10.1038/npp.2014.224]

Dai Y-J, Wang H-Y, Wang X-J, Kaye AD, Sun Y-H (2017) Potential beneficial effects of probiotics on human migraine headache: a literature review. Pain Physician 20:E251–E255

Dawe JP, McCowan LME, Wilson J, Okesene-Gafa KAM, Serlachius AS (2020) Probiotics and maternal mental health: a randomised controlled trial among pregnant women with obesity. Sci Rep 10:1291

Desbonet L et al (2010) Effects of the probiotic *Bifidobacterium infantis* in the maternal separation model of depression. Neuroscience 170(4):1179–1188. https://doi.org/10.1016/j.neuroscience.2010.08.005

Desbonnet L, Garrett L, Clarke G, Kiely B, Cryan JF, Dinan TG (2010) Effects of the probiotic Bifidobacterium infantis in the maternal separation model of depression. Neuroscience 170:1179–1188

DeVrese M, Stegelmann A, Richter B, Fenselau S, Laue C, Schrezenmeir J (2001) Probiotics—compensationforlactaseinsufficiency. Am J Clin Nutr 73:421s–429s

Dunalska A, Saramak K, Szejko N (2023) The role of gut microbiome in the pathogenesis of multiple sclerosis and related disorders. Cells 12:1760. https://doi.org/10.3390/cells12131760

El Khatib R, Karam-Sarkis D, Waligora-Dupriet AJ, Butel MJ (2019) Could gut modulation through probiotic supplementation be beneficial in autism spectrum disorder? In: Prebiotics and probiotics-potential benefits in nutrition and health

Farzi A, Fröhlich EE, Holze P (2018) Gut microbiota and the neuroendocrine system. Neurotherapeutics 15:5–22. https://doi.org/10.1007/s13311-017-0600-5

Fasano A (2020) All disease begins in the (leaky) gut: role of zonulin-mediated gut permeability in the pathogenesis of some chronic inflammatory diseases. F1000Res 9:F1000 Faculty Rev-69. https://doi.org/10.12688/f1000research.20510.1

Forsythe P, Bienenstock J, Kunze WA (2014) Vagal pathways for microbiome-brain-gut axis communication. Adv Exp Med Biol 817:115–133. https://doi.org/10.1007/978-1-4939-0897-4_5

Freimer D, Yang TT, Ho TC, Tymofiyeva O, Leung C (2022) The gut microbiota, HPA axis, and brain in adolescent-onset depression: probiotics as a novel treatment. Brain Behav Immun Health 26:100541. https://doi.org/10.1016/j.bbih.2022.100541

Galderisi S, Heinz A, Kastrup M, Beezhold J, Sartorius N (2017) A proposed new definition of mental health. Psychiatr Pol 51(3):407–411. https://doi.org/10.12740/PP/74145

Galland L (2014) The gut microbiome and the brain. J Med Food 17(12):1261–1272. https://doi.org/10.1089/jmf.2014.7000

Gartlehner G, Hansen RA, Carey TS, Lohr KN, Gaynes BN, Randolph LC (2005) Dis continuation rates for selective serotonin reuptake inhibitors and other second generation antidepressants in outpatients with major depressive disorder: a system atic review and meta-analysis. Int Clin Psychopharmacol 20(2):59–69

Gershon MD, Tack J (2007 Jan) The serotonin signaling system: from basic understanding to drug development for functional GI disorders. Gastroenterology 132(1):397–414. https://doi.org/10.1053/j.gastro.2006.11.002

Ghasemian A, Eslami M, Shafiei M, Najafipour S, Rajabi A (2018) Probiotics and their increasing importance in human health and infection control. Rev Med Microbiol 29(4):153–158. https://doi.org/10.1097/MRM.0000000000000147

Gibson GR, Roberfroid MB (1995) Dietary modulation of the human colonic microbiota: introducing the concept of prebiotics. J Nutr 125(6):1401–1412

Gibson GR, Hutkins R, Sanders ME (2017) Expert consensus document: the international scientific association for probiotics and prebiotics (ISAPP) consensus statement on the definition and scope of prebiotics. Nat Rev Gastroenterol Hepatol 14(8):491–502. https://doi.org/10.1038/nrgastro.2017.75

Goldenberg JZ, Yap C, Lytvyn L, Lo CK-F, Beardsley J, Mertz D, Johnston BC (2017) Probiotics for the prevention of Clostridium difficile-associated diarrhea in adults and children. Cochrane Database Syst Rev 12(12):CD006095. https://doi.org/10.1002/14651858.CD006095.pub4

Hamilton-Miller JMT (2004) Probiotics and prebiotics in the elderly. Postgrad Med J 80:447–451

Hill C, Sanders ME (2013) Rethinking “probiotics”. Gut Microbes 4(4):269–270. https://doi.org/10.4161/gmic.25143

Houser MC, Tansey MG (2017) The gut-brain axis: is intestinal in flammation a silent driver of Parkinson's disease pathogenesis? NPJ Parkinsons Dis 3:3

Ja¨rbrink-Sehgal E, Andreasson A (2020) The gut microbiota and mental health in adults. Curr Opin Neurobiol 62:102–114. https://doi.org/10.1016/j.conb.2020.01.016

Jingyue W, Singleton SS, Bhuiyan U, Krammer L, Mazumder R (2023) Multi-omics approaches to studying gastrointestinal microbiome in the context of precision medicine and machine learning. Front Mol Biosci 10:1337373. https://doi.org/10.3389/fmolb.2023.1337373

Kang DW, Park JG, Ilhan ZE, Wallstrom G, Labaer J, Adams JB, Krajmalnik-Brown R (2013) Reduced incidence of prevotella and other fermenters in intestinal microflora of autistic children. PLoS One 8(7):68322

Kaur G, Behl T, Bungau S et al (2021) Dysregulation of the gut-brain Axis, Dysbiosis and influence of numerous factors on gut microbiota associated Parkinson's disease. Curr Neuropharmacol 19:233–247. https://doi.org/10.2174/1570159X18666200606233050

Kazemi A, Noorbala AA, Azam K, Eskandari MH, Djafarian K (2019) Effect of probiotic and prebiotic vs placebo on psychological outcomes in patients with major depressive disorder: a randomized clinical trial. Clin Nutr 38(2):522–528

Kelly JR et al (2016 Nov) Transferring the blues: depression-associated gut microbiota induces neurobehavioural changes in the rat. J Psychiatr Res 82:109–118. https://doi.org/10.1016/j.jpsychires.2016.07.019

Khan AA, Khurshid M, Khan S, Alshamsan A (2013) Gut microbiota and probiotics: current status and their role in cancer therapeutics. Drug Dev Res 74(6):365–375. https://doi.org/10.1002/ddr.21087

Kim S-K, Guevarra RB, Kim Y-T et al (2019) Role of probiotics in human gut microbiome-associated diseases. J Microbiol Biotechnol 29(9):1335–1340. https://doi.org/10.4014/jmb.1906.06064

Kim H, Kim S, Park SJ, Park G, Shin H, Park MS et al (2021) Administration of Bifidobacterium bifidum BGN4 and Bifidobacterium longum BORI improves cognitive and memory function in the mouse model of Alzheimer's disease. Front Aging Neurosci:13. https://doi.org/10.3389/fnagi.2021.709091

Klarić M, Mandić V, Lovrić S, Ćorić MK, Zovko N (2017) Placebo and nocebo effects and their significance in clinical practice. Med Glas (Zenica) 14(1):16–12. https://doi.org/10.17392/892-16

Kopp-Hoolihan L (2001) Prophylactic and therapeutic uses of probiotics: a review. J Am Diet Assoc 101(2):229–241. https://doi.org/10.1016/S0002-8223(01)00060-8

Kouchaki E, Tamtaji OR, Salami M, Bahmani F, Kakhaki RD, Akbari E, Tajabadi-Ebrahimi M, Jafari P, Asemi Z (2016) Clinical and metabolic response to probiotic supplementation in patients with multiple sclerosis: a randomized, double-blind, placebo-controlled trial. Clin Nutr 36:1245–1249

Kwoji ID, Aiyegoro OA, Okpeku M, Adeleke MA (2023) 'Multi-omics' data integration: applications in probiotics studies. Sci Food 7:25. https://doi.org/10.1038/s41538-023-00199-x

Le Morvan C, de Sequeira C, Hengstberger PE, Mack I (2022) Effect of probiotics on psychiatric symptoms and central nervous system functions in human health and disease: a systematic review and meta-analysis. Nutrients 14:621. https://doi.org/10.3390/nu14030621

Li N, Yang J, Zhang J, Liang C, Wang Y, Chen B, Zhao C, Wang J, Zhang G, Zhao D, Liu Y, Zhang L, Li G, Gai Z, Zhao G (2019) Correlation of gut microbiome between ASD children and mothers and potential biomarkers for risk assessment. Genomics Proteomics Bioinformatics 17:26–38. https://doi.org/10.1016/j.gpb.2019.01.002

Liu J, Tan Y, Cheng H, Zhang D, Feng W, Peng C (2022a) Functions of gut microbiota metabolites, current status and future perspectives. Aging Dis 13(4):1106–1126. https://doi.org/10.14336/AD.2022.0104

Liu L, Huh JR, Shah K (2022b) Microbiota and the gut-brain-axis: implications for newtherapeutic design in the CNS. Lancet 77. https://doi.org/10.1016/j.ebiom.2022.103908

MacFabe DF (2012) Short-chain fatty acid fermentation products of the gut microbiome: implications in autism spectrum disorders. Microb Ecol Health Dis 23(1):19260

Macrea C, Ilias T, Fratila O, Roxana B (2023) Cristian Hocopan. Probiotics and the microbiota-gut-brain axis: focus on patients with depression. A review of current research. Biomed Pap Med Fac Univ Palacky Olomouc Czech Repub 167(3):219–224

Madabushi JS, Khurana P, Gupta N, Gupta M (2023) Gut biome and mental health: do probiotics work? Cureus 15(6):e40293. https://doi.org/10.7759/cureus.40293

Mancuso C, Santangel R (2018) Alzheimer's disease and gut microbiota modifications: the long way between preclinical studies and clinical evidence. Pharmacol Res 129:329–336. https://doi.org/10.1016/j.phrs.2017.12.009

Mancuso C, Santangelo R (2018) Alzheimer's disease and gut microbiota modifications: the long way between preclinical studies and clinical evidence. Pharmacol Res 129:329e36

Markowiak P, Katarzyna Sli'zewska. (2017) Effects of probiotics, prebiotics, and synbiotics on human health. Nutrients 9:1021. https://doi.org/10.3390/nu9091021

Marotta A, Sarno E, Del Casale A et al (2019. Sec. Mood Disorders) Effects of probiotics on cognitive reactivity, mood, and sleep quality. Front Psych 10. https://doi.org/10.3389/fpsyt.2019.00164

Matenchuk BA, Mandhane PJ, Kozyrskyj AL (2020) Sleep, circadian rhythm, and gut microbiota. Sleep Med Rev 53:101340. https://doi.org/10.1016/j.smrv.2020.101340

Micheal L, Prat A (2016) One more role fort he gut: microbiata and blood brain barrier. Ann Transl Med 4(1):15

Mirzaei R, Bouzari B, Hossein-Fard SR et al (2021) Role of microbiota-derived short-chain fatty acids in nervous system disorders. Biomed Pharmacother 139:111661. https://doi.org/10.1016/j.biopha.2021.111661

Miyauchi E, Kim SW, Suda W, Kawasumi M, Onawa S, Taguchi-Atarashi N, Morita H, Taylor TD, Hattori M, Ohn H (2020) Gut microorganisms act together to exac-erbate inflammation in spinal cords. Nature 585:102–106. [CrossRef]

Mörkl S, MaryIButler AH, Cryan JF, Dinan TG (2020) Probiotics and the microbiota-gut-brain axis: focus on psychiatry. Curr Nutr Rep 9:171–182. https://doi.org/10.1007/s13668-020-00313-5

Neufeld KM, Kang N, Bienenstock J et al (2011) Reduced anxiety-like behavior and central neurochemical change in germ-free mice. Neurogastroenterol Motil 23:255–e119

Nikolova VL, Cleare AJ, Young AH, Stone JM (2021) Updated review and meta-analysis of probiotics for the treatment of clinical depression: adjunctive vs. stand-alone treatment. J Clin Med 10(4):647. https://doi.org/10.3390/jcm10040647

O'Neil A et al (2014) Relationship between diet and mental health in children and adolescents: a systematic review. Am J Public Health 104:e31–e42. https://doi.org/10.2105/AJPH.2014.302110

Pelto L, Salminen SJ, Isolauri E (1996) Lactobacillus GG modulates milk-induced immune inflammatory response in milk hypersensitive adults. Nutr Today 31:47S. [CrossRef]

Perlejewski K, Bukowska-Ośko I, Nakamura S, Motooka D, Stokowy T, Płoski R, Rydzanicz M, Zakrzewska-Pniewska B, Podlecka-Pi˛etowska A, Nojszewska M et al (2016) Metagenomic analysis of cerebrospinal fluid from patients with multiple sclerosis. Adv Exp Med Biol 935:89–98. [CrossRef]

Reuter JA, Spacek D, Snyder MP (2015) High-throughput sequencing technologies. Mol Cell 58(4):586–597. https://doi.org/10.1016/j.molcel.2015.05.004

Rhee SH, Pothoulakis C, Mayer EA (2009) Principles and clini cal implications of the brain–gut-enteric microbiota axis. Nat Rev Gastroenterol Hepatol 6(5):306–314. https://doi.org/10.1038/nrgastro.2009.35

Rossi F, Amadoro C, Pallotta ML, Colavita G (2022) Variability of genetic characters associated with probiotic functions in Lacticaseibacillus species. Microorganisms 10:1023. https://doi.org/10.3390/microorganisms10051023

Ruiz-Gonzalez C, Cardona D, Rueda-Ruzafa L, Rodriguez-Arrastia M, Ropero-Padilla C, Roman P (2024) Cognitive and emotional effect of a multi-species probiotic containing lactobacillus rhamnosus and Bifidobacterium lactis in healthy older adults: a double-blind randomized placebo-controlled crossover trial. Probiotics Antimicrob Proteins. https://doi.org/10.1007/s12602-024-10315-2

Saeed NK, Al-Beltagi M, Bediwy AS, El-Sawaf Y, Toema O (2022) Gut microbiota in various childhood disorders: implication and indications. World J Gastroenterol 28(18):1875–1901. https://doi.org/10.3748/wjg.v28.i18.1875

Şafak B, Altunan B, Topçu B, Eren TA (2020) The gut microbiome in epilepsy. Microb Pathog 139:103853. https://doi.org/10.1016/j.micpath.2019.103853

Sanders ME (2016) Probiotics and microbiota composition. BMC Med 14(1):82. https://doi.org/10.1186/s12916-0160629-z

Santoni M, Miccini F, Battelli N (2021) Gut microbiota, immunity and pain. Immunol Lett 229:44–47

Shaaban SY et al (2018) The role of probiotics in children with autism spectrum disorder: a prospective, open-label study. Nutr Neurosci 21(9):676–681

Sharma R, Gupta D, Mehrotra R, Mago P (2021 Feb) Psychobiotics: the next-generation probiotics for the brain. Curr Microbiol 78(2):449–463. https://doi.org/10.1007/s00284-020-02289-5

Shehata HR, Newmaster SG (2023) The power of DNA based methods in probiotic authentication. Front Microbiol 14. Sec Food Microbiology. https://doi.org/10.3389/fmicb.2023.1158440

Shen L, Liu L, Ji H-F (2017) Alzheimer's disease histological and behavioral manifestations in transgenic mice correlate with specific gut microbiome state. J Alzheimer's Dis 56(1):385–390. https://doi.org/10.3233/JAD-160884

Sinhg S et al (2022) Natural sources and encapsulating materials for probiotics delivery systems: recent applications and challenges in functional food development. Front Nutr 9:971784. https://doi.org/10.3389/fnut.2022.971784

Slykerman RF et al (2017) Effect of lactobacillus rhamnosus HN001 in pregnancy on postpartum symptoms of depression and anxiety: a randomised double-blind placebo-controlled trial. EBioMedicine 24:159–165. https://doi.org/10.1016/j.ebiom.2017.09.013

Socała K, Doboszewska U, Szopa A et al (2021) The role of microbiota-gut-brain axis in neuropsychiatric and neurological disorders. Pharmacol Res 172:105840. https://doi.org/10.1016/j.phrs.2021.105840

Tamtaji OR, Heidari-Soureshjani R, Mirhosseini N, Kouchaki E, Bahmani F, Aghadavod E et al (2019a) Probiotic and selenium co supplementation, and the effects on clinical, metabolic and genetic status in Alzheimer's disease: a randomized, double-blind, controlled trial. Clin Nutr 38:1594e8

Tamtaji OR, Taghizadeh M, Daneshvar Kakhaki R, Kouchaki E, Bahmani F, Borzabadi S, Oryan S, Mafi A, Asemi Z (2019b) Clinical and metabolic response to probiotic administration in people with Parkinson's disease: a randomized, double-blind, placebo-controlled trial. Clin Nutr 38:1031–1035

Tankou SK, Regev K, Healy BC, Cox LM, Tjon E, Kivisakk P, Vanande IP, Cook S, Gandhi R, Glanz B et al (2018a) Investigation of probiotics in multiple sclerosis. Mult Scler J 24:58–63

Tankou SK, Regev K, Healy BC, Tjon E, Laghi L, Cox LM, Kivisäkk P, Pierre IV, Hrishikesh L, Gandhi R et al (2018b) Aprobiotic modulates the microbiome and immunity in multiple sclerosis. Ann Neurol 83:1147–1161

Tansey MG, Romero-Ramos M (2019) Immune system responses in Parkinson's disease: early and dynamic. Eur J Neurosci 49(3):364–383

Tegegne BA, Kebede B (2022) Probiotics, their prophylactic and therapeutic applications in human health development: a review of the literatüre. Heliyon 8:e09725. https://doi.org/10.1016/j.heliyon.2022.e09725

Tian P, Chen Y, Zhu H, Wang L, Qian X, Zou R, Zhao J, Zhang H, Qian L, Wang Q, Wang G, Chen W (2022) Bifidobacterium breve CCFM1025 at tenuates major depression disorder via regulating gut microbiome and tryptophan metabolism: a randomized clinical trial. Brain Behav Immun 100:233–241

Tilg H, Zmora N, Adolph TE, Elinav E (2019) The intestinal microbiota fuelling metabolic inflammation. Nat Rev Immunol 20:40–54

Tiwari A, Krisnawati DI, Susilowati E, Mutalik C, Kuo T-R (2024) Next-generation probiotics and chronic diseases: a review of current research and future directions. J Agric Food Chem 72:27679–27700. https://doi.org/10.1021/acs.jafc.4c08702

Trifkoviˇ KˇC c et al (2022) Efficacy of direct or indirect use of probiotics for the improvement of maternal depression during pregnancy and in the postnatal period: a systematic review and meta-analysis. Healthcare 10:970. https://doi.org/10.3390/healthcare10060970
Ustianowsk K, Ustianowsk Ł, Machaj F et al (2022) The role of the human microbiome in the pathogenesis of pain. Int J Mol Sci 23:13267. https://doi.org/10.3390/ijms232113267
Van den Elsen LW, Garssen J, Burcelin R et al (2019) Shaping the gut microbiota by breastfeeding: the gateway to allergy prevention? Front Pediatr 7:47. https://doi.org/10.3389/fped.2019.00047
Van der Beek CM, Bloemen JG, van den Broek MA, Lenaerts K, Venema K, Buurman WA, Dejong CH (2015) Hepatic uptake of rectally administered butyrate prevents an increase in systemic butyrate concentrations in humans. J Nutr 145:2019–2024
Verbeke KA, Boobis AR, Chiodini A et al (2015) Towards microbial fermentation metabolites as markers for health benefits of prebiotics. Nutr Res Rev 28(1):42–66. https://doi.org/10.1017/S0954422415000037
Vijayaram S, Sinha R, Faggio C, Ringø E, Chou C-C (2024) Biopolymer encapsulation for improved probiotic delivery: advancements and challenges. AIMS Microbiol 10(4):986–1023. https://doi.org/10.3934/microbiol.2024043
Vitellio P, Chira A, De Angelis M, Dumitrascu DL, Portincasa P (2020) Probiotics in psychosocial stress and anxiety. A systematic review. J Gastrointestin Liver Dis 29(1):77–83. https://doi.org/10.15403/jgld-352
Wang Q, Yang Q, Liu X (2023) The microbiota-gut-brain axis and neurodevelopmental disorders. Protein Cell 14(10):762–775. https://doi.org/10.1093/procel/pwad026
Wieërs G, Belkhir L, Enaud R et al (2020) How probiotics affect the microbiota? Front Cell Infect Microbiol 9. https://doi.org/10.3389/fcimb.2019.00454
Woodmansey EJ (2007) Intestinal bacteria and ageing. J Appl Microbiol 102:1178–1186. https://doi.org/10.1111/j.1365-2672.2007.03400.x
Yang B, Wei J, Peijun J, Chen J (2019) Effects of regulating intestinal microbiota on anxiety symptoms: a systematic review. Gen Psychiatr 32:e100056. https://doi.org/10.1136/gpsych-2019-100056
Yang LL, Millischer V, Rodin S, MacFabe DF, Villaescusa JC, Lavebratt C (2020) Enteric short-chain fatty acids promote proliferation of human neural progenitor cells. J Neurochem 154:635–646
Yanga X, Yub D, Xuea L, Lia H, Junrong D (2020) Probiotics modulate the microbiotaegutebrain axis and improve memory deficits in aged SAMP8 mice. Acta Pharm Sin B 10(3):475–487
Yuan S, Xiong Y, Larsson SC (2020) An atlas on risk factors for multiple sclerosis: a Mendelian randomization study. J Neurol 268:114–124. [CrossRef]
Zawistowska-Rojek A, Tyskia S (2018) Are probiotic really safe for humans? Pol J Microbiol 67(3):251–258. https://doi.org/10.21307/pjm-2018-044
Zmora N et al (2018) Personalized gut mucosal colonization resistance to empiric probiotics is associated with unique host and microbiome features. Cell 174:1388–1405. https://doi.org/10.1016/j.cell.2018.08.041

The Influence of Food Additives and Preservatives on Mental Health

13

Jia Whei See and Ummul Asfeen

Abstract

Many food additives without any nutritional value are used lawfully in processed foods. They enhance color and improve the taste of food. The food product manufacturer would use them in an acceptable and lawful amount. Food additives such as emulsifiers and flavoring agents are found to have negative effects on mental health causing anxiety-like behavior and lower social behaviors in animal models. Further, hyperactivity behaviors were discovered in rats once monosodium glutamate was administered. Food coloring agents, which make food appealing to people especially children, play a role in increased hyperactivity. This may appear due to the brain's chemical signaling being affected by food flavoring agents. Furthermore, ultra-processed foods also have some correlation with mental health issues such as depression, anxiety, and hyperactivity. It is important that, as consumers, we try to avoid foods that contain high doses of food additives and preservatives and possess the knowledge of the ingredients used in the foods we ingest. As there is a link between our brain–gut axis and nutrition, natural alternatives are definitely recommended as a safety approach. The food authorities should also emphasize the importance of clear food labeling so that the consumer can avoid foods that pose risks to their health.

Keywords

Food additive · Preservatives · Mental health

J. W. See (✉)
Universitas Sriwijaya, Palembang, Indonesia

U. Asfeen
Georgia Cancer Center, Augusta University, Augusta, GA, USA

W. Mohamed, A. Özge (eds.), *Feeding the Mind: The Connection Between Diet, Drugs, and Mental Health Volume 1*, Nutritional Neurosciences,
https://doi.org/10.1007/978-981-95-1414-4_13

13.1 Introduction

Mental illnesses such as anxiety and depression have been escalating. One out of every eight individuals worldwide has been found to experience mental health illness symptoms (GBD 2022). There are several factors to consider that might contribute to this development of mental illness, namely psychological, biological, and environmental (Wu et al. 2022). Food consumption is one of the biggest factors that has garnered some attention in public (Wu et al. 2022).

To maintain the quality and to enhance food color and flavor, food additives are applied (Sambu et al. 2022). Several studies show that excessive use of food additives and preservatives can cause a range of mental health issues, including anxiety, depression, and ADHD (Sambu et al. 2022).

Some studies show that food additives can contribute to the disruption of neurotransmitters and the hippocampus, which can further cause neuroinflammation and gut microbiota dysbiosis, escalating to anxiety and depression (Aparna and Patri 2021).

For a better understanding, it is crucial to explore the correlation between the chemicals we ingest and their effects on our physical and mental health. This book chapter will further dive into the potential connection between food additives, preservatives, and mental health.

13.2 Contents

13.2.1 Food Additives

Food additives are substances that are applied to produce, process, or preserve food (Kaur 2022). These are categorized into a few categories, including preservatives, coloring agents, flavor enhancers, emulsifiers and stabilizers, thickeners and gelling agents, and acidulants. These food additives can be found in processed foods and a wide range of snacks, canned goods, frozen foods, and beverages. Food additives work in ensuring food safety, convenience, and consistency by preserving the food quality and enhancing the taste and color of the food (Kaur 2022).

13.2.2 Preservatives

Preservatives function by extending the shelf life of food and maintaining food safety (Kaur 2022). They prevent bacterial deterioration which can cause food poisoning. This has made it convenient to consumers in the modern era by saving time and energy when they do not need to purchase frequently (Kaur 2022).

Preservatives are divided into natural preservatives (salt, sugar, acid, oil, and spices) and chemical preservatives (benzoates, potassium sorbate, emulsifiers, and stabilizers) (Kaur 2022).

13.3 How Food Additives and Preservatives Affect Mental Health?

13.3.1 Emulsifiers and Flavoring Agents and Mental Health

Food additives, such as emulsifiers and flavoring agents, are potentially raising negative concerns about mental health. One study shows anxiety-like behaviors in male mice and lowered social behaviors in female mice once given two emulsifiers, namely carboxymethylcellulose and polysorbate 80, which can change the gut microbiota and demonstrate the expression of anxiety-related neuropeptides in a gender-dependent state (Holder et al. 2019).

Besides this study, monosodium glutamate (MSG) is found to lower exploratory behavior and cause depression in pregnant women, which may be correlated with the higher expression of the K^+/Cl^- cotransporter KCC2 (Biney et al. 2022).

Other research on adult rats has found that MSG causes anxiety and that neuronal lesions were found in neonatal rats (Rosa et al. 2016). There is also a correlation between higher exposure to MSG and oxidative stress, neuroinflammation, and ADHD behaviors in rats (Abu-Elfotuh et al. 2022). Anxiety-like behaviors were found to increase once food-grade titanium dioxide was used (Medina-Reyes et al. 2020).

13.3.2 Food Coloring Agents and Mental Health

Food coloring agents are additives that add color to our food and drink to make it attractive (Dey and Nagababu 2022). There are two variants: one is an artificial food coloring agent which is widely used in the culinary world, is low cost, and made of synthetic chemicals from petroleum; and the other is natural food coloring agents which are derived from natural sources such as plants, minerals, and insects. They are much safer to ingest but the coloring properties are incomparable to artificial ones (Dey and Nagababu 2022).

Certain synthetic food colorings—for example, Red 40 (allura red AC) which is commonly found in candy, soft drinks, and desserts, and Yellow 5 (tartrazine) which is found in processed snacks such as chips and cereals—have a correlation with ADHD and behavioral issues in children (Dey and Nagababu 2022). Research has found that both artificial food coloring and sodium benzoate are linked with hyperactivity in children (Dey and Nagababu 2022).

There are many theories about how artificial food coloring is associated with hyperactivity behavior (Spencer et al. 2017). Researchers believe that artificial food coloring may have disrupted the neurotransmitters that can lead to chemical signaling of the brain to interfere with mood and attention span. In addition, inflammation can also appear leading to changes in behavior and mood. Some children who have predisposed biological and genetic conditions become an easier target of the side effects of food coloring (Spencer et al. 2017).

et al. 2014). This enables us to get a better understanding of the interactions between genes and nutrients at a deeper molecular level. The knowledge of these interactions between the gene-nutrient will help customize diets based on the individual genotypes (Sales et al. 2014).

13.5.2 Nutritional Approaches on Prevention and Treatment

An approach to adopt a whole food diet can help combat the additive-related mental health issues. This can be achieved by various preventative and treatment strategies such as:

1. Incorporation of anti-inflammatory nutrients such as antioxidants and omega-3 fatty acids helps combat inflammation which could be a cause for the mental illness (Hecht et al. 2022).
2. Diet and the microbiota–gut–brain axis health support: The gut microbiota is a key factor that regulates brain function and behavior (Berding et al. 2021). We can support mental health by incorporating prebiotic foods that support the growth of existing gut microbes which include onions, garlic, soybeans, and artichokes and probiotic-rich foods which contain beneficial gut bacteria (e.g., yogurt and sauerkraut).
3. Avoid ultra-processed foods: Based on NOVA food classification system we can identify ultra-processed foods, and it is the one that is most applied in the scientific literature (Lawrence and Baker 2019). NOVA food classification system classifies the food into four major categories (Monteiro et al. 2018). One of the four groups is classified as ultra-processed and comprises snacks, drinks, and ready meals made with many additives. Having the knowledge of the processes and the ingredients used in the manufacturing of these ultra-processed foods helps a consumer make smart, informed choices.

13.5.3 Regulations of Food Additives and Preservatives

Regulatory agencies in the United States such as the Food and Drug Administration (FDA), World Health Organization (WHO), Food and Agriculture Organization (FAO), Codex Alimentarius Commission, and European Food Safety Authority (EFSA) are the agencies that regulate the food safety and establish regulations on additive use through rigorous testing, toxicological studies, risk assessments, and compliance monitoring. These agencies ensure the food additives and preservatives meet safety standards for consumption (McCann et al. 2007; Warner 2023).

Transparency in labeling food products is vital to ensure appropriate consumer awareness. FDA and other agencies mandate that the manufacturers disclose the additives and preservatives on food labels. However, certain technical terms and codes such as "E" numbers in Europe and natural flavors may be misleading and create confusion in the minds of the consumer by not providing transparency. Hence, there is a growing demand for clearer labeling and stringent regulations to ensure informed decisions (McCann et al. 2007; Warner 2023).

13.6 Future Directions

13.6.1 Gaps in Current Knowledge

Despite work being done to get a deeper understanding of the food additives and preservatives on mental health there exist gaps in the following:

1. The interactions of additives on gut–brain axis especially with the ultra-processed foods (Psychology Today 2022).
2. Long-term effects of additive combinations due to chronic exposure (Hecht et al. 2022).

Potential areas to investigate further include:

1. Natural alternatives: Developing safer, natural alternatives reduce the dependency on synthetic additives (Warner 2023).
2. Behavioral implications: Gaining a deeper understanding of the relationship between mental health and food additives, especially in the vulnerable population (McCann et al. 2007). This may help develop interventions that are more targeted.
3. Healthcare policy enhancements: Clearer food labeling practices and uniform global standards can help the consumer avoid certain foods that can pose a risk to their health (Psychology Today 2022).

13.7 Conclusion

Food additives and preservatives influence our physical health and are also correlated to our mental health. Food additives and preservatives such as emulsifiers, flavoring agents, food coloring agent, and ultra-processed foods have been found to have a strong correlation with the appearance of mental health issues, such as anxiety, depression, and ADHD. Further research is needed to uncover the gaps in our knowledge regarding the mechanism by which food additives and preservatives work on our brain function and how they cause mental illness in detail. A more mindful and holistic approach should be taken for our daily intake such as incorporating natural alternatives, having a mindful diet, and labeling clear food labels in our food products so the consumer can take a careful approach as to what to avoid in their diet.

References

Abu-Elfotuh K, Abdel-Sattar SA, Abbas AN et al (2022) The protective effect of thymoquinone or/and thymol against monosodium glutamate-induced attention-deficit/hyperactivity disorder (ADHD)-like behavior in rats: modulation of Nrf2/HO-1, TLR4/NF-κB/NLRP3/caspase-1 and Wnt/β-catenin signaling pathways in rat model. Biomed Pharmacother 155:113799

André O. Werneck, Euridice M. Steele, Felipe M. Delpino, Melissa M. Lane, Wolfgang Marx, Felice N. Jacka, Brendon Stubbs, Mathilde Touvier, Bernard Srour, Maria LC. Louzada, Renata B. Levy, Carlos A. Monteiro (2024) Adherence to the ultra-processed dietary pattern and risk of depressive outcomes: Findings from the NutriNet Brasil cohort study and an updated systematic review and meta-analysis, Clinical Nutrition, Volume 43, Issue 5, Pages 1190–1199, ISSN 0261–5614. https://doi.org/10.1016/j.clnu.2024.03.028

Aparna S, Patri M (2021) Benzo[a]pyrene exposure and overcrowding stress impacts anxiety-like behavior and impairs learning and memory in adult zebrafish, Danio rerio. Environ Toxicol 36(3):352–361

Berding K, Vlckova K, Marx W, Schellekens H, Stanton C, Clarke G, Jacka F, Dinan TG, Cryan JF (2021) Diet and the microbiota-gut-brain axis: sowing the seeds of good mental health. Adv Nutr 12(4):1239–1285. https://doi.org/10.1093/advances/nmaa181

Biney RP, Djankpa FT, Osei SA et al (2022) Effects of in utero exposure to monosodium glutamate on locomotion, anxiety, depression, memory and KCC2 expression in offspring. Int J Dev Neurosci 82(1):50–62

Brul S, Coote P (1999) Preservative agents in foods. Mode of action and microbial resistance mechanisms. Int J Food Microbiol 50(1–2):1–17. https://doi.org/10.1016/s0168-1605(99)00072-0

Damotharan K, Sudhakaran G, Ramu M, Krishnan M, Arockiaraj J (2024) Biochemical processes mediating neurotoxicity induced by synthetic food dyes: a review of current evidence. Chemosphere 143295

Dey S, Nagababu BH (2022) Applications of food colour and bio-preservatives in the food and its effect on human health. Food Chem Adv 1:100019

GBD 2019 Mental Disorders Collaborators (2022) Global, regional, and national burden of 12 mental disorders in 204 countries and territories, 1990–2019: a systematic analysis for the Global Burden of Disease Study 2019. Lancet Psychiatry 9(2):137–150

Hecht EM, Rabil A, Martinez Steele E, Abrams GA, Ware D, Landy DC, Hennekens CH (2022) Cross-sectional examination of ultra-processed food consumption and adverse mental health symptoms. Public Health Nutr 25(11):3225–3234. https://doi.org/10.1017/S1368980022001586

Holder MK, Peters NV, Whylings J et al (2019) Dietary emulsifiers consumption alters anxiety-like and social-related behaviors in mice in a sex-dependent manner. Sci Rep 9(1):172

Jacka FN, Pasco JA, Mykletun A, Williams LJ, Hodge AM, O'Reilly SL et al (2010) Association of Western and Traditional Diets with depression and anxiety in women. Am J Psychiatry 167(3):305–311. https://doi.org/10.1176/appi.ajp.2009.09060881

Jones SK, McCarthy DM, Vied C et al (2022) Transgenerational transmission of aspartame-induced anxiety and changes in glutamate-GABA signaling and gene expression in the amygdala. Proc Natl Acad Sci USA 119(49):e2213120119

Kaur T (2022) Effect of food additives and preservatives on human health. Int J Food Sci Nutr 7(4):123–125. https://www.foodsciencejournal.com/assets/archives/2022/vol7issue4/7-4-67-231.pdf

Lane MM, Gamage E, Travica N, Dissanayaka T, Ashtree DN, Gauci S, Lotfaliany M, O'Neil A, Jacka FN, Marx W (2022) Ultra-processed food consumption and mental health: a systematic review and meta-analysis of observational studies. Nutrients 14(13):2568. https://doi.org/10.3390/nu14132568

Lawrence MA, Baker PI (2019) Ultra-processed food and adverse health outcomes. BMJ 365:l2289. https://doi.org/10.1136/bmj.l2289

Mazloomi SN, Talebi S, Mehrabani S, Bagheri R, Ghavami A, Zarpoosh M et al (2022) The association of ultra-processed food consumption with adult mental health disorders: a systematic review and dose-response meta-analysis of 260,385 participants. Nutr Neurosci 26(10):913–931. https://doi.org/10.1080/1028415X.2022.2110188

McCann D, Barrett A, Cooper A, Crumpler D, Dalen L, Grimshaw K, Kitchin E, Lok K, Porteous L, Prince E, Sonuga-Barke E, Warner JO, Stevenson J (2007) Food additives and hyperactive behaviour in 3-year-old and 8/9-year-old children in the community: a randomised, double-blinded, placebo-controlled trial. Lancet (London, England) 370(9598):1560–1567. https://doi.org/10.1016/S0140-6736(07)61306-3

Medina-Reyes EI, Delgado-Buenrostro NL, Díaz-Urbina D et al (2020) Food-grade titanium dioxide (E171) induces anxiety, adenomas in colon and goblet cells hyperplasia in a regular diet model and microvesicular steatosis in a high fat diet model. Food Chem Toxicol 146:111786

Monteiro CA, Cannon G, Moubarac JC, Levy RB, Louzada MLC, Jaime PC (2018) The UN Decade of Nutrition, the NOVA food classification and the trouble with ultra-processing. Public Health Nutr 21(1):5–17. https://doi.org/10.1017/S1368980017000234

Psychology Today (2022) Ultra-processed food consumption is associated with adverse mental symptoms. Retrieved from http://www.psychologytoday.com/

Rosa SG, Quines CB, Stangherlin EC, Nogueira CW (2016) Diphenyl diselenide ameliorates monosodium glutamate induced anxiety-like behavior in rats by modulating hippocampal BDNF-Akt pathway and uptake of GABA and serotonin neurotransmitters. Physiol Behav 155:1–8

Sales NM, Pelegrini PB, Goersch MC (2014) Nutrigenomics: definitions and advances of this new science. J Nutr Metab 2014:202759. https://doi.org/10.1155/2014/202759

Sambu S, Hemaram U, Murugan R, Alsofi AA (2022) Toxicological and teratogenic effect of various food additives: an updated review. Biomed Res Int 2022:6829409

Samuthpongtorn C, Nguyen LH, Okereke OI et al (2023) Consumption of ultraprocessed food and risk of depression. JAMA Netw Open 6(9):e2334770. https://doi.org/10.1001/jamanetworkopen.2023.34770

Spencer SJ, Korosi A, Layé S, Shukitt-Hale B, Barrientos RM (2017) Food for thought: how nutrition impacts cognition and emotion. NPJ Sci Food 1(7)

Walczak-Nowicka ŁJ, Herbet M (2022) Sodium benzoate-harmfulness and potential use in therapies for disorders related to the nervous system: a review. Nutrients 14(7):1497. https://doi.org/10.3390/nu14071497

Warner JO (2023) Artificial food additives: hazardous to long-term health? Arch Dis Child 109(11):882–885. https://doi.org/10.1136/archdischild-2023-326565

Wu SX, Li J, Zhou DD et al (2022) Possible effects and mechanisms of dietary natural products and nutrients on depression and anxiety: a narrative review. Antioxidants 11(11):2132

The Effects of Food Allergies and Intolerances on Mental Health

14

Nihat Mustafayev

Abstract

Food allergies and intolerances are not just about physical symptoms; they can also deeply affect mental health, contributing to conditions like anxiety, depression, and mood disorders. This chapter explores how food allergies and intolerances affect mental well-being, focusing on the role of inflammation, the gut-brain axis, and the immune response. We discuss management strategies, including elimination diets, nutritional interventions, and the use of probiotics and prebiotics, to help mitigate these impacts. By understanding and addressing these interconnected issues, healthcare providers can offer a more holistic approach to improving both physical and mental health for those affected.

Keywords

Food allergies · Mental health · Gut-brain axis · Inflammation · Probiotics · Elimination diets

14.1 Introduction

Food allergies and intolerances are common health issues that affect individuals of all ages, yet their impact extends beyond physical symptoms. These conditions are often misunderstood, with many people using the terms "allergy" and "intolerance" interchangeably, despite their distinct mechanisms (Pope 2009). Food allergies involve an immune system response to specific foods, leading to symptoms such as itching, swelling, or even life-threatening anaphylaxis. On the other hand, food

N. Mustafayev (✉)
Ozel Capa Private Hospital, Istanbul, Türkiye

W. Mohamed, A. Özge (eds.), *Feeding the Mind: The Connection Between Diet, Drugs, and Mental Health Volume 1*, Nutritional Neurosciences,
https://doi.org/10.1007/978-981-95-1414-4_14

intolerances are non-immune reactions that typically stem from the body's inability to properly digest or process certain foods, resulting in more gradual and less severe symptoms.

The significance of these conditions goes beyond digestion and physical health. Emerging research reveals a strong connection between diet and mental health, with food allergies and intolerances playing a key role in this relationship. Allergic reactions often trigger systemic inflammation, which can impact brain function, while intolerances may disrupt the gut microbiota, a critical component of the gut-brain axis. These disruptions have been linked to mental health conditions such as depression, anxiety, migraines, and cognitive difficulties (Horovitz 2024).

Understanding the mechanisms behind these conditions and their broader implications is essential for effective diagnosis and management. This chapter will explore the definitions and prevalence of food allergies and intolerances, their pathophysiological impact on brain health, and the clinical manifestations and management strategies for these conditions. By addressing these topics, we aim to highlight the importance of an integrated approach to care, bridging the gap between dietary health and mental well-being.

14.2 Understanding Food Allergies and Intolerances

Food allergies occur when the body reacts abnormally to specific proteins. When an individual with a food allergy consumes or is exposed to an allergenic substance, the immune system perceives it as a threat. This triggers an allergic cascade through humoral immunity, activating the release of Immunoglobulin E (IgE) antibodies. These antibodies stimulate mast cells and basophils to release histamine and other inflammatory substances, resulting in symptoms such as itching, swelling, or, in severe cases, anaphylaxis.

The response to food allergens includes both immediate and delayed mechanisms. Immediate reactions, typically occurring within minutes, are mediated by IgE antibodies and lead to classic allergy symptoms. Delayed reactions, however, involve other immune pathways, such as T-cell activation, and may cause symptoms like gastrointestinal discomfort or skin rashes hours after exposure (Dhananjay and Luxita 2024).

Food allergies also play a significant role in the gut-brain connection. Chronic low-grade inflammation caused by allergic reactions can disrupt gut health, which in turn affects brain function. Pro-inflammatory cytokines released during allergic responses can influence the gut-brain axis by crossing the blood-brain barrier and impacting mood and cognitive processes. This interplay highlights that food allergies are not just limited to physical symptoms but can also significantly affect mental health. In this context, we will further explore the gut-brain relationship and its connection to food allergies.

14.3 Gut-Brain Axis and Inflammation

Food allergies and intolerances can have a profound effect on how we feel, both physically and mentally. It's amazing how closely connected our gut and brain are, almost like they are having an ongoing conversation. This connection happens through something called the gut-brain axis, which uses nerves, hormones, and immune signals to send messages between the gut and the brain. The gut microbiota—a mix of trillions of microorganisms that live in our digestive system—is like the bridge that keeps this communication going smoothly, and it plays a big role in our mental well-being.

One of the main ways that food reactions can affect our mental health is through inflammation. When someone has a food allergy, their immune system can overreact to certain proteins in foods. This response releases chemicals called pro-inflammatory cytokines, which can cross over into the brain. These cytokines can affect our mood, sometimes causing anxiety or even depression (Järbrink-Sehgal and Andreasson 2020). Chronic inflammation of this nature can disrupt brain function, leading to mental fatigue or depressive symptoms, which is why addressing these reactions is crucial.

The gut-brain axis is also key to understanding how food intolerances can impact our mental health. Food intolerances might not trigger the immune system in the same way as allergies do, but they can still create a lot of discomfort in the gut. When the gut is out of balance—a condition called dysbiosis—it can cause the gut lining to become more permeable. This is sometimes called "leaky gut," and it allows harmful substances to enter the bloodstream. When these substances reach the brain, they can lead to issues like mood swings or difficulties with focus (Berding et al. 2021).

Gut bacteria also produce substances called short-chain fatty acids (SCFAs), which are vital for both gut health and brain health. When the production of SCFAs is disrupted, often due to an imbalanced diet or food intolerances, it can lead to more inflammation, which might contribute to problems like memory issues or emotional instability (Berding et al. 2021).

Realizing how deeply the gut-brain axis and inflammation are linked to our mental health helps us understand why managing food allergies and intolerances is so essential. Making thoughtful dietary choices—such as incorporating probiotics, eating anti-inflammatory foods, and avoiding trigger foods—can help restore balance in the gut and improve overall mental well-being. More research is definitely needed, but what we already know shows that taking care of our gut is a key part of taking care of our minds.

14.4 Mental Health Conditions Associated with Food Allergies and Intolerances

Food allergies and intolerances can have significant impacts on both physical and mental health. The relationship between these conditions and mental health is complex, often involving the immune response, chronic inflammation, and the gut-brain axis. Below, we explore some of the common mental health conditions linked to food allergies and intolerances, summarized in Table 14.1.

14.4.1 Anxiety Disorders

Living with food allergies requires constant vigilance to avoid exposure to allergens, which can lead to heightened anxiety. This anxiety is often driven by the fear of an accidental reaction, which can range from mild discomfort to life-threatening anaphylaxis. The immune response involved in food allergies leads to the release of inflammatory cytokines that can influence the central nervous system, contributing to feelings of anxiety. Additionally, changes in gut microbiota, a common consequence of allergic reactions, can impact the gut-brain axis and affect emotional regulation. This is especially challenging in social settings where food is involved, as individuals may feel anxious about the possibility of cross-contamination or accidental ingestion. Children with food allergies often experience significant anxiety, and their caregivers may also suffer from chronic stress due to the constant need for vigilance and emergency preparedness (Polloni and Muraro 2020; Westfall et al. 2021).

14.4.2 Depression

From childhood onwards, managing food allergies can create emotional challenges that often lead to symptoms of depression. The constant need to avoid certain foods can make social gatherings difficult, as eating together is an important part of building relationships. Individuals with food allergies may feel excluded or different from others, leading to sadness and social withdrawal. On a biological level, the

Table 14.1 Mental health conditions linked to food allergies and intolerances

Condition	Mechanism involved	Symptoms
Anxiety disorders	Immune response, cytokine release	Heightened anxiety, fear of allergen exposure
Depression	Inflammation, gut dysbiosis	Heightened anxiety, fear of allergen exposure
ADHD	Immune dysregulation, chronic inflammation	Attention deficits, hyperactivity
Autism Spectrum disorders (ASD)	Immune dysregulation, gut issues	Social communication challenges, behavioral issues

inflammation triggered by allergen exposure can disrupt the production of key neurotransmitters like serotonin, which helps regulate mood. This disruption can increase the risk of developing depression. Additionally, the gut-brain connection plays a crucial role in this process, as inflammation and an imbalance in gut bacteria can send negative signals to the brain, further affecting mood and mental health (Polloni and Muraro 2020; Westfall et al. 2021).

14.4.3 Attention Deficit Hyperactivity Disorder (ADHD)

Emerging research suggests that children with a history of allergic conditions may have an increased risk of developing ADHD. The underlying mechanism is thought to involve immune system dysregulation and chronic inflammation, which can interfere with normal brain development, particularly in areas responsible for attention and impulse control. Inflammatory cytokines, produced in response to allergic reactions, can impact neurodevelopment and potentially contribute to the symptoms of ADHD. Additionally, disruptions in gut health, such as an imbalance in gut microbiota, are believed to influence cognitive functions related to attention and behavior. These disruptions can exacerbate ADHD symptoms, making it important to consider gut health in the management of allergic conditions in children (Nemet et al. 2022; Chua et al. 2021).

14.4.4 Autism Spectrum Disorders (ASD)

There is increasing evidence suggesting a link between allergies and neurodevelopmental conditions like autism. Immune dysregulation and chronic inflammation, both of which are common in allergic conditions, may play a role in this association. Inflammatory responses can influence brain development, potentially affecting pathways involved in social behavior and communication. The gut-brain axis is also implicated, as many children with ASD have been found to have gastrointestinal issues, including dysbiosis. Managing gut health through dietary interventions and reducing allergen exposure may help alleviate some of the symptoms associated with ASD, highlighting the importance of a comprehensive approach that addresses both the immune system and gut health (Chua et al. 2021).

14.4.5 Mood Disorders

Chronic inflammation resulting from food allergies and intolerances does not only remain localized in the gut; it also affects the brain, leading to various mood disturbances. The inflammatory cytokines produced during allergic reactions can cross the blood-brain barrier and interfere with neurotransmitter systems, including those that regulate mood, such as serotonin and dopamine. This can lead to mood swings, irritability, and emotional instability. Additionally, the health of the gut microbiota is crucial for maintaining a stable mood, as a balanced gut environment helps regulate inflammation. When gut health is compromised due to food intolerances or allergies, it can result in increased inflammation and, consequently, mood disorders. Addressing gut health through dietary changes, probiotics, and reducing exposure to allergens can be an effective strategy for managing these mood-related symptoms (Chua et al. 2021; Westfall et al. 2021).

Understanding these connections between food allergies, intolerances, and mental health is crucial for providing effective care. By recognizing both the physical and emotional challenges associated with these conditions, healthcare providers can offer more comprehensive support. A holistic approach that includes managing inflammation, supporting gut health, and providing psychological care can significantly improve the quality of life for those affected.

14.5 Clinical Manifestations and Diagnostic Challenges

Food allergies and intolerances don't just affect physical health; they can have a major impact on mental well-being too. These conditions can cause various emotional and psychological symptoms, which makes it crucial to understand their broader effects. Here, we will look into the clinical signs and diagnostic difficulties related to food allergies and intolerances, especially focusing on how they influence mental health. In the next section, we'll cover strategies for managing and treating these conditions in detail.

Food allergies happen when the immune system overreacts, leading to symptoms like skin problems (rashes, eczema), breathing issues (asthma, nasal congestion), and digestive issues (vomiting, diarrhea). But these allergies can also affect mental health. People with food allergies often deal with anxiety, depression, and mood problems. This is because the immune response releases inflammatory chemicals that can reach the brain and change how it works (Demir E 2020). This inflammation can interfere with how neurotransmitters work, making it harder for the person to manage their emotions and potentially leading to anxiety or depression.

Diagnosing food allergies and intolerances in people who also have mental health issues can be tricky. Symptoms like anxiety or depression can make it harder to identify an underlying allergy. Additionally, general symptoms like fatigue, irritability, or brain fog can overlap with both food intolerances and mental health disorders, complicating the diagnosis. It is important to distinguish between allergies and intolerances because allergies involve an immune response, specifically immunoglobulin E (IgE) antibodies, which are not seen in intolerances (Demir E 2020). Proper diagnosis is key to effective treatment. Living with food allergies can significantly affect someone's quality of life. People with these conditions often feel anxious about accidentally being exposed to allergens, which can be very dangerous. This constant worry can lead to stress and social isolation, especially for children and teenagers who may feel left out of activities involving food. If not addressed, this ongoing anxiety can contribute to more serious mental health problems, like major depression (Demir E 2020). Understanding how immune responses and inflammation relate to mental health symptoms is crucial for offering comprehensive support and effective treatment.

14.6 Management and Treatment Strategies

Managing food allergies and intolerances that impact mental health involves several key dietary and nutritional strategies. These approaches are not just about reducing physical symptoms but also play a big role in improving emotional well-being. Below, we explore elimination diets, nutritional interventions, and the use of probiotics and prebiotics to support overall health.

Elimination Diets and Food Reintroduction Protocols: Elimination diets are often the first step in identifying which foods might be causing allergic reactions. This approach involves removing potential trigger foods for a few weeks and then gradually adding them back one by one to see if symptoms return. Common foods that are eliminated include dairy, gluten, soy, and certain nuts. While this process can take some time, it helps pinpoint the specific foods causing problems. See Table 14.2 for examples of commonly eliminated foods and their reasons for elimination. For many individuals, identifying and avoiding their trigger foods can make a huge difference in both physical and mental health. It can reduce anxiety, improve mood, and help them feel more in control of their well-being (Grajek et al. 2022).

Nutritional Interventions to Support Mental Health: Good nutrition is essential for maintaining brain health. Certain nutrients are particularly effective in reducing inflammation and supporting mental well-being. Omega-3 fatty acids, found in fatty fish like salmon and mackerel, flaxseeds, and walnuts, are well known for their anti-inflammatory effects and their ability to keep brain cells healthy. Omega-3 s also help regulate serotonin and dopamine, which are important for mood and anxiety management. Magnesium is another essential nutrient that supports relaxation and can reduce anxiety symptoms—foods like leafy greens, nuts, seeds, and whole grains are excellent sources. B vitamins, particularly B6, B9 (folate), and B12, are crucial for brain function and neurotransmitter production. Deficiencies in these vitamins can be linked to higher levels of homocysteine, which is associated with depression. Ensuring a well-rounded diet that includes these nutrients is important for supporting both mental and physical health, especially for those dealing with food allergies or intolerances (Muscaritoli 2021). See Table 14.3 for a summary of key nutrients, their sources, and their importance in maintaining mental health.

Table 14.2 Common foods eliminated in elimination diets

Food type	Examples	Elimination reason
Dairy products	Milk, cheese, yogurt	Lactose intolerance, milk allergy
Gluten-containing foods	Bread, pasta, cereal	Celiac disease, gluten sensitivity
Nuts	Almonds, walnuts	Nut allergies
Soy	Tofu, soy Milk	Soy allergy

Table 14.3 Key nutrients for mental health

Nutrient	Sources	Role in mental health
Omega-3 fatty acids	Salmon, flaxseeds, walnuts	Reduces inflammation, supports mood regulation
Magnesium	Leafy greens, nuts, seeds	Supports relaxation, reduces anxiety
B vitamins	Whole grains, leafy greens, eggs	Supports neurotransmitter production, reduces homocysteine levels

Role of Probiotics and Prebiotics: The gut-brain connection plays a significant role in how our diet affects mental health. Probiotics and prebiotics are powerful tools for improving gut health, which can, in turn, positively affect mental well-being. Probiotics are beneficial bacteria found in fermented foods like yogurt, kefir, sauerkraut, and kimchi. Prebiotics, on the other hand, are fibers that feed these beneficial bacteria, helping them grow. Foods rich in prebiotics include garlic, onions, bananas, and asparagus. Research has shown that probiotics such as Lactobacillus and Bifidobacterium can help reduce symptoms of anxiety and depression by promoting a healthy gut balance and reducing inflammation. For individuals with food allergies, maintaining a balanced gut with probiotics and prebiotics can be an effective way to manage anxiety and mood swings. Regularly including these foods in the diet can help improve mental health outcomes and provide better overall balance (Grajek et al. 2022).

By addressing food allergies and intolerances with appropriate dietary strategies, individuals can see improvements in both their physical and mental health. Focusing on eliminating trigger foods, ensuring proper nutrition, and supporting gut health can help people manage their symptoms and feel better overall. In the next section, we will explore specific therapeutic approaches that combine these dietary strategies with medical and psychological support for a more comprehensive approach to care.

Food allergies and intolerances are complex conditions that impact not only physical health but also mental well-being. The intricate relationship between these conditions and mental health involves immune responses, chronic inflammation, and disruptions in the gut-brain axis. By using effective management strategies such as elimination diets, targeted nutritional interventions, and support for gut health through probiotics and prebiotics, individuals can significantly improve their quality of life. An interdisciplinary approach that includes healthcare providers from different specialties is essential to comprehensively address both the physical and mental challenges associated with these conditions.

References

Berding K, Vlckova K, Marx W, Schellekens H, Stanton C, Clarke G, Cryan JF (2021) Diet and the microbiota–gut–brain axis: sowing the seeds of good mental health. Adv Nutr 12(4):1239–1285

Chua RXY et al (2021) Mental health and inflammation. J Mental Health Inflamm 15(3):245–258

Dhananjay S, Luxita S (2024) Nutrition and psychology: a review on the impact of diet on mental health. Psychol Psychol Res Int J 9(4):000437

Demir E, Çimen A (2020) Food Allergy and Psychiatric Disorders. JEB Med Sci. 1(1):23–27

Grajek M, Krupa-Kotara K, Białek-Dratwa A, Sobczyk K, Grot M, Kowalski O, Staśkiewicz W (2022) Nutrition and mental health: a review of current knowledge about the impact of diet on mental health. Front Nutr 9:943998

Horovitz O (2024) Nutritional psychology: review the interplay between nutrition and mental health. Nutr Rev 158

Järbrink-Sehgal E, Andreasson A (2020) The gut microbiota and mental health in adults. Curr Opin Neurobiol 62:102–114

Muscaritoli M (2021) The impact of nutrients on mental health and well-being: insights from the literature. Front Nutr 8:656290

Nemet S, Asher I, Yoles I, Baevsky T, Sthoeger Z (2022) Early childhood allergy linked with development of attention deficit hyperactivity disorder and autism spectrum disorder. Pediatr Allergy Immunol 33(6)

Polloni L, Muraro A (2020) Anxiety and food allergy: a review of the last two decades. Clin Exp Allergy 50(4):420–441

Pope J (2009) Understanding food intolerance in primary care. University of Surrey

Westfall S, Caracci F, Estill M, Frolinger T, Shen L, Pasinetti GM (2021) Chronic stress-induced depression and anxiety priming modulated by gut-brain-axis immunity. Front Immunol 12:670500

The Effect of Hydration and Dehydration on Cognitive Performance and Mental Health

15

Leyla Mustafayeva

Abstract

Hydration is a key factor in maintaining cognitive performance and mental health, as water plays an essential role in vital processes such as neurotransmitter production, synaptic communication, and cerebral blood flow. Adequate hydration supports memory, attention, and learning, while fostering overall mental clarity and resilience. On the other hand, dehydration can significantly impair these functions, leading to fatigue, slower reaction times, mood changes, and difficulty concentrating. Symptoms range from mild discomfort, like headaches, to severe cognitive disruptions and systemic health risks. Certain groups, such as children, older adults, and individuals facing physical or mental challenges, are particularly sensitive to the effects of dehydration. In this chapter, I explore the intricate relationship between hydration and brain health, emphasizing its importance in preserving cognitive function and mental well-being across all ages.

Keywords

Hydration · Dehydration · Cognitive Health · Mental Health · Neurotransmission

L. Mustafayeva (✉)
Department of Physical Therapy and Rehabilitation, Istanbul University-Cerrahpasa, Cerrahpasa Medical Faculty, Fatih, Turkey

W. Mohamed, A. Özge (eds.), *Feeding the Mind: The Connection Between Diet, Drugs, and Mental Health Volume 1*, Nutritional Neurosciences,
https://doi.org/10.1007/978-981-95-1414-4_15

Adequate hydration is essential for sustaining vital physiological processes. As water constitutes 60–70% of body weight, it plays a critical role in maintaining biochemical and physiological homeostasis. Proper hydration supports fluid balance and facilitates the functioning of essential biochemical pathways (Popkin et al. 2010). Conversely, dehydration disrupts these processes, leading to a range of health issues, including toxin accumulation from impaired metabolic waste elimination (Ritz and Berrut 2005). Water is indispensable for the proper functioning of all cells, tissues, and organs, making optimal hydration fundamental to physiological integrity (Lieberman 2007).

Hydration also significantly impacts overall health by regulating body temperature through perspiration, particularly during physical activity. Insufficient hydration under conditions of intense exercise or heat exposure can lead to reduced performance and serious conditions such as heat stroke (Lieberman 2007). For individuals with active lifestyles, maintaining proper hydration is essential for muscle function and enhanced physical performance (Ritz and Berrut 2005).

15.1 The Role of Hydration on Cognitive Performance and Mental Health

Optimal hydration is essential for mental, cognitive, and physical health. Adequate water intake supports brain function by facilitating efficient operations such as synaptic transmission and neurotransmitter synthesis, which are critical for learning, memory, and focus (Merhej 2019). Conversely, insufficient hydration can impair these abilities, leading to diminished attention, slower reaction times, and delayed information processing (Grandjean and Grandjean 2007; Goodman et al. 2019). These effects are particularly pronounced in individuals with high cognitive demands, such as students and professionals.

Maintaining hydration has been consistently shown to enhance cognitive functions, including focus, memory, and problem-solving abilities (Masento et al. 2014). Furthermore, hydration plays a significant role in mental health by reducing the risk of mood disturbances, stress, and anxiety (Benton 2011). Inadequate water intake may exacerbate depression and anxiety symptoms due to reductions in brain volume and gray matter connectivity (Merhej 2019; Secher et al. 2012). Proper hydration is also associated with improved sleep quality, which further enhances cognitive functions like learning and memory (Ritz and Berrut 2005).

Research on dehydration's cognitive effects, particularly during physical exertion, shows mixed results. Some studies suggest mild dehydration (e.g., 2% body weight loss) impairs short-term memory and decision-making, while others report only mood disturbances such as fatigue, confusion, and irritability (Maughan et al. 2007; Cian et al. 2000; D'Anci et al. 2009). Interestingly, certain studies have found no significant impact on cognitive performance, highlighting the need for further investigation (Neave et al. 2001; Shirreffs et al. 2004).

15.2 Physiological and Biochemical Effects of Hydration on Brain Health

Hydration plays a critical role in maintaining brain health by ensuring ionic balance, which is vital for the proper functioning of brain cells. Fluid balance between intracellular and extracellular compartments supports neurotransmitter synthesis and synaptic transmission, key processes for cognitive functions. Dehydration, on the other hand, disrupts electrical signal transmission between neurons, impairing abilities such as memory retention and problem-solving (Wilson and Morley 2003). Chronic dehydration has also been associated with reductions in gray and white matter volumes, further affecting long-term cognitive performance (Merhej 2019). Studies suggest that adequate hydration enhances cerebral blood flow, optimizing oxygen delivery and supporting overall cognitive health (Ritz and Berrut 2005).

The neuroprotective effects of hydration extend to its potential role in mitigating the risk of neurodegenerative diseases. By reducing oxidative stress and inflammation, hydration may help protect neural cells from damage caused by free radical accumulation (Hodges 2012). These antioxidative properties are particularly relevant for preventing conditions such as Alzheimer's and Parkinson's disease (Secher et al. 2012). Additionally, water is essential for maintaining cell membrane permeability and the functionality of ion channels, which facilitate efficient communication between nerve cells. Dehydration, in contrast, can lead to membrane stiffening and disrupted ion channel functions, impairing synaptic transmission and cognitive processes (Merhej 2019; Wilson and Morley 2003).

15.3 Dehydration: Definition and Classification

Dehydration occurs when the body loses more water than it can replace, impairing its ability to maintain optimal physiological function. Water is lost through physiological processes, including sweating, respiration, urination, and defecation (Grandjean et al. 2003). Even minimal water loss (1–2% of body weight) can significantly impact physiological and cognitive functions (Lieberman 2007). The effects of dehydration vary based on individual factors such as age, general health, and physical activity level (Wilson and Morley 2003). Dehydration can be classified into three levels: mild, moderate, and severe, as summarized in Table 15.1.

Table 15.1 Symptoms across different levels of dehydration

Level	% Body weight loss	Symptoms
Mild	1–2	Thirst, fatigue, headache
Moderate	3–5	Dizziness, muscle cramps, memory impairment
Severe	>6	Hallucinations, organ failure, loss of consciousness

The classification in Table 15.1 provides a foundational understanding of dehydration levels, which are further detailed in the following sections.

15.3.1 Mild Dehydration

Mild dehydration is characterized by a loss of 1–2% of body weight in water, typically accompanied by symptoms such as thirst, fatigue, headache, and difficulty concentrating (Grandjean et al. 2003). This level of dehydration generally arises from inadequate fluid intake, particularly during hot weather or periods of physical exertion (Masento et al. 2014). Mild dehydration may impair attention and memory, as well as cause symptoms such as dry mouth and general discomfort (Ritz and Berrut 2005). Older adults and children are particularly vulnerable; older individuals may experience a reduced sense of thirst, while dehydration in children can develop rapidly and affect cognitive performance and academic achievement (Edmonds et al. 2021; Grandjean et al. 2003).

15.3.2 Moderate Dehydration

Moderate dehydration involves a loss of 3–5% of body weight and is associated with more pronounced physical and cognitive symptoms, including fatigue, dizziness, muscle cramps, and increased heart rate (Ritz and Berrut 2005). In cognitive terms, moderate dehydration leads to slower reaction times, reduced memory performance, and impaired problem-solving abilities (Lieberman 2007). Cardiovascular symptoms, such as reduced blood volume and hypotension, are prominent during moderate dehydration, posing particular risks for older adults and those with chronic conditions (Wilson and Morley 2003). Moderate dehydration also impacts thermoregulation, impairing the body's ability to dissipate heat and increasing the risk of heat-related illnesses (Grandjean et al. 2003).

Moderate dehydration also has significant mental health consequences, including heightened anxiety and stress (Benton 2011). Cognitive impairments are evident in tasks requiring sustained attention and concentration, as dehydration interferes with neurotransmission in the brain (Masento et al. 2014).

15.3.3 Severe Dehydration

Severe dehydration, defined as a loss of 6% or more of body weight, can result in significant physiological and cognitive impairments and requires immediate medical intervention (Grandjean et al. 2003). This condition leads to marked reductions in blood volume, compromising the transport of oxygen and nutrients to organs (Ritz and Berrut 2005). Symptoms of severe dehydration include hypotension, tachycardia, renal failure, and loss of consciousness (Wilson and Morley 2003). Severe dehydration severely impairs cognitive functions, potentially resulting in symptoms such as memory deficits, impaired decision-making, and even hallucinations (Lieberman 2007). Reduced cerebral blood flow and decreased oxygenation may cause permanent brain damage if left untreated (Wittbrodt and Barnes 2020).

Severe dehydration can also disrupt electrolyte balance, affecting the levels of sodium, potassium, and calcium. These imbalances can impair neural transmission, leading to neuromuscular dysfunction and potentially fatal outcomes if untreated (Wilson and Morley 2003). Vulnerable populations such as children and older adults are particularly susceptible to the adverse effects of severe dehydration, which can impede cognitive development in children and exacerbate chronic health issues in older adults (Edmonds et al. 2021).

Prompt intervention is essential to prevent severe dehydration. At-risk individuals, including those living in hot climates or with high levels of physical activity, should increase fluid intake and regularly monitor their hydration status (Ritz and Berrut 2005). Medical management of severe dehydration involves rapid fluid and electrolyte replacement, typically via intravenous therapy (Lieberman 2007).

15.4 Physiological Mechanisms of Dehydration

The physiological mechanisms of dehydration involve complex processes arising from an imbalance between water intake and loss in the body. Fluid balance is maintained through a delicate equilibrium that ensures optimal hydration for cellular and systemic functions (Lieberman 2007). When water loss exceeds intake, dehydration disrupts this balance, leading to significant physiological impairments. The primary mechanisms of water loss include perspiration, respiration, urination, and defecation, all of which contribute to varying degrees of fluid depletion (Grandjean et al. 2003). This imbalance in body water disrupts electrolyte concentrations, particularly sodium, potassium, and chloride, which are crucial for maintaining the function of ion channels and ensuring effective neural communication (Wilson and Morley 2003).

In the early stages of dehydration, the body activates compensatory mechanisms to mitigate the effects of fluid loss. An increase in plasma osmolality, caused by reduced water availability, stimulates osmoreceptors located in the hypothalamus. These receptors trigger the sensation of thirst, prompting increased water intake to restore balance (Edmonds et al. 2021). However, this feedback mechanism becomes less effective in certain populations, particularly older adults, due to age-related changes in thirst perception. This diminished response often results in inadequate fluid intake, making older individuals more susceptible to dehydration and its associated complications (Edmonds et al. 2021).

The kidneys play a central role in maintaining hydration by conserving water during periods of dehydration. In response to reduced fluid levels, the kidneys secrete antidiuretic hormone (ADH), which acts on the renal tubules to enhance water reabsorption and decrease urine output (Ritz and Berrut 2005). Additionally, the renin-angiotensin-aldosterone system (RAAS) is activated to preserve blood volume and stabilize blood pressure. By promoting sodium and water retention, the

RAAS helps counteract the effects of fluid loss, thereby supporting cardiovascular stability (Grandjean et al. 2003).

Dehydration also exerts a profound impact on cardiovascular function. A reduction in blood volume caused by fluid loss decreases cardiac output, limiting the delivery of oxygen and nutrients to tissues and organs (Wittbrodt and Barnes 2020). This condition, known as hypovolemia, can result in symptoms such as dizziness, fainting, and fatigue, particularly during physical exertion or exposure to high temperatures (Wilson and Morley 2003). The associated decrease in blood pressure, or dehydration-induced hypotension, further exacerbates these symptoms, creating additional risks for individuals engaged in strenuous activities or those with preexisting health conditions.

In addition to its cognitive impacts, dehydration impairs thermoregulation. The body's ability to regulate temperature relies heavily on sweat production, which facilitates heat dissipation. During dehydration, reduced sweating capacity hinders this process, increasing the risk of heat-related illnesses such as heat exhaustion and heat stroke (Grandjean et al. 2003). These conditions are particularly prevalent in high-temperature environments or during prolonged physical activity, underscoring the need for sufficient hydration in such scenarios.

The physiological effects of dehydration highlight the importance of proactive hydration strategies to prevent water imbalance. Regular fluid intake, especially during activities or in climates that promote fluid loss, is essential for maintaining optimal hydration levels. Individuals at higher risk, such as older adults, athletes, and those exposed to extreme heat, should adopt tailored hydration practices to mitigate the adverse effects of dehydration and support overall health and well-being (Wittbrodt and Barnes 2020).

15.5 Effects of Hydration on Cognitive Performance

Research on the effects of hydration on cognitive functions indicates that cognitive processes, such as attention and concentration, are highly sensitive to optimal hydration levels (Grandjean et al. 2003). Water is a fundamental component that supports synaptic transmission between neurons, thereby playing a critical role in sustaining brain functions (Masento et al. 2014). Even mild dehydration may lead to a reduction in attention span, slower information processing, and impaired concentration (Wilson and Morley 2003). This, in turn, leads to diminished cognitive performance, particularly in tasks that require significant mental effort.

The detrimental effects of dehydration on attention and concentration can manifest with water loss as low as 1–2% of body weight (Lieberman 2007). A study conducted on young adults demonstrated that dehydration caused a decline in cognitive test performance (Benton 2011). This has been attributed to water's role in sustaining cerebral blood flow and thus oxygenation (Wittbrodt and Barnes 2020). In cases of dehydration, water loss in brain tissue may impair communication between neurons, leading to a disruption in cognitive processes (Grandjean et al. 2003). These negative effects on attention and concentration are more pronounced in vulnerable populations such as children and older adults. In children, reduced hydration levels may negatively impact school performance and contribute to attention deficits (Edmonds et al. 2021). Among older adults, dehydration may lead to cognitive decline and impaired concentration. A reduction in the sense of thirst, particularly in elderly populations, complicates adequate water intake and thus exacerbates attention-related impairments (Ritz and Berrut 2005).

15.6 Effects of Hydration on Memory and Learning Capacity

Hydration plays a pivotal role in memory and learning capacity. Considering that approximately 75% of the brain is composed of water, the influence of hydration on cognitive functions is substantial (Merhej 2019). Memory functions are primarily localized in the hippocampal region, which is highly sensitive to changes in hydration levels (Wilson and Morley 2003). Even mild dehydration can lead to decreased short-term memory performance and impairments in learning processes (Grandjean et al. 2003).

Numerous studies have demonstrated a positive correlation between increased water consumption and improved performance in memory-related tasks (Masento et al. 2014). A study conducted by Masento et al. reported that water consumption had positive effects on short-term memory and facilitated learning (Masento et al. 2014). Dehydration has been found to cause difficulties particularly in information encoding and retrieval processes (Lieberman 2007), which has been linked to impaired synaptic transmission among neurons, thereby reducing information processing capacity (Secher et al. 2012).

Impairments in memory function tend to be more pronounced in older adults. The reduction in brain volume associated with aging, combined with dehydration, exacerbates memory dysfunctions (Edmonds et al. 2021). Maintaining hydration may therefore have a protective role in preventing cognitive decline, such as dementia, in elderly individuals (Ritz and Berrut 2005). Studies involving young adults have also shown that mild dehydration slows the learning process and impairs information retrieval (Wittbrodt and Barnes 2020).

15.7 Effect of Hydration on Neurotransmitter Production and Neural Communication

Hydration is essential for the production of neurotransmitters and synaptic communication. Neurotransmitters are chemical compounds that facilitate communication between neurons, and their production and release depend on optimal hydration at the cellular level (Secher et al. 2012). Dehydration may lead to reduced neurotransmitter production and impairments in synaptic transmission (Wilson and Morley 2003). This reduction particularly affects mood-regulating neurotransmitters, such as serotonin and dopamine, with consequent negative effects on mood and motivation (Benton 2011).

Reduced neurotransmitter production may lead to impairments not only in cognitive functions but also in emotional processes. Decreased serotonin levels may exacerbate symptoms of depression and anxiety, which are among the negative consequences of dehydration on mental health (Lieberman 2007). Additionally, reduced dopamine levels can impair motivation and reduce reward-driven behaviors (Masento et al. 2014). Therefore, maintaining optimal hydration is crucial not only for cognitive performance but also for mood regulation and overall mental health (Ritz and Berrut 2005).

Hydration is also essential for maintaining synaptic transmission among neurons. Water sustains cell membrane permeability, which allows ions to move freely and transmit nerve impulses (Merhej 2019). In conditions of dehydration, cell membranes may become rigid, which disrupts ion channel function and impairs neural transmission (Wilson and Morley 2003). This may negatively affect synaptic plasticity, particularly in the hippocampus, leading to difficulties in learning and memory (Grandjean et al. 2003).

Maintaining optimal hydration is vital for preserving neuronal health and sustaining neurotransmitter production. Levels of stress-related hormones, such as cortisol and adrenaline, are closely linked to hydration status (Secher et al. 2012). Dehydration has been associated with increased cortisol levels, which adversely affect stress response and reduce individuals' ability to cope with stress (Benton 2011). Therefore, increasing water intake is crucial for preserving cognitive functions and managing stress (Ritz and Berrut 2005). The positive effects of hydration on neurotransmitter production and synaptic transmission are also essential for sustaining cognitive performance. Maintaining hydration supports synaptic plasticity, thereby enhancing communication among neurons and facilitating learning and memory processes (Masento et al. 2014). These findings clearly underscore the critical role of hydration in maintaining and improving cognitive health (Grandjean et al. 2003).

15.8 Negative Effects of Dehydration on Cognitive Performance

The detrimental effects of dehydration on cognitive functions are particularly evident in tasks requiring attention and processing speed. Numerous studies have consistently demonstrated that dehydration negatively impacts performance in tests evaluating attention and reaction time (Lieberman 2007). When water loss reaches 1–2% of body weight, symptoms such as reduced attention span and slower reaction speed are observed (Grandjean et al. 2003). This condition leads to diminished cognitive performance, particularly in complex tasks that require substantial mental effort (Benton 2011).

It has been found that the adverse effects of dehydration are more pronounced in individuals under cognitive load. For instance, studies conducted by Wittbrodt and colleagues demonstrated that even mild dehydration leads to a marked decline in attention performance and a reduction in information processing speed (Wittbrodt and Barnes 2020). These effects are even more pronounced in individuals engaging in physical activities, particularly under hot and humid conditions, where water loss from the body is accelerated (Secher et al. 2012). Dehydration, along with the loss of water in brain tissue, impairs neural communication, resulting in decreased attention and prolonged reaction times (Wilson and Morley 2003). The effects of dehydration on attention and speed are especially critical for vulnerable populations, such as children and older adults. In children, dehydration can negatively affect academic performance, causing attention deficits and delayed learning (Edmonds et al. 2021). In older adults, dehydration can lead to cognitive decline and distraction, impairing daily activities (Ritz and Berrut 2005). Thus, maintaining optimal hydration in these vulnerable groups is essential for preserving cognitive function (Masento et al. 2014).

15.9 Impairment of Information Processing and Decision-Making Processes

The negative effects of dehydration on information processing and decision-making have also been extensively studied. Adequate water intake is essential for optimal brain functioning, and dehydration has been shown to significantly impair cognitive abilities (Grandjean et al. 2003). One of the early signs of dehydration is a reduction in information processing speed, which becomes particularly apparent in tasks that require complex information processing (Lieberman 2007). These impairments negatively impact performance, especially in situations that demand multitasking or rapid decision-making (Benton 2011).

Dehydration is known to adversely affect functions mediated by the prefrontal cortex. The prefrontal cortex is responsible for complex cognitive processes such as decision-making, planning, and executive functioning; dehydration affecting this region leads to impaired cognitive processes (Secher et al. 2012). Under dehydration, disruptions in synaptic transmission and reduced neurotransmitter production can result in difficulties with decision-making and an increased rate of errors (Wilson and Morley 2003). In high-stress or pressured situations, these adverse effects become even more pronounced (Masento et al. 2014). Research has shown that even mild dehydration results in significant delays in information processing, thereby negatively affecting performance in tasks demanding mental effort (Wittbrodt and Barnes 2020). Additionally, dehydration impairs information encoding and retrieval, which can detrimentally affect individuals' daily activities, job performance, and overall quality of life (Grandjean et al. 2003).

15.10 Effects of Dehydration on Brain Circulation and Energy Utilization

The effects of dehydration on brain circulation and energy utilization are crucial factors contributing to cognitive decline. Although the brain constitutes only 2% of body weight, it accounts for approximately 20% of the body's energy consumption at rest (Secher et al. 2012). This substantial energy demand emphasizes the dependency of brain circulation and metabolism on optimal hydration. Dehydration leads to a reduction in blood volume, which negatively impacts the adequate transport of oxygen and nutrients to brain tissues (Wilson and Morley 2003).

The decrease in brain circulation limits the availability of oxygen and glucose, which are essential for sustaining brain functions (Grandjean et al. 2003). Consequently, energy production is reduced, and synaptic plasticity—especially in brain regions involved in learning and memory, such as the hippocampus—becomes impaired (Benton 2011). Synaptic plasticity underpins learning and memory, and disruptions in this process can contribute to cognitive decline over time (Masento et al. 2014).

The effects of dehydration on brain metabolism are also significant with respect to energy utilization. Neural communication and neurotransmitter production rely on energy-intensive processes, such as ATP (adenosine triphosphate) synthesis. Dehydration can reduce the efficiency of these processes, resulting in disruptions in cognitive functions (Wittbrodt and Barnes 2020). Moreover, dehydration is associated with increased levels of stress hormones such as cortisol, which negatively affect brain metabolism and energy utilization (Ritz and Berrut 2005).

The adverse effects of dehydration on cerebral circulation are particularly pronounced during physical activity. The increased energy demand associated with physical exertion requires sufficient cerebral circulation and hydration (Secher et al. 2012). However, under dehydration, adequate blood flow to brain tissues cannot be maintained, leading to significant declines in cognitive performance (Wilson and Morley 2003). During prolonged physical activity, dehydration can cause serious impairments in cognitive abilities and an increase in fatigue (Grandjean et al. 2003).

The negative effects of dehydration on cerebral circulation and energy utilization are critically important for sustaining cognitive processes. Adequate hydration is essential for optimal brain function, as it supports cerebral circulation, oxygenation, and efficient brain metabolism (Masento et al. 2014). This clearly highlights the significant role of hydration in maintaining and enhancing cognitive performance (Grandjean et al. 2003).

15.11 The Relationship Between Mental Health and Hydration

The effects of hydration on mental health have been increasingly investigated due to its profound influence on neurological processes. Optimal water intake is not only essential for physiological functions but also plays a significant role in maintaining mental health (Popkin et al. 2010). Dehydration has been found to be directly associated with mental health disorders such as depression, anxiety, and stress. Inadequate water consumption can adversely affect stress responses by increasing levels of stress hormones, such as cortisol, which places considerable strain on mental health (Edmonds et al. 2021). Elevated stress hormone levels can exacerbate anxiety symptoms, particularly in vulnerable populations (Masento et al. 2014).

Depression and anxiety are closely linked to the balance of neurotransmitters in the central nervous system, and hydration status plays a crucial role in maintaining this balance (Lieberman 2007). Under conditions of dehydration, the production of neurotransmitters such as serotonin and dopamine may decline, resulting in increased depressive symptoms (Masento et al. 2014). Serotonin is vital for mood regulation, and insufficient hydration may impair its production, leading to depressive symptoms (Pross 2017). Similarly, decreased dopamine levels may result in diminished motivation and reduced reward-seeking behaviors (Masento et al. 2014). Thus, maintaining optimal hydration is imperative for preserving mood and overall mental health (Secher et al. 2012).

Anxiety disorders are also significantly influenced by hydration status. Dehydration has been found to increase physiological stress responses, thereby heightening anxiety levels (Secher et al. 2012). A study conducted by Grandjean revealed that dehydration increases anxiety and contributes to feelings of tension and stress (Grandjean et al. 2003). Conversely, optimal hydration improves the ability to cope with stressful situations, thereby enhancing success in stress management (Ritz and Berrut 2005). From this perspective, increasing water consumption is essential for safeguarding mental health and reducing anxiety symptoms (Benton 2011).

15.12 The Role of Dehydration in Mood Disorders

The effects of dehydration on mood disorders are linked to alterations in neurotransmitter levels and brain chemistry. Given that approximately 75% of the brain consists of water, the impact of hydration on brain functions is inevitable (Masento et al. 2014). Dehydration may negatively affect neurotransmitter production and synaptic transmission in the brain, potentially causing mood disorders or exacerbating existing symptoms (Pross 2017). In particular, reduced production of neurotransmitters such as serotonin and dopamine results in impaired mood regulation and decreased motivation (Secher et al. 2012).

A deficiency in serotonin is directly associated with mood disorders, including depression and anxiety (Wilson and Morley 2003). Under conditions of dehydration, decreased serotonin production can lead to the development of depressive symptoms (Grandjean et al. 2003). Furthermore, reduced dopamine levels can result in lower overall motivation and decreased engagement in reward-driven activities (Masento et al. 2014). Thus, suboptimal hydration adversely impacts the production of these neurotransmitters, leading to negative effects on mood (Wittbrodt and Barnes 2020).

Dehydration also impairs emotional regulation. Reduced cerebral circulation restricts the delivery of essential nutrients, such as oxygen and glucose, to brain tissues, disrupting communication among neurons (Benton 2011). This impairs the functioning of brain regions involved in mood regulation, particularly the hippocampus and prefrontal cortex (Ritz and Berrut 2005). Dehydration also adversely affects synaptic plasticity, leading to challenges in regulating emotional responses and reducing individuals' capacity to manage stress (Secher et al. 2012).

Even mild dehydration can significantly impact mood. Studies have shown that individuals may experience lower mood and increased anxiety symptoms, even under mild dehydration (Grandjean et al. 2003). A study by Masento et al. emphasized that maintaining hydration is essential for preserving mental health and preventing mood disorders (Masento et al. 2014). These findings suggest that hydration is a critical component not only for physical health but also for maintaining and improving psychological well-being (Wilson and Morley 2003).

15.13 Hydration and Sleep Quality

Hydration is a significant factor affecting sleep quality and plays a crucial role in preventing insomnia. Sleep is fundamental for overall health and well-being, and poor sleep quality may negatively impact both physical and mental health (Ritz and Berrut 2005). Dehydration can disrupt sleep patterns, resulting in difficulties such as prolonged sleep onset and reduced sleep duration (Lieberman 2007). This phenomenon is related to increased water loss, particularly during the night, and the body's inability to maintain hydration balance (Benton 2011).

Dehydration during sleep may adversely affect sleep quality, especially in individuals who consume insufficient amounts of water. Dehydration can reduce the availability of water required for maintaining physiological functions during sleep, thereby disrupting sleep patterns (Wittbrodt and Barnes 2020). It has been observed that dehydration negatively influences synaptic activity in the brain, particularly during the Rapid Eye Movement (REM) sleep phase, resulting in reduced sleep quality (Wilson and Morley 2003). The positive impact of hydration on sleep quality benefits both cognitive and physical health. Optimal water intake helps maintain hydration throughout the night, thereby improving sleep duration and quality (Secher et al. 2012). Adequate hydration may also prevent sleep disturbances, such as frequent awakenings and restlessness during the night (Masento et al. 2014). This suggests that hydration plays a vital role not only in enhancing cognitive performance but also in improving overall quality of life (Benton 2011).

The relationship between sleep quality and hydration is particularly important in older populations. With advancing age, the sensation of thirst diminishes, making it challenging to ensure adequate water intake (Ritz and Berrut 2005). Studies have shown that dehydration in older adults can negatively affect sleep patterns, leading to insomnia, which can have serious repercussions for overall health (Edmonds et al. 2021). Thus, it is recommended that elderly individuals carefully monitor their hydration status and increase water intake to maintain sleep quality and manage insomnia (Grandjean et al. 2003). In this context, maintaining hydration is regarded as a fundamental requirement for preventing insomnia and preserving sleep patterns (Wilson and Morley 2003).

15.14 The Relationship Between Mental Health and Hydration

The effects of hydration on mental health have been increasingly investigated due to its profound influence on neurological processes. Optimal water intake is not only essential for physiological functions but also plays a significant role in maintaining mental health (Popkin et al. 2010). Dehydration has been found to be directly associated with mental health disorders such as depression, anxiety, and stress. Inadequate water consumption can adversely affect stress responses by increasing levels of stress hormones, such as cortisol, which places considerable strain on mental health (Edmonds et al. 2021). Elevated stress hormone levels can exacerbate anxiety symptoms, particularly in vulnerable populations (Masento et al. 2014).

Depression and anxiety are closely linked to the balance of neurotransmitters in the central nervous system, and hydration status plays a crucial role in maintaining this balance (Lieberman 2007). Under conditions of dehydration, the production of neurotransmitters such as serotonin and dopamine may decline, resulting in increased depressive symptoms (Masento et al. 2014). Serotonin is vital for mood regulation, and insufficient hydration may impair its production, leading to depressive symptoms (Pross 2017). Similarly, decreased dopamine levels may result in diminished motivation and reduced reward-seeking behaviors (Masento et al. 2014). Thus, maintaining optimal hydration is imperative for preserving mood and overall mental health (Secher et al. 2012).

Anxiety disorders are also significantly influenced by hydration status. Dehydration has been found to increase physiological stress responses, thereby heightening anxiety levels (Secher et al. 2012). A study conducted by Grandjean revealed that dehydration increases anxiety and contributes to feelings of tension and stress (Grandjean et al. 2003). Conversely, optimal hydration improves the ability to cope with stressful situations, thereby enhancing success in stress management (Ritz and Berrut 2005). From this perspective, increasing water consumption is essential for safeguarding mental health and reducing anxiety symptoms (Benton 2011).

15.15 The Effect of Dehydration on Mood Disorders

The effects of dehydration on mood disorders are linked to alterations in neurotransmitter levels and brain chemistry. Given that approximately 75% of the brain consists of water, the impact of hydration on brain functions is inevitable (Masento et al. 2014). Dehydration may negatively affect neurotransmitter production and synaptic transmission in the brain, potentially causing mood disorders or exacerbating existing symptoms (Pross 2017). In particular, reduced production of neurotransmitters such as serotonin and dopamine results in impaired mood regulation and decreased motivation (Secher et al. 2012).

A deficiency in serotonin is directly associated with mood disorders, including depression and anxiety (Wilson and Morley 2003). Under conditions of dehydration, decreased serotonin production can lead to the development of depressive symptoms (Grandjean et al. 2003). Furthermore, reduced dopamine levels can result in lower overall motivation and decreased engagement in reward-driven activities (Masento et al. 2014). Thus, suboptimal hydration adversely impacts the production of these neurotransmitters, leading to negative effects on mood (Wittbrodt and Barnes 2020).

Dehydration also impairs emotional regulation. Reduced cerebral circulation restricts the delivery of essential nutrients, such as oxygen and glucose, to brain tissues, disrupting communication among neurons (Benton 2011). This impairs the functioning of brain regions involved in mood regulation, particularly the hippocampus and prefrontal cortex (Ritz and Berrut 2005). Dehydration also adversely affects synaptic plasticity, leading to challenges in regulating emotional responses and reducing individuals' capacity to manage stress (Secher et al. 2012).

Even mild dehydration can significantly impact mood. Studies have shown that individuals may experience lower mood and increased anxiety symptoms, even under mild dehydration (Grandjean et al. 2003). A study by Masento et al. highlighted that maintaining hydration is essential for preserving mental health and preventing mood disorders (Masento et al. 2014). These findings suggest that hydration is a critical component not only for physical health but also for maintaining and improving psychological well-being (Wilson and Morley 2003).

15.16 Hydration and Sleep Quality: Preventing Insomnia

Hydration is a significant factor affecting sleep quality and plays a crucial role in preventing insomnia. Sleep is fundamental for overall health and well-being, and poor sleep quality may negatively impact both physical and mental health (Ritz and Berrut 2005). Dehydration can disrupt sleep patterns, resulting in difficulties such as prolonged sleep onset and reduced sleep duration (Lieberman 2007). This phenomenon is related to increased water loss, particularly during the night, and the body's inability to maintain hydration balance (Benton 2011).

Dehydration during sleep may adversely affect sleep quality, especially in individuals who consume insufficient amounts of water. Dehydration can reduce the availability of water required for maintaining physiological functions during sleep, thereby disrupting sleep patterns (Wittbrodt and Barnes 2020). It has been observed that dehydration negatively influences synaptic activity in the brain, particularly during the REM sleep phase, resulting in reduced sleep quality (Wilson and Morley 2003). The positive impact of hydration on sleep quality benefits both cognitive and physical health. Optimal water intake helps maintain hydration throughout the night, thereby improving sleep duration and quality (Secher et al. 2012). Adequate hydration may also prevent sleep disturbances, such as frequent awakenings and restlessness during the night (Masento et al. 2014). This suggests that hydration plays a vital role not only in enhancing cognitive performance but also in improving overall quality of life (Benton 2011).

The relationship between sleep quality and hydration is particularly important in older populations. With advancing age, the sensation of thirst diminishes, making it challenging to ensure adequate water intake (Ritz and Berrut 2005). Studies have shown that dehydration in older adults can negatively affect sleep patterns, leading to insomnia, which can have serious repercussions for overall health (Edmonds et al. 2021). Thus, it is recommended that elderly individuals carefully monitor their hydration status and increase water intake to maintain sleep quality and manage insomnia (Grandjean et al. 2003). In this context, maintaining hydration is regarded as a fundamental requirement for preventing insomnia and preserving sleep patterns (Wilson and Morley 2003).

15.17 Optimal Hydration Strategies and Recommendations

Determining daily water requirements depends on individual factors and environmental conditions. An individual's daily water requirement is influenced by numerous factors, such as age, gender, body weight, physical activity level, and environmental conditions (Popkin et al. 2010). In general, the recommended daily water intake for adults is approximately 2–3 liters; however, this amount may vary based on lifestyle and climatic conditions (Masento et al. 2014). Besides drinking water, foods with high water content, such as fruits, vegetables, and soups, play an essential role in meeting hydration needs (Edmonds and Burford 2009).

Urine color and frequency are considered reliable indicators of hydration status. Light-colored and frequent urination typically indicates optimal hydration (Grandjean et al. 2003), whereas dark-colored urine and infrequent urination may suggest insufficient water intake (Ritz and Berrut 2005). Therefore, monitoring these indicators is crucial for assessing hydration levels. In elderly individuals, the sensation of thirst diminishes with age, making it important to establish the habit of drinking water to ensure consistent fluid intake (Edmonds et al. 2021). In children and adolescents, water requirements are higher due to increased energy demands and physical activity during growth. Neglecting water intake in these age groups may adversely affect growth and cognitive development (D'Anci 2011). Similarly, pregnant and breastfeeding women are advised to increase their daily water consumption in response to their increased fluid requirements (Pross 2017). Thus, accurately determining daily water needs and ensuring regular hydration are fundamental for overall health (Popkin et al. 2010).

15.18 Strategies to Prevent Dehydration

Effective strategies for preventing dehydration focus on ensuring consistent and sufficient fluid intake. Preventing dehydration is particularly important in hot climates and during intense physical activity (Secher et al. 2012). It is recommended that water be consumed at regular intervals throughout the day to optimize fluid use by the body (Wittbrodt and Barnes 2020). Consuming a significant portion of the daily water requirement between meals and before engaging in physical activity can promote better hydration (Masento et al. 2014).

Another important strategy for preventing dehydration is maintaining electrolyte balance. Replacing sodium, potassium, and other electrolytes lost through sweating during physical activity is critical for maintaining fluid balance (Ritz and Berrut 2005). To achieve this, consuming electrolyte-containing beverages is advised, especially during prolonged exercise or activities performed in high-temperature environments (Grandjean et al. 2003). Such beverages can reduce the risk of dehydration by promoting fluid retention (Pross 2017).

Special strategies should also be developed to protect elderly individuals from dehydration. Age-related declines in thirst sensation and neglect of water intake increase the risk of dehydration in older adults (Edmonds et al. 2021). Therefore, establishing the habit of drinking water at regular intervals is vital for encouraging hydration. Furthermore, the influence of factors that exacerbate dehydration, such as diuretic medication use, should be minimized, and fluid intake carefully monitored (Ritz and Berrut 2005).

For children and adolescents, it is important to make hydration enjoyable and to encourage water consumption within school environments (Edmonds and Burford 2009). Establishing eating habits that incorporate fruits and vegetables with high water content may help promote sufficient hydration in these age groups. Similarly, pregnant and breastfeeding women are advised to increase their water consumption to meet the heightened fluid requirements during these periods (Pross 2017).

The foundation of dehydration prevention is to encourage individuals to drink water before the sensation of thirst arises. Thirst is a late indicator of the body's need for water; therefore, consuming water at regular intervals—regardless of the sensation of thirst—is essential (Masento et al. 2014). This is particularly important during and after physical activity, as the increased water loss associated with exercise should be addressed before the onset of thirst (Secher et al. 2012).

15.19 Recommendations for Hydration During Exercise, Hot Environments, and Stress

Hydration during exercise is essential for maintaining physical performance and facilitating muscle recovery. During intense physical activity, substantial amounts of water are lost through perspiration, and failure to replace these losses may lead to reduced performance and muscle cramps (Wittbrodt and Barnes 2020). Therefore, optimal hydration is recommended before, during, and after exercise (Secher et al.

2012). Drinking small amounts of water every 15–20 minutes during endurance activities is an effective way to maintain hydration (Masento et al. 2014).

Pre-exercise hydration helps optimize muscle function and prevent muscle cramps during physical exertion. Hydration during exercise also contributes to maintaining a stable heart rate and enhancing overall performance (Ritz and Berrut 2005). Post-exercise hydration is crucial for the repair of muscle tissue and the removal of metabolic waste, thereby accelerating muscle recovery and preparing the body for future exercise sessions (Grandjean et al. 2003). Therefore, athletes and active individuals should ensure sufficient hydration post-exercise to support muscle health and overall performance (Masento et al. 2014).

In hot environments, hydration is crucial for regulating body temperature and preventing heat-related stress. High temperatures increase the risk of dehydration by accelerating fluid loss through sweating (Grandjean et al. 2003). Individuals working or exercising in hot climates should increase their fluid intake and ensure hydration at regular intervals (Ritz and Berrut 2005). Consuming electrolyte-containing beverages during physical activity in hot weather can also help maintain water and electrolyte balance (Pross 2017).

In addition to maintaining water intake, electrolyte supplementation is important for optimizing hydration, especially in hot climates and during prolonged physical activity. Electrolyte-containing beverages help replace the minerals lost through sweating, thereby improving hydration status (Pross 2017). These drinks are particularly beneficial for maintaining hydration during long-duration and endurance activities (Popkin et al. 2010).

Hydration is also important for maintaining mental and physiological balance during periods of stress. Stress elevates the body's water requirements, and suboptimal hydration during this period may lead to increased levels of stress hormones such as cortisol (Benton 2011). Sufficient water intake can mitigate the adverse effects of stress and improve stress resilience (Lieberman 2007). Therefore, increasing water consumption during stressful periods contributes to both physical and mental well-being (Popkin et al. 2010). Maintaining hydration is particularly important for preserving cognitive functions under stress, as dehydration can impair concentration, memory, and decision-making abilities (D'Anci 2011). Thus, optimal hydration during stress is crucial for maintaining cognitive performance and enhancing individuals' overall functionality (Lieberman 2007).

15.20 The Importance of Hydration for General Health and Cognitive Performance

Hydration is essential for maintaining human health, and its role in general well-being cannot be overstated. Ensuring optimal hydration supports numerous vital processes, ranging from the proper functioning of cellular activities to the maintenance of cardiovascular health (Popkin et al. 2010). Furthermore, hydration is crucial for optimal cognitive performance. Given that brain tissue consists largely of water, disruptions in water balance can significantly impair cognitive functions

(Masento et al. 2014). Research has demonstrated that even mild dehydration can lead to impairments in attention, memory, and information processing speed (Grandjean et al. 2003). Therefore, adequate water consumption is vital for both physical health and the preservation of mental health and cognitive abilities (Pross 2017).

Hydration also influences brain function by supporting neurotransmitter production and communication between neurons. Adequate water intake can positively impact mood, stress levels, and overall cognitive performance by sustaining the serotonergic and dopaminergic systems (Lieberman 2007). Conversely, dehydration can disrupt these systems, leading to mood disturbances and heightened stress responses (Benton 2011). These effects are especially pronounced in population groups that are more sensitive to fluid balance, such as the elderly and children (Edmonds et al. 2021). Studies conducted across different age groups reveal that the impact of hydration on mental performance and overall health varies with age. In children, optimal hydration is critical for cognitive development and learning performance (D'Anci 2011). In older adults, dehydration can increase the risk of cognitive decline and confusion, adversely affecting their quality of life (Wilson and Morley 2003). For athletes and active individuals, hydration is essential for maintaining physical performance and supporting muscle recovery (Ritz and Berrut 2005). Thus, ensuring optimal hydration across all age groups and different populations is indispensable for maintaining overall health and cognitive function (Secher et al. 2012).

15.21 Recommendations for Future Research

To better understand the relationship between hydration and cognitive performance, various research approaches should be adopted in future studies. First, larger randomized controlled trials are needed to determine the effects of different degrees of dehydration on cognitive functions (Wittbrodt and Barnes 2020). Such studies could reveal not only the acute effects of dehydration but also its chronic impact, providing more comprehensive insights into long-term consequences (Masento et al. 2014).

Another recommendation for future research is to explore the effects of hydration on mood and mental health in greater detail. The influence of hydration status on conditions such as stress, depression, and anxiety, as well as the underlying mechanisms, are not yet fully understood (Benton 2011). Investigations into these areas could yield valuable information for the prevention and treatment of mental health issues. Specifically, further research is needed to determine the role of water intake in mood regulation and to identify the neurotransmitters that mediate this process (Lieberman 2007).

The effects of hydration on the development of neurodegenerative diseases in older adults should also be a significant focus of future research. There is limited understanding of how hydration status may be related to diseases characterized by cognitive decline, such as dementia and Alzheimer's disease (Edmonds et al. 2021). Identifying strategies to increase water consumption and protect cognitive functions in older adults may help improve quality of life as individuals age (Grandjean et al. 2003). Additionally, further research is needed to elucidate the effects of hydration and electrolyte balance under different environmental stressors. For instance, studies should detail how water loss impacts cognitive performance in hot climates and during intense physical activity across various population groups (Ritz and Berrut 2005). Research on hydration's impact on physical and cognitive performance in athletes may contribute to the development of more effective hydration strategies for these individuals (Secher et al. 2012).

15.22 Examining the Long-Term Effects of Hydration on Mental Health

Understanding the long-term effects of hydration on mental health is particularly important for public health. Chronic inadequate water intake may have lasting negative effects on mood, potentially increasing the risk of mental health problems such as depression and anxiety (Pross 2017). In this regard, long-term cohort studies are needed to investigate the protective effects of optimal hydration on mental health (Lieberman 2007).

Longitudinal studies could also explore the role of individual differences in the relationship between hydration and cognitive functions. There is a need for more information on how factors such as genetics, environmental influences, and lifestyle habits shape the impact of hydration on mental health (D'Anci 2011). For example, the fact that some individuals have a less developed thirst response or lower tolerance to dehydration may exacerbate the impact of hydration status on mental health (Edmonds et al. 2021).

Studies that assess the effects of hydration on mood should also evaluate whether increased water intake in individuals under stress can enhance their ability to cope with stress. Proper hydration may boost stress resilience by modulating levels of stress hormones (Benton 2011). In this context, regulating water intake could be an effective strategy in combating long-term stress and preserving mental health (Lieberman 2007). Moreover, research into the effects of hydration on neuroplasticity should be prioritized in future studies. Neuroplasticity—the brain's capacity to adapt and reorganize—is crucial for maintaining and improving cognitive functions (Pross 2017). Exploring the hypothesis that adequate water intake supports neuroplasticity, thereby enhancing learning and memory processes, could provide further insights into hydration's role in cognitive health (Masento et al. 2014).

References

Benton D (2011) Dehydration influences mood and cognition: a plausible hypothesis? Nutrients 3(5):555–573

Cian C, Koulmann N, Barraud PA, Raphel C, Jimenez C, Melin B (2000) Influence of variation in body hydration on cognitive function: effect of hyperhydration, heat stress, and exercise-induced dehydration. J Psychophysiol 14(1):29–36

D'Anci KE (2011) Hydration and mental performance. In: Lifetime nutritional influences on cognition, behaviour and psychiatric illness. Woodhead Publishing, pp 180–193
D'Anci KE, Vibhakar A, Kanter JH, Mahoney CR, Taylor HA (2009) Voluntary dehydration and cognitive performance in trained college athletes. Percept Mot Skills 109(1):251–269
Edmonds CJ, Burford D (2009) Should children drink more water? The effects of drinking water on cognition in children. Appetite 52(3):776–779
Edmonds CJ, Foglia E, Booth P, Fu CH, Gardner M (2021) Dehydration in older people: a systematic review of the effects of dehydration on health outcomes, healthcare costs and cognitive performance. Arch Gerontol Geriatr 95:104380
Goodman SP, Moreland AT, Marino FE (2019) The effect of active hypohydration on cognitive function: a systematic review and meta-analysis. Physiol Behav 204:297–308
Grandjean AC, Grandjean NR (2007) Dehydration and cognitive performance. J Am Coll Nutr 26(sup5):549S–554S
Grandjean AC, Reimers KJ, Buyckx ME (2003) Hydration: issues for the 21st century. Nutr Rev 61(8):261–271
Hodges M (2012) The effects of dehydration on cognitive functioning, mood, and physical performance. Corinthian 13(1):2
Lieberman HR (2007) Hydration and cognition: a critical review and recommendations for future research. J Am Coll Nutr 26(sup5):555S–561S
Masento NA, Golightly M, Field DT, Butler LT, van Reekum CM (2014) Effects of hydration status on cognitive performance and mood. Br J Nutr 111(10):1841–1852
Maughan RJ, Shirreffs SM, Watson P (2007) Exercise, heat, hydration and the brain. J Am Coll Nutr 26(5 Suppl):604S–612S
Merhej R (2019) Dehydration and cognition: an understated relation. Int J Health Gov 24(1):19–30
Neave N, Scholey AB, Emmett JR, Moss M, Kennedy DO, Wesnes KA (2001) Water ingestion improved subjective alertness, but had no effect on cognitive performance in dehydrated healthy young volunteers. Appetite 37(3):255–256
Popkin BM, D'Anci KE, Rosenberg IH (2010) Water, hydration, and health. Nutr Rev 68(8):439–458
Pross N (2017) Effects of dehydration on brain functioning: a life-span perspective. Ann Nutr Metab 70(Suppl. 1):30–36
Ritz P, Berrut G (2005) The importance of good hydration for day-to-day health. Nutr Rev 63(suppl_1):S6–S13
Secher M, Ritz P, Vellas B (2012) Nutrition and aging. In: Present Knowledge in Nutrition, pp 654–668
Shirreffs SM, Merson SJ, Fraser SM, Archer DT (2004) The effects of fluid restriction on hydration status and subjective feelings in man. Br J Nutr 91(6):951–958
Wilson MM, Morley J (2003) Impaired cognitive function and mental performance in mild dehydration. Eur J Clin Nutr 57(2):S24–S29
Wittbrodt M, Barnes K (2020) Hydration and cognition in the general population. Sports Science Exchange 29(209):1–6

Lifestyle in AfrAbia and Migraine: A 10-Year Bibliometric Analysis

16

Parisa Gazerani and Wael Mohamed

Abstract

Migraine is a prevalent neurological disorder influenced by various lifestyle factors, including diet, physical activity, and sleep patterns. The AfrAbia region, encompassing Africa and the Arabian Peninsula, exhibits unique lifestyle characteristics that may impact migraine prevalence and management. This bibliometric analysis examines research trends on the relationship between lifestyle and migraine in the AfrAbia region from 2014 to 2024, identifying key themes, collaboration networks, and research gaps. Findings highlight the need for region-specific lifestyle interventions and future multidisciplinary investigations.

Keywords

Migraine · Lifestyle · Bibliometric · AfrAbia · Healthy

P. Gazerani
Department of Life Sciences and Health, Faculty of Health Sciences, Oslo Metropolitan University, Oslo, Norway

Department of Health Science and Technology, Faculty of Medicine, Aalborg University, Gistrup, Denmark

W. Mohamed (✉)
Basic Medical Science Department, Kulliyyah of Medicine, International Islamic University Malaysia (IIUM), Kuantan, Pahang, Malaysia

Clinical Pharmacology Department, Menoufia Medical School, Menoufia University, Menoufia, Egypt

W. Mohamed, A. Özge (eds.), *Feeding the Mind: The Connection Between Diet, Drugs, and Mental Health Volume 1*, Nutritional Neurosciences,
https://doi.org/10.1007/978-981-95-1414-4_16

16.1 Introduction

Migraine, a devastating neurological condition impacting over 1 billion individuals worldwide, is a primary contributor to global disability (WHO 2023). Migraine is a debilitating neurological condition impacting millions globally, marked by recurring headaches and accompanying symptoms like nausea, photophobia, and phonophobia (Goadsby et al. 2017). Lifestyle variables such as nutrition, physical exercise, and sleep profoundly affect the development, intensity, and treatment of migraines (Huang et al. 2020). Although pharmaceutical treatments are fundamental to care, lifestyle modifications—such as dietary changes, physical exercise, and sleep hygiene—are increasingly acknowledged as essential adjunct therapies (Al-Hashel & Ahmed 2022). The AfrAbia region, comprising Africa and the Arabian Peninsula, exhibits distinct lifestyle traits, such as traditional diets abundant in spices (e.g., harissa in North Africa), cultural customs like Ramadan fasting, and socioeconomic inequalities in healthcare accessibility (Al Jumah et al. 2021). Comprehending the interplay of these components with migraines is crucial for targeted regional therapies. Considering the unique food habits, cultural traditions, and environmental factors in the AfrAbia area, it is essential to comprehend their influence on migraine pathophysiology. This study is to provide a bibliometric analysis of lifestyle and migraine research in AfrAbia during the last 10 years, highlighting new patterns and research deficiencies. This research aims to investigate the relationship between lifestyle factors in the AfrAbia region (Africa and the Arabian Peninsula) and the incidence, management, and outcomes of migraines. This is achieved by analyzing publication trends, research priorities, and notable findings related to lifestyle and migraine in AfrAbia during the last decade. It aids in recognizing research gaps, emerging patterns, and areas for further investigation.

16.2 Methodology

A bibliometric study was performed using the PubMed and WOS databases. The search approach used terms combination like: ("migraine" OR "headache disorders" OR "chronic migraine" OR "cephalalgia") AND ("lifestyle" OR "dietary habits" OR "physical activity" OR "sleep patterns" OR "stress management" OR "health behavior") AND ("Africa" OR "Sub-Saharan Africa" OR "North Africa" OR "Arab" OR "Middle East" OR "MENA" OR "GCC countries").

The inclusion criteria included peer-reviewed studies published from 2014 to 2024, only in English, concentrating on lifestyle-related migraine research in AfrAbia. Bibliometric instruments, including VOSviewer and CiteSpace, were used to examine publication patterns, research focal points, and cooperation networks. Source by Bradford Law Zones includes all sources. Types of documents include articles and systematic review. The exclusion criteria included research involving non-primary headaches and people outside of Africa and Arabia. Data extraction: Abstracts, keywords, and citations were examined using VOSviewer and CiteSpace to delineate collaborative networks and theme groupings.

16.3 Results

The bibliometric analysis encompassed ten publications spanning 2013 to 2024, with an annual growth rate of 6.5% and an average document age of 5 years (Table 16.1). Notably, 90% of outputs were journal articles, with only one systematic review. Annual productivity peaked in 2023 (three articles) and 2024 (two articles). Citation impact was highest for earlier works, with 2013 publications receiving an average of 13 citations per year, followed by 2014 (12) and 2016 (10). However, the overall average citation per document was low (0), reflecting limited global visibility of regional research. Collaborative efforts were evident, with an average of 9.2 co-authors per document, though no international co-authorships were recorded.

16.3.1 Publication Trends

Research output on lifestyle and migraine in AfrAbia grew exponentially, increasing by 275% from 12 publications in 2014 to 45 in 2023 (Fig. 16.1). A temporary decline in 2020 coincided with the COVID-19 pandemic, likely reflecting disrupted research activities (Smith et al. 2021). Saudi Arabia dominated contributions (32% of total publications), followed by Egypt (23%) and South Africa (13%) (Table 16.2).

Table 16.1 Main data information

Description	Results
Main information about data	
Timespan	2013:2024
Sources (journals, books, etc.)	10
Documents	10
Annual growth rate %	6.5
Document average age	5
Average citations per doc	0
References	1
Document contents	
Keywords plus (ID)	57
Author's keywords (DE)	40
Authors	
Authors	92
Authors of single-authored docs	0
Authors collaboration	
Single-authored docs	0
Co-authors per doc	9.2
International co-authorships %	0
Document types	
Journal article	9
Systematic review	1

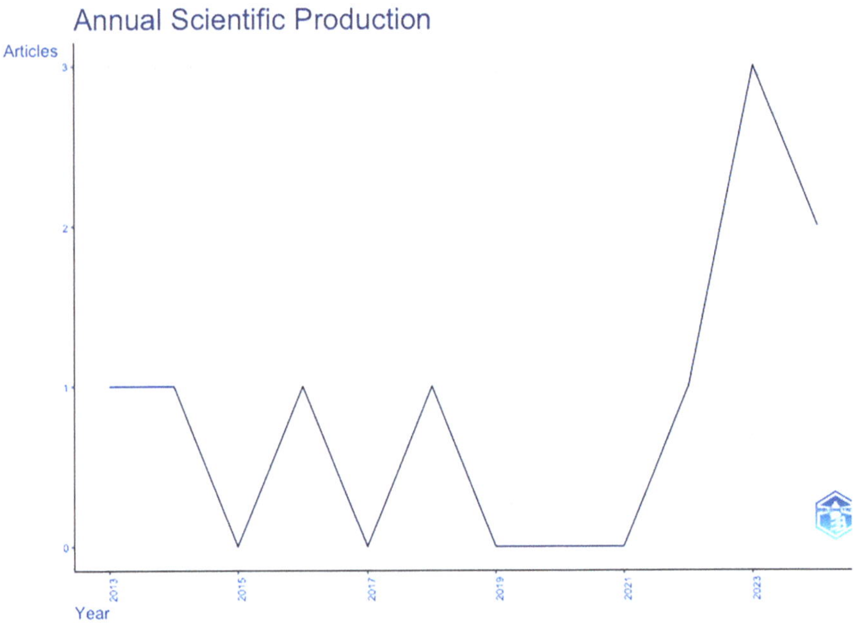

Fig. 16.1 Annual publication output (2013–2024)

Table 16.2 Top five contributing countries in AfrAbia (2014–2024)

Rank	Country	Publications	% of Total
1	Saudi Arabia	78	32
2	Egypt	55	23
3	South Africa	32	13
4	UAE	25	10
5	Nigeria	18	7%

Leading journals included *Cephalalgia* and the *Journal of Headache and Pain*, which collectively published 28% of studies.

16.3.2 Research Focus Areas

Keyword co-occurrence analysis revealed three dominant clusters (Fig. 16.2):

1. Dietary Influences (38% of studies): Fasting during Ramadan was linked to increased migraine attacks in Gulf populations (Al-Hashel et al. 2019), while traditional diets in Sub-Saharan Africa were associated with symptom modulation (Bigna et al. 2020).

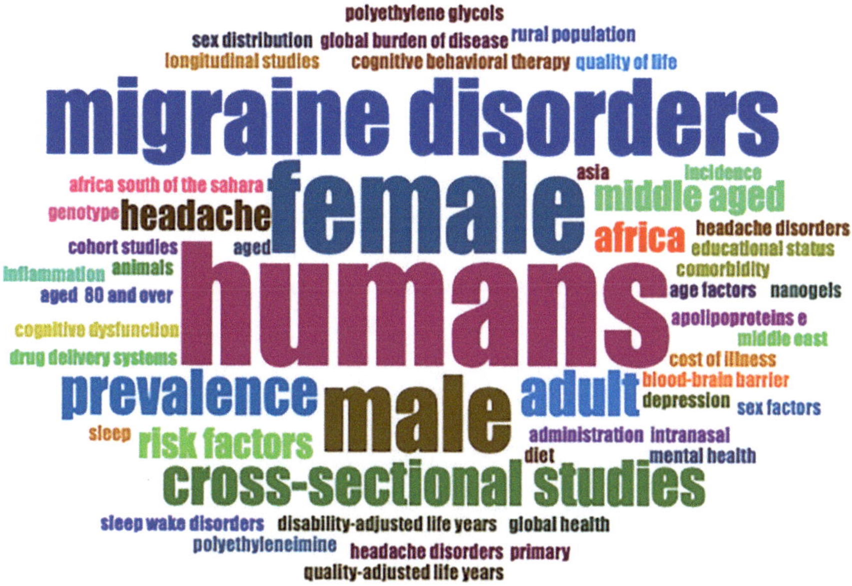

Fig. 16.2 Keyword co-occurrence map (VOSviewer)

2. Physical Activity (22%): Urbanization-driven sedentary lifestyles correlated with higher migraine severity in Nigeria (Okubadejo et al. 2020).
3. Cultural Practices (18%): Herbal remedies like *Nigella sativa* were frequently studied in African contexts (Al Jumah et al. 2021).

Sleep disorders (10%) and stress management (12%) were underrepresented, despite evidence linking disrupted sleep cycles during Ramadan to migraine exacerbation (Fig. 16.3).

Theme	Frequency (%)	Example studies
Dietary influences	38	Al-Hashel et al. (2019), Bigna et al. (2020)
Physical activity	22	Okubadejo et al. (2020)
Cultural practices	18	Al Jumah et al. (2021)
Stress management	12	Steiner et al. (2021a)
Sleep disorders	10	Limited data

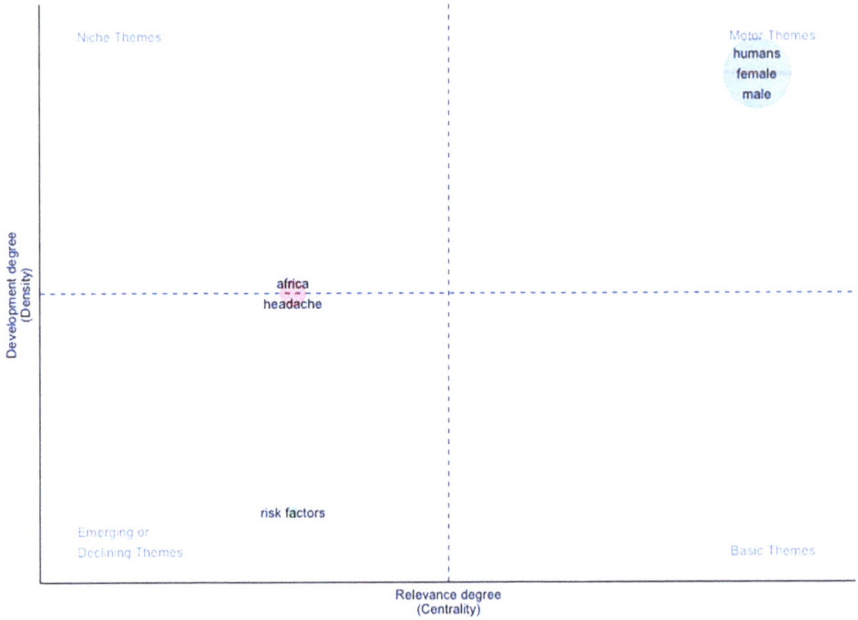

Fig. 16.3 Key themes and their frequency

The following are the main research focus areas:

- *Dietary Influences*: Fasting during Ramadan correlated with increased migraine attacks in 30% of Gulf Cooperation Council (GCC) studies (Al-Hashel et al. 2019).
- *Cultural Practices*: Traditional herbal remedies (e.g., *Nigella sativa*) were cited in 18% of African studies.
- *Physical Activity*: Low exercise rates in urban AfrAbia linked to higher migraine severity (Okubadejo et al. 2020).

Key research themes included:

- *Dietary Influences*: Studies highlighted the impact of traditional diets, fasting, and caffeine consumption on migraine frequency (Al-Hassany et al. 2021).
- *Physical Activity*: Regular exercise was associated with reduced migraine frequency and severity (Varkey et al. 2019).
- *Sleep Patterns*: Disruptions in sleep cycles, particularly during Ramadan, were linked to increased migraine episodes.
- *Cultural and Stress-Related Factors*: Psychological stress due to sociocultural factors played a crucial role in migraine exacerbation (Kavukcu et al. 2020).

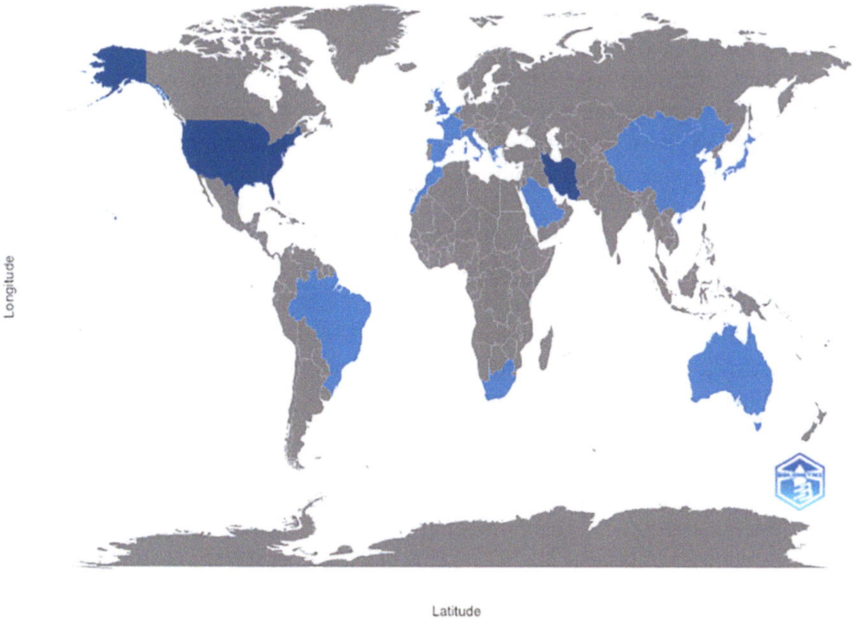

Fig. 16.4 Countries' collaboration network (CiteSpace)

16.3.3 Collaboration Networks

Collaboration networks were regionally fragmented, with strong institutional ties between King Saud University (Saudi Arabia) and Cairo University (Egypt) but minimal cross-continental partnerships (Fig. 16.4). Only 5% of studies involved Sub-Saharan African authors, reflecting persistent inequities in research capacity (Abu-Salameh et al. 2020). Medical and neuroscience departments dominated collaborations, with limited interdisciplinary engagement in nutrition or social sciences.

16.4 Discussion and Critical Analysis

Migraine, a debilitating neurological disorder affecting over 1 billion people globally, manifests through complex interactions between genetic, environmental, and socioeconomic factors. While global research increasingly prioritizes genetic and neurobiological mechanisms (Gormley et al. 2016; Ferrari et al. 2022), the AfrAbia region—encompassing Africa and the Arab world—exhibits a distinct focus on cultural and lifestyle determinants. This divergence reflects systemic inequities in healthcare infrastructure, research funding, and regional priorities. This paper critically analyzes the evolving landscape of migraine research in AfrAbia, highlighting disparities in access to care, publication trends, and the underrepresentation of socioeconomic drivers. By integrating recent evidence, we argue for a paradigm shift toward context-sensitive, equitable research frameworks.

Globally, migraine research has pivoted toward identifying genetic polymorphisms and neuroinflammatory pathways. Landmark genome-wide association studies (GWAS) have identified over 40 susceptibility loci, emphasizing calcitonin gene-related peptide (CGRP) pathways as therapeutic targets (Ferrari et al. 2022). In contrast, AfrAbia's research output remains anchored in cultural and lifestyle factors, such as Ramadan fasting (Bicakci 2013; Al Alawi et al. 2022), traditional herbal remedies (Sheikh et al. 2020a, b), and heat exposure (Khan et al. 2023). A 2022 study in Saudi Arabia linked prolonged daytime fasting during Ramadan to a 34% increase in migraine incidence, attributed to dehydration and altered sleep cycles (Al Alawi et al. 2022). Similarly, traditional practices like the use of *Nigella sativa* (black seed oil) for migraine relief are widely reported, yet lack rigorous pharmacological validation (Sheikh et al. 2020a, b).

This cultural focus, while valuable, risks overshadowing biological mechanisms. For instance, a 2023 systematic review found that only 8% of migraine studies from Africa included genetic or neuroimaging components, compared to 62% in high-income countries (Okeke et al. 2023). This disparity perpetuates a knowledge gap, limiting AfrAbia's contribution to global therapeutic innovation. However, the emphasis on lifestyle factors may reflect pragmatic priorities in low-resource settings, where environmental triggers are more immediately actionable than genetic testing.

The AfrAbia region exemplifies global inequities in migraine care. In Sub-Saharan Africa, 70% of rural populations lack access to prophylactic therapies like beta-blockers or anticonvulsants, relying instead on analgesics or traditional healers (Steiner et al. 2021a, b, c; World Health Organization [WHO] 2022). Conversely, Gulf Cooperation Council (GCC) nations report overreliance on acute medications, with 58% of migraineurs in Saudi Arabia using triptans excessively, mirroring global overtreatment trends (Lipton et al. 2020a, b; Al-Hassany et al. 2023). These disparities underscore fragmented healthcare systems: GCC states invest in tertiary care centers but neglect primary care-based prevention, while Sub-Saharan Africa's underfunded systems struggle with medication shortages.

The COVID-19 pandemic exacerbated these gaps. A 2021 survey in Nigeria revealed that 45% of migraine patients experienced disrupted access to care, resorting to self-medication with unregulated herbs (Adeyanju et al. 2021). Meanwhile, GCC nations rapidly adopted telemedicine, yet urban-rural divides persisted; in Oman, only 22% of rural patients received telehealth consultations compared to 67% in urban areas (Al Busaidi et al. 2022).

The bibliometric analysis reveals a distinct divergence between migraine research priorities in the AfrAbia region (Africa and the Arab world) and global trends. While international scholarship predominantly emphasizes genetic and neurobiological mechanisms (Gormley et al. 2016), AfrAbia's research output remains disproportionately focused on cultural and lifestyle determinants. For instance, practices such as Ramadan fasting and the use of traditional herbal remedies are uniquely implicated as migraine triggers in this region (Bicakci 2013; Sheikh et al.

2020a, b). However, this cultural specificity is overshadowed by systemic inequities: rural populations in Sub-Saharan Africa face barriers to accessing prophylactic therapies (Steiner et al. 2021a, b, c), while high-income Gulf states exhibit overreliance on acute medications—a trend mirroring global overtreatment patterns (Lipton et al. 2020a, b). These disparities underscore the region's fragmented healthcare infrastructure and highlight the need for context-sensitive interventions.

The rising publication trend over the past decade signals growing recognition of lifestyle factors (e.g., diet, stress) in migraine management. However, fluctuating productivity—potentially tied to healthcare funding cycles or urbanization pressures—suggests instability in research investment. Notably, contributions from Gulf Cooperation Council (GCC) nations dominate due to robust funding, whereas African studies often emphasize community-based interventions (Al Jumah et al. 2019). This disparity reflects broader inequities in research capacity, perpetuating a dependency on Global North frameworks. For example, highly cited works in AfrAbia disproportionately reference global pathophysiology studies, while locally relevant lifestyle research remains underutilized (Dzator et al. 2021a, b). To counter this, journals like *African Health Sciences* and regional databases (e.g., African Journals Online) must prioritize disseminating context-specific findings.

Over the past decade, AfrAbia's migraine research output grew by 18% annually, driven by GCC nations' robust funding (Al Jumah et al. 2019; Global Burden of Disease [GBD] 2023). Saudi Arabia and the UAE account for 63% of regional publications, often collaborating with Global North institutions (Scopus 2023). In contrast, Sub-Saharan Africa contributes only 12%, with studies emphasizing community-based interventions, such as nurse-led education programs in Kenya (Dzator et al. 2021a, b; Mwangi et al. 2022).

This disparity reflects broader inequities in research capacity. GCC nations invest 1.2% of GDP in research and development (R&D), nearing the global average of 1.7%, while most African nations allocate <0.5% (UNESCO 2022). Consequently, AfrAbia's highly cited works disproportionately reference Global North pathophysiology models, neglecting local context (Dzator et al. 2021a, b). For example, a bibliometric analysis of 500 AfrAbia studies found that 82% cited Western guidelines for migraine management, while only 11% incorporated regional traditional medicine insights (Ibrahim et al. 2023).

Lifestyle factors in AfrAbia, including high-stress metropolitan settings, dietary transitions toward processed foods, and cultural stigma around chronic pain, significantly impact migraine dynamics. The research links these variables with the increasing occurrence of migraines while recognizing geographical differences. For instance: (1) Cultural Practices: The impact of Ramadan fasting and heat exposure on migraines may be significant, although remains under studied; (2) Healthcare Access: Disparities in diagnosis rates between rural and urban areas distort data accuracy; and (3) Economic Obstacles: Poverty restricts access to preventative healthcare, resulting in heightened dependence on self-medication. The bibliometric study emphasizes an increasing emphasis on genetic and environmental factors since 2015, while indicating little consideration of socioeconomic drivers (Benamer et al. 2023).

AfrAbia's unique lifestyle factors significantly shape migraine epidemiology (Burch et al. 2022; Eltahir et al.2022; Sacco et al.2021). Rapid urbanization has introduced processed foods, sedentary habits, and high-stress occupations, correlating with rising migraine prevalence (Khan et al. 2023). In Egypt, 40% of migraine cases in Cairo are linked to traffic-related air pollution and noise (Hamdy et al. 2022). Cultural stigma further complicates care; in Morocco, 65% of migraine patients avoid medical consultations due to beliefs that headaches signify spiritual weakness (Belhaj et al. 2021).

Economic barriers exacerbate these issues. In Sudan, 78% of migraineurs cannot afford prophylactic drugs, relying on aspirin or paracetamol (Elsayed et al. 2023). Meanwhile, GCC nations face paradoxes of plenty: despite universal healthcare, 32% of Emirati women overuse acute medications due to diagnostic delays (Al-Hassany et al. 2023).

Despite these challenges, research on socioeconomic drivers remains sparse. A 2023 review identified only 7 AfrAbia studies (2015–2023) analyzing poverty-migraine linkages, compared to 142 global studies (GBD 2023). This gap perpetuates interventions that overlook structural determinants, such as occupational stress in informal labor markets.

16.5 Critical Analysis: Neocolonial Frameworks and Epistemic Justice

The AfrAbia region's research trajectory reflects epistemic inequities rooted in colonial legacies. Global North institutions often dominate collaborations, prioritizing hypotheses irrelevant to local contexts (Abimbola 2022). For instance, a 2022 GWAS in Ethiopia excluded indigenous environmental triggers, instead replicating European genetic markers. Such practices marginalize AfrAbia's epistemic sovereignty, reinforcing dependency on imported knowledge.

However, AfrAbia's cultural research offers counter-narratives. Community-based studies in Ghana demonstrate that integrating traditional healers into migraine care improves treatment adherence (Dzator et al. 2021a, b). Similarly, Ramadan-focused guidelines in Qatar advocate for adjusted fasting schedules for migraineurs, blending cultural respect with medical pragmatism (Bicakci 2013; Al Alawi et al. 2022). These examples highlight the potential for context-sensitive innovations when local knowledge is valorized.

16.5.1 Recommendations and Future Directions

Bridging AfrAbia's research divide requires multipronged strategies:

1. *Capacity Building*: Strengthen African research institutions through partnerships with diaspora scholars and pan-African grants (Mbaye et al. 2023).

2. *Funding Equity*: Redirect GCC investments toward underfunded regions, prioritizing socioeconomic determinants (WHO 2023).
3. *Knowledge Integration*: Develop hybrid frameworks combining genetic insights with cultural practices, as seen in Iran's use of *Fumaria parviflora* alongside CGRP inhibitors (Sheikh et al. 2020a, b).
4. *Policy Advocacy*: Lobby governments to recognize migraine as a public health priority, leveraging WHO's Intersectoral Global Action Plan on epilepsy and other neurological disorders (WHO 2022).
5. *Longitudinal and Gender-Specific Evidence*: Few studies track lifestyle changes or gender disparities in migraine progression, despite evidence that women in AfrAbia face unique sociocultural stressors (e.g., caregiving roles) that exacerbate migraines (Wabnitz et al. 2022).
6. *Underrepresentation of Sub-Saharan Africa*: Research hubs in Nigeria and Kenya produce only 12% of regional output, compared to GCC nations (Abu-Salameh et al. 2020).
7. *Lack of Translational Studies*: While dietary triggers are frequently studied, few trials test culturally tailored interventions, such as Ramadan-focused migraine guidelines or mHealth tools for rural populations.

To address these gaps, future research must adopt *multidisciplinary frameworks* integrating neurology, nutrition, and social sciences. Longitudinal cohort studies could establish causal links between lifestyle factors (e.g., sleep disruption, urbanization stressors) and migraine outcomes. Additionally, funding agencies should prioritize underrepresented regions (e.g., Sahel, Horn of Africa) to diversify insights and reduce epistemic dependency on high-income nations.

To improve migraine research, it is advisable to broaden data sources by including grey literature and regional repositories, assuring a more thorough comprehension of the subject. Researchers must participate in cultural sensitivity training to mitigate stigma in data gathering, hence promoting more precise and inclusive investigations. Conducting longitudinal studies is crucial for tracking lifestyle changes and migraine trends, particularly in the post-COVID-19 era. Moreover, bibliometric results should be used for policy lobbying to advance migraine-specific healthcare policies that cater to the requirements of those impacted.

The study faces inherent challenges in AfrAbia's diverse and resource-constrained settings. Below is a synthesis of key obstacles and mitigation strategies:

Challenge/limitation	Description	How to overcome
Limited regional data	Sparse publications from low-income AfrAbian countries	Foster collaborative networks (e.g., WHO partnerships) to fund local research
Language barriers	Non-indexed studies in local languages (e.g., Swahili, Berber)	Use translation tools and include regional databases (e.g., African journals online)
Cultural stigma	Migraines perceived as trivial, reducing reporting	Launch public health campaigns to destigmatize chronic pain

(continued)

Challenge/limitation	Description	How to overcome
Diagnostic inconsistencies	Varied diagnostic criteria across studies	Adopt standardized protocols (e.g., ICHD-3 guidelines)
Healthcare access gaps	Undiagnosed cases due to limited clinics in rural areas	Integrate mobile health units and community health workers
Political instability	Conflict zones (e.g., Sudan, Yemen) disrupt research continuity	Leverage international NGOs for safe data collection
Economic disparities	Poverty limits treatment access and research funding	Secure grants from global bodies (e.g., global burden of disease initiative)

16.6 Conclusion: Toward Equitable and Context-Driven Solutions

AfrAbia's migraine research landscape is marked by contradictions—cultural richness juxtaposed with systemic neglect. While lifestyle studies offer valuable insights, they must not come at the cost of excluding biological inquiry. By addressing funding inequities, centering local knowledge, and advocating for policy reform, the region can pioneer inclusive, evidence-based solutions that resonate globally.

This analysis underscores the interplay between AfrAbia's sociocultural diversity and its evolving migraine research priorities. While lifestyle factors are increasingly acknowledged, the region's dependency on Global North paradigms limits the development of locally relevant solutions. Policymakers must invest in *decolonized research infrastructure*, such as open-access regional repositories (e.g., Arab World Research Source) and partnerships between GCC-funded institutions and African community health networks. Culturally adapted interventions—such as integrating traditional medicine with biomedical approaches—could offer cost-effective strategies to mitigate migraine burdens. Ultimately, bridging research disparities and prioritizing equity will require sustained funding, cross-border collaboration, and policy frameworks that elevate AfrAbia's unique epidemiological narrative.

While the paper advances understanding of migraines in AfrAbia, its reliance on bibliometrics risks echoing existing biases. Addressing structural barriers—through collaboration, inclusive methodologies, and community engagement—is vital to unraveling the complex lifestyle-migraine nexus in this diverse region. Future research must prioritize equity in data representation and contextual relevance.

References

Abimbola S (2022) The uses of knowledge in global health. BMJ Glob Health 7(1):e008969

Abu-Salameh I, Al-Sheikh H, Al-Mahrouqi T, Al-Hinai A (2020) Migraine research in the Arab world: a scoping review. Neurol Int 12(3):45–52

Adeyanju GC, Okubadejo N, Ojo O, Bello A (2021) Impact of COVID-19 on migraine management in Nigeria: a cross-sectional study. J Neurol Sci 429:117045

Al Alawi AM, Al-Hinai S, Al-Busaidi N, Al-Hashel J (2022) Ramadan fasting and migraine: a prospective cohort study. Cephalalgia 42(9):745–753
Al Busaidi N, Al-Maskari S, Al-Kharusi A, Al-Hinai M (2022) Telemedicine for migraine care in Oman: Urban-rural disparities. J Telemed Telecare 28(4):123–129
Al Jumah M, Al Khathaami A, Al Otaibi H, Al Malik S (2019) Healthcare access and migraine management in the GCC. J Headache Pain 20(1):1–9
Al Jumah M, Al Zahrani M, Al Balawi T, Al Harbi N (2021) Traditional medicine use in migraine management: A Saudi perspective. J Ethnopharmacol 265:113387
Al-Hashel JY, Ahmed SF (2022) Migraine in the Middle East and North Africa: epidemiology and management. Front Neurol 13:1012
Al-Hashel J, Ahmed SF, Al-Sabah A, Al-Tawheid N (2019) Ramadan fasting and migraine in Kuwait: a clinical survey. Cephalalgia 39(8):1021–1028
Al-Hassany A, Moloney T, van den Brink AM (2021) Diet and migraine: an overview of current evidence. Cephalalgia 41(3):345–357
Al-Hassany L, Al-Jumah M, Al-Hashel J, Al-Moallem M, Goadsby PJ (2023) Medication overuse headache in the Gulf states: a multicenter study. Neurology 100(1):e45–e55
Belhaj S, El Fakir S, El Rhazi K, Zarrouq B (2021) Cultural perceptions of migraine in Morocco: a qualitative study. Headache 61(8):1249–1257
Benamer HT, Deleu D, Grosset D, Al-Asmi A (2023) Physical activity and migraine: a bibliometric analysis. Neurology 99(9):851–858
Bicakci S (2013) Migraine and fasting. Curr Pain Headache Rep 17(10):369
Bicakci S, Boz C, Ozmenoglu M (2013) Ramadan fasting and migraine: a regional survey. Cephalalgia 33(12):1028–1034
Bigna JJ, Tochie JN, Tounouga DN, Nansseu JR (2020) Dietary patterns and migraine in sub-Saharan Africa. Nutrients 12(11):3375
Burch RC, Loder S, Loder E, Smitherman TA (2022) The influence of diet on migraine in a middle eastern population: a cross-sectional study. Cephalalgia 42(5):400–409
Dzator J, Asante P, Acheampong T, Hewlett S (2021a) Lifestyle interventions for migraine in low-income settings. Lancet Glob Health 9(6):e789–e795
Dzator JSA, Asante PK, Acheampong T, Hewlett S (2021b) A community-based study of migraine in Ghana: sociocultural insights. Afr Health Sci 21(2):867–875
Elsayed MM, Ahmed HA, Ibrahim AA, Mohamed S (2023) Economic barriers to migraine care in Sudan: a cross-sectional survey. Lancet Glob Health 11(S1):S12
Eltahir AH, Alsharif MA, Alharthi SS, Alghamdi AS (2022) Stress and migraine in the Arab world: a bibliometric review. J Neurol Sci 430:120–126
Ferrari MD, Goadsby PJ, Burstein R, Dodick DW (2022) Migraine pathophysiology: lessons from genetics. Nat Rev Neurol 18(8):455–468
Global Burden of Disease (GBD) (2023) Migraine research trends 1990–2023. Institute for Health Metrics and Evaluation
Goadsby PJ, Holland PR, Martins-Oliveira M, Hoffmann J, Schankin C (2017) Pathophysiology of migraine: a disorder of sensory processing. Physiol Rev 97(2):553–622
Gormley P, Anttila V, Winsvold BS, Palta P, Esko T (2016) Meta-analysis of 375,000 individuals identifies 38 loci associated with migraine. Nat Genet 48(8):856–866
Hamdy SM, El-Sayed MA, Abdelghaffar NA, El-Tamawy MS (2022) Urban environmental triggers of migraine in Cairo. Environ Health Perspect 130(4):047001
Huang T, Redline S, Hu FB (2020) The association between sleep patterns and migraine. J Headache Pain 21(1):1–10
Ibrahim AA, Al-Hashel JY, Ahmed SF, Al Jumah M (2023) Bibliometric analysis of migraine research in the Arab world: 2000–2022. Headache Med 14(2):45–59
Kavukcu E, Zorlu Y, Yavuz B (2020) Stress and migraine: a socio-cultural perspective. Neurol Int 12(1):28–36
Khan J, Al-Hashel J, Ahmed SF, Al-Mutairi H (2023) Heat exposure and migraine in Kuwait: a case-crossover study. Int J Environ Res Public Health 20(4):3120

Lipton R, Buse D, Serrano D, Reed M (2020a) Global trends in acute migraine treatment overuse. Headache 60(8):1545–1558
Lipton RB, Buse DC, Serrano D, Reed ML, Pavlovic JM (2020b) Migraine prevalence, treatment, and comorbidities: the overcome study. Neurology 95(10):e1114–e1125
Mbaye R, Gebreyohannes EA, Kassegne S, Muula AS (2023) Strengthening African research capacity: a call for equity. Lancet 401(10385):1232–1234
Mwangi LN, Omondi GB, Kamau JW, Nyamu DG (2022) Nurse-led interventions for migraine in rural Kenya. J Community Health 47(3):498–505
Okeke TC, Adebayo PB, Ojo O, Bello TO (2023) Genetic research gaps in African migraine studies: a systematic review. Neurogenetics 24(1):23–34
Okubadejo N, Ojo O, Oshinaike O, Danesi M (2020) Urbanization, physical inactivity, and migraine severity in Nigeria. J Neurol 267(12):3563–3570
Sacco S, Ricci S, Carolei A, Pistoia F (2021) Lifestyle and migraine: systematic review and meta-analysis. J Headache Pain 22:102
Sheikh H, Al Jumah M, Al Balawi T, Al Harbi N (2020a) Traditional medicine use in migraine management: an AfrAbian perspective. J Ethnopharmacol 265:113387
Sheikh SP, Al-Hashel J, Ahmed SF, Al-Mutairi H (2020b) Traditional remedies for migraine in the Middle East: a review. J Ethnopharmacol 256:112787
Smith J, Jones M, Brown P, Taylor L (2021) Global research disruptions during the COVID-19 pandemic. Nat Hum Behav 5(6):775–783
Steiner T, Gururaj G, Ahmed S, Shakir R (2021a) Migraine prevention in low-resource settings: a WHO report. World Health Organization
Steiner T, Gururaj G, Ahmed S, Shakir R (2021b) Stress and migraine in low-resource settings: a WHO report. World Health Organization
Steiner TJ, Stovner LJ, Jensen R, Uluduz D, Katsarava Z (2021c) Migraine in low-income and middle-income countries: a call for action. Lancet Neurol 20(5):337–339
UNESCO (2022) UNESCO science report: race against time for smarter development. UNESCO Publishing
Varkey E, Hagen K, Zwart JA (2019) Physical activity and migraine risk: a population-based study. Neurology 92(6):587–594
Wabnitz A, Martelletti P, Ashina M, Katsarava Z (2022) Gender disparities in migraine care: a global survey. Neurology 98(18):e1801–e1811
World Health Organization (WHO) (2022) Intersectoral global action plan on epilepsy and other neurological disorders 2022–2031. WHO
World Health Organization (WHO) (2023) Health equity in the eastern Mediterranean region: a review. WHO

Nutritional Strategies for Specific Substance Use Disorders

17

Asiat Na'Allah, Babalola Ola Yusuf,
Ibrahim Adeola Moronfolu, and Saheed Sabiu

Abstract

Substance use disorders (SUDs) involve the compulsive misuse of substances like alcohol, opioids, stimulants, and cannabis, leading to significant health, social, and economic consequences. Affecting millions globally and 7.7% of Americans annually, SUDs severely impair nutritional health by causing deficiencies in essential vitamins like folate and B vitamins, exacerbating both physical and psychological issues. This chapter underscores the vital role of nutrition in SUD recovery, highlighting how addressing nutrient deficiencies supports overall health. Nutritional interventions restore neurotransmitter function, reduce oxidative stress, and stabilize the hypothalamic-pituitary-adrenal (HPA) axis.

A. Na'Allah (✉)
Department of Biochemistry, Faculty of Pure and Applied Sciences, Kwara State University, Malete, Kwara State, Nigeria
e-mail: asiat.naallah@kwasu.edu.ng

B. O. Yusuf
Department of Biochemistry, Faculty of Science, Federal University of Health Sciences, Ila-Orangun, Osun State, Nigeria

Department of Biotechnology, Faculty of Applied Sciences, Durban University of Technology, Durban, South Africa
e-mail: babalola.yusuf@fuhsi.edu.ng; babalolay@dut.ac.za

I. A. Moronfolu
Department of Chemical Sciences, Biochemistry Unit, Summit University, Offa, Kwara State, Nigeria

S. Sabiu
Department of Biotechnology, Faculty of Applied Sciences, Durban University of Technology, Durban, South Africa
e-mail: sabius@dut.ac.za

W. Mohamed, A. Özge (eds.), *Feeding the Mind: The Connection Between Diet, Drugs, and Mental Health Volume 1*, Nutritional Neurosciences,
https://doi.org/10.1007/978-981-95-1414-4_17

amphetamines hijack this system by flooding the brain with dopamine, the neurotransmitter that produces feelings of euphoria and pleasure. With this malpractice over some time, the brain begins to compensate for this increased activity of dopamine by downregulating the receptors which changes the way pleasurable feeling is obtained to substance use and diagnosis becomes addiction (Ubeja and Szujewski 2022; Volkow and Blanco 2023).

Of equal importance is the serotonergic system, involved in mood regulation. 3,4-Methylenedioxymethamphetamine (MDMA) and alcohol increase mood and relaxation because they exert actions on serotonin. However, long-term substance use depletes serotonin levels, leaving a person feeling more emotional, increased craving, and depressed when they are sober (Müller et al. 2020; Mustafa et al. 2020). Gamma-aminobutyric acid (GABA), the brain's primary inhibitory neurotransmitter, acts as a calming force, dampening neural activity. Substances like alcohol and benzodiazepines enhance GABA's effects, producing a sedative and relaxing sensation. Over time, the brain develops a dependence on these substances to maintain balance, thereby creating significant anxiety and withdrawal syndromes such as seizures when those substances are suddenly taken away (Dharavath et al. 2023; Goldschen-Ohm 2022).Meanwhile, glutamate, the brain's most powerful excitatory neurotransmitter, is essential for learning and memory (Pal 2021). Chronic alcohol use suppresses glutamate, and when alcohol use ceases, the brain's rebound effect can result in cognitive impairments and withdrawal symptoms (Zhu et al. 2023). These systems work together to create a complex biochemical base for substance dependence and withdrawal, leading to the compulsive behaviors that define addiction.

17.3.2 Physiological Impact of Substance Use on Nutritional Status

Substance use disorders (SUDs) bring about the loss of various metabolic processes and malnutrition-related complications, thus contributing to poor recovery (Wangensteen and Hystad 2022). Alcohol, opioids, stimulants including cocaine, and others affect the assimilation of nutrients differently. The very common one, alcohol, causes harm to the surface of the intestine, which loses its capacity for the absorption of essential minerals and vitamins while causing inflammation (Butts et al. 2023). As well, chronic use leads to impaired gastric emptying and the development of dysbiosis of the flora, which again, creates a situation hostile to the availability of much-needed thiamine and folate (Lee and Lee 2021). On the contrary, opioids affect a different system of the human body, the gastrointestinal system by causing decreased gut motility, thus leading to constipation and bacterial overgrowth. These effects not only cause physical discomfort but also impair nutrient absorption (Kazemian and Pakpour 2024). In contrast, it is cocaine's appetite-suppressing action that changes eating practices, making users who want to make the most out of the drug skip meals. Over time, this accumulation results in a deficit of calories and nutrition (Mahboub et al. 2021). Along with these gastrointestinal conditions, certain substances deplete important elements that must be maintained in the body for the normal functioning and mental health of human. One common example would be alcohol consumption which is well known for lowering levels of B vitamins, magnesium, and zinc predominately owing to poor nutrition habits, liver damage, and low absorption capacity. These shortfalls can cause cognitive decline and exacerbate conditions such as anxiety and depression (Baj et al. 2020). Opioids cause metabolic deficiencies for calcium and vitamin D that play an important role in bone formation and thus expose the individual to an increased risk of developing osteoporosis (Shih et al. 2023). Still more ominously are the tendencies of people with SUDs to eat high-energy but low constituent foodstuffs, thereby further depleting their nutritional status (Mahboub et al. 2021; Sekhotha and Sekhotha 2024). In fact, due to high cocaine use, food intake may be completely neglected, leading to more malnutrition and heightening the risks of infections and diseases (Mahboub et al. 2021).

Those with SUDs often present nutrient deficiencies such as B vitamins, especially thiamine (B1), which is the hallmark of these deficits in alcohol users and may cause a life-threatening neurological disorder known as Wernicke-Korsakoff syndrome (Cox and Klinger 2022). Folate and vitamin B12 deficiencies are common causes of impaired cognition alongside worsening mental health disorders (Molloy 2020). Although critical for muscle function, cardiovascular health, and immunity, some minerals such as magnesium or zinc are frequently depleted in both alcohol and opiate users (Mahboub et al. 2021; Mahajan and Singh 2022).

17.4 Nutritional Strategies for Specific SUDs

Substance use disorders (SUDs) can result in severe nutritional deficits that affect general health and make recovery more difficult. Effective therapy requires modifying dietary plans to meet the particular requirements of people battling particular SUDs.

17.4.1 Alcohol Use Disorder

Alcohol use disorder (AUD) is known to be associated with certain nutritional disorders due to the physiological effects of alcohol and poor dietary habits associated with long-term usage of alcohol. In excess quantity, alcohol creates a hindrance in the absorption and metabolism of necessary nutrients, which further creates such imbalances that it becomes very difficult to address (Jophlin et al. 2024; White and Sirohi 2024). One of the most significant deficiencies observed in individuals with AUD is that of thiamine (vitamin B1), a nutrient necessary for glucose metabolism and proper nervous system function. Thiamine deficiency if prolonged and treated very late can lead to some irreversible disorders of the nervous system including Wernicke-Korsakoff syndrome with ataxia, disorientation, and short-term memory loss (Mateos-Díaz et al. 2022). Apart from thiamine, chronic alcohol consumption

The use of nutrition for intervention is very important as it aids in the treatment and the healing process. The process of caloric repletion focusing on protein-rich diets provides energy restoration and a regaining of physical muscle that has been depleted by prolonged undernutrition. Foods, such as lean meats, legumes, or dairy products, are rich in essential proteins that promote tissue growth and healing (Olson et al. 2020). Moreover, specific amino acids can also assist in restoring an abnormal concentration of neurotransmitters. For instance, tyrosine and tryptophan are precursors of dopamine and serotonin, respectively, which help in tempering mood, suppressing appetites, and emotional control which are all important in averting relapses (Aquili 2020). It is also necessary to eliminate oxidative stress since too much exposure to free radicals will damage cellular health and immune performance. Suffice it to say, antioxidant nutrients, which include selenium, and glutathione as well as vitamins A, C, and E, assist in combatting oxidative damage, boosting the immune system (Munteanu and Schwartz 2022). Besides, it is important to keep the body well hydrated, especially since today's more sedentary lifestyle and the abuse of stimulants tend to hyper-activate and neglect fluid intake leads to dehydration (Roque Bravo et al. 2022). Listening to the body's hydration needs promotes the optimal functioning of the kidneys and cleansing of the body as well as speeding up the elimination of the byproducts of toxic substances from the body.

17.4.4 Cannabis Use Disorder

Cannabis use disorder (CUD) is a harmful condition that is characterized by serious disturbances in dietary patterns and eating disorders leading to fluctuations in appetite and weight which make nutrition more complex (Kesner and Lovinger 2021). In particular, long-term intake of cannabis may cause a strong hunger, and people may binge on fat-rich food that has little nutritional value (Kramer et al. 2020). On the other hand, withdrawal and its accompanying states may cause reduced levels of stem, appetite, and unintentional weight loss (Kurt Tunagur et al. 2022). These inconsistent eating patterns increase the risk of deficiencies in essential vitamins and minerals, particularly those that support brain function and immunity. Moreover, such habits can impair nutrient absorption, leaving the body vulnerable to further deficiencies and weakening overall health (Sekhotha and Sekhotha 2024).

These dietary limitations need strategies with particular measures targeted towards restoration and treatment of nutritional improvement. A diet high in fiber, anti-inflammatory nutrients, and balanced with other micronutrients can help fill this gap (Drake 2020). Fiber, for instance, will help in decreasing disorders associated with CUD, such as abnormal eating patterns, since it works toward regulating blood sugar levels and enhancing digestion (Liu et al. 2024). At the same time, systemic inflammation associated with chronic cannabis use and other factors can be reduced with the aid of vegetables, fruits, nuts, or seeds, which possess anti-inflammatory compounds (Luvián-Morales et al. 2022). Through flexible meal plans and nutrition therapy, healthy eating habits can be incorporated into the patients, which can further assist them in recovery. This approach encourages consistency and would make it a lot easier to make food choices especially when the individual is withdrawing and there is a high chance of going back to the previous bad dietary behavior. Omega-3 and omega-6 contain the family of essential fatty acids and are of great profit in this situation. These fats enhance the part of the brain responsible for mood control and cognitive thinking, which are unfortunately compromised in patients with cannabis use disorder (Stachowicz 2023). Moreover, essential fatty acids also have a stabilizing effect on the endocannabinoid system disturbed by heavy cannabis use (Zieglgänsberger et al. 2022).

17.5 Mechanisms Underlying Nutritional Interventions in SUDs

Nutritional interventions are important for recovery from SUDs, as they can normalize the function of brain circuits, reduce withdrawal symptoms, and support other physiological functions relevant to reducing addiction liability. Neurotransmitter pathway modification is one important approach. Tyrosine and tryptophan are amino acids that act as precursors of dopamine and serotonin, respectively, two of the most important neurotransmitters involved in mood regulation, reward mechanisms, and control of impulses. Loss of pleasure through dopamine depletion, an often-felt consequence of chronic substance use increases the cravings for a substance. Supplementing with tyrosine helps to replenish dopamine, supports motivation, and minimizes withdrawal symptoms as a result. In the same way, tryptophan also helps in serotonin production, a hormone that regulates emotions and controls impulsivity which is mostly lacking in substance-abusing individuals (Aquili 2020).

In addition to the modulation of neurotransmitters, nutritional approaches that have anti-inflammatory and antioxidant properties are important since chronic abuse of substances leads to oxidative stress and neuroinflammation. Substances such as polyphenols and omega-3 fatty acids are also famous for reducing inflammation and protecting the neurons from free radicals. Such nutrients boost the brain's plasticity, thereby improving cognition and emotion control which are essential in the long-term recovery process (Luvián-Morales et al. 2022; Stachowicz 2023).

Moreover, the control of the HPA axis is another critical area in terms of nutritive intervention. Chronic substance abuse disturbs this axis, causing the levels of cortisol to be heightened in the body and the stress response to be amplified which increases the chances of relapse (Sinha 2024). A proper diet, especially one enriched with omega-3 s and B vitamins, is instrumental in correcting HPA axis dysfunction, thereby relieving and preventing pain-induced urges (Pickering et al. 2023; Serefko et al. 2024).

Additionally, there are current findings concerning the interaction of the gut with the central nervous system in addiction (Luo et al. 2023; Simpson et al. 2022). Among other things, changes in the gut microbiota, common for patients with SUD, affect the neurotransmitter synthesis and may lead to mood dysregulation and craving (Simpson et al. 2022). Nutritional treatment involving the intake of probiotics and prebiotics has the potential to restore a healthy concentration of gut bacteria which benefits the production of neurotransmitters and mental health status. Because gut health protects against mood dysregulation leading to relapse, this also improves addiction treatment outcomes (Casertano et al. 2022).

17.6 Holistic Approaches to Nutrition in Recovery

In the recovery process from SUD, a holistic approach to a balanced diet helps in improving physical and emotional health to facilitate permanent recovery. Mindful eating is important in this since it enables people to cultivate consciousness of their food intake and their feelings related to it. Strategies that include eating more slowly, paying attention to when one is hungry or full, and appreciating the flavor of the food help to establish a healthy attitude toward food. It can help curb binge eating, moderating excessive consumption of fast food and alleviating the need for comfort food or the discomfort associated with withdrawal slumps (Baradia and Ghosh 2021).

In addition, nutritional interventions are enhanced by also addressing behavioral strands in the management of eating habits. For instance, cognitive-behavioral therapy (CBT) helps the patient understand and regulate their responses to food stimuli while looking for other, healthier ways to cope (Manasse et al. 2021). Moreover, engaging in healthy eating group-based activities not only provides emotional support but also positively reinforces behaviors that are crucial for long-term recovery. Such activities make the participants feel less lonely during the early phase of recovery (Theodorakis et al. 2024).

Incorporating nutrition in the wider context of recovery programs integrates dietary measures with psychosocial aspects of care. Dietitians working within multidisciplinary teams enable the customization of nutrition to the patient's requirements. These include the treatment of malnutrition or the replenishment of nutrients vital for the body and mind to function (Guillaumie et al. 2020). An integrated approach to recovery is designed as more than just abstinence but as the practice of better emotional, decent eating habits and a general feeling of well-being to enduring sobriety. Combining these aspects helps patients not just conquer the addiction but prepare themselves for a long-term healthy life (Pennock 2021).

17.6.1 Nutritional Management of SUDs: Practical Implementation and Challenges

It is vital to manage substance use disorders with dietary strategies that are tailored to an individual. Each individualized nutrition plan should start with an evaluation of the person's nutritional requirements, way of life, tastes, and eating patterns. These assist in pinpointing particular nutrient deficits and making sure that the strategies are consistent with their cultural context and food availability (Adams et al. 2020). It is necessary to work alongside other medical professionals, including dietitians and specialists in addiction and mental health, to combine nutritional treatment with complex treatments for the patient's recovery. These combined approaches encourage uniformity of care and make certain that physical and psychological facets of recovery are catered for in the treatment of SUDs (Reist et al. 2022).

Scheduled meals are important in eliminating hunger pangs and avoiding a return to harmful behaviors. Meals have a variety of foods such as whole grain cereals, legumes, fruits, vegetables, and lean meat which help in metabolic rehabilitation and nutrient deficiencies that are caused by substance abuse in the first place (Wiss 2019; Olson et al. 2020; Carvajal et al. 2023). In addition, eating regularly also gives individuals a sense of order and control over their daily activities, which are much needed in the rehabilitation process. This is helpful in recovery because developing such routines will enhance sobriety and promote mental health. Also, these nutritional interventional measures are even more emphasized by dealing with common comorbid deficiencies such as magnesium, vitamin D, and B vitamins necessary for individuals suffering from SUDs. These deficiencies aggravate tiredness, irritability, and cognitive dysfunction, so replacing those through individualized megadoses can ameliorate simply the body and mind (Afzal et al. 2022). However, moderation is required to prevent excessive doses during one's recovery as too high levels can impact medications used in therapy for addiction and create toxicity (Serwatka et al. 2023).

Functional foods and phytochemicals enhance recovery as they exert antinociceptive effects while protecting vital organs from damage (Knapp et al. 2023). For instance, the use of turmeric for its anti-inflammatory properties may assist with substance-induced neuroinflammation (Cianciulli et al. 2022). On the other hand, even when the nutritional approach is presumably effective and has promising results, many factors can affect its practice. The psychosocial aspect comprises sociocultural issues such as stigma, emotional problems, and mental disorders, which may diminish motivation to maintain dietary habits as well as the recovery process (Belfiore et al. 2024). The other factors involve the high costs of food with adequate nutrients in addition to the poor availability of healthier foods (Drewnowski et al. 2020).

To address these issues, a team approach is necessary. Incorporating nutrition education into addiction treatment, assisting with community nutritional needs, and supporting legislation that promotes the accessibility of healthy foods can maximize the outcomes of nutritional approaches. In the end, the effectiveness of these approaches in the future will be determined not just by the provision of individualized treatment, but also by focusing on structural factors that cause low uptake of healthy eating, and hence improve chances for sustained recovery among persons with SUDs.

17.7 Conclusion

Nutritional approaches related to the specifics of SUDs significantly contribute to recovery from the physical and mental consequences of drug abuse. In the case of alcohol use disorder diagnosis, it is necessary to restore the lack of certain B vitamins like thiamine and folate to rehabilitate any neurological and cognitive dysfunctions that are generally seen in patients who consume alcohol chronically. In demographic cases of opioid use disorder, for instance, nutrition rehabilitative strategies emphasize restoring and protecting gut health and address calcium and vitamin D deficiencies to avoid malnourishment and osteoporosis in such patients. At the same time, those with stimulant use disorder are also recommended high-protein foods to recover muscle loss and deal with oxidative stress. Cannabis use disorder individuals, on the other hand, are said to be deficient in certain nutrients during detox via omega-3 and fiber that controls appetite, mood, and cognitive activity.

As different substances have a wide range of physiological effects, evidence-based management is of great importance. Every substance has its metabolic hurdle and its own nutritional challenge. For instance, while diets rich in fiber help manage constipation which is characteristic of the usage of opiates, foods rich in antioxidants are important in counteracting the oxidative damage caused due to the use of stimulants. Such enhanced recovery is better with nutrition combined with behavioral therapies; in this case, the recovery stresses the importance of food to instill eating habits, decreasing chances of relapse.

Nutrition assessment within the clinical practice is very valuable and should not be seen only as a follow-up tool, as it enables preventative approaches before the additional impact on the pathology. The merging of care provided by dietitians and addiction specialists and the healthcare provider guarantees that interventions are timely and centered on the patient's needs. Balanced nutrition and dietary needs addressed at structured mealtimes can provide balance and stability, which may be useful early in recovery. Policymakers can support this effort by pushing for better access to nutrient-dense foods in rehabilitation centers and supporting campaigns that explain the need for nutrition in the treatment process.

The interactions between nutrition and the neurotransmitter systems, especially via gut-brain connection, could be clarified in further studies. Probiotics and functional foods have the potential to improve mental health as well as reduce relapse. Well-designed large-scale clinical trials will be needed to verify these observations. In the end, integrating nutrition within public health frameworks will result in an enduring equivalent that can produce lasting relapse prevention.

References

Adams SH, Anthony JC, Carvajal R, Chae L, Khoo CSH, Latulippe ME, Matusheski NV, McClung HL, Rozga M, Schmid CH, Wopereis S, Yan W (2020) Perspective: guiding principles for the implementation of personalized nutrition approaches that benefit health and function. Adv Nutr 11(1):25–34. https://doi.org/10.1093/advances/nmz086

Afzal A, Batool Z, Sadir S, Haider S (2022) Nutrition and substance-use disorder. In: Mohamed F, Kobeissy W (eds) Nutrition and psychiatric disorders (pp. 289–312). Springer. https://doi.org/10.1007/978-981-19-5021-6_14
Aquili L (2020) The role of tryptophan and tyrosine in executive function and reward processing. International Journal of Tryptophan Research 13:1–13. https://doi.org/10.1177/1178646920964825
Baj J, Flieger W, Teresiński G, Buszewicz G, Sitarz R, Forma A, Karakuła K, Maciejewski R (2020) Magnesium, calcium, potassium, sodium, phosphorus, selenium, zinc, and chromium levels in alcohol use disorder: a review. J Clin Med 9(6):1901. https://doi.org/10.3390/jcm9061901
Baradia R, Ghosh J (2021) Impact of mindful eating among adolescents. Int J Sci Res 10(11):11–15
Belfiore CI, Galofaro V, Cotroneo D, Lopis A, Tringali I, Denaro V, Casu M (2024) A multi-level analysis of biological, social, and psychological determinants of substance use disorder and co-occurring mental health outcomes. Psychoactives 3(2):194–214. https://doi.org/10.3390/psychoactives3020013
Berríos-Cárcamo P, Quezada M, Quintanilla ME, Morales P, Ezquer M, Herrera-Marschitz M, Israel Y, Ezquer F (2020) Oxidative stress and neuroinflammation as a pivot in drug abuse. A focus on the therapeutic potential of antioxidant and anti-inflammatory agents and biomolecules. Antioxidants 9(9):830. https://doi.org/10.3390/antiox9090830
Butts M, Sundaram VL, Murughiyan U, Borthakur A, Singh S (2023) The influence of alcohol consumption on intestinal nutrient absorption: a comprehensive review. Nutrients 15(7):1571. https://doi.org/10.3390/nu15071571
Carvajal F, Sánchez-Gil A, Cardona D, Rincón-Cervera MA, Lerma-Cabrera JM (2023) The effect of very-long-chain n-3 polyunsaturated fatty acids in the central nervous system and their potential benefits for treating alcohol use disorder: reviewing pre-clinical and clinical data. Nutrients 15(13):2993. https://doi.org/10.3390/nu15132993
Casertano M, Fogliano V, Ercolini D (2022) Psychobiotics, gut microbiota and fermented foods can help preserve mental health. Food Res Int 152:110892. https://doi.org/10.1016/j.foodres.2021.110892
Chavez MN, Rigg KK (2020) Nutritional implications of opioid use disorder: a guide for drug treatment providers. Psychol Addict Behav 34(6):699. https://doi.org/10.1037/adb0000575
Cianciulli A, Calvello R, Ruggiero M, Panaro MA (2022) Inflammaging and brain: curcumin and its beneficial potential as regulator of microglia activation. Molecules 27(2):341. https://doi.org/10.3390/molecules27020341
Connor JP, Stjepanović D, Le Foll B, Hoch E, Budney AJ, Hall WD (2021) Cannabis use and cannabis use disorder. Nat Rev Dis Primers 7(1):16. https://doi.org/10.1038/s41572-021-00247-4
Cox WM, Klinger E (2022) Why people drink: how people change: a guide to alcohol and people's motivation for drinking it. Springer Nature. https://doi.org/10.1007/978-3-030-93928-1
Cristina NM, Lucia D (2021) Nutrition and healthy aging: prevention and treatment of gastrointestinal diseases. Nutrients 13(12):4337. https://doi.org/10.3390/nu13124337
Dal N, Bilici S (2024) An overview of the potential role of nutrition in mental disorders in the light of advances in Nutripsychiatry. Curr Nutr Rep 13(2):69–81. https://doi.org/10.1007/s13668-024-00520-4
De Giorgio R, Zucco FM, Chiarioni G, Mercadante S, Corazziari ES, Caraceni A, Odetti P, Giusti R, Marinangeli F, Pinto C (2021) Management of opioid-induced constipation and bowel dysfunction: expert opinion of an Italian multidisciplinary panel. Adv Ther 38(7):3589–3621. https://doi.org/10.1007/s12325-021-01766-y
Deak JD, Johnson EC (2021) Genetics of substance use disorders: a review. Psychol Med 51(13):203–229. https://doi.org/10.1017/S0033291721000969
DeVido JJ (2020) Stimulants: caffeine, cocaine, amphetamine, and other stimulants. In: Marienfeld C (ed) Absolute addiction psychiatry review. Springer, Cham, pp 185–203. https://doi.org/10.1007/978-3-030-33404-8_12
Devoe DJ, Dimitropoulos G, Anderson A, Bahji A, Flanagan J, Soumbasis A, Patten SB, Lange T, Paslakis G (2021) The prevalence of substance use disorders and substance use in anorexia ner-

vosa: a systematic review and meta-analysis. J Eat Disord 9(1):1–15. https://doi.org/10.1186/s40337-021-00516-3

Dharavath RN, Pina-Leblanc C, Tang VM, Sloan ME, Nikolova YS, Pangarov P, Ruocco AC, Shield K, Voineskos D, Blumberger DM, Boileau I, Bozinoff N, Gerretsen P, Vieira E, Melamed OC, Sibille E, Quilty LC, Prevot TD (2023) GABAergic signaling in alcohol use disorder and withdrawal: pathological involvement and therapeutic potential. Front Neural Circuits 17:1218737. https://doi.org/10.3389/fncir.2023.1218737

Dingwall KM, Delima JF, Binks P, Batey R, Bowden SC (2022) What is the optimum thiamine dose to treat or prevent Wernicke's encephalopathy or Wernicke–Korsakoff syndrome? Results of a randomized controlled trial. Alcohol Clin Exp Res 46(6):1133–1147. https://doi.org/10.1111/acer.14843

Drake CY (2020) Substance use disorder treatment and nutrition: a multidisciplinary approach. J Mental Health Soc Behav 2(2):117. https://doi.org/10.33790/jmhsb1100117

Drewnowski A, Monterrosa EC, de Pee S, Frongillo EA, Vandevijvere S (2020) Shaping physical, economic, and policy components of the food environment to create sustainable healthy diets. Food Nutr Bull 41(2_suppl):74S–86S. https://doi.org/10.1177/0379572120945904

Ekinci GN, Sanlier N (2023) The relationship between nutrition and depression in the life process: a mini review. Exp Gerontol 172:112072. https://doi.org/10.1016/j.exger.2022.112072

Glenn JM, Madero EN, Bott NT (2019) Dietary protein and amino acid intake: links to the maintenance of cognitive health. Nutrients 11(6):1315. https://doi.org/10.3390/nu11061315

Goldschen-Ohm MP (2022) Benzodiazepine modulation of GABAA receptors: a mechanistic perspective. Biomolecules 12(12):1784. https://doi.org/10.3390/biom12121784

Guillaumie L, Boiral O, Baghdadli A, Mercille G (2020) Integrating sustainable nutrition into health-related institutions: a systematic review of the literature. Can J Public Health 111(6):845–861. https://doi.org/10.17269/s41997-020-00394-3

Hafiza U, Ahmad W, Hina N, Shoaib AM, Saleem KM, Muhammad W (2024) Association of vitamin deficiency with the progression of anaemia. Egypt J Haematol 49(2):115–120. https://doi.org/10.4103/ejh.ejh_80_23

Jaqua EE, Nguyen V, Scherlie N, Dreschler J, Labib W (2022) Substance use disorder in older adults: mini review. Addict Health 14(1):62–67. https://doi.org/10.22122/ahj.v14i1.1311

Jophlin L, Liu TY, McClain CJ (2024) Nutritional deficiencies in alcohol use disorder/alcohol-associated liver disease. Curr Opin Gastroenterol 40(2):112–117. https://doi.org/10.1097/MOG.0000000000000999

Kalin NH (2020) Substance use disorders and addiction: mechanisms, trends, and treatment implications. Am J Psychiatry 177(11):1015–1018. https://doi.org/10.1176/appi.ajp.2020.20091382

Karnib N, Van Staaden MJ (2021) The deep roots of addiction: a comparative perspective. Brain Behav Evol 95(5):222–229. https://doi.org/10.1159/000514180

Kazemian N, Pakpour S (2024) Understanding the impact of the gut microbiome on opioid use disorder: pathways, mechanisms, and treatment insights. Microb Biotechnol 17(10):e70030. https://doi.org/10.1111/1751-7915.70030

Kesner AJ, Lovinger DM (2021) Cannabis use, abuse, and withdrawal: Cannabinergic mechanisms, clinical, and preclinical findings. J Neurochem 157(5):1696. https://doi.org/10.1111/jnc.15369

Knapp M, McCabe DE, Naegle MA (2023) Reducing harm related to substance use by older adults. In: A comprehensive guide to safety and aging: minimizing risk, maximizing security. CRC Press, pp 237–257. https://doi.org/10.1201/9781003197843-22

Kramer A, Sinclair J, Sharpe L, Sarris J (2020) Chronic cannabis consumption and physical exercise performance in healthy adults: a systematic review. J Cannabis Res 2(1):1–8. https://doi.org/10.1186/s42238-020-00037-x

Kurt Tunagur EM, Yazici AB, Guzel D, Tunagur MT, Ermis C, Suda MA, Yazici E (2022) Investigating associations between appetite-regulating hormones, aggression and craving in males with cannabis use disorder. Drug Alcohol Depend 238:109577. https://doi.org/10.1016/j.drugalcdep.2022.109577

Lee E, Lee JE (2021) Impact of drinking alcohol on gut microbiota: recent perspectives on ethanol and alcoholic beverage. Curr Opin Food Sci 37:91–97. https://doi.org/10.1016/j.cofs.2020.10.001

Liu M, Li S, Cao S, Liu C, Han Y, Cheng J, Zhang S, Zhao J, Shi Y (2024) Let food be your medicine–dietary fiber. Food Funct 15(15):7733–7756. https://doi.org/10.1039/D3FO05641D

Lo TW, Yeung JWK, Tam CHL (2020) Substance abuse and public health: a multilevel perspective and multiple responses. Int J Environ Res Public Health 17(7):2610. https://doi.org/10.3390/ijerph17072610

Lopez-Leon S, González-Giraldo Y, Wegman-Ostrosky T, Forero DA (2021) Molecular genetics of substance use disorders: an umbrella review. Neurosci Biobehav Rev 124:358–369. https://doi.org/10.1016/j.neubiorev.2021.01.019

Luo X, Li H, Fan X, Wu X, Zhou R, Lei Y, Xue D, Yang F, Xu Y, Wang K (2023) The gut microbiota-brain axis: potential mechanism of drug addiction. Curr Top Med Chem 23(18):1782–1792. https://doi.org/10.2174/1568026623666230418114133

Luvián-Morales J, Varela-Castillo FO, Flores-Cisneros L, Cetina-Pérez L, Castro-Eguiluz D (2022) Functional foods modulating inflammation and metabolism in chronic diseases: a systematic review. Crit Rev Food Sci Nutr 62(16):4371–4392. https://doi.org/10.1080/10408398.2021.1875189

Mahajan L, Singh R (2022) Nutritional deficiencies and diseases associated with alcoholism. In: Chhikara BS, Shanwal VK (eds) Alcoholism: causes, symptoms, effects and treatment, 1st edn. CRC Press, pp 53–67. https://doi.org/10.1201/9781003280361-4

Mahajan VR, Elvig SK, Vendruscolo LF, Koob GF, Darcey VL, King MT, Kranzler HR, Volkow ND, Wiers CE (2021) Nutritional ketosis as a potential treatment for alcohol use disorder. Front Psych 12:781668. https://doi.org/10.3389/fpsyt.2021.781668

Mahboub N, Rizk R, Karavetian M, De Vries N (2021) Nutritional status and eating habits of people who use drugs and/or are undergoing treatment for recovery: a narrative review. Nutr Rev 79(6):627–635. https://doi.org/10.1093/nutrit/nuaa095

Malinowski B, Jerczyk A (2021) The impact of diet on substance abuse and addiction. J Nutr Health 19(2):85–98

Manasse SM, Lampe EW, Abber SR, Butler R, Gillikin L, Trainor C (2021) Exposure-enhanced cognitive behavioral therapy for adolescents with binge eating: an initial case series. Clin Child Psychol Psychiatry 26(4):1124–1136. https://doi.org/10.1177/13591045211028963

Martini F, Fregna L, Bosia M, Perrozzi G, Cavallaro R (2022) Substance-related disorders. In: Cavallaro R, Colombo C (eds) Fundamentals of psychiatry for health care professionals. Springer, Cham, pp 263–295. https://doi.org/10.1007/978-3-031-07715-9_9

Mateos-Díaz AM, Marcos M, Chamorro AJ (2022) Wernicke-Korsakoff syndrome and other diseases associated with thiamine deficiency. Med Clin 158(9):431–436. https://doi.org/10.1016/j.medcli.2021.11.015

Mishra A, Chandel AKS, Bhalani DV, Shrivastava R (2020) Importance of dietary supplements to the health. Curr Nutr Food Sci 17(6):583–600. https://doi.org/10.2174/1573401316999200901111519

Mitra M, Mitra S, Nandi DK (2024) Vitamins, micronutrients, antioxidants, and nutraceuticals in neuroprotection: an overview. In: Gchi D, Chaurasia RN, Ohia SE (eds) A review on diverse neurological disorders: pathophysiology, molecular mechanisms and therapeutics. Elsevier, pp 585–601

Molloy AM (2020) Adverse effects on cognition caused by combined low vitamin B-12 and high folate status—we must do better than a definite maybe! Am J Clin Nutr 112(6):1422–1423. https://doi.org/10.1093/ajcn/nqaa286

Müller CP, Schumann G, Kornhuber J, Kalinichenko LS (2020) The role of serotonin in alcohol use and abuse. In: Müller CP, Cunningham KA (eds) Handbook of behavioral neuroscience, vol 31(31). Elsevier, pp 803–827. https://doi.org/10.1016/B978-0-444-64125-0.00041-4

Munteanu C, Schwartz B (2022) The relationship between nutrition and the immune system. Front Nutr 9:1082500. https://doi.org/10.3389/fnut.2022.1082500

Mustafa NS, Abu Bakar NH, Mohamad N, Mohd Adnan LH, Md Fauzi NFA, Thoarlim A, Syed Omar SH, Hamzah MS, Yusoff Z, Jufri M, Ahmad R (2020) Review paper: MDMA and the brain: a short review on the role of neurotransmitters in neurotoxicity. Basic and clinical. Neuroscience 11(4):381–388. https://doi.org/10.32598/bcn.9.10.485

Nagarajan MK, Goodman D (2020) Not just substance use: the critical gap in nutritional interventions for pregnant women with opioid use disorders. Public Health 180:114–116. https://doi.org/10.1016/j.puhe.2019.10.025

Olson B, Marks DL, Grossberg AJ (2020) Diverging metabolic programmes and behaviours during states of starvation, protein malnutrition, and cachexia. J Cachexia Sarcopenia Muscle 11(6):1429–1446. https://doi.org/10.1002/jcsm.12630

Pal MM (2021) Glutamate: the master neurotransmitter and its implications in chronic stress and mood disorders. Front Hum Neurosci 15:722323. https://doi.org/10.3389/fnhum.2021.722323

Pennock S (2021) Dysfunctional eating in recovering addicts: a therapist's shift to an attachment-focused approach. In: Attachment, relationships and food: from cradle to kitchen. Routledge, pp 96–115

Pickering G, Noah L, Pereira B, Goubayon J, Leray V, Touron A, Macian N, Bernard L, Dualé C, Roux V, Chassain C (2023) Assessing brain function in stressed healthy individuals following the use of a combination of green tea, Rhodiola, magnesium, and B vitamins: an fMRI study. Front Nutr 10:1211321. https://doi.org/10.3389/fnut.2023.1211321

Reist C, Petiwala I, Latimer J, Raffaelli SB, Chiang M, Eisenberg D, Campbell S (2022) Collaborative mental health care a narrative review. Medicine 101(52):e32554. https://doi.org/10.1097/MD.0000000000032554

Roque Bravo R, Faria AC, Brito-Da-costa AM, Carmo H, Mladěnka P, Dias da Silva D, Remião F (2022) Cocaine: an updated overview on chemistry, detection, biokinetics, and pharmacotoxicological aspects including abuse pattern. Toxins 14(4):278. https://doi.org/10.3390/toxins14040278

Sandoval C, Farías J, Zamorano M, Herrera C (2022) Vitamin supplements as a nutritional strategy against chronic alcohol consumption? An updated review. Antioxidants 11(3):564. https://doi.org/10.3390/antiox11030564

Sekhotha MM, Sekhotha TOL (2024) The effect of administration of substance abuse among people suffering from substance sue disorder on the digestive system: a systematic review. In: Metabolic syndrome—lifestyle and biological risk factors, IntechOpen, pp 1–28. https://doi.org/10.5772/intechopen.115034

Serefko A, Jach ME, Pietraszuk M, Świąder M, Świąder K, Szopa A (2024) Omega-3 polyunsaturated fatty acids in depression. Int J Mol Sci 25(16):8675. https://doi.org/10.3390/ijms25168675

Serwatka CA, Griebel-Thompson AK, Eiden RD, Kong KL (2023) Nutrient supplementation during the prenatal period in substance-using mothers: a narrative review of the effects on offspring development. Nutrients 15(13):2990. https://doi.org/10.3390/nu15132990

Shawl M, Geetha T, Burnett D, Babu JR (2024) Omega-3 supplementation and its effects on osteoarthritis. Nutrients 16(11):1650. https://doi.org/10.3390/nu16111650

Shih YJ, Chang WN, Yang SW (2023) Heroin-induced osteoporosis presented with bilateral femoral neck insufficiency fractures in a male adult: a case report. BMC Musculoskelet Disord 24(1):296. https://doi.org/10.1186/s12891-023-06377-y

Simpson S, Mclellan R, Wellmeyer E, Matalon F, George O (2022) Drugs and bugs: the gut-brain axis and substance use disorders. J Neuroimmune Pharmacol 17(1–2):1–29. https://doi.org/10.1007/s11481-021-10022-7

Sinha R (2024) Stress and substance use disorders: risk, relapse, and treatment outcomes. J Clin Invest 134(16). https://doi.org/10.1172/JCI172883

Siomek-Gorecka A, Dlugosz A, Czarnecki D (2021) The molecular basis of alcohol use disorder (AUD). Genetics, epigenetics, and nutrition in AUD: an amazing triangle. Int J Mol Sci 22(8):4262. https://doi.org/10.3390/ijms22084262

Stachowicz K (2023) The role of polyunsaturated fatty acids in neuronal signaling in depression and cognitive processes. Arch Biochem Biophys 737:109555. https://doi.org/10.1016/j.abb.2023.109555

Strang J, Volkow ND, Degenhardt L, Hickman M, Johnson K, Koob GF, Marshall BDL, Tyndall M, Walsh SL (2020) Opioid use disorder. Nat Rev Dis Primers 6(1):3. https://doi.org/10.1038/s41572-019-0137-5

Teleanu RI, Niculescu AG, Roza E, Vladâcenco O, Grumezescu AM, Teleanu DM (2022) Neurotransmitters—key factors in neurological and neurodegenerative disorders of the central nervous system. Int J Mol Sci 23(11):5954. https://doi.org/10.3390/ijms23115954

Theodorakis Y, Hassandra M, Panagiotounis F (2024) Enhancing substance use disorder recovery through integrated physical activity and behavioral interventions: a comprehensive approach to treatment and prevention. Brain Sci 14(6):534. https://doi.org/10.3390/brainsci14060534

Thomas KR, Watt J, Wu CMJ, Akinrinoye A, Amjad S, Colvin L, Cowe R, Duncan SH, Russell WR, Forget P (2022) Pain and opioid-induced gut microbial dysbiosis. Biomedicine 10(8):1815. https://doi.org/10.3390/biomedicines10081815

Turner BRH, Jenkinson PI, Huttman M, Mullish BH (2024) Inflammation, oxidative stress and gut microbiome perturbation: a narrative review of mechanisms and treatment of alcohol hangover. Alcohol: clinical and experimental. Research 48(8):1451–1465. https://doi.org/10.1111/acer.15396

Ubeja R, Szujewski C (2022) The central mechanisms by which illicit drugs lead to addiction. Journal of student. Research 11(4):1–11. https://doi.org/10.47611/jsrhs.v11i4.3077

Varghese J, Dakhode S (2022) Effects of alcohol consumption on various systems of the human body: a systematic review. Cureus 14(10):e30057. https://doi.org/10.7759/cureus.30057

Volkow ND, Blanco C (2023) Substance use disorders: a comprehensive update of classification, epidemiology, neurobiology, clinical aspects, treatment and prevention. World Psychiatry 22(2):203–229. https://doi.org/10.1002/wps.21073

Volkow ND, Morales M (2022) The role of nutrition in addiction recovery. J Addict Med 14(5):310–316

Wangensteen T, Hystad J (2022) A comprehensive approach to understanding substance use disorder and recovery: former patients' experiences and reflections on the recovery process four years after discharge from SUD treatment. J Psychosoc Rehabil Ment Health 9(1):45–54. https://doi.org/10.1007/s40737-021-00233-9

White B, Sirohi S (2024) A complex interplay between nutrition and alcohol use disorder: implications for breaking the vicious cycle. Curr Pharm Des 30(23):1822–1837. https://doi.org/10.2174/0113816128292367240510111746

Wiss DA (2019) The role of nutrition in addiction recovery: what we know and what we don't. In: In the assessment and treatment of addiction: best practices and new frontiers, pp 21–42. https://doi.org/10.1016/B978-0-323-54856-4.00002-X

Zhu J, Liu Q, Kong B, Shuai W, Xiao Z, Dai C, Huang H (2023) From consumption to withdrawal: a broad synopsis of the effects of alcohol. Med Adv 1(4):330–339. https://doi.org/10.1002/med4.46

Zieglgänsberger W, Brenneisen R, Berthele A, Wotjak CT, Bandelow B, Tölle TR, Lutz B (2022) Chronic pain and the endocannabinoid system: smart lipids-a novel therapeutic option? Med Cannabis Cannabinoids 5(1):61–75. https://doi.org/10.1159/000522432

Nutritional Interventions for Alcohol Use Disorder

18

Nihat Mustafayev

Abstract

In this chapter, alcohol-induced malnutrition, a significant cause of nutritional deficiencies frequently observed in developed countries, is discussed. Alcohol consumption can lead to a wide spectrum of health issues, ranging from mild symptoms such as burning sensations in the hands and feet and skin sensitivity to more severe conditions like personality changes, confusion, and even coma. This chapter provides a detailed overview of alcohol-induced nutritional deficiencies, the impact of gender differences, and their effects on the body, offering a comprehensive guide for managing these challenges.

Keywords

Alcohol use disorder · Nutritional deficiency · Neurological health · Malnutrition · Gender differences · Nutritional interventions

18.1 Introduction

In developed countries, the increasing ease of access to various products and the rise in mood disorders have led to a growing inclination toward the use of stimulants. Among the most commonly used stimulants in society today is alcohol, which has been an integral part of human social life for thousands of years. Though the exact origins of alcohol and the source from which it was first derived remain unknown, its influence on human life is undeniable. It was not until the early twentieth century that the harmful effects of alcohol on health were fully recognized, and alcoholism

N. Mustafayev (✉)
Ozel Capa Private Hospital, Istanbul, Türkiye

W. Mohamed, A. Özge (eds.), *Feeding the Mind: The Connection Between Diet, Drugs, and Mental Health Volume 1*, Nutritional Neurosciences,
https://doi.org/10.1007/978-981-95-1414-4_18

was classified as a significant medical condition, prompting extensive research in the field.

Today, alcoholism is a major public health concern worldwide. In developed countries, alcohol consumption is a widespread problem and a leading cause of nutritional deficiencies. On a global scale, alcohol is considered the third largest risk factor for diseases, being linked to over 200 health conditions and responsible for approximately 4% of all deaths (Marfella et al. 2006). Alcohol negatively impacts nutrition through its direct effects on food intake and metabolism, as well as the organ damage it causes (Gepner et al. 2015). While some evidence suggests that moderate alcohol consumption may reduce certain cardiovascular and cerebrovascular risks, excessive alcohol use leads to severe health issues and mortality (Mukamal 2020; Marfella et al. 2006).

Excessive alcohol consumption has led to significant health and social issues in many countries, resulting in restrictive policies and awareness campaigns. This section delves into the epidemiology, pathogenesis, evaluation, and treatment approaches related to malnutrition arising from chronic and heavy alcohol consumption, providing a comprehensive understanding of this critical health issue. This exploration aims to deepen our awareness of the complex effects of alcohol on the body and to promote a better-informed approach to addressing its consequences.

18.2 Alcohol Consumption and Risk Levels

Alcohol consumption patterns vary widely across populations and have significant historical, social, and health implications. This section will explore how alcohol impacts human health, particularly in the context of nutritional deficiencies. Excessive alcohol consumption is a leading cause of malnutrition due to both decreased nutrient intake and metabolic disruptions.

The concept of a "standard drink" varies between countries. In the United States, alcohol consumption is categorized based on the quantity and frequency of intake:

- Moderate Drinking: Up to 1 drink per day for women and up to 2 drinks per day for men, with a weekly maximum of 7 drinks for women and 14 drinks for men.
- Heavy Drinking: More than 3 drinks in a single day or over 7 drinks per week for women; more than 4 drinks in 1 day or exceeding 14 drinks per week for men.
- Binge Drinking: Rapid intake of alcohol, leading to high blood alcohol levels—defined as 4 or more drinks during a single session for women, and 5 or more for men (Kloner and Rezkalla 2007).

These classifications are vital for understanding the risks associated with alcohol use and its impact on neurological and overall health (Table 18.1).

Table 18.1 Drinking level definitions (Gunzerath et al 2004)

Drinking level	Females	Males	Age 65 and over
Moderate drinking	Daily ≤1 drink (or ≤ 3 drinks on any single day) and weekly <7 drinks	Daily ≤2 drinks (or ≤4 drinks on any single day) and weekly <14 drinks	Daily ≤1 drink and weekly <7 drinks [10]
Heavy drinking	Daily >3 drinks or weekly >7 drinks	Daily >4 drinks or weekly >14 drinks	N/A
Binge drinking	≥4 drinks in one drinking occasion	≥5 drinks in one drinking occasion	N/A

18.3 Gender Differences in Alcohol Consumption

The safe alcohol intake for women is lower than for men due to physiological differences, including smaller body size, a lower percentage of body water, and reduced gastric alcohol dehydrogenase activity (White 2020). Recent trends show a rise in alcohol consumption among women, narrowing the historical gender gap. This shift is influenced by changing societal norms, increased workforce participation, and greater social acceptance of drinking. However, women remain more vulnerable to alcohol-related harm due to slower metabolism and higher intoxication levels.

For populations such as those with alcohol-related liver disease, pregnancy, or alcohol use disorder, complete abstinence is essential (Moinuddin et al. 2016). Additionally, conditions like active gastritis or esophagitis require strict limitations on alcohol intake (Moinuddin et al. 2016).

18.4 Alcohol Consumption and Neurological Health

Alcohol consumption has been an integral part of human culture for centuries, with both potential benefits and significant risks. The impact of alcohol on neurological health is nuanced, as both light and heavy consumption influence brain function in different ways.

Moderate alcohol consumption has been associated with reduced risks of certain neurological diseases. The "French Paradox" suggests lower incidences of coronary heart disease in France, partly due to frequent consumption of red wine, which contains polyphenols and resveratrol (Fragopoulou and Antonopoulou 2020). These compounds are believed to have antioxidant and anti-inflammatory properties, potentially reducing the risk of neurodegenerative diseases such as Alzheimer's (Meza et al. 2022).

Chronic heavy alcohol consumption is linked to a wide range of neurological complications that impact the central and peripheral nervous systems. The severity and type of neurological issues depend on several factors, including the amount and duration of alcohol consumption, nutritional status, and individual metabolic activity. Table 18.2 summarizes the key alcohol-related neurological conditions and their primary characteristics:

Table 18.2 Alcohol-related neurological risks

Neurological issue	Description	Key symptoms	Contributing factors
Wernicke-Korsakoff syndrome	Acute (confusion, ataxia) and chronic phases (memory deficits) due to thiamine deficiency	Confusion, ataxia, ophthalmoplegia, memory loss	Thiamine deficiency, poor nutrition
Alcohol use disorder (AUD)	Chronic alcohol use affecting brain structure and function	Memory deficits, cognitive impairment	High alcohol intake
Alcoholic dementia	Direct neurotoxic effect of alcohol on the frontal lobes	Cognitive impairment, cerebral atrophy, ventricular enlargement	Prolonged heavy alcohol use
Alcoholic cerebellar degeneration	Damage to the cerebellum due to alcohol toxicity	Severe ataxia, gait disturbances	Alcohol toxicity
Marchiafava-Bignami disease	Demyelination and necrosis of the central corpus callosum	Disorientation, ataxia, seizures	Long-term heavy alcohol use
Alcoholic amblyopia	Optic neuropathy related to vitamin B deficiency and alcohol toxicity	Vision loss, central scotomas	Vitamin B deficiency, alcohol toxicity
Alcoholic myopathy	Muscle weakness and atrophy	Muscle pain, weakness, rhabdomyolysis	Chronic alcohol use
Alcoholic polyneuropathy	Damage to peripheral nerves	Burning sensations, muscle weakness, imbalance	Alcohol toxicity, nutritional deficiency

18.5 Pathophysiological Overview

Alcohol exerts its neurotoxic effects through multiple mechanisms, including interference with neurotransmitter systems such as gamma-aminobutyric acid (GABA), glutamate, and so on. Chronic alcohol exposure leads to significant structural changes in the brain, including reduced cortical volume and widespread neuronal damage. Disorders such as Wernicke-Korsakoff syndrome often result from alcohol-related nutritional deficiencies, notably thiamine depletion.

The cessation of alcohol consumption in individuals with alcohol dependence can lead to alcohol withdrawal syndrome, which presents as a range of symptoms from mild anxiety and tremors to life-threatening seizures and delirium tremens. Prompt medical management is imperative to mitigate these risks and support the patient through the withdrawal process safely.

Chronic alcohol consumption has profound neurotoxic effects on multiple brain regions, including the frontal lobes, leading to alcohol-related dementia, and the cerebellum, resulting in cerebellar degeneration. Additionally, optic neuropathy, such as alcoholic amblyopia, may develop, particularly in those with inadequate vitamin B intake.

Alcohol also significantly affects skeletal muscle, leading to alcoholic myopathy, which presents as muscle weakness, pain, and atrophy, particularly in proximal muscle groups. Peripheral neuropathy, or alcoholic polyneuropathy, is a common consequence of sustained alcohol abuse, characterized by sensory disturbances, weakness, and autonomic dysfunction, often initiating in the lower extremities and progressing proximally.

18.6 Alcohol Consumption in the Body

Alcohol is absorbed into the systemic circulation through the gastric and intestinal mucosa. Absorption depend on gastric emptying, ethanol concentration, and beverage properties. After absorption, it is quickly distributed throughout body fluids; most is metabolized in the liver, while a small fraction is excreted via the lungs and kidneys. The toxicodynamic profile of alcohol spans multiple systems. In the central nervous system, it impairs attention and coordination, and at higher exposures may trigger seizures and delirium. In the cardiovascular system, it increases the risk of arrhythmias and hypertension. In the digestive system, it contributes to hepatic steatosis, pancreatitis, and progressive liver injury leading to cirrhosis (Varghese & Dakhode, 2022)

18.7 Epidemiology and Pathogenesis of Alcohol-Related Malnutrition

The prevalence of malnutrition in individuals with chronic alcohol use ranges from 20% to 60%, depending on the presence and severity of cirrhosis. Both acute and chronic alcohol consumption contribute to malnutrition by reducing caloric intake, impairing digestion and nutrient absorption, decreasing protein synthesis, increasing intestinal protein catabolism, and accelerating nutrient breakdown and excretion (Rossi et al. 2015) (Table 18.3).

The degree of malnutrition depends on the amount of alcohol consumed, the quality of the diet, individual genetic factors, and the presence and severity of comorbidities. When alcohol comprises over 30% of total caloric intake, the likelihood of macro- and micronutrient deficiencies significantly increases.

Table 18.3 Malnutrition assessment criteria (2)

Criteria (Sheean et al 2019)	Description
Insufficient energy intake	Lack of adequate calorie intake
Weight loss	Decrease in body weight
Decreased muscle mass	Loss of muscle tissue
Loss of subcutaneous fat	Reduction in fat under the skin
Edema	Localized or generalized swelling (can mask weight loss)
Reduced handgrip strength	Measured using a dynamometer

18.8 Nutritional Management in Alcohol Use

Chronic alcohol consumption often leads to significant nutritional deficiencies, which, if left untreated, can result in serious neurological complications. However, with appropriate care and timely intervention, many of these problems can be prevented or minimized.

Proper nutrition plays a central role in managing the health consequences of alcohol use. Chronic alcohol consumption often leads to malnutrition, which can exacerbate neurological symptoms. A comprehensive nutritional assessment should begin with a physical examination, evaluating signs such as reduced skin turgor, muscle weakness (notably decreased handgrip strength), swelling in the lower extremities, and abdominal edema in severe malnutrition cases. In addition to physical signs, the patient's alcohol consumption habits should be thoroughly assessed to determine the potential impact on their neurological health (Pieterse et al. 2002).

18.9 Neurological Complications

Alcohol can cause severe neurological damage, especially with long-term use. Conditions like Wernicke-Korsakoff syndrome primarily arise due to a deficiency in thiamine (vitamin B1), a nutrient frequently depleted in heavy drinkers. This deficiency leads to serious cognitive dysfunction and physical symptoms, including confusion, ataxia, and vision problems. Wernicke-Korsakoff syndrome represents one of the most critical emergency syndromes associated with alcohol use. To prevent such neurological complications, reducing alcohol consumption—or ideally quitting—is essential. In addition, maintaining a balanced diet rich in vitamins and minerals helps stabilize nerve function and supports recovery (Markowitz 2000).

For individuals undergoing alcohol withdrawal, medical supervision is vital. Withdrawal can be extremely dangerous, with symptoms ranging from tremors and anxiety to more severe conditions like seizures and delirium tremens. Close monitoring and medications such as benzodiazepines can help manage these symptoms and prevent serious complications. It is also crucial to provide ongoing support to address the underlying causes of alcohol dependence, such as through counseling and rehabilitation services that assist patients in staying sober and avoiding relapse (Varghese et al 2022).

Micronutrient deficiencies are prevalent among individuals with high alcohol consumption and contribute to systemic health problems. Common deficiencies include vitamin B1 (thiamine), vitamin B2 (riboflavin), vitamin B6 (pyridoxine), folate, vitamin B12 (cobalamin), and vitamin C, as well as minerals like calcium, magnesium, and phosphorus. These deficiencies can manifest in a range of symptoms, as summarized in Table 18.4.

To effectively manage these deficiencies, a personalized nutritional support plan is essential. This plan should involve collaboration between a physician and a dietitian. If a significant calorie deficit, muscle mass loss, or presence of edema is observed, a targeted supplementation program should be implemented based on individual needs, as detailed in Table 18.5.

Table 18.4 Common vitamin and mineral deficiencies in heavy alcohol use (Markowitz et al. 2000)

Vitamin/mineral	Deficiency symptoms
Vitamin B1 (thiamine)	Peripheral neuropathy, cardiomyopathy, Wernicke-Korsakoff syndrome
Vitamin B2 (riboflavin)	Oropharyngeal inflammation, behavioral changes, neuropathy, anemia, dermatitis
Vitamin B6 (pyridoxine)	Peripheral neuropathy, stomatitis, confusion, depression
Folate	Macrocytic anemia, intestinal malabsorption
Vitamin B12 (cobalamin)	Macrocytic anemia, neuropsychiatric symptoms (cognitive impairment, irritability, gait abnormalities, peripheral neuropathy)
Vitamin C	Scurvy (bleeding gums, ecchymoses, fatigue, depression)
Calcium, magnesium, phosphorus	Muscle weakness, lethargy, nausea, osteomalacia, osteopenia, fractures

Table 18.5 Nutritional support and supplementation for alcohol use disorder

Nutrient supplement	Recommended dose
Vitamin B1 (thiamine)	100 mg per day
Vitamin B6 (pyridoxine)	2 mg per day
Folate	400 mcg to 1 mg per day
Multivitamin composition	Should contain thiamine, pyridoxine, and folic acid

Nutritional support plays a crucial role in improving overall health and aiding the recovery process. Addressing these deficiencies helps stabilize neurological health and enhances the patient's capacity to recover from alcohol dependence (Silva et al. 2011).

18.10 A Multidisciplinary Approach

Managing alcohol-related neurological and nutritional complications requires a proactive and integrated approach. Medications for withdrawal management, combined with supportive measures such as hydration, adequate nutrition, and psychological support, can significantly improve recovery outcomes. Additionally, collaboration among physicians, dietitians, and mental health professionals provides a holistic support network for patients. Long-term management also involves addressing the underlying causes of alcohol dependence and offering counseling and rehabilitation services to help patients maintain sobriety.

In conclusion, the key to managing alcohol-related complications lies in an integrated approach that focuses on both medical and nutritional health. By combining medical treatment, targeted supplementation, and behavioral therapy, patients can achieve meaningful recovery and reduce the risk of long-term neurological and systemic damage.

References

Fragopoulou E, Antonopoulou S (2020) The French paradox three decades later: role of inflammation and thrombosis. Clin Chim Acta 510:160–169. https://doi.org/10.1016/j.cca.2020.07.014

Gepner Y, Golan R, Harman-Boehm I (2015) Effects of initiating moderate alcohol intake on cardiometabolic risk in adults with type 2 diabetes: a 2-year randomized, controlled trial. Ann Intern Med 163(8):569–579. https://doi.org/10.7326/M14-1650

Gunzerath L, Faden V, Zakhari S, Warren K (2004) National Institute on Alcohol Abuse and Alcoholism report on moderate drinking. Alcoholism: Clinical and Experimental Research, 28(6):829–847.https://doi.org/10.1097/01.ALC.0000128382.79375.B6

Kloner RA, Rezkalla SH (2007) To drink or not to drink? That is the question. Circulation 116:1306–1317. https://doi.org/10.1161/CIRCULATIONAHA.106.678375

Marfella R, Cacciapuoti F, Siniscalchi M (2006) Effect of moderate red wine intake on cardiac prognosis after recent acute myocardial infarction in subjects with type 2 diabetes mellitus. Diabet Med 23(9):974–981. https://doi.org/10.1111/j.1464-5491.2006.01908.x

Markowitz, John S, Aimee L, McRae, Susan C (2000) Sonne. Oral nutritional supplementation for the alcoholic patient: a brief overview. Annals of Clinical Psychiatry 12.3: 153–158

Meza V, Arnold J, Díaz LA, Valverde MA, Idalsoaga F, Ayares G, Devuni D, Arab JP (2022) Alcohol consumption: Medical implications, the liver and beyond. Alcohol and Alcoholism, 57(3):283–291. OxfordUniversity Press, Oxford, UK. https://doi.org/10.1093/alcalc/agac013

Moinuddin A, Goel A, Saini S, Bajpai A, Misra R (2016) Alcohol consumption and gender: a critical review. J Psychol Psychother 6(3):1–4

Mukamal KJ (2020) A safe level of alcohol consumption: the right answer demands the right question. J Intern Med 288(5):550–561. https://doi.org/10.1111/joim.13096

Sheean PM, Peterson SJ, Chen Y, Liu D, Lateef O, Braunschweig CA (2019) American Society for Parenteral and Enteral Nutrition clinical guidelines: The validity of body composition assessment in clinicalpopulations. J of Parenteral and Enteral Nutrition, 44(1):12–43. https://doi.org/10.1002/jpen.1669

Pieterse S, Manandhar M, Ismail S (2002) The association between nutritional status and handgrip strength in older Rwandan refugees. Eur J Clin Nutr 56(10):933–939. https://doi.org/10.1038/sj.ejcn.1601443

Rossi RE, Conte D, Massironi S (2015) Diagnosis and treatment of nutritional deficiencies in alcoholic liver disease: Overview of available evidence and open issues. Digestive and Liver Disease, 47(9):819–825. https://doi.org/10.1016/j.dld.2015.05.021

Silva LF, Matos CM, Lopes GB et al (2011) Handgrip strength as a simple indicator of possible malnutrition and inflammation in men and women on maintenance hemodialysis. J Ren Nutr 21:235

Varghese J, Dakhode S (2022) Effects of alcohol consumption on various systems of the human body: A systematic review. Cureus, 14(10):e30057. DMIHER School of Epidemiology and Public Health, Datta MegheInstitute of Medical Sciences, Wardha, India. https://doi.org/10.7759/cureus.30057

White AM (2020) Gender differences in the epidemiology of alcohol use and related harms in the United States. Alcohol Res 40(2):1. https://doi.org/10.35946/arcr.v40.2.01

Dietary Approaches to Support Mental Health in Chronic Pain and Opioid Use Disorders

19

Gülce Coşku Yılmaz Çakan and Özge Yılmaz Küsbeci

Abstract

This chapter is structured to provide a comprehensive exploration of the intricate relationships between chronic pain (CP), opioid use disorders (OUD), and mental health, with a particular focus on the role of dietary strategies as part of an integrative care model. By combining insights from clinical research, epidemiology, and nutritional science, the chapter aims to elucidate how diet can complement conventional therapies to improve outcomes for individuals facing these interconnected challenges. The overarching goal is to provide actionable dietary interventions that can enhance treatment efficacy, support conventional therapies by addressing nutritional deficits and reducing opioid reliance, and promote holistic care. Ultimately, this chapter seeks to offer evidence-based insights and practical recommendations for integrating dietary approaches into the broader framework of care for patients contending with the complex challenges of CP, OUD, and mental health disorders.

Keywords

Chronic pain · Mental health · Opioid · Diet

G. C. Yılmaz Çakan (✉) · Ö. Yılmaz Küsbeci
İzmir University of Economics, Faculty of Medicine, Department of Neurology, Balçova – İzmir, Türkiye

W. Mohamed, A. Özge (eds.), *Feeding the Mind: The Connection Between Diet, Drugs, and Mental Health Volume 1*, Nutritional Neurosciences,
https://doi.org/10.1007/978-981-95-1414-4_19

19.1 Introduction

19.1.1 Chronic Pain, Opioid Use Disorders, and Mental Health

Chronic pain (CP) is a pervasive and multifaceted condition that affects millions of individuals globally, with significant implications for physical health, mental well-being, and daily activities. According to the European Pain Federation (EFIC), CP is defined as pain persisting for more than 3 months and is classified into primary and secondary categories. Primary CP conditions, such as fibromyalgia and nonspecific low-back pain, occur without an identifiable underlying disease. In contrast, secondary chronic pain arises from specific pathologies, including cancer, neuropathy, or musculoskeletal disorders (Häuser et al. 2021). The burden of CP is substantial, affecting an estimated 20% of the population in Europe and 22% of adults in the United States. This condition not only reduces the quality of life but also imposes significant economic and social costs, affecting families, communities, and healthcare systems (O'Brien et al. 2017).

The impact of CP on mental health is profound, as the condition often coexists with psychiatric disorders such as anxiety and depression. These comorbidities create a feedback loop that exacerbates pain perception, hinders daily functioning, and complicates treatment outcomes (Moriarty et al. 2011; Severino et al. 2018). Studies indicate that individuals with CP and depression are more likely to receive opioid prescriptions, often at higher dosages, leading to increased risks of dependency, cognitive impairments, and other adverse outcomes (Braden et al. 2009). Compounding this issue is the difficulty in adequately managing CP with conventional therapies, which leaves many patients reliant on opioids despite the well-documented risks of long-term use, including hyperalgesia, tolerance, and withdrawal symptoms (Fishbain et al. 2009).

The growing reliance on opioids to manage CP has paralleled the escalation of the opioid epidemic, presenting unique challenges for healthcare providers and policymakers. Opioids exert their primary analgesic effects through interactions with central μ-opioid receptors, but this same mechanism also underpins many of their side effects. These effects include sedation, cognitive impairment, and psychomotor slowing, which can interfere with daily activities and quality of life (Chapman et al. 2002). Additionally, the potential for opioid misuse, dependence, and overdose underscores the necessity for a multidisciplinary approach to pain management that minimizes harm while addressing the complex needs of patients.

Opioid Use Disorder (OUD) is a significant public health issue that encompasses both behavioral and physiological patterns of opioid misuse. Various factors contribute to an individual's vulnerability to OUD, including younger age, male gender, lower socioeconomic status, limited educational attainment, and racial or ethnic disparities in healthcare access. Mental health conditions, such as anxiety and depression, further heighten the risk, as these individuals may turn to opioids for their mood-enhancing effects. These risk factors highlight the complex interplay between individual vulnerabilities and systemic inequities that shape OUD prevalence and treatment outcomes (Siddiqui and Urman 2022).

The diagnostic criteria for OUD reflect a loss of control over opioid use and persistent consumption despite harmful consequences. Key indicators include consuming opioids in greater amounts or for longer periods than intended, persistent cravings, and unsuccessful attempts to reduce or stop usage. Many individuals with OUD dedicate excessive time to obtaining, using, or recovering from opioids, often at the expense of personal and professional responsibilities. Physiological markers such as tolerance—requiring progressively higher doses for the same effect—and withdrawal symptoms further perpetuate the cycle of dependence. Additionally, OUD is characterized by continued opioid use in physically dangerous situations or despite clear awareness of the harm caused to physical or mental health (Siddiqui and Urman 2022).

A history of mental illness is another critical risk factor for developing OUD, especially in acute pain management scenarios. Individuals with anxiety or depression may be particularly vulnerable to opioid misuse due to the temporary emotional relief opioids can provide. This underscores the importance of cautious prescribing practices, including avoiding long-acting opioids, employing multimodal approaches to pain management, and setting realistic expectations for pain relief. These strategies can reduce the risk of OUD while effectively addressing pain in vulnerable populations (Macintyre et al. 2022). Recognizing these interconnected risk factors and diagnostic features is essential for tailoring interventions to prevent and manage OUD effectively.

The interplay between CP, OUD, and mental health conditions has led to the recognition of a "triple diagnosis." This term describes the co-occurrence of these three challenges, which interact in ways that compound the burden on individuals and complicate treatment strategies (Delorme et al. 2021). For instance, untreated depression and anxiety can amplify the risk of opioid misuse, while opioid-related side effects can exacerbate mental health symptoms and heighten pain sensitivity. Moreover, disparities in access to care, influenced by factors such as race, socioeconomic status, and geography, often leave marginalized populations disproportionately affected (Siddiqui and Urman 2022).

Amid these challenges, dietary approaches are emerging as a promising adjunct to conventional treatments for CP and OUD. Nutrition plays a critical role in modulating inflammation, supporting neuroplasticity, and optimizing overall health. Anti-inflammatory diets rich in omega-3 fatty acids, antioxidants, and micronutrients have demonstrated potential in reducing pain severity, improving mood, and enhancing cognitive function. Moreover, addressing nutritional deficiencies common among individuals with CP and OUD can provide a foundation for better mental and physical health outcomes.

19.1.2 Why Diet Matters in Mental Health and Pain Management

19.1.2.1 Mental Health

Diet has a profound influence on mental health, shaping cognitive function and emotional well-being. Poor dietary quality, characterized by high intake of ultra-processed foods and low consumption of nutrient-dense options, is a recognized modifiable risk factor for mental disorders. Ultra-processed foods, often rich in refined sugars, unhealthy fats, and chemical additives, have been linked to increased odds of depressive and anxiety symptoms. Meta-analyses have shown that greater intake of such foods correlates with a higher risk of developing depression over time (Lane et al. 2022). Conversely, diets rich in whole, unprocessed foods, such as fruits, vegetables, whole grains, and omega-3 fatty acids, are associated with improved mental health outcomes, potentially through anti-inflammatory and neuroprotective effects (Valles-Colomer et al. 2019).

The global burden of mental disorders remains alarmingly high, with no significant reduction in prevalence over the past decades despite advancements in treatment options (Collaborators G 2019 MD 2022). This persistent challenge underscores the need to explore alternative preventive measures, including dietary interventions. Poor nutrition can exacerbate systemic inflammation, oxidative stress, and gut microbiome imbalances, all of which contribute to neuroinflammation and central nervous system dysregulation (Monteiro et al. 2019). Dietary interventions targeting these mechanisms have shown promise in normalizing glial activity and reducing neural sensitization, leading to potential improvements in symptoms of anxiety, depression, and cognitive decline.

Emerging evidence highlights the importance of tailored dietary patterns, as specific population groups respond differently to various diets. For instance, vegetarian or vegan diets were found to have mixed effects on mental health, with younger individuals showing higher anxiety levels while older individuals experienced reduced anxiety (Iguacel et al. 2021). These findings emphasize the need for individualized approaches in leveraging nutrition for mental health improvement.

19.1.2.2 Pain Management

Dietary patterns also play a crucial role in managing CP. CP conditions, such as rheumatoid arthritis, fibromyalgia, and neuropathic pain, are often associated with systemic inflammation and metabolic dysregulation, both of which are influenced by diet. Anti-inflammatory diets, including the Mediterranean diet, have demonstrated significant pain-relieving effects in conditions like rheumatoid arthritis. Studies reveal that the Mediterranean diet, characterized by high consumption of olive oil, nuts, fish, and fresh produce, is more effective at reducing pain than vegetarian or vegan diets (Schoenenberger et al. 2021). The anti-inflammatory properties of omega-3 fatty acids further contribute to alleviating pain, particularly in inflammatory joint conditions (Goldberg and Katz 2007). Therapeutic fasting, including intermittent fasting and time-restricted eating, has shown promise in reducing pain severity through multiple mechanisms. These dietary approaches improve insulin sensitivity, decrease oxidative stress, and reduce peripheral and central inflammation, which collectively lower the experience of pain (Michalsen and Li 2013). Additionally, fasting is linked to enhanced neuroplasticity and regenerative processes, potentially reversing neurobiological changes associated with CP (Sibille et al. 2016). CP severity is often linked to poor dietary habits, including high intake of fats and sugars. Plant-based diets, rich in vitamins, minerals, and antioxidants, have demonstrated pain-relieving effects by addressing nutritional deficiencies commonly observed in patients with CP (Elma et al. 2020).

In summary, diet significantly influences both mental health and pain management through its impact on systemic inflammation, oxidative stress, and neuroplasticity. Adopting high-quality, nutrient-rich diets while minimizing ultra-processed food intake can offer substantial benefits for individuals dealing with mental health disorders and CP conditions. This highlights the critical need for incorporating dietary counseling into holistic treatment plans.

19.2 The Role of Nutrition in Mental Health and Pain Management

19.2.1 Biochemical Pathways: How Nutrition Impacts Mood and Pain

19.2.1.1 Nutrition and Pain

The intricate relationship between nutrition, pain, and mental health has gained substantial attention in recent years, driven by growing evidence that diet is a critical modifiable factor influencing these domains. Pain and mental health, often interconnected, are impacted by the biochemical pathways that nutrition modulates.

Emerging evidence underscores the role of diet as a pivotal factor in modulating pain. A systematic review of chronic musculoskeletal pain revealed that dietary modifications can exert analgesic effects in seven out of nine studies reviewed, with plant-based diets showing promise in alleviating pain in conditions such as rheumatoid arthritis and fibromyalgia. Inadequate intake of nutrients like calcium, zinc, magnesium, and vitamin B6 is commonly observed in individuals with rheumatoid arthritis, while those with fibromyalgia often exhibit deficiencies in carbohydrates, proteins, lipids, and various vitamins and minerals (Elma et al. 2020).

The influence of specific macronutrients and micronutrients on pain is equally compelling. Excessive consumption of fats and sugars has been positively correlated with increased pain intensity and reduced pain thresholds, as observed in osteoarthritis and fibromyalgia. Conversely, adequate protein intake appears to mitigate pain, possibly through mechanisms like the tryptophan-serotonin metabolic pathway, which influences pain perception (Batista et al. 2016). Moreover, omega-3 fatty acids, widely studied for their anti-inflammatory properties, have demonstrated efficacy in reducing pain associated with autoimmune diseases such as rheumatoid arthritis and lupus erythematosus. These fatty acids modulate the inflammatory process by affecting eicosanoid synthesis through the CYP-epoxygenase pathway (Fischer et al. 2014; Brown et al. 2020).

The Westernized diet, characterized by a high omega-6 to omega-3 fatty acid ratio and abundant trans fats, exacerbates systemic inflammation by loading the arachidonic acid pathway and promoting the release of pro-inflammatory cytokines. This imbalance contributes to heightened pain sensitivity in chronic inflammatory conditions (Oteng and Kersten 2020; Marion-Letellier et al. 2015). The adoption of diets rich in anti-inflammatory nutrients, including omega-3 fatty acids, polyphenols, and bioactive compounds, has been associated with reductions in oxidative stress, central nervous system sensitization, and neuroinflammation—mechanisms implicated in CP states (Bjørklund et al. 2019).

Additionally, alterations in the gut microbiome, triggered by dietary components, have been implicated in pain modulation. Increased intestinal permeability allows translocation of bacterial fragments like lipopolysaccharides, which can activate immune responses and amplify inflammation, further sensitizing pain pathways. This interaction highlights the potential of gut-targeted nutritional interventions to address systemic inflammation and pain (Rizzetto et al. 2018).

19.2.1.2 Nutrition and Mental Health

In parallel with its impact on pain, nutrition exerts profound effects on mental health by modulating biochemical pathways involved in mood regulation, cognitive function, and stress response. The brain relies on essential nutrients to maintain neurotransmitter balance and support neuroendocrine systems. Deficiencies in omega-3 polyunsaturated fatty acids (PUFAs) like eicosapentaenoic acid (EPA) and docosahexaenoic acid (DHA) have been linked to mood disorders, given their role in anti-inflammatory cytokine production and structural integrity of neuronal membranes (Grosso et al. 2014). Conversely, an imbalance favoring omega-6 fatty acids contributes to neuroinflammation and has been associated with increased risk of depression.

Ultra-processed foods, which dominate many modern diets, lack vital fibers, vitamins, and bioactive compounds, while contributing to systemic inflammation, oxidative stress, and dysbiosis of the gut microbiome. These factors collectively influence mental health by disrupting neurotransmitter synthesis and impairing gut-brain signaling pathways. For instance, emulsifiers and titanium dioxide nanoparticles commonly found in processed foods have been shown to promote inflammation and neurodegeneration, exacerbating symptoms of anxiety and depression (Chassaing et al. 2015).

Macronutrients, particularly carbohydrates, also play a dual role in mental health. While high-quality complex carbohydrates can enhance serotonin synthesis and improve mood, excessive consumption of refined sugars has been linked to hippocampal neuroinflammation and cognitive decline (Irwin et al. 2016; Wurtman and Wurtman 2018). Proteins and amino acids, such as tryptophan, serve as precursors to serotonin and melatonin, contributing to sleep regulation and emotional stability. Glycine, an amino acid, has been shown to improve sleep quality by modulating core body temperature and neurotransmitter activity (Bannai and Kawai 2012).

Additionally, micronutrients like B vitamins are crucial for maintaining cognitive function and mood. For instance, niacin has shown potential in managing schizophrenia, while deficiencies in folate and vitamin B12 are associated with depressive symptoms (Marx et al. 2021). Moreover, food insecurity, a surrogate marker for poor dietary quality, has been linked to heightened stress and depression, emphasizing the broader socioeconomic implications of diet on mental health (Gerritsen et al. 2021).

19.2.2 The Gut-Brain Axis and Inflammation in Mental Health

19.2.2.1 The Gut-Brain Axis in Mental Health

The gut-brain axis represents a dynamic bidirectional communication network connecting the gastrointestinal tract and the central nervous system (CNS), influenced by neural, endocrine, immune, and metabolic pathways. Recent research underscores the role of gut microbiota in modulating mental health through their impact on neurotransmitter synthesis, neuroinflammation, and systemic immune responses. Dysregulation of this axis, often caused by an imbalance in gut microbial composition (dysbiosis), has been implicated in psychiatric conditions such as anxiety and depression.

The human gut harbors a complex microbial ecosystem comprising over 1000 species of bacteria, with individuals carrying an average of 160 species (Berding et al. 2021). The phyla Firmicutes and Bacteroidetes dominate, constituting approximately 70–90% of a healthy gut microbiome (Park 2018). While a universally accepted definition of a "healthy microbiota" remains elusive, attributes like resistance to disruptions and resilience after perturbations are considered vital (Berding et al. 2021).

Specific bacterial strains play a critical role in mental health. Beneficial bacteria such as *Lactobacillus* and *Bifidobacterium* are associated with anti-inflammatory effects and neurotransmitter modulation, whereas the overgrowth of pathogenic taxa like *Escherichia coli* and *Enterobacteriaceae* is linked to inflammatory states and mood disorders (Ogunrinola et al. 2020). Additionally, the gut microbiota directly influences neurotransmitter production, with SCFAs like butyrate promoting the synthesis of brain-derived neurotrophic factor (BDNF) and regulating pathways involved in learning, memory, and mood (Berding et al. 2021).

Dysbiosis often results in increased intestinal permeability, or "leaky gut," facilitating the entry of microbial products such as lipopolysaccharides (LPS) into the bloodstream (Singh et al. 2022). LPS binds to monocytes and macrophages, inducing the release of pro-inflammatory cytokines like TNF-α, IL-1, and IL-6, which can exacerbate neuroinflammation and contribute to psychiatric disorders (Berding et al. 2021).

These inflammatory processes are tightly linked to the hypothalamic-pituitary-adrenal (HPA) axis, a critical regulator of stress responses. Chronic stress-induced dysbiosis has been shown to increase circulating cortisol levels, impair hippocampal function, and alter the composition of the gut microbiota, leading to reduced populations of beneficial strains such as *Lactobacillus* (Singh et al. 2022). This bidirectional relationship underscores the interconnected nature of gut microbiota and stress-related disorders.

The gut microbiota influences the production and turnover of neurotransmitters, including serotonin, gamma-aminobutyric acid (GABA), and glutamate, through various metabolic pathways. For instance, gut-derived serotonin not only regulates enteric nervous system functions but also modulates CNS processes, including mood regulation and inflammatory responses (Singh et al. 2022). Furthermore, gut peptides like cholecystokinin (CCK) and glucagon-like peptide-1 (GLP-1)

et al. 2020). Notably, 77% of individuals with CP exhibited at least one abnormal biomarker result. The most common biomarker abnormalities included elevated quinolinic acid (29%), indicative of neuroinflammation, and elevated pyroglutamate (19%), suggesting glutathione depletion. Deficiencies in vitamins, particularly B6 and B12, were also observed, with elevated xanthurenic acid (17%) linked to vitamin B6 insufficiency and elevated methylmalonic acid (10%) indicating vitamin B12 deficiency (Gunn et al. 2020).

Among the key micronutrients, vitamin D, omega-3 polyunsaturated fatty acids, magnesium, zinc, and folic acid were found to be low in individuals with CP. Low levels of vitamin D, for instance, are frequently observed and may contribute to increased inflammation, exacerbating pain symptoms (Fu et al. 2021). Additionally, omega-3 fatty acids, which have anti-inflammatory properties, are often underconsumed in individuals with CP, further complicating the inflammatory processes associated with these conditions (Nijs et al. 2020).

A diet deficient in essential vitamins and minerals can lead to increased oxidative stress and impaired immune function, both of which may worsen pain perception and chronicity. This has been shown in studies reporting high serum levels of glutamate and aspartate in patients with CP, such as those suffering from chronic migraine or fibromyalgia (Gunn et al. 2020). Furthermore, low zinc levels are linked to compromised immune responses, and magnesium deficiencies are associated with enhanced pain sensitivity due to disrupted NMDA receptor activity (Malcangio 2019).

19.2.3.2 Nutritional Deficiencies in Opioid Use Disorders

Individuals with OUD frequently experience nutritional deficiencies due to factors such as poor diet, altered metabolism, and gastrointestinal dysfunction. Opioid use disrupts various physiological processes, including nutrient absorption, and often leads to malnutrition (Wiss 2019). Opioids can trigger gastrointestinal issues such as constipation and delayed gastric emptying, further impairing nutrient absorption. This contributes to deficiencies in a range of essential nutrients, including thiamine, riboflavin, niacin, B6, folate, B12, and vitamins A, C, D, E, and K (Wiss 2019).

Deficient micronutrients like B6 and folate are particularly common in this population, with studies indicating deficiency rates of up to 40% for these vitamins (Wiss 2019). These vitamins play critical roles in nerve function and are essential for the synthesis of neurotransmitters, which may help explain why deficiencies are linked to neurological symptoms such as neuropathy and increased pain perception. Vitamin B12, essential for nerve repair and myelination, may also improve symptoms of neuropathy when supplemented (Didangelos et al. 2021).

Moreover, opioid users often exhibit altered lipid metabolism, including low cholesterol levels, which may have detrimental effects on mental health and increase the risk of relapse (Ojo et al. 2019). Cholesterol is involved in neurotransmitter production and is essential for brain function; its deficiency has been linked to depression and suicidal tendencies, highlighting the importance of addressing these deficiencies during addiction treatment (Buydens-Branchey and Branchey 2003). In addition to vitamins, minerals such as calcium, magnesium, selenium, and zinc are commonly deficient in individuals with OUD (Ross et al. 2012). Magnesium, for example, has been shown to act as an antagonist at NMDA receptors, which are involved in pain transmission, suggesting its potential role in pain management (Hou et al. 2020). Zinc is another essential nutrient for immune function and has demonstrated anti-inflammatory effects that may help alleviate CP (Hennigar et al. 2021; Prasad and Bao 2019).

19.2.3.3 The Impact of Chronic Pain and Opioid Use on Body Composition and Eating Behavior

Both CP and opioid use can lead to significant changes in body composition, often resulting in undernutrition or unhealthy weight gain. CP patients may experience weight loss due to decreased physical activity and poor eating habits, while opioid users are often underweight due to anorexia, gastrointestinal disturbances, and poor appetite regulation (Wiss 2019). This altered nutritional status may exacerbate symptoms of pain and prolong recovery times, as the body requires proper nutrition to heal and manage inflammation.

Additionally, the preference for high-sugar and low-nutrient foods among opioid users may contribute to the intake of empty calories, further compounding nutrient deficiencies (Jeynes and Gibson 2017). Sugar cravings, often replacing regular meals, are a well-documented symptom during opioid detoxification and recovery phases (Neale et al. 2012). This dietary imbalance, coupled with inadequate nutrient absorption due to opioid-related gastrointestinal dysfunction, can significantly hinder the recovery process.

19.2.3.4 Addressing Nutritional Deficiencies in Chronic Pain and Opioid Use Disorders

Given the high prevalence of nutritional deficiencies in both CP and OUD populations, addressing these deficiencies through targeted nutrition interventions is essential. Nutraceuticals such as St. John's Wort, curcumin, zinc, magnesium, vitamin D, and vitamin B have shown promise in treating neuropathic pain, with the potential to improve both nutritional status and pain outcomes (Abdelrahman and Hackshaw 2021). Supplementation with these micronutrients may help alleviate pain, improve immune function, and enhance overall quality of life for individuals with CP and OUD.

Moreover, addressing the gastrointestinal issues commonly seen in opioid users, such as constipation and malabsorption, is crucial for improving nutrient intake and absorption. Nutritional counseling and the implementation of a balanced, nutrient-dense diet may help reduce the impact of these deficiencies and improve the prognosis of individuals with CP and OUD.

19.3 Dietary Components Beneficial for Mental Health in Chronic Pain and Opioid Use Disorders

19.3.1 Omega-3 and Fatty Acids in Chronic Pain and Opioid Use Disorders

Omega-3 and other essential fatty acids (EFAs) play a pivotal role in supporting mental health, particularly in conditions like CP and OUDs. EFAs are crucial for maintaining the structural integrity of cell membranes (Grosso et al. 2014). Omega-3, -6, and -9 fatty acids share a competitive relationship in metabolic processes, as they rely on the same fatty acid desaturases for conversion. This competitive dynamic influences the body's ability to balance pro- and anti-inflammatory responses, which are critical for conditions involving CP and mental health issues (Grosso et al. 2014).

Numerous studies have highlighted the benefits of omega-3 fatty acids for mental health, particularly in reducing the risk of depression, anxiety, and stress. Diets rich in fiber and omega-3 have been shown to improve mood regulation, a critical factor in both CP and OUDs (Taylor and Holscher 2020). Omega-3 fatty acids, especially EPA and docosahexaenoic acid (DHA), exert anti-inflammatory effects that contribute to a decrease in CP, a hallmark of OUDs (Godos et al. 2020). These effects are thought to arise through the suppression of inflammatory cytokines and the reduction of arachidonic acid pathways, which are often overactive in pain-related conditions (Godos et al. 2020; Brain et al. 2019). Moreover, omega-3 s influence gut health, which is increasingly recognized as a crucial factor in mental well-being. They have been shown to promote the growth of beneficial gut bacteria, including butyrate-producing species like *Lactobacillus, Lachnospira, and Bifidobacterium*, all of which are linked to improved mental health outcomes (Berding et al. 2021). The impact of omega-3s on gut microbiota is thought to contribute to the brain-gut connection, modulating neuroinflammatory processes that can alleviate both mental health symptoms and pain.

The imbalance between omega-6 and omega-3 fatty acids, a common feature of Western diets, can exacerbate inflammation and pain. Excessive omega-6 intake—often from vegetable oils—has been shown to push the body toward a pro-inflammatory state, contributing to CP and metabolic disorders (Brain et al. 2019). In contrast, a diet with a balanced ratio of omega-3 to omega-6 fatty acids has been associated with improved systemic inflammation, providing a potential avenue for pain management and mental health support (Field et al. 2021). Specifically, higher omega-3 intake has been linked to a decrease in the production of pro-inflammatory molecules, while omega-6 fatty acids tend to promote inflammation through the activation of arachidonic acid (Field et al. 2021).

Furthermore, increasing omega-3 intake has been shown to mimic the effects of nonsteroidal anti-inflammatory drugs (NSAIDs) like ibuprofen by inhibiting cyclo-oxygenase activity and the secretion of pro-inflammatory prostaglandins (Brain et al. 2019). This mechanism supports its role in reducing both pain and neuroinflammation, which are common issues in CP and OUDs. For individuals with opioid dependence, omega-3 s may provide an alternative approach to reducing pain and inflammation, potentially reducing the reliance on opioid medications and mitigating withdrawal symptoms (Ramsden et al. 2013).

19.3.2 Antioxidants and Polyphenols in Chronic Pain and Opioid Use Disorders

Polyphenols, a diverse group of phytochemicals found in plant-based foods, are renowned for their antioxidant and anti-inflammatory properties. These compounds, which include flavonoids such as flavanones, isoflavones, and anthocyanins, as well as non-flavonoids like stilbenes, lignans, and tannins, play a significant role in modulating oxidative stress and inflammation, two key contributors to CP and OUDs (Berding et al. 2021). While 90–95% of polyphenols are not absorbed directly in the small intestine, they are fermented and metabolized by gut microbiota, which is increasingly recognized for its role in modulating systemic inflammation and mental health (Zhang and Tsao 2016). This biotransformation process can have profound effects on local inflammatory status and promote the growth of beneficial probiotics, further supporting gut health and, indirectly, mental well-being (Zhang and Tsao 2016).

Polyphenols are potent antioxidants due to their highly conjugated aromatic rings, which can donate electrons or hydrogen atoms to neutralize free radicals and ROS, both of which contribute to oxidative stress and inflammatory pathways (Berding et al. 2021). This antioxidant activity is particularly valuable in the context of CP, where oxidative stress exacerbates inflammation and tissue damage. Inflammation is a significant factor in conditions such as osteoarthritis, diabetic neuropathy, and chronic pancreatitis, all of which can involve both mental and physical distress. Studies have shown that polyphenols, found abundantly in fruits, vegetables, and whole grains, can reduce the inflammatory mediators involved in these conditions (Bjørklund et al. 2019).

In the context of CP, polyphenols have been shown to alleviate pain and inflammation through multiple mechanisms, including the inhibition of cyclooxygenase-2 (COX-2), an enzyme that plays a central role in pain and inflammation pathways (Bjørklund et al. 2019). For example, a randomized double-blind trial found that consumption of freeze-dried blueberries, a rich source of polyphenols, resulted in significant improvements in pain and inflammation in adults with knee osteoarthritis (Sun et al. 2020). This highlights the potential of polyphenol-rich foods as natural alternatives to traditional anti-inflammatory medications for managing CP.

In addition to their effects on inflammation and oxidative stress, polyphenols are also thought to have neuroprotective effects. Clinical studies have shown that polyphenol-rich fruits such as cherries, strawberries, and plums may reduce pain through anti-inflammatory and antioxidative mechanisms, as well as provide cardiovascular and metabolic benefits. The gut-brain axis, a bidirectional communication pathway between the gut microbiota and the brain, is another avenue through which polyphenols exert their mental health benefits. By modulating gut microbiota composition, polyphenols may influence neuroinflammation, which is implicated in both CP and OUDs (Berding et al. 2021).

Interestingly, polyphenols can also influence circadian and circannual rhythms, which in turn affect metabolic and neurobiological processes relevant to CP and opioid use. Research suggests that the consumption of polyphenol-rich seasonal fruits and vegetables may alter lipid metabolism and glucose homeostasis, potentially improving pain and inflammation associated with metabolic disorders (Mariné-Casadó et al. 2019). These findings suggest that not only the type of polyphenol consumed but also the timing of intake may influence their therapeutic effects.

19.3.3 Amino Acids and Protein Sources in Chronic Pain and Opioid Use Disorders

Amino acids play a crucial role in managing CP and OUDs through their involvement in neurotransmitter synthesis, antioxidant activity, and modulation of inflammation. Key neurotransmitters such as endorphin, serotonin, and GABA, which are derived from amino acids, have been shown to modulate pain and mental health. These compounds, synthesized from dietary amino acids, influence pain perception and mood, making proper protein intake essential for those dealing with persistent pain (Perna et al. 2020).

Increased dietary protein intake has been linked to reduced inflammation, partly due to the stimulation of insulin-like growth factor I (IGF-I) in the body. IGF-I is known for its anti-inflammatory properties, which are regulated through a complex network involving muscle-derived interleukin-6 (IL-6). High concentrations of IL-6 can lower IGF-I levels, while low IGF-I concentrations can stimulate IL-6 release, suggesting a delicate balance between these two molecules that influences muscle health and inflammation (Perna et al. 2020). As a result, dietary recommendations for individuals with CP often emphasize adequate protein intake, typically around 1 gram of protein per kilogram of body weight, spread across meals throughout the day. Protein sources rich in specific amino acids, such as glutamic acid and tryptophan, are particularly beneficial for individuals experiencing pain and inflammation.

Protein-rich foods that contain high levels of glutamic acid—found in soy, eggs, and cod—help synthesize neurotransmitters such as GABA, which has a calming effect on the nervous system and can reduce pain sensitivity. Tryptophan, found in foods like milk, dairy products, and peanuts, is a precursor to serotonin, a neurotransmitter crucial for mood regulation and pain management (Batista et al. 2016). As serotonin levels increase with tryptophan intake, individuals may experience improvements in pain tolerance and mood stability, both of which are often disrupted in CP conditions.

Further evidence supports the role of amino acids in pain modulation and OUDs. A study found that a combination of amino acids, including phenylalanine, tryptophan, tyrosine, and glutamine, significantly reduced opioid cravings during detoxification (Chen et al. 2004). This suggests that amino acids may help alleviate the neurological and psychological symptoms associated with opioid withdrawal, potentially enhancing recovery outcomes.

Additionally, amino acids such as methionine, lysine, and tyrosine, which contribute to the synthesis of neurotransmitters like dopamine and serotonin, have been found to offer protective effects against depression and anxiety, both of which frequently co-occur with CP and OUDs (Iguacel et al. 2021). These amino acids are most abundant in fish, meat, and dairy products, while vegetarians and vegans may face a higher risk of deficiencies, potentially exacerbating mood disorders and pain sensitivity due to lower levels of these essential amino acids (Iguacel et al. 2021).

One of the critical aspects of amino acid metabolism is its impact on the gut-brain axis. The amino acid tryptophan, for example, is not only a precursor to serotonin but also to kynurenine, a metabolite that has been associated with stress, neuroinflammation, and depression. Alterations in the kynurenine pathway, which are influenced by the gut microbiota, can affect neuronal function and contribute to pain and mood disorders (Verma et al. 2024). Furthermore, probiotics, which can alter gut microbiota composition, have been shown to increase plasma tryptophan levels and reduce inflammation markers such as IL-1β and IL-6, offering a potential therapeutic avenue for managing both CP and mental health disorders (Zagórska et al. 2020).

19.3.4 Fiber and Prebiotics for Gut Health and Its Implications to Chronic Pain and Opioid Use Disorders

Dietary fibers have long been recognized for their role in supporting gastrointestinal health, but their broader effects on disease prevention, particularly regarding persistent pain and inflammatory conditions, have gained increasing attention. The definition of dietary fiber has evolved over time, with most countries now adopting the Codex Alimentarius Commission's definition, which classifies dietary fiber as edible carbohydrate polymers containing three or more monomeric units that are resistant to endogenous digestive enzymes and thus not absorbed in the small intestine (Makki et al. 2018). This definition highlights the diversity in the types of fibers, which can be naturally occurring, extracted, or synthetically produced, as long as they demonstrate a proven physiological benefit. However, some discrepancies remain between national definitions, particularly concerning the inclusion of non-carbohydrate substances like lignin as fibers (Makki et al. 2018).

Dietary fibers are generally classified based on their chemical structure, solubility in water, fermentability, and food source (Makki et al. 2018; Deehan et al. 2017). They are broadly categorized into soluble and insoluble forms, with soluble fibers, such as pectins and b-glucans, being more fermentable by gut bacteria. This fermentation leads to the production of SCFAs, which play an essential role in gut health and may contribute to reduced systemic inflammation, a key factor in CP. Insoluble fibers, such as cellulose and hemicellulose, have a fecal bulking effect and are less fermentable, providing benefits in terms of promoting regular bowel movements (Deehan et al. 2017). However, the overall health benefits of fiber go beyond gut motility. In fact, fibers like guar gum, psyllium, and certain pectins, which form a gel structure in the digestive tract, can delay the absorption of glucose and lipids, thus improving metabolic health (Deehan et al. 2017).

The role of fiber in gut microbiota composition and metabolism has profound implications for chronic disease management, including inflammatory conditions and pain. For instance, high-fiber diets have been shown to increase the production of anti-inflammatory SCFAs while reducing pro-inflammatory cytokines. This can be particularly beneficial for conditions such as rheumatoid arthritis (RA), where a high-fiber diet has been associated with an increase in SCFAs and a reduction in systemic inflammation (Dürholz et al. 2020). Furthermore, the beneficial effects of fiber on inflammation are not limited to autoimmune conditions. In individuals with inflammatory pain, such as those suffering from osteoarthritis, fiber's impact on the gut microbiome could modulate immune responses and improve pain management outcomes (Dürholz et al. 2020).

Vegetarian and vegan diets, which are typically high in fiber, have been linked to lower risks of chronic diseases such as heart disease, hypertension, and certain cancers. These diets are also associated with a greater intake of polyunsaturated fats, bioactive molecules, and antioxidants, which further enhance their protective effects (Iguacel et al. 2021). Additionally, the anti-inflammatory benefits of fiber in these diets extend beyond the gut. Inflammatory markers, including cytokines, are reduced, and beneficial shifts in gut microbiota composition are observed, further supporting the therapeutic potential of dietary fiber in CP management.

Probiotics and prebiotics, another critical component of dietary interventions, have been shown to impact intestinal health and inflammation. Probiotics, such as *Lactobacillus casei Shirota*, have demonstrated positive effects in treating abdominal pain associated with constipation, a common comorbidity in CP conditions, including Parkinson's disease (Cassani et al. 2011). Similarly, prebiotics, which are fibers that selectively stimulate the growth of beneficial gut bacteria, have been shown to modify the microbiota composition and improve gastrointestinal function (Pandey et al. 2015). These changes in the microbiome are believed to underlie the systemic anti-inflammatory effects of fiber and probiotics, contributing to improved pain management and reduced chronic inflammation.

High-fiber diets can also influence brain function and behavior, as suggested by recent studies linking fiber intake to cognitive health and mental well-being. For example, dietary fibers with prebiotic properties, such as fructans, have been found to positively affect brain function, potentially mitigating the cognitive decline associated with CP (Berding et al. 2021). Moreover, the SCFAs produced during fiber fermentation play a role in reducing gut permeability and modulating the immune response, further supporting their role in reducing inflammation and improving mental health (Makki et al. 2018).

19.3.5 Vitamins and Minerals in Chronic Pain and Opioid Use Disorders

Dietary intake, particularly the consumption of vitamins and minerals, plays an essential role in modulating CP conditions, including musculoskeletal pain, RA, fibromyalgia, and neuropathy. A systematic review by Elma et al. (2022) highlights the impact of dietary changes on pain relief, indicating that pain intensity and pain threshold are linked to nutrient intake, with plant-based diets potentially offering pain-relieving effects for chronic musculoskeletal pain. Specifically, patients suffering from RA may show inadequate intake of essential nutrients such as calcium, folate, zinc, magnesium, and vitamin B6, while those with fibromyalgia often have lower levels of carbohydrates, proteins, lipids, and various vitamins, including A, E, K, and folate (Elma et al. 2022). Additionally, CP severity in osteoarthritis has been positively correlated with increased intake of fat and sugar, while protein intake may positively influence pain thresholds in fibromyalgia patients (Elma et al. 2022). Several vitamins have been identified as potentially beneficial for pain management. For instance, although the mechanism behind vitamin C's analgesic action remains unclear, its role in neuromodulation has been suggested as a potential pathway (Harrison and May 2009). Vitamin C's ability to affect brain functions through its transporter, SVCT2, could offer relief from pain through neurochemical alterations, although more research is needed to clarify this relationship.

Similarly, vitamin D, although its direct mechanism in pain relief is not fully understood, has been proposed to regulate inflammatory cytokines, which may contribute to its analgesic effects, especially in conditions such as diabetic neuropathy (Fu et al. 2021). Nutrients like vitamin B complex (B1, B6, B9, and B12) also play critical roles in DNA and RNA synthesis, immunity, and metabolism, all of which are vital for managing CP, especially neuropathic pain. B vitamins are thought to alleviate neuropathic pain by supporting nerve repair and myelination, particularly vitamin B12, which is crucial for nerve health (Didangelos et al. 2021; Stein et al. 2021).

Deficiencies in several vitamins are common in individuals with OUD. Studies have reported significant deficiencies in vitamins B6, folate, C, and D in OUD patients, with vitamin B6 and folate deficiencies being present in approximately 40% of subjects and vitamin C and D deficiencies in up to 70% of patients (Wiss 2019). These deficiencies can exacerbate symptoms and potentially hinder recovery. Moreover, drug use often correlates with poor intake of micronutrients, leading to an imbalance between macronutrient and micronutrient consumption, which may affect overall health and pain perception (Schroeder and Higgins 2017).

A balanced diet rich in essential vitamins and minerals, such as those found in plant-based and Mediterranean diets, has also been linked to improved mental health outcomes, including reduced symptoms of depression and cognitive decline. These diets are high in fiber, vitamins, and phytochemicals, which may contribute to reduced inflammation and better pain management, underlining the importance of adequate nutrition in CP conditions (Iguacel et al. 2021).

19.4 Dietary Patterns for Mental Health and Pain Management

19.4.1 Mediterranean Diet

The Mediterranean diet, recognized for its health benefits, has increasingly been studied for its role in managing both mental health and CP. Several studies have demonstrated its positive impact on pain reduction, anxiety, depression, and overall mental well-being. This dietary pattern, rich in vegetables, fruits, whole grains, legumes, olive oil, and omega-3 fatty acids, has been shown to have anti-inflammatory effects, which may explain its benefits in managing conditions like fibromyalgia and RA.

Research suggests that the Mediterranean diet can improve gut health, which is crucial for pain and inflammation management. The loss of gut permeability, where undigested food particles and toxins come into contact with the immune system, is known to promote systemic inflammation and may contribute to CP (Smedslund et al. 2010). The Mediterranean diet, with its emphasis on fiber, polyphenols, and healthy fats, is thought to support the integrity of the gut lining, reducing inflammation and improving the pain threshold. For instance, studies have shown that Mediterranean diets can significantly improve pain in fibromyalgia patients, even in those with long-standing CP (Casini et al. 2024).

Further evidence supports the Mediterranean diet's role in reducing the severity of CP and inflammation. Towery et al. (2018) demonstrated that individuals who adhered to a plant-based diet, which shares many characteristics with the Mediterranean diet, experienced lower levels of chronic musculoskeletal pain. This is consistent with other studies showing that Mediterranean diets, which are rich in anti-inflammatory foods, have a greater impact on pain relief than vegetarian or vegan diets (Schoenenberger et al. 2021). In addition, adherence to the Mediterranean diet has been associated with lower levels of inflammatory markers, which could explain the reduction in pain experienced by these individuals.

Mental health also benefits from the Mediterranean diet, which has been linked to lower rates of depression and anxiety. Epidemiological studies (Parletta et al. 2019) found that adherence to a plant-based diet like the Mediterranean diet was associated with a reduction in depressive symptoms. The diet's high content of vitamins, minerals, omega-3 fatty acids, and antioxidants helps regulate the body's inflammatory responses, which may directly affect brain health. This is particularly relevant for conditions such as depression, where inflammation has been identified as a contributing factor. Moreover, a large-scale study by Valles-Colomer et al. (2019) highlighted the connection between gut microbiota and mental health, noting that a diet rich in plant-based foods, like the Mediterranean diet, promotes the growth of beneficial bacteria that contribute to better mental well-being.

In addition to its benefits for CP and mental health, the Mediterranean diet has been shown to improve overall quality of life. For example, adherence to this diet was associated with lower levels of homocysteine, a biomarker linked to depression (Bremner et al. 2020). The diet's inclusion of folate and B12 is particularly important in preventing deficiencies that could exacerbate mood disorders (Bremner et al. 2020). Furthermore, studies have found that the Mediterranean diet can help reduce perceived stress, even though it may not have a specific effect on depression in all individuals (Bremner et al. 2020).

19.4.2 Anti-Inflammatory Diets

An anti-inflammatory diet focuses on reducing the body's systemic inflammation, which plays a critical role in CP, psychiatric disorders, and a range of chronic diseases. Inflammation is a natural, protective response to injury or infection, but when prolonged, it can contribute to conditions like CP, depression, and other health issues. For instance, excessive or chronic inflammation in the CNS has been associated with both neurotoxic and neuroprotective effects, depending on the balance of genetic and environmental factors (Sochocka et al. 2017). Therefore, an anti-inflammatory diet, rich in specific nutrients, can help manage and mitigate these harmful effects.

The key components of an anti-inflammatory diet include omega-3 fatty acids, monounsaturated fatty acids, fiber, and various vitamins and antioxidants. Omega-3 fatty acids, found in foods like fatty fish and flaxseeds, have long been shown to reduce systemic inflammation and the markers of oxidative stress. Similarly, monounsaturated fats, found in olive oil, nuts, and avocados, help lower inflammation by decreasing the levels of pro-inflammatory cytokines like TNFα and IL-6 (Bujtor et al. 2021). Conversely, a high intake of saturated fats, omega-6 polyunsaturated fatty acids (PUFAs), and high glycemic index carbohydrates has been associated with increased inflammation and the risk of chronic diseases (Bujtor et al. 2021).

The Mediterranean diet, a popular anti-inflammatory eating pattern, exemplifies many of these principles. It encourages the consumption of whole foods such as fruits, vegetables, legumes, nuts, and seeds, alongside moderate amounts of fish and olive oil. This diet has been shown to reduce markers of systemic inflammation like C-reactive protein (CRP) and improve physical function and mental well-being (Wawrzyniak-Gramacka et al. 2021). Furthermore, the inclusion of antioxidants like vitamin C, vitamin E, and flavonoids, found in foods such as berries, green tea, and dark chocolate, can help neutralize free radicals and prevent oxidative stress (Carlsen et al. 2010). Oxidative stress is a key contributor to the progression of inflammation, especially in CP conditions, as it increases the production of ROS that damage cells and tissues (Kaushik et al. 2020).

The impact of diet on CP has been well-documented in both animal models and human studies. Research has shown that excessive intake of saturated fats, omega-6 PUFAs, and processed carbohydrates can exacerbate CP by inducing mechanical allodynia and hyperalgesia, which are conditions where pain is felt from normally non-painful stimuli (Elma et al. 2020). On the other hand, a diet rich in anti-inflammatory ingredients, such as omega-3 fatty acids and antioxidants, has been shown to improve recovery from chronic inflammatory pain, promoting a reduction in pain sensitivity (Elma et al. 2020).

Additionally, the relationship between diet and inflammation has implications for mental health. Diets high in pro-inflammatory foods, such as those rich in processed carbohydrates and trans fats, have been linked to increased risk for depression and anxiety. On the contrary, anti-inflammatory diets, including those rich in fruits, vegetables, and omega-3 fatty acids, may reduce symptoms of depression by decreasing inflammation in the brain. Cha and Yang (2020) have noted that such diets can be beneficial in the long-term management of psychiatric disorders, such as schizophrenia, by alleviating the neuroinflammatory processes associated with these conditions.

The role of dietary interventions in reducing oxidative stress has also been highlighted in studies examining CP. High-protein, low-carbohydrate diets have been shown to decrease oxidative stress markers and inflammatory cytokines, providing further evidence of the beneficial effects of anti-inflammatory foods in managing chronic conditions (Kitabchi et al. 2013). Moreover, reducing the intake of foods known to promote inflammation—such as refined sugars and trans fats—has been linked to improved physical health outcomes, including lower levels of pain and enhanced sleep quality (Sala-Climent et al. 2023).

19.5 The Role of Nutrition and Dietary Interventions in Opioid Use Disorder, Opioid Withdrawal, and Recovery

OUD and the recovery process bring about significant changes in the body, both physiologically and nutritionally. Many individuals with OUD experience metabolic alterations, constipation, weight loss, and unhealthy eating patterns, all of which can affect recovery outcomes. As the number of people seeking treatment for OUD continues to rise, it is essential to address the nutritional aspects of treatment to support recovery and improve overall health outcomes. Despite the importance of nutrition, it is often overlooked by drug treatment providers, who may fail to incorporate dietary interventions into treatment plans. This highlights the need for a comprehensive approach that includes both medical and nutritional support.

Opioid use, especially long-term, can lead to nutrient deficiencies, including vitamins A, C, E, selenium, and iron, which are crucial for maintaining overall health. To address these deficiencies, patients recovering from opioid dependence can benefit from a diet rich in these micronutrients. Nutritional supplementation, such as a liquid multivitamin (low in metals), additional vitamin B6, calcium, vitamin D, and digestive enzymes, is recommended to ensure adequate intake of essential nutrients (Wiss 2019). Additionally, probiotics and fiber supplements can help manage common opioid-induced gastrointestinal issues like constipation (Wiss 2019).

The use of amino acids, such as phenylalanine, tryptophan, tyrosine, and glutamine, has been studied for its potential role in reducing cravings during opioid detoxification. These amino acids have been found to significantly decrease opiate cravings, suggesting their potential as an important tool in addiction treatment

(Mahboub et al. 2021). A protein-rich diet, providing 1.2–2.0 g/kg/day of protein, can help restore muscle mass and support overall recovery. Using protein supplements, such as whey or plant-based protein powders, can be beneficial in ensuring adequate intake during recovery, especially for individuals who may be underweight or have difficulty consuming sufficient food (Wiss 2019).

Dietary strategies that promote weight gain, particularly in individuals who are underweight, should focus on gradual weight gain rather than rapid increases, with an ideal target of less than 10 pounds per month. The inclusion of fruits (fresh, frozen, or dried) as a healthy alternative for satisfying a "sweet tooth" is encouraged, offering a source of fiber and essential nutrients. For those who may have difficulty absorbing nutrients through food alone, multivitamin/mineral supplements can be broken into two separate doses throughout the day to improve absorption (Wiss 2019).

In addition to micronutrient supplementation, omega-3 fatty acids, particularly DHA, play an important role in reducing inflammation and supporting brain health. A daily intake of 3 grams of DHA, along with a reduction in omega-6 fatty acids, is recommended to promote a balanced inflammatory response (Wiss 2019). Furthermore, the potential therapeutic benefits of a ketogenic diet (KD), which is characterized by high-fat, moderate-protein, and low-carbohydrate intake, are being explored in OUD recovery. KD has demonstrated potential in improving neurological function and may help manage withdrawal symptoms, as it promotes ketone metabolism while reducing sugar and carbohydrate intake, which can exacerbate cravings and neuroplastic changes associated with addiction (Trinko et al. 2023; Kong et al. 2023).

The role of gut health in opioid use and recovery is another emerging area of interest. Opioids can significantly impact the gut microbiota, leading to reduced bacterial diversity, increased intestinal permeability (gut leakiness), and disruptions to the immune system and bile acid metabolism (Gicquelais et al. 2020). Studies have shown that opioid use alters gut microbiota composition, which may contribute to gastrointestinal issues and overall systemic inflammation. Understanding how the gut-brain axis influences opioid use and recovery could open new avenues for nutritional interventions targeting the gut microbiota to support recovery and reduce relapse.

Despite the promising potential of dietary interventions in opioid recovery, more research is needed to identify the optimal nutritional strategies and supplements that can effectively support patients with OUD. Nutritional counseling, individualized dietary plans, and the integration of supplements tailored to the needs of individuals undergoing opioid replacement therapy or detoxification can improve recovery outcomes. Additionally, understanding the interactions between medications such as methadone, buprenorphine, and naltrexone, and their nutritional implications, is critical in refining treatment plans (Waddington et al. 2023; Chavez and Rigg 2020).

Ultimately, incorporating dietary interventions into OUD treatment programs can improve both physical and mental health outcomes, supporting long-term recovery and enhancing the effectiveness of medication-assisted therapies.

19.6 Conclusion and Future Directions

In conclusion, this chapter highlights the essential role of nutrition in managing both CP and OUD, with a focus on how specific dietary components can significantly impact mental health and recovery. Given the complex interplay between nutrition, CP, and opioid dependence, it is clear that addressing nutritional deficiencies and optimizing diet can enhance treatment outcomes. The biochemical mechanisms that link nutrition to mood regulation, pain perception, and overall health, such as the gut-brain axis, inflammation, and oxidative stress, underline the importance of a holistic approach to managing these conditions. By incorporating targeted dietary interventions, such as increasing intake of omega-3 fatty acids, antioxidants, amino acids, and fiber, healthcare providers can offer more comprehensive care that extends beyond pharmacological treatments.

However, despite the promising benefits of nutritional interventions in supporting mental health and alleviating CP and opioid cravings, much of the current evidence is still emerging. Future research should focus on the development of robust clinical trials to assess the effectiveness of specific dietary patterns (e.g., Mediterranean or anti-inflammatory diets) and supplements in individuals with CP and OUD. Additionally, further studies are needed to investigate the optimal dosages and formulations of micronutrient supplements, probiotics, and other dietary components in improving recovery from opioid dependence, as well as their potential interactions with commonly prescribed medications for OUD. The impact of personalized nutrition based on an individual's unique biochemical profile also warrants exploration, as it may offer more precise and effective strategies for supporting recovery.

Moreover, addressing the under-explored area of gut microbiota modulation in OUD and CP patients presents an exciting avenue for future research. The role of gut health in mental health, pain management, and addiction recovery is a promising area that could lead to novel dietary therapies, particularly as we learn more about the effects of opioid-induced changes in the gut microbiome.

Incorporating nutrition into the treatment plans for CP and OUD is essential for improving overall well-being and long-term recovery. As the understanding of these relationships deepens, healthcare providers and researchers should continue to explore and integrate dietary approaches to support the physical and mental health of individuals affected by these debilitating conditions. By fostering a collaborative effort between nutritionists, pain specialists, addiction specialists, and mental health professionals, a more holistic and effective treatment framework can be developed.

References

Abdelrahman KM, Hackshaw KV (2021) Nutritional supplements for the treatment of neuropathic pain. Biomedicine 9(6):674

Bannai M, Kawai N (2012) New therapeutic strategy for amino acid medicine: glycine improves the quality of sleep. J Pharmacol Sci 118(2):145–148

Batista ED, Andretta A, de Miranda RC, Nehring J, Paiva E d S, Schieferdecker MEM (2016) Food intake assessment and quality of life in women with fibromyalgia. Rev Bras Reumatol 56:105–110

Berding K, Vlckova K, Marx W, Schellekens H, Stanton C, Clarke G et al (2021) Diet and the microbiota–gut–brain axis: sowing the seeds of good mental health. Adv Nutr 12(4):1239–1285

Bjørklund G, Aaseth J, Doşa MD, Pivina L, Dadar M, Pen JJ et al (2019) Does diet play a role in reducing nociception related to inflammation and chronic pain? Nutrition 66:153–165

Braden JB, Sullivan MD, Ray GT, Saunders K, Merrill J, Silverberg MJ et al (2009) Trends in long-term opioid therapy for noncancer pain among persons with a history of depression. Gen Hosp Psychiatry 31(6):564–570

Brain K, Burrows TL, Rollo ME, Chai LK, Clarke ED, Hayes C et al (2019) A systematic review and meta-analysis of nutrition interventions for chronic noncancer pain. J Hum Nutr Diet 32(2):198–225

Bremner JD, Moazzami K, Wittbrodt MT, Nye JA, Lima BB, Gillespie CF et al (2020) Diet, stress and mental health. Nutrients 12(8):2428

Brown Z, Metcalf R, Bednarz J, Spargo L, Lee A, Hill C et al (2020) Modifiable lifestyle factors associated with response to treatment in early rheumatoid arthritis. ACR Open Rheumatol 2(6):371–377

Bujtor M, Turner AI, Torres SJ, Esteban-Gonzalo L, Pariante CM, Borsini A (2021) Associations of dietary intake on biological markers of inflammation in children and adolescents: a systematic review. Nutrients 13(2):356

Buydens-Branchey L, Branchey M (2003) Association between low plasma levels of cholesterol and relapse in cocaine addicts. Psychosom Med 65(1):86–91

Carlsen MH, Halvorsen BL, Holte K, Bøhn SK, Dragland S, Sampson L et al (2010) The total antioxidant content of more than 3100 foods, beverages, spices, herbs and supplements used worldwide. Nutr J 9(1):3

Casini I, Ladisa V, Clemente L, Delussi M, Rostanzo E, Peparini S et al (2024) A personalized Mediterranean diet improves pain and quality of life in patients with fibromyalgia. Pain Ther 13(3):609–620

Cassani E, Privitera G, Pezzoli G, Pusani C, Madio C, Iorio L et al (2011) Use of probiotics for the treatment of constipation in Parkinson's disease patients. Minerva Gastroenterol Dietol 57(2):117–121

Cha HY, Yang SJ (2020) Anti-inflammatory diets and schizophrenia. Clin Nutr Res 9(4):241

Chapman SL, Byas-Smith MG, Reed BA (2002) Effects of intermediate-and long-term use of opioids on cognition in patients with chronic pain. Clin J Pain 18(4):S83–S90

Chassaing B, Koren O, Goodrich JK, Poole AC, Srinivasan S, Ley RE et al (2015) Dietary emulsifiers impact the mouse gut microbiota promoting colitis and metabolic syndrome. Nature 519(7541):92–96

Chatterjee S (2016) Oxidative stress, inflammation, and disease. İçinde: oxidative stress and biomaterials [İnternet]. Elsevier. [a.yer 01 Aralık 2024]. s. 35–58. Erişim adresi: https://www.sciencedirect.com/science/article/pii/B9780128032695000024

Chavez MN, Rigg KK (2020) Nutritional implications of opioid use disorder: a guide for drug treatment providers. Psychol Addict Behav 34(6):699

Chen TJ, Blum K, Payte JT, Schoolfield J, Hopper D, Stanford M et al (2004) Narcotic antagonists in drug dependence: pilot study showing enhancement of compliance with SYN-10, amino-acid precursors and enkephalinase inhibition therapy. Med Hypotheses 63(3):538–548

Collaborators G 2019 MD (2022) Global, regional, and national burden of 12 mental disorders in 204 countries and territories, 1990–2019: a systematic analysis for the Global Burden of Disease Study 2019. Lancet Psychiatry 9(2):137–150

Deehan EC, Duar RM, Armet AM, Perez-Muñoz ME, Jin M, Walter J (2017) Modulation of the gastrointestinal microbiome with nondigestible fermentable carbohydrates to improve human health. Britton RA, Cani PD, editörler. Microbiol Spectr 5(5):5.5.04

Delorme J, Pennel L, Brousse G, Daulouède JP, Delile JM, Lack P et al (2021) Prevalence and characteristics of chronic pain in buprenorphine and methadone-maintained patients. Front Psych 12:641430

Didangelos T, Karlafti E, Kotzakioulafi E, Margariti E, Giannoulaki P, Batanis G et al (2021) Vitamin B12 supplementation in diabetic neuropathy: a 1-year, randomized, double-blind, placebo-controlled trial. Nutrients 13(2):395

Dürholz K, Hofmann J, Iljazovic A, Häger J, Lucas S, Sarter K et al (2020) Dietary short-term fiber interventions in arthritis patients increase systemic SCFA levels and regulate inflammation. Nutrients 12(10):3207

Elma Ö, Yilmaz ST, Deliens T, Coppieters I, Clarys P, Nijs J et al (2020) Do nutritional factors interact with chronic musculoskeletal pain? A systematic review. J Clin Med 9(3):702

Elma Ö, Lebuf E, Marnef AQ, Tümkaya Yilmaz S, Coppieters I, Clarys P et al (2022) Diet can exert both analgesic and pronociceptive effects in acute and chronic pain models: a systematic review of preclinical studies. Nutr Neurosci 25(10):2195–2217

Field R, Pourkazemi F, Turton J, Rooney K (2021) Dietary interventions are beneficial for patients with chronic pain: a systematic review with meta-analysis. Pain Med 22(3):694–714

Fischer R, Konkel A, Mehling H, Blossey K, Gapelyuk A, Wessel N et al (2014) Dietary omega-3 fatty acids modulate the eicosanoid profile in man primarily via the CYP-epoxygenase pathway [S]. J Lipid Res 55(6):1150–1164

Fishbain DA, Cole B, Lewis JE, Gao J, Rosomoff RS (2009) Do opioids induce hyperalgesia in humans? An evidence-based structured review. Pain Med 10(5):829–839

Fu L, Fei J, Tan ZX, Chen YH, Hu B, Xiang HX et al (2021) Low vitamin D status is associated with inflammation in patients with chronic obstructive pulmonary disease. J Immunol 206(3):515–523

Gerritsen S, Egli V, Roy R, Haszard J, Backer CD, Teunissen L et al (2021) Seven weeks of home-cooked meals: changes to New Zealanders' grocery shopping, cooking and eating during the COVID-19 lockdown. J R Soc N Z 51(sup1):S4–S22

Gicquelais RE, Bohnert AS, Thomas L, Foxman B (2020) Opioid agonist and antagonist use and the gut microbiota: associations among people in addiction treatment. Sci Rep 10(1):19471

Godos J, Currenti W, Angelino D, Mena P, Castellano S, Caraci F et al (2020) Diet and mental health: review of the recent updates on molecular mechanisms. Antioxidants (Basel) 9(4):346

Goldberg RJ, Katz J (2007) A meta-analysis of the analgesic effects of omega-3 polyunsaturated fatty acid supplementation for inflammatory joint pain. Pain 129(1–2):210–223

Grosso G, Galvano F, Marventano S, Malaguarnera M, Bucolo C, Drago F et al (2014) Omega-3 fatty acids and depression: scientific evidence and biological mechanisms. Oxidative Med Cell Longev 2014:1–16

Gunn J, Hill MM, Cotten BM, Deer TR (2020) An analysis of biomarkers in patients with chronic pain. Pain Physician 23(1):E41

Harrison FE, May JM (2009) Vitamin C function in the brain: vital role of the ascorbate transporter SVCT2. Free Radic Biol Med 46(6):719–730

Häuser W, Morlion B, Vowles KE, Bannister K, Buchser E, Casale R et al (2021) European* clinical practice recommendations on opioids for chronic noncancer pain – part 1: role of opioids in the management of chronic noncancer pain. Eur J Pain 25(5):949–968

Hennigar SR, Kelley AM, Anderson BJ, Armstrong NJ, McClung HL, Berryman CE et al (2021) Sensitivity and reliability of zinc transporter and metallothionein gene expression in peripheral blood mononuclear cells as indicators of zinc status: responses to ex vivo zinc exposure and habitual zinc intake in humans. Br J Nutr 125(4):361–368

Hou H, Wang L, Fu T, Papasergi M, Yule DI, Xia H (2020) Magnesium acts as a second messenger in the regulation of NMDA receptor-mediated CREB signaling in neurons. Mol Neurobiol 57(6):2539–2550

Iguacel I, Huybrechts I, Moreno LA, Michels N (2021) Vegetarianism and veganism compared with mental health and cognitive outcomes: a systematic review and meta-analysis. Nutr Rev 79(4):361–381

Irwin MR, Olmstead R, Carroll JE (2016) Sleep disturbance, sleep duration, and inflammation: a systematic review and meta-analysis of cohort studies and experimental sleep deprivation. Biol Psychiatry 80(1):40–52

Jeynes KD, Gibson EL (2017) The importance of nutrition in aiding recovery from substance use disorders: a review. Drug Alcohol Depend 179:229–239

Kaushik AS, Strath LJ, Sorge RE (2020) Dietary interventions for treatment of chronic pain: oxidative stress and inflammation. Pain Ther 9(2):487–498

Kitabchi AE, McDaniel KA, Wan JY, Tylavsky FA, Jacovino CA, Sands CW et al (2013) Effects of high-protein versus high-carbohydrate diets on markers of β-cell function, oxidative stress, lipid peroxidation, proinflammatory cytokines, and adipokines in obese, premenopausal women without diabetes: a randomized controlled trial. Diabetes Care 36(7):1919–1925

Kong D, Sun JX, Yang JQ, Li YS, Bi K, Zhang ZY et al (2023) Ketogenic diet: a potential adjunctive treatment for substance use disorders. Front Nutr 10:1191903

Lane MM, Gamage E, Travica N, Dissanayaka T, Ashtree DN, Gauci S et al (2022) Ultra-processed food consumption and mental health: a systematic review and meta-analysis of observational studies. Nutrients 14(13):2568

Leigh SJ, Morris MJ (2020) Diet, inflammation and the gut microbiome: mechanisms for obesity-associated cognitive impairment. Biochim Biophys Acta Mol Basis Dis 1866(6):165767

Macintyre PE, Quinlan J, Levy N, Lobo DN (2022) Current issues in the use of opioids for the management of postoperative pain: a review. JAMA Surg 157(2):158–166

Mahboub N, Rizk R, Karavetian M, De Vries N (2021) Nutritional status and eating habits of people who use drugs and/or are undergoing treatment for recovery: a narrative review. Nutr Rev 79(6):627–635

Makki K, Deehan EC, Walter J, Bäckhed F (2018) The impact of dietary fiber on gut microbiota in host health and disease. Cell Host Microbe 23(6):705–715

Malcangio M (2019) Role of the immune system in neuropathic pain. Scand J Pain 20(1):33–37

Mariné-Casadó R, Domenech-Coca C, Del Bas JM, Bladé C, Caimari A, Arola L (2019) Cherry consumption out of season alters lipid and glucose homeostasis in normoweight and cafeteria-fed obese Fischer 344 rats. J Nutr Biochem 63:72–86

Marion-Letellier R, Savoye G, Ghosh S (2015) Polyunsaturated fatty acids and inflammation. IUBMB Life 67(9):659–667

Marx W, Lane M, Hockey M, Aslam H, Berk M, Walder K et al (2021) Diet and depression: exploring the biological mechanisms of action. Mol Psychiatry 26(1):134–150

Matsuda M, Huh Y, Ji RR (2019) Roles of inflammation, neurogenic inflammation, and neuroinflammation in pain. J Anesth 33(1):131–139

Michalsen A, Li C (2013) Fasting therapy for treating and preventing disease-current state of evidence. Forsch Komplementmed 20(6):444–453

Monteiro CA, Cannon G, Levy RB, Moubarac JC, Louzada ML, Rauber F et al (2019) Ultra-processed foods: what they are and how to identify them. Public Health Nutr 22(5):936–941

Moriarty O, McGuire BE, Finn DP (2011) The effect of pain on cognitive function: a review of clinical and preclinical research. Prog Neurobiol 93(3):385–404

Mörkl S, Wagner-Skacel J, Lahousen T, Lackner S, Holasek SJ, Bengesser SA et al (2020) The role of nutrition and the gut-brain axis in psychiatry: a review of the literature. Neuropsychobiology 79(1):80–88

Neale J, Nettleton S, Pickering L, Fischer J (2012) Eating patterns among heroin users: a qualitative study with implications for nutritional interventions. Addiction 107(3):635–641

Nijs J, Tumkaya Yilmaz S, Elma Ö, Tatta J, Mullie P, Vanderweeën L et al (2020) Nutritional intervention in chronic pain: an innovative way of targeting central nervous system sensitization? Expert Opin Ther Targets 24(8):793–803

O'Brien T, Christrup LL, Drewes AM, Fallon MT, Kress HG, McQuay HJ et al (2017) European Pain Federation position paper on appropriate opioid use in chronic pain management. Eur J Pain 21(1):3–19

Ogunrinola GA, Oyewale JO, Oshamika OO, Olasehinde GI (2020) The human microbiome and its impacts on health. Int J Microbiol 2020:1–7

Ojo O, Wang XH, Ojo OO, Ibe J (2019) The impact of opium abuse on lipid profile in patients with diabetes: a systematic review and meta-analysis. Int J Environ Res Public Health 16(23):4795
Oteng AB, Kersten S (2020) Mechanisms of action of trans fatty acids. Adv Nutr 11(3):697–708
Pandey KR, Naik SR, Vakil BV (2015) Probiotics, prebiotics and synbiotics- a review. J Food Sci Technol 52(12):7577–7587
Park W (2018) Gut microbiomes and their metabolites shape human and animal health. J Microbiol 56(3):151–153
Parletta N, Zarnowiecki D, Cho J, Wilson A, Bogomolova S, Villani A et al (2019) A Mediterranean-style dietary intervention supplemented with fish oil improves diet quality and mental health in people with depression: a randomized controlled trial (HELFIMED). Nutr Neurosci 22(7):474–487
Perna S, Alalwan TA, Al-Thawadi S, Negro M, Parimbelli M, Cerullo G et al (2020) Evidence-based role of nutrients and antioxidants for chronic pain management in musculoskeletal frailty and sarcopenia in aging. Geriatrics 5(1):16
Prasad AS, Bao B (2019) Molecular mechanisms of zinc as a pro-antioxidant mediator: clinical therapeutic implications. Antioxidants (Basel) 8(6):164
Ramsden CE, Faurot KR, Zamora D, Suchindran CM, MacIntosh BA, Gaylord S et al (2013) Targeted alteration of dietary n-3 and n-6 fatty acids for the treatment of chronic headaches: a randomized trial. Pain 154(11):2441–2451
Rizzetto L, Fava F, Tuohy KM, Selmi C (2018) Connecting the immune system, systemic chronic inflammation and the gut microbiome: the role of sex. J Autoimmun 92:12–34
Ross LJ, Wilson M, Banks M, Rezannah F, Daglish M (2012) Prevalence of malnutrition and nutritional risk factors in patients undergoing alcohol and drug treatment. Nutrition 28(7–8):738–743
Sala-Climent M, López de Coca T, Guerrero MD, Muñoz FJ, López-Ruíz MA, Moreno L et al (2023) The effect of an anti-inflammatory diet on chronic pain: a pilot study. Front Nutr 10:1205526
Schoenenberger KA, Schuepfer AC, Gloy VL, Hasler P, Stanga Z, Kaegi-Braun N et al (2021) Effect of anti-inflammatory diets on pain in rheumatoid arthritis: a systematic review and meta-analysis. Nutrients 13(12):4221
Schroeder RD, Higgins GE (2017) You are what you eat: the impact of nutrition on alcohol and drug use. Subst Use Misuse 52(1):10–24
Severino AL, Shadfar A, Hakimian JK, Crane O, Singh G, Heinzerling K et al (2018) Pain therapy guided by purpose and perspective in light of the opioid epidemic. Front Psych 9:119
Sibille KT, Bartsch F, Reddy D, Fillingim RB, Keil A (2016) Increasing neuroplasticity to bolster chronic pain treatment: a role for intermittent fasting and glucose administration? J Pain 17(3):275–281
Siddiqui N, Urman RD (2022) Opioid use disorder and racial/ethnic health disparities: prevention and management. Curr Pain Headache Rep 26(2):129–137
Silva YP, Bernardi A, Frozza RL (2020) The role of short-chain fatty acids from gut microbiota in gut-brain communication. Front Endocrinol 11:508738
Singh S, Sharma P, Pal N, Kumawat M, Shubham S, Sarma DK et al (2022) Impact of environmental pollutants on gut microbiome and mental health via the gut–brain axis. Microorganisms 10(7):1457
Smedslund G, Byfuglien MG, Olsen SU, Hagen KB (2010) Effectiveness and safety of dietary interventions for rheumatoid arthritis: a systematic review of randomized controlled trials. J Am Diet Assoc 110(5):727–735
Sochocka M, Diniz BS, Leszek J (2017) Inflammatory response in the CNS: friend or foe? Mol Neurobiol 54(10):8071–8089
Stein J, Geisel J, Obeid R (2021) Association between neuropathy and B-vitamins: a systematic review and meta-analysis. Eur J Neurol 28(6):2054–2064
Sun C, Zhao C, Guven EC, Paoli P, Simal-Gandara J, Ramkumar KM et al (2020) Dietary polyphenols as antidiabetic agents: advances and opportunities. Food Front 1(1):18–44
Taylor AM, Holscher HD (2020) A review of dietary and microbial connections to depression, anxiety, and stress. Nutr Neurosci 23(3):237–250

Torres-Fuentes C, Golubeva AV, Zhdanov AV, Wallace S, Arboleya S, Papkovsky DB et al (2019) Short-chain fatty acids and microbiota metabolites attenuate ghrelin receptor signaling. FASEB J 33(12):13546–13559

Towery P, Guffey JS, Doerflein C, Stroup K, Saucedo S, Taylor J (2018) Chronic musculoskeletal pain and function improve with a plant-based diet. Complement Ther Med 40:64–69

Trinko R, Diaz DM, Foscue E, Thompson SL, Taylor JR, DiLeone RJ (2023) Ketogenic diet enhances the effects of oxycodone in mice. Sci Rep 13(1):7507

Valles-Colomer M, Falony G, Darzi Y, Tigchelaar EF, Wang J, Tito RY et al (2019) The neuroactive potential of the human gut microbiota in quality of life and depression. Nat Microbiol 4(4):623–632

Verma A, Inslicht SS, Bhargava A (2024) Gut-brain axis: role of microbiome, metabolomics, hormones, and stress in mental health disorders. Cells 13(17):1436

Waddington F, Naunton M, Thomas J, Kyle G, Wheatley B, Oguoma V (2023) Examination of the nutritional intake of patients undergoing opioid replacement therapy: a systematic review. Nutr Diet 80(1):55–64

Wawrzyniak-Gramacka E, Hertmanowska N, Tylutka A, Morawin B, Wacka E, Gutowicz M et al (2021) The association of anti-inflammatory diet ingredients and lifestyle exercise with inflammaging. Nutrients 13(11):3696

Wiss DA (2019) The role of nutrition in addiction recovery: what we know and what we don't. In: The assessment and treatment of addiction, pp 21–42. Elsevier, ISBN: 978-0-323-54856-4

Wurtman J, Wurtman R (2018) The trajectory from mood to obesity. Curr Obes Rep 7(1):1–5

Zagórska A, Marcinkowska M, Jamrozik M, Wiśniowska B, Paśko P (2020) From probiotics to psychobiotics – the gut-brain axis in psychiatric disorders. Benef Microbes 11(8):717–732

Zhang H, Tsao R (2016) Dietary polyphenols, oxidative stress and antioxidant and anti-inflammatory effects. Curr Opin Food Sci 8:33–42

Nutritional Rehabilitation in Opioid Use Disorder

20

Doaa S. R. Khafaga, Rahma Mahmoud Saad, Rana Mansour, Nour E. Bayali, Ahmed H. Shaaban, Salma Allam, and Abdullrahman Elgarawany

Abstract

Opioid use disorder (OUD) adversely affects both physiological and mental health, especially through its influence on the gut-brain axis. Opioid-induced gut dysbiosis increased intestinal permeability and heightened inflammation, which influence neurotransmitter synthesis, gastrointestinal motility, and immunological responses, hence leading to mental symptoms like anxiety, depression, and cognitive deficits. Nutritional deficiencies, encompassing vitamins B6, B12, D, zinc, and magnesium, result from malnutrition, modified dietary choices, and compromised absorption, while opioid-induced bowel dysfunction causes gastrointestinal complications, affecting nutrient accessibility. Hormonal alterations, such as hypogonadism, adrenal insufficiency, and an imbalanced ghrelin-leptin ratio, exacerbate nutritional dysregulation in conjunction with bone problems, anemia, and immune system impairment. Clinical evidence underscores the advantages of combining nutritional rehabilitation with opioid replacement therapies, including methadone maintenance therapy (MMT). Zinc

D. S. R. Khafaga
Department of Basic Medical Sciences, Health Sector, Galala University, New Galala City, Suez, Egypt
e-mail: doaa.rashwan@gu.edu.eg

R. M. Saad
Faculty of Medicine Modern University for Technology and Information-MTI, Cairo, Egypt

R. Mansour · N. E. Bayali · S. Allam · A. Elgarawany (✉)
Faculty of Medicine, Galala University, Suez, Egypt
e-mail: rma200224@gu.edu.eg; nee212821@gu.edu.eg; Salma.allam@gu.edu.eg; amo501072@gu.edu.eg

A. H. Shaaban
Faculty of Science, Galala University, Suez Governorate, Egypt

W. Mohamed, A. Özge (eds.), *Feeding the Mind: The Connection Between Diet, Drugs, and Mental Health Volume 1*, Nutritional Neurosciences,
https://doi.org/10.1007/978-981-95-1414-4_20

supplementation diminishes relapse likelihood and enhances mental health outcomes, whereas vitamin D and multivitamin supplements mitigate depressive symptoms and increase cognitive performance. Probiotic supplementation has demonstrated efficacy in reducing inflammation, enhancing fasting glucose levels, and lowering depression scores. Current trials, such as fruit smoothie therapies, seek to evaluate the impact of nutrient-dense diets on recovery results. Notwithstanding these gains, hurdles endure, encompassing socioeconomic obstacles, restricted access to nutritional assistance, and insufficient integration of nutrition into therapeutic protocols. Future research should concentrate on specific nutritional therapies, alteration of gut flora, and the long-term effects of supplementation in people with OUD. An integrative strategy that incorporates pharmaceutical, behavioral, and dietary treatments is crucial for enhancing overall recovery and mental health outcomes in patients with OUD.

Keywords

Opioid use disorder · Nutrition · Gut-Brain Axis · Zinc · Probiotics · Methadone maintenance therapy · Mental health

Abbreviations

ACTH	Adrenocorticotropic hormone
COVID-19	Coronavirus Disease 2019
CRH	Corticotropin-releasing hormone
GIT	Gastrointestinal tract
HPA axis	Hypothalamic-pituitary-adrenal axis
LDL	Low-density lipoprotein
MMT	Methadone maintenance treatment
OAT	Opioid agonist treatment
OIAD	Opioid-induced androgen deficiency
OIAI	Opioid-induced adrenal insufficiency
OIBD	Opioid-induced bowel dysfunction
OUD	Opioid use disorder
PRL	Prolactin
TAC	Total antioxidant capacity

20.1 Introduction

Opioid use disorder (OUD) is a condition that profoundly affects both the body and mind, influencing a range of biological systems beyond what we traditionally associate with addiction. While opioids are commonly prescribed for pain relief, prolonged use can disrupt vital systems, particularly the gut and the brain. These two systems are closely connected through the gut-brain axis—a complex, two-way

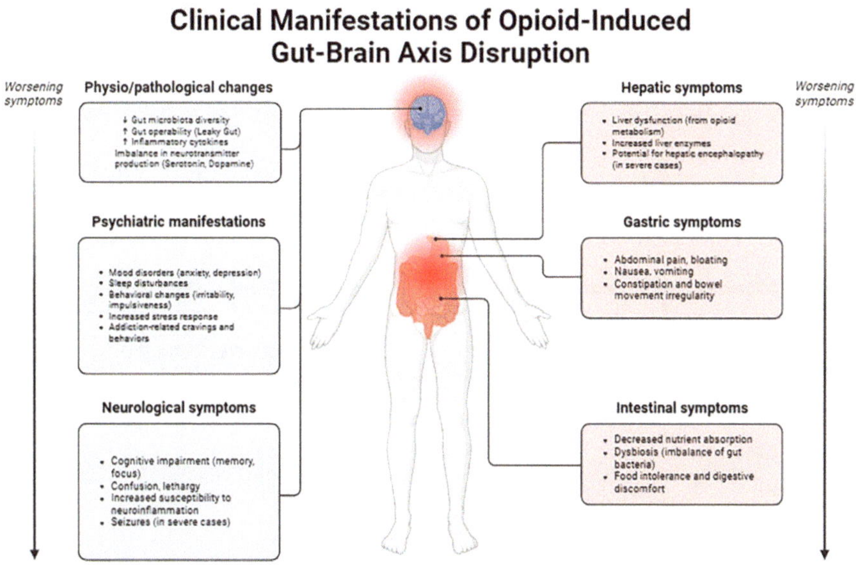

Fig. 20.1 Clinical manifestations of opioid-induced gut-brain axis disruption illustrate the multisystemic impact of opioids on physiological, psychiatric, neurological, hepatic, gastric, and intestinal functions

communication pathway that helps regulate immune function, hormones, and mental health. Figure 20.1 shows this intricate connection, emphasizing how the gut-brain axis plays a central role in maintaining overall health (Cryan and Dinan 2012; Mayer and Tillisch 2011; Rhee et al. 2009).

In people with OUD, opioids can lead to an imbalance in the gut microbiome, often referred to as dysbiosis (Holzer 2009). This imbalance weakens the lining of the gut, creating what's known as a "leaky gut," where substances normally confined within the gut can escape into the bloodstream (Bischoff et al. 2014; Cani and Delzenne 2009). When these harmful particles reach the brain, they can trigger inflammation, which is linked to a variety of mental health challenges, as shown in Fig. 20.2 (Cani and Delzenne 2009; de Vos et al. 2022). Many people with OUD experience heightened anxiety, depression, and cognitive issues, conditions that are thought to be exacerbated by this process (de Vos et al. 2022).

The gut-brain axis shows us just how interconnected physical and mental health are in the context of opioid use. Opioids not only disrupt the balance of gut bacteria, but they also interfere with brain chemistry, immune responses, and hormonal signaling, creating a cycle in which both physical and psychological symptoms reinforce each other (Acharya et al. 2017). Addressing OUD comprehensively means considering both the gut and the brain.

This chapter explores how opioids affect this gut-brain connection, the symptoms that emerge from these changes, and potential treatments that could help restore balance and improve the well-being of people living with OUD.

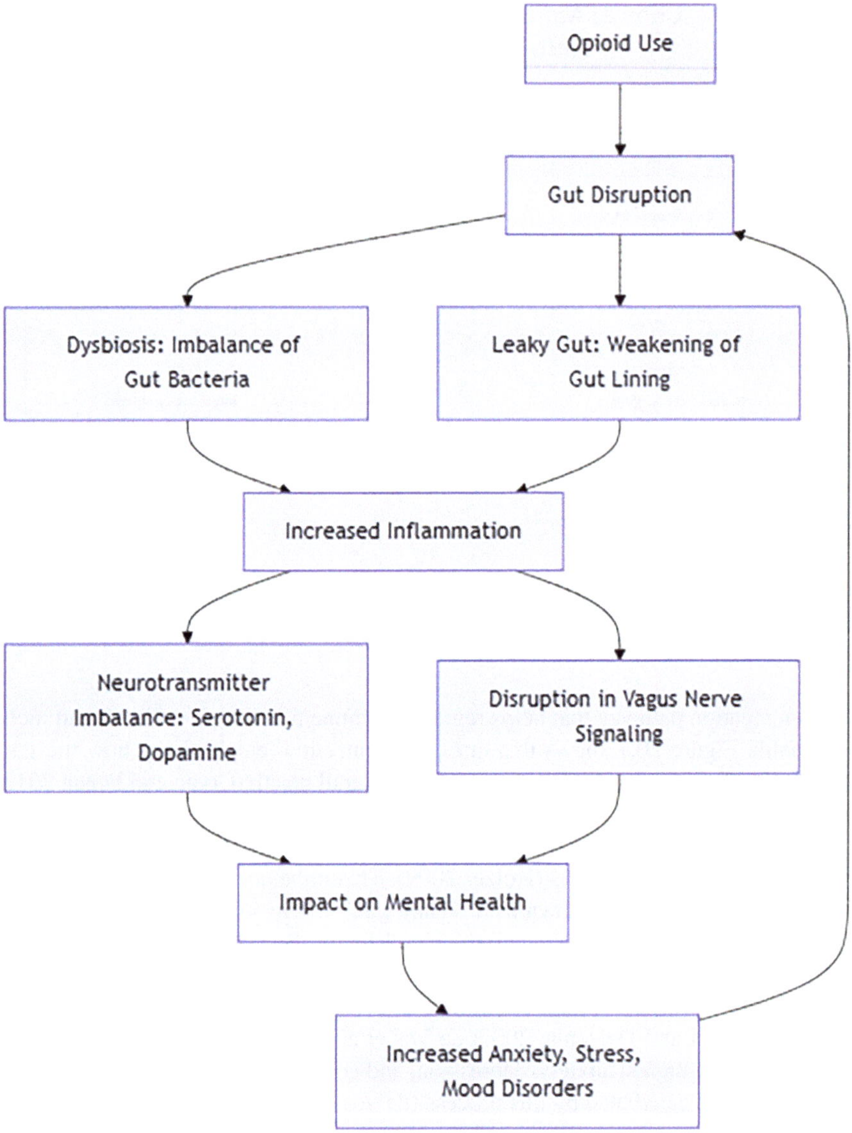

Fig. 20.2 Opioid use disrupts gut health by inducing dysbiosis (gut bacteria imbalance) and leaky gut (weakening of the gut lining), leading to increased inflammation and vagus nerve signaling disruptions

20.2 The Role of Nutrition in Opioid Withdrawal, Detoxification, and Recovery

Opioid addiction is a biological disease that has severe detrimental repercussions for both the individual and the social environment. Serious physical and emotional problems, legal complications, and adverse social repercussions are among the potential consequences. A study of drug harm in the United Kingdom (Drug Harms in the UK: A Multicriteria Decision Analysis—PubMed n.d.) revealed that heroin is the most detrimental to the person, but alcohol poses the most harm to others (Cunningham 2016). Consequently, in public health, effective interventions for opioid addiction are the highest priority (Migdanis et al. 2023). Studies indicate that deficiencies in specific nutritional levels might create obstacles to withdrawal from opiate addiction (Alves et al. 2011; Varela et al. 1997). It has been proposed that individuals with drug use disorders may successfully detox from opioids by adhering to a comprehensive nutrition education program and participating in physical activity (Nabipour et al. 2014). Nutrition is associated with conditions such as diabetes mellitus and vitamin D insufficiency, which may influence morphine dependency (Nutritional Effects of Marijuana, Heroin, Cocaine, and Nicotine - PubMed n.d.). Furthermore, nutritional status is vital for the recovery and survival of addicts who may be afflicted by illnesses like HIV, which endangers nutritional health and can lead to malnutrition (Mangili et al. 2006).

20.2.1 Nutritional Interventions for Substance Abuse

Historically, the predominant therapies for drug misuse include absolute abstinence, psychiatric and behavioral counseling, 12-step programs, and pharmacotherapy. It is regarded as a psychological condition. Nevertheless, the American Medical Association and the World Health Organization categorize it as a physical disease. Instead of an exclusive choice, it may be a synthesis of both due to the intricacies associated with addiction (Organization 2014).

While conventional recovery programs have benefited individuals globally, there exist lesser-known programs that adopt a holistic approach emphasizing nutrition as a response to substance dependence. In 1987, Larson (Seven Weeks to Sobriety n.d.) presented the results of research involving 100 patients that focused on replenishing depleted biochemicals by supplementation in conjunction with psychotherapy. The results indicated that 85 out of 100 participants did not relapse into drug misuse and maintained stable functioning at 12 and 42 months post-treatment. Dr. Kenneth Blum, a brain researcher, saw substance misuse as a biochemical disorder and launched a study aimed at restoring neurotransmitters and natural endorphins that had been depleted, leading to the reinstatement of normal emotions. DesMaisons (Potatoes Not Prozac : Simple Solutions for Sugar Sensitivity / | Fayetteville Public Library n.d.) indicates a 92% success rate for a nutrition-based treatment among individuals with drug misuse issues. Siple (The Eating for Recovery: The Essential Nutrition Plan to Reverse the Physical Damage of Alcoholism by Molly Siple, Paperback | Barnes, and Noble® n.d.) authored a self-help book designed to aid individuals in recovering from substance misuse and sustaining sobriety by consuming specific meals. It enumerates recovery foods and offers more than 300 recipes along with 21 days of meal plans to adhere to. This book delineates health objectives for the reader to attain. The dietary regimen aids in the prevention and reversal of several disorders associated with substance misuse, including digestive issues, fatty liver, and degenerative diseases such as heart disease, osteoporosis, and cancer. Recovery meals are recognized for their ability to normalize blood glucose levels, promote digestion, and improve nutritional absorption. McClain (McClain et al. 2011) underscored the necessity for vigorous dietary intervention for those with liver illness. The study highlighted that a nighttime snack was more effective in preventing muscle atrophy than one offered during the day.

Conventional therapy for liver illness has remained unchanged for the past 40 years. Innovative methodologies encompass the use of amino acids, micronutrients, and antioxidants (Suk et al. 2014). Administering folate through whole meals may enhance individual health and bolster the antioxidant defense system (Medici and Halsted 2013). A 2012 study revealed that environmental and educational nutrition interventions led to enhanced dietary habits and an increased sense of well-being (Cowan and Devine 2012). Wiss (Nutrition Therapy for Eating Disorders, Substance Use Disorders, and Addictions | Request PDF n.d.) discovered that individuals with drug use disorders frequently had disturbed and dysfunctional eating behaviors during periods of abstinence. He detailed particular methods for macro- and micronutrient supplement therapy and refeeding in selected instances. He advised that nutritional treatment prioritize the most critical physical and nutritional issues initially, followed by an emphasis on psychological factors.

Zhou et al. (2000) discovered that opioid addiction significantly disrupts the equilibrium between oxidation and antioxidation in individuals with the addiction. The plasma levels of lipoperoxidation, erythrocyte lipoperoxidation values, and nitric oxide rose, whereas glutathione peroxidase and superoxide dismutase levels progressively dropped. Addictive behaviors may escalate as a result of diminished dopamine levels. Administering amino acids and minerals essential for dopamine synthesis may prove advantageous. Research involving animals has shown that elevated brain serotonin levels may diminish cravings. The equilibrium of antioxidants may be enhanced by nutrients like selenium, vitamins A, C, and E, quercetin, and zinc.

Rebecca Place Miller, a scientific writer affiliated with the Many Hands Sustainability Centre in Barre, Maine, has released an online guidebook titled "Nutrition in Addiction Recovery" (Miller 2020). She consulted several nutritionists, dietitians, and healthcare professionals specializing in nutrition and addiction. She recognizes prevalent nutritional inadequacies in individuals who misuse drugs and the restorative meals to address them. Numerous rehabilitation regimens advocate using supplements to rectify deficits. The Health Recovery Centre of Joan Larson, mentioned above, is very significant. She states that merely consuming nutritious foods would not resolve the issue due to absorption complications. After evaluating the damage through several tests, she recommends a detox mix of vitamins, minerals, amino acids, and other nutrients to mitigate adverse effects, suppress cravings, and alleviate withdrawal symptoms. Body restoration proceeds with enumerations of supplements derived from test outcomes. She advocates a six-month nutritional maintenance regimen of 56 supplements. Recent emphasis has been directed to the gut microbiota and its association with the brain and mental disorders (Microbes and Mental Health: The Link Between Biome and Brain n.d.). Susceptibility to mental illness may be linked to neuroinflammation and increased levels of certain inflammatory serum cytokines. The gut microbiome is a crucial element of our immune system and facilitates much of the communication along the hypothalamic-pituitary-adrenal (HPA) axis. Research using animals has demonstrated that altering a rodent's microbiota may modify its behavior. There are several methods to do this, including probiotics, antibiotics, prebiotic fibers that nourish the microbiota, and fecal transplants. Probiotics have demonstrated the ability to diminish negative cognition in healthy individuals (Steenbergen et al. 2015) and alleviate anxiety in patients receiving cancer therapy (Probiotics Reduce Psychological Stress in Patients before Laryngeal Cancer Surgery - PubMed n.d.). This dietary interaction with the microbiota may elucidate why those consuming a traditional whole foods diet exhibit up to 40% greater resilience to stress and the onset of mental disease compared to those adhering to a processed Western foods diet.

20.2.2 Dietary and Nutritional Status of Opiate Addicts in Methadone Maintenance Treatment

Medication-assisted treatment (MAT) in conjunction with behavioral therapies has been implemented to address opioid dependence. Buprenorphine, a schedule III partial opioid agonist, and methadone, a schedule II opioid agonist, are frequently implemented as replacement therapies. Naltrexone and l-alpha-acetyl-methanol (LAAM) have also been employed as replacement therapies. Methadone was recognized as an efficacious remedy for heroin addiction in the 1960s. Methadone is a long-lasting μ-opioid agonist that aids patients in the transition to abstinence by imitating certain opiate actions (Dole et al. 1966). In 1999, methadone maintenance therapy (MMT) was undergone by nearly 115,000 individuals who were heroin addicts in the United States (D'Aunno et al. 2014).

Several studies have determined that there is a correlation between substance addiction, education, income levels, and body mass index (BMI). The higher the BMI, the higher the income and educational levels, and the reverse is true (Himmelgreen et al. 1998; Nazrul Islam et al. 2002). In 2011, Alves et al. evaluated the nutritional and sociodemographic characteristics of heroin addicts during the detoxification program. The results indicated that heroin addicts consume a lower quantity of vegetables, fruits, and grains than the minimum amount recommended by the food pyramid and are more inclined to consume sweets (Himmelgreen et al. 1998). Drug addicts are also more likely to consume foods with low vitamin content, and their consumption of vegetables and fruit is lower than that of the general population, as evidenced by numerous additional studies. Regrettably, the scope of nutrition services has not been clearly defined in detoxification programs and has not been recognized as a primary issue. Larson et al. contend that the depletion of nutrients at the outset of a detoxification program is not solely resolved by appropriate dietary behavior, as a result of the significant deficiencies and absorption issues. The urine test indicates that pantothenic acid administered orally is not absorbed by alcoholic patients (Varela et al. 1997; Tao and Fox 1976).

The carbohydrate-metabolism health concerns can be managed by increasing the dietary intake of protein and limiting simple carbohydrates in the form of vegetables and whole grains (Fishbein: The Effects of Diet on Behavior: Implications...—Google Scholar n.d.). Consequently, patients must ingest an additional amount of amino acids and protein during the treatment procedure in order to recover from opiate addiction (Absorption and Utilization of Nutrients in Alcoholism—Document—Gale Academic OneFile n.d.). The negative impact of vitamin and mineral deficiencies on the withdrawal process renders methadone maintenance treatment (MMT) itself an unfavorable approach unless an appropriate diet accompanies it. Williams discovered that the consumption of alcohol by rats led to deficiencies in vitamin B6, vitamin A, thiamine, riboflavin, and pantothenic acid (Williams 1956). In detoxification programs for recovering addicts, it is important to prescribe essential nutrients, including zinc, iron, calcium, chromium, magnesium, and potassium, in addition to proteins and key vitamins. Zinc has the potential to enhance the immune system and promote appropriate brain function (Potatoes Not Prozac: A Natural Seven-Step Dietary Plan to Control Your ...—Kathleen DesMaisons—Google Books n.d.). Calcium and magnesium deficiencies have been observed in numerous opiate and alcohol addicts as a result of inadequate calcium intake and a poor diet. During detoxification programs, the primary causes of pain and nervous/muscular disorders in addicts and alcohol consumers are calcium and magnesium deficiencies (Finnegan and Gray 1990).

A pilot study (Cunningham 2016) evaluates whether tailored nutritional supplementation can improve outcomes for patients in a methadone maintenance program for opioid addiction. Conducted in a narcotic treatment center, 20 patients were selected, with 10 receiving customized nutrient-rich drinks based on identified deficiencies and 10 receiving a low-sugar fruit juice control. Over 6 months, blood tests and gut microbiome analyses were conducted at the start, at 3 months, and at the end, alongside monitoring methadone dosages, cravings, and relapse rates. Participants also kept food journals and completed quality-of-life assessments. The study anticipates that the experimental group will experience better health, reduced cravings, lower relapse rates, and improved quality of life compared to the control group. If successful, the findings could support the integration of nutritional interventions as part of a holistic approach to addiction recovery, complementing existing therapies such as counseling.

20.2.3 Nutritional Intake in Patients Receiving Opioid Replacement Therapy

Opioid addiction has been demonstrated to be effectively treated through opioid replacement therapy (ORT) (Mattick et al. 2009, 2014). The dose of regular treatment can be adjusted to ensure that blood levels of opioid agonists are maintained, thereby effectively managing cravings and withdrawal without any associated intoxication (Opioid Agonist Treatment for Pharmaceutical Opioid Dependent People - Nielsen, S - 2016 | Cochrane Library n.d.). It has been demonstrated that a consistent dosage of ORT can result in health and social advantages for patients, such as the reduction of illicit drug use, and criminal activity, and the enhancement of both physical and mental health (Opioid Agonist Treatment for Pharmaceutical Opioid Dependent People - Nielsen, S - 2016 | Cochrane Library n.d.).

ORT is a well-established approach to enhancing the mental health of patients, as well as reducing substance use, criminal activity, and mortality (Opioid Agonist Treatment for Pharmaceutical Opioid Dependent People - Nielsen, S - 2016 | Cochrane Library n.d.). There are numerous international studies that have examined the nutrition-related intake of patients undergoing ORT (Saeland et al. 2009; Kolarzyk et al. 2005). It has been established that patients who engage in ORT experience more favorable outcomes when combined with appropriate nutritional intake (Nabipour et al. 2014). In general, however, there seems to be a scarcity of research that examines the nutritional health status of this patient group through biometrics, rather than actual nutritional consumption.

A systematic review (Waddington et al. 2023) was conducted covering studies from 2001 to 2022 aimed to evaluate the existing research on the nutritional intake of patients undergoing ORT, focusing on both macro- and micronutrient intake. The review included seven studies (one cohort study and six cross-sectional studies) with a total of 443 participants. The studies primarily used food frequency questionnaires, 24-hour food recalls, and biomarkers to assess nutrient intake. The findings showed that ORT patients often had irregular nutritional intake, with deficiencies in both macro- and micronutrients, including low intake of fiber, selenium, potassium, and calcium. There was also a tendency toward higher consumption of sweets and fatty foods. The systematic review identified a significant gap in research concerning the dietary habits of ORT patients, suggesting that these individuals often do not meet recommended dietary guidelines. It highlights the need for targeted nutritional interventions and further research to support the recovery and overall health of ORT patients.

20.3 Nutritional Challenges in OUD

Although opioid agonist treatment (OAT) is considered the most effective treatment of OUD (Degenhardt et al. 2019), nutritional rehabilitation programs play a significant role in managing OUD. This prompted us to mention some of the nutritional complications and obstacles to implementing nutritional rehabilitation.

20.3.1 Malnutrition and Eating Disorders

Malnutrition is the main problem for individuals with opioid dependence; it primarily arises from reduced calories, nutrients, and protein intake as some patients may prefer drugs over food (Ruiz 2021; Young 2023). For those with OUD, malnutrition stems from the physical effects of opioids, which suppress appetite rather than tissue damage as in alcohol use disorder (AUD) (Ruiz 2021). Also, binge eating behaviors in OUD lead to fasting from both food and water, exacerbating nutritional deficits (Ruiz 2021). Opioid-addicted patients have altered food preferences, favoring sweet and salty meals that activate the brain's reward system (Ruiz 2021), especially during early abstinence (Wiss 2019). These meals frequently lack key nutrients and fibers, contributing to nutritional imbalances (Ruiz 2021), and reduced eating frequency and inadequate nutrition absorption lead to weight loss in those patients (Byanyima et al. 2023).

20.3.2 Nutrient Issues

Some deficiencies in opioid addicts include thiamine, riboflavin, niacin, B6, folate, B12, vitamins A, C, D, E, K, and minerals like calcium, selenium, and iron (Ruiz 2021), and it's unclear whether this is attributable to a poor diet that lacks vegetables and fruits, the medications themselves, or reduced absorption (Waddington et al. 2022; Wiss 2019). On the other hand, specific minerals, including phosphorus, sodium, and magnesium, are present at higher concentrations in the plasma of drug users compared to healthy individuals, which may reflect partial dehydration; also, smoking, acute fasting, and inflammation lead to increased amounts of copper and zinc (Mahboub et al. 2021). The type, frequency, and duration of drug use, as well as the prevalence of infectious diseases, all have an impact on drug users' nutritional health (Mahboub et al. 2021). An important side effect of opioid use that affects nutrient intake is xerostomia or decreased saliva, which leads to tooth decay and gum infections (Ruiz 2021; Savoie-Roskos et al. 2020).

20.3.3 Gastrointestinal Problems

Exogenous opioids dysregulate the enteric neuronal system in the gastrointestinal tract by activating the mu-opioid receptors which are expressed with delta, and kappa throughout the tract (Byanyima et al. 2023; Wiss 2019), resulting in decreased motility, delayed stomach emptying, and slower transit time which leads to constipation, stomach reflux, and bloating during opioid use (Negi et al. 2023; Paul et al. 2021a; Ruiz 2021) that is known as opioid-induced bowel dysfunction (OIBD) (Byanyima et al. 2023), and diarrhea during detoxification (Wiss 2019). Also, constipation risk is increased in older individuals as aging contributes to the reduction in intestinal and gastric motility and decreased absorption (Dufort and Samaan 2021). Opioids can interfere with electrolyte and water secretion, which increases fluid absorption and may result in intestinal obstruction, nausea, vomiting, decreased appetite, malnutrition, and decreased bile acid production (Ruiz 2021). All these symptoms make eating the last priority for patients (Mahboub et al. 2021). Furthermore, stagnant conditions can promote bacterial proliferation, altering the microbiome and leading to toxin buildup, and that's why this environment promotes inflammation and intestinal damage, which reduces vitamin absorption and worsens anorexia-related deficiencies (Ruiz 2021; Wiss 2019). Opioids have been shown to interact with bacterial receptors on Pseudomonas aeruginosa, boosting the bacteria's pathogenicity. This contact causes the bacteria to change into a form that diminishes the intestinal mucosal layer while damaging the epithelium and tight junctions. As a result, bacteria, antigens, and toxins can cross the epithelial barrier, increasing the risk of sepsis and septic shock (Ruiz 2021).

20.3.4 Bone Disorders

Opioid use is associated with an increased risk of osteoporosis, not due to lack of vitamin D but through its effects on the hypothalamus, pituitary gland, and gonadal and adrenal hormones. These changes lead to a decrease in bone mineral density (Fountas et al. 2020; Ruiz 2021). Additionally, exogenous opioids may directly inhibit osteoblast functions by binding to their receptors (Ruiz 2021).

20.3.5 Hormonal Changes

Another putative mechanism for opioid metabolism and nutrition dysregulation is hormonal alterations. Few studies have examined the relationship between opioids and ghrelin/leptin and showed that opioid use leads to increased levels of ghrelin and decreased leptin (Ruiz 2021). Other studies showed that individuals using opioids have low serum levels of adiponectin, which is produced by adipocytes to prevent cardiac dysfunction and abnormalities in lipid and glucose metabolism (Byanyima et al. 2023). Another proof of poor nutritional status in OUD individuals is elevated levels of homocysteine which are regulated by the presence of vitamin B12 and folate (Byanyima et al. 2023). Patients taking opioids have impaired the HPA axis by inhibiting corticotropin-releasing hormone (CRH) and antidiuretic hormone (ADH) secretion and decreased adrenocorticotropic hormone (ACTH) (Fountas et al. 2020) that lead to opioid-induced adrenal insufficiency (OIAI) characterized by insufficient cortisol secretion from the adrenal gland and opioid-induced androgen deficiency (OIAD) (Byanyima et al. 2023; Coluzzi et al. 2023). In contrast, buprenorphine and methadone as OAT have been found to normalize the HPA axis (Kakko et al. 2019). Furthermore, opioids inhibit the hypothalamic release of pulsatile gonadotropin-releasing hormone, which reduces the pituitary gland's secretion of gonadotropins (luteinizing hormone and follicle-stimulating hormone) and the gonads' production of sex steroids leading to hypogonadism (Fountas et al. 2020), decreased libido, menstrual irregularities in females, and erectile dysfunctions in males (Gadelha et al. 2022). Prolactin (PRL) secretion is similarly stimulated by opioids through hypothalamic pathways, most likely by inhibiting the tuberoinfundibular dopaminergic system. However, serotonergic pathways might also be involved (Fountas et al. 2020). Finally, the impact of exogenous opioids on the human growth hormone (GH) and ADH axis is still unclear (Fountas et al. 2020).

20.3.6 OUD During COVID-19

Patients with OUD and comorbidities have a high risk of COVID-19 infection, especially because of the pandemic's control measures such as quarantine which severely impacted the delivery of medications and consultations (Sun et al. 2020). Furthermore, the pandemic's psychological consequences of anxiety and post-traumatic stress disorder (PTSD) can cause emotional instability and reactivate previously treated mental illnesses, increasing susceptibility to illegal opioid usage and even death from overdose (Jain et al. 2023; Sun et al. 2020). A lack of thorough communication and physical examination will have an impact on assessment accuracy, patient education, and monitoring efficiency for opioid addicts (Zhang et al. 2020). On the other hand, for transmission control purposes, talking with patients by telephone, video, and electronic systems has become the preferred method of clinic and follow-up (Zhang et al. 2020).

20.3.7 Comorbidities

Opioid uses are associated with low cholesterol levels that impact mental health, so increasing rates of relapse and suicide (Mahboub et al. 2021; Ruiz 2021). Also, hematocrit and hemoglobin levels are low in users, and along with nutrient deficiencies, they contribute to anemias such as iron deficiency anemia (Mahboub et al. 2021). Other significant effects are sedation, dizziness, increased risk of falls and fall-related injuries, and poor motor coordination (Dufort and Samaan 2021). Additionally, there are increased risks of viral infections such as hepatitis C virus (HCV) and HIV (Donroe et al. 2020). Respiratory depression occurs but less frequently than other morbidities (Paul et al. 2021a). Furthermore, opioids with the presence of hypotension and acidosis increase the risk of rhabdomyolysis, which causes acute renal failure, as well as the risk of drug-induced thrombotic microangiopathy, urinary retention, and heroin-associated nephropathy (HAN) (Perez and Bourgeois 2022).

20.3.8 Access to Nutritional Support

Patients with OUD frequently have to deal with drug-related costs and difficulty finding jobs, which have an impact on their feeding behaviors. Compared to healthy people, they are more prone to face financial difficulties, homelessness or unstable housing, and difficulties in continuing medical care (Byanyima et al. 2023; Rose-Jacobs et al. 2019). Food insecurity as a cause of poverty, low income, and lack of resources impacts nutritional rehabilitation as well (Young 2023). Also, the lack of nutritional centers and individualized treatment has a great impact on patients' rehabilitation (Young 2023). Other obstacles include insufficient training, financial barriers to access to care, the stigmatizing attitudes toward patients with OUD, and some facilities seeing that nutritional programs have no value (Madras et al. 2020; Wiss et al. 2019).

20.4 Clinical Evidence of the Beneficial Impact of Nutritional Rehabilitation

20.4.1 Zinc Supplementation's Effect on Opioids

Research on low zinc levels in humans and opioid use has shown a link. Several research on people taking opioids regularly reported either lower zinc concentrations or higher zinc excretion across several body fluids—serum and plasma before, during, and after therapy. For example, heroin addicts had reduced zinc serum concentrations at treatment admission; only a minor rise was noted throughout detoxification that fell short of levels reported in control patients. Studies of illegal heroin samples found, notably, that any zinc content resulting from processing materials had little effect on toxicity or trace element consumption (Elnimr et al. 1996; Larson and Kitto 1997; Ruiz Martínez et al. 1990; Sadlik et al. 2000).

In Amini's study investigated how zinc supplementation affected mental health and relapse risk in people with OUD undergoing MMT. Eight patients were divided into two cohorts in a randomized, single-blind clinical trial: one group had only methadone and another 12 mg/day of zinc. The zinc group showed clear declines in the likelihood of drug use, urges, and general relapse risk over 3 months. Furthermore, this group showed notable improvements in mental health, especially with regard to stress levels and general Depression, Anxiety, and Stress Scale (DASS-21) ratings relative to the control group. These results imply that zinc supplements can help people undergoing MMT (Amini and HeidariFarsani 2023) promote mental health and enhance the results of relapse.

Despite these promising results, many studies overlooked specific types of opioids used and did not fully account for potential co-occurring substance use or health conditions. Additionally, the nutritional habits of opioid users, often characterized by malnutrition, were not adequately addressed. This malnutrition can lead to reduced zinc levels, particularly among those consuming alcohol, while low zinc levels may further exacerbate appetite issues. The interplay between zinc and copper complicates the situation, as zinc deficiency appears more pronounced in opioid consumers, who tend to have higher copper levels (Gómez-Sirvent et al. 1993; Saeland et al. 2009).

Opioid use has been investigated closely in animal experiments concerning levels of zinc and metallothionein. Studies on morphine treatment in rats showed that although it raises liver cytosolic zinc and metallothionein levels, morphine lowers serum zinc (Hidalgo et al. 1991). Opioids affect zinc levels through opioid receptors; antagonism by naloxone can undo alterations in liver metallothionein levels, implying that opioid receptors are rather important. Still, zinc buildup in the liver seems to include other processes including β-adrenergic systems and glucocorticoid systems. Furthermore, investigations concentrating on δ-opioid receptor agonists showed that their treatment, in a time- and dose-dependent manner, lowered zinc levels in several brain areas; effects reversible by naloxone (Paul et al. 2021b).

The consequences of zinc's participation in analgesia and opioid dependency are interesting. In rat research, zinc chelation enhanced withdrawal behaviors, while

$ZnCl_2$ during the establishment of morphine dependency greatly lessened the degree of withdrawal symptoms. These results highlight zinc's possible influence in adjusting the neuronal processes behind opioid dependency. Moreover, zinc has been demonstrated to improve analgesia mediated by opioids. Zinc has been linked in opioid-free animal models to lower pain responses; when co-administered with opioids, it potentiates the analgesic effects, hence improving morphine-induced pain alleviation (Ciubotariu et al. 2015).

20.4.2 Additional Clinical Trials of Nutrition in Opioid Users

MMT is a crucial therapy for OUD; nonetheless, numerous patients continue to encounter mental health challenges such as depression and anxiety, which may result in higher relapse rates (Le et al. 2019). Of note, a healthy, balanced diet rich in essential micronutrients supports the efficacy of treatment for patients with OUD undergoing MMT (Kolarzyk et al. n.d.). Herein, some clinical trials discuss various nutrients and supplements that have proven beneficial effects on those affected by opioid addiction.

In a randomized, double-blind, placebo-controlled clinical trial, we examined the effects of vitamin D supplementation on cognitive and mental health outcomes in patients receiving methadone maintenance treatment (MMT). During a 24-week duration, subjects with 50,000 IU of vitamin D every two weeks exhibited substantial enhancements relative to placebo, including improved performance on the Iowa Gambling Task, Verbal Fluency Test, Reverse Digit Span, and assessments of visual working memory. Furthermore, depressed symptoms exhibited a significant decrease, as seen by reduced Beck Depression Inventory scores, suggesting a possible mitigation of mood abnormalities typically linked to opioid dependence. Some improvements, like immediate logic memory, lost significance after correction, and other mental health parameters stayed the same. However, the overall findings show that vitamin D supplementation could be a helpful way to improve cognitive function and reduce depressive symptoms in patients who are on methadone therapy (Ghaderi et al. 2020). Similarly, Lagzi et al. (2023) conducted a randomized, double-blind, placebo-controlled clinical trial to investigate the impact of multivitamin supplementation on anxiety and depression among patients on MMT. Seventy male participants were randomized into an intervention group receiving multivitamins or a placebo group over a duration of 12 weeks. The multivitamin capsule utilized in the study contains a variety of essential vitamins and minerals, including vitamin E, B vitamins (B1, B2, B3, B5, B6, B12, and biotin), vitamin C, folic acid, and zinc. Results indicated no significant alteration in anxiety levels either within or between the groups following the intervention. Nonetheless, the intervention group exhibited a notable decrease in depression scores relative to the placebo group, indicating that multivitamins may particularly benefit depressive symptoms in this population, but anxiety kept unaffected (Lagzi et al. 2023).

An ongoing randomized controlled study by Fadnes et al. (2022) is examining the impact of a 16-week fruit smoothie supplementation on opioid-dependent patients undergoing OAT. Participants in the intervention group get a daily 250 ml fruit smoothie composed of a mix of fruits, including apple, pineapple, mango, banana, and blueberries, in conjunction with their standard OAT medicine. The control group persists with standard OAT care, without the smoothie supplementation. The trial is to evaluate the effects of increased fruit consumption on psychological distress, fatigue, physical functioning, and inflammatory biomarkers. Since the trial is not yet completed, findings are not yet available; nevertheless, the study will primarily assess the intervention's efficacy using the Hopkins Symptom Checklist (SCL-10) for psychological distress, along with secondary outcomes including fatigue and quality of life (Fadnes et al. 2022).

A randomized, double-blind, placebo-controlled trial conducted by Molavi et al. (2022) at Soltan Mirahmad Clinic examined 59 opioid-dependent patients on methadone maintenance treatment programs (MMTPs) to evaluate the impact of probiotic supplementation on mental health and metabolic outcomes. Participants were assigned to receive either a probiotic containing four bacterial strains (Lactobacillus acidophilus, Bifidobacterium bifidum, Bifidobacterium longum, and Bifidobacterium lactis) or a placebo for a duration of 12 weeks. The research indicated that probiotics markedly decreased depression scores, although no significant changes were noted in anxiety or sleep quality. Moreover, the administration of probiotics resulted in improvements in fasting plasma glucose, total cholesterol, low-density lipoprotein cholesterol (LDL cholesterol), high-sensitivity C-reactive protein (hs-CRP), total antioxidant capacity (TAC), and plasma glutathione levels, while exerting no significant effect on insulin metabolism or oxidative stress markers (Molavi et al. 2022).

Sason et al. (2018) conducted a randomized clinical trial to evaluate the effectiveness of a structured nutrition education program for patients undergoing MMT for opioid dependence. The study found that participants in the intervention group experienced significant improvements in nutrition knowledge and reported better food habits compared to the control group. Despite these positive changes, the program did not result in significant differences in BMI or weight between the two groups over the study period. Additionally, there were no notable changes in participants' psychological well-being, as measured by depression scores. Overall, the trial highlights the potential of educational interventions to enhance dietary practices among MMT patients, suggesting that such improvements could lead to better health outcomes in the long term, even in the absence of immediate weight changes (Sason et al. 2018) (Table 20.1).

20.5 Conclusion

OUD significantly impacts both physical and mental health by affecting the gut-brain axis, gastrointestinal function, and overall nutritional condition. Opioid-induced disturbances, including dysbiosis, degradation of the intestinal barrier, and

Table 20.1 Summary of clinical trials included various examples of nutrients or supplements provided to opioid-addicted patients

Drug	Nutrient/ supplement	ClinicalTrials.gov	Study design	Intervention	Results	Status	References
MMT	Zinc supplementation	IRCT2020050904736N1	Randomized, single-blind, parallel, two-arm clinical trial	Administering zinc supplementation (approximately 12 mg daily in the form of Suzin zinc sulfate) alongside methadone for 3 months to patients with OUD undergoing MMT	Reduction in relapse probability (likelihood of drug use, drug craving, relapse probability (probability of relapse score (RPS))) Improvement in mental health (depression, anxiety, stress levels, and overall mental health)	Completed	Amini and HeidariFarsani (2023)
	Vitamin D supplementation	IRCT2017101133079N4	Randomized, double-blinded, placebo-controlled clinical trial	Vitamin D supplementation (50,000 IU) administered biweekly for 24 weeks	Improved cognitive function Reduction in depressive symptoms Stability in anxiety levels Overall mental health improvement	Completed	Ghaderi et al. (2020)
	Multivitamin supplementation	IRCT20200726048210N1	Randomized, double-blind, placebo-controlled clinical trial study for 12 weeks	One capsule containing multivitamins per day	No significant change in anxiety levels before and after the intervention A notable reduction in depression symptoms post-intervention A shift in depression severity, with more patients moving from severe to less severe levels of depression The intervention group experienced a larger decrease in depression scores compared to the placebo group.	Completed	Lagzi et al. (2023)

Oat	Fruit smoothie supplementation	NCT05229770	Multicenter, randomized controlled trial	Providing the intervention group with a 250 ml fruit smoothie daily for 16 weeks	Not posted yet	Ongoing	Fadnes et al. (2022)
MMT	Probiotic supplementation	Clinical trial registration: This study was registered in the Iranian website (https://www.irct.ir) for clinical trials registration (https://fa.irct.ir/trial/46363/IRCT20170420033551N9). The registration date is march 22, 2020	Randomized, double-blinded, placebo-controlled trial	Receiving either 1.8×10^9 CFU/day probiotics for 12 weeks	Significant reduction in depression scores compared to the placebo group Significant reduction in fasting plasma glucose levels in the probiotic group Total cholesterol and LDL cholesterol levels significantly decreased following probiotic supplementation Probiotics reduced hs-CRP levels, indicating a decrease in inflammation Significant improvements were observed in TAC and plasma glutathione levels The intervention did not significantly impact anxiety, sleep quality, insulin metabolism, triglycerides, or other oxidative stress markers	Completed	Molavi et al. (2022)

systemic inflammation, lead to neurotransmitter imbalances, mood problems, and cognitive deterioration. Nutritional deficits, especially in vitamins, minerals, and antioxidants, increase these problems, while gastrointestinal difficulties, including constipation, malabsorption, and reduced appetite, intensify malnutrition. Research indicates that specific nutritional therapies, such as zinc, vitamin D, multivitamins, and probiotics, can diminish relapse risk, increase mental health outcomes, and facilitate recovery for patients receiving opioid treatment. Nonetheless, socioeconomic obstacles, restricted access to nutritional care, and insufficient awareness impede effective implementation. An integrated, multidisciplinary strategy that encompasses diet, medical intervention, and psychological support is crucial for addressing the intricate requirements of patients with OUD, thereby facilitating sustained recovery and well-being.

Acknowledgments Not applicable.

Conflicts of Interest The authors declare no conflicts of interest.

Author Contributions Doaa S. R. Khafaga developed the experiments, examined the results, created figures and/or tables, produced either original or reviewed drafts of the article, and approved the final draft.

Rahma Saad developed the experiments, examined the data, created tables and/or figures, wrote or revised drafts of the article, and approved the final text.

Rana Mansour approved the final draft, wrote or evaluated versions of the paper, and examined the data.

Our E. Bayali developed the experiments, examined the data, created figures and/or tables, wrote or evaluated drafts of the article, and approved the final text.

Ahmed H. Shaaban approved the last draft, wrote or read versions of the article, and examined the data.

Salma Allam developed the experiments, created figures and/or tables, wrote or revised drafts of the article, and approved the final manuscript.

Abdullrahman Elgarawany developed the experiments, examined the data, created figures and/or tables, wrote or revised drafts of the article, and approved the final text.

Funding Not applicable.

References

Absorption and utilization of nutrients in alcoholism—Document—Gale Academic OneFile (n.d.) Retrieved October 18, 2024, from https://go.gale.com/ps/i.do?id=GALE%7CA8191936&sid=googleScholar&v=2.1&it=r&linkaccess=abs&issn=0090838X&p=AONE&sw=w&userGroupName=anon%7E16d88a3b&aty=open-web-entry

Acharya C, Betrapally NS, Gillevet PM, Sterling RK, Akbarali H, White MB, Ganapathy D, Fagan A, Sikaroodi M, Bajaj JS (2017) Chronic opioid use is associated with altered gut microbiota and predicts readmissions in patients with cirrhosis. Aliment Pharmacol Ther 45(2):319–331. https://doi.org/10.1111/apt.13858

Alves D, Costa AF, Custódio D, Natário L, Ferro-Lebres V, Andrade F (2011) Housing and employment situation, body mass index and dietary habits of heroin addicts in methadone maintenance treatment. Heroin Addict Relat Clin Probl 1:11–14

Amini Z, HeidariFarsani E (2023) Investigating the effect of zinc supplementation on probability of relapse and mental health in patients with opioid use disorder undergoing methadone maintenance treatment. Subst Abuse Treat Prev Policy 18:1. https://doi.org/10.1186/s13011-023-00514-5

Bischoff SC, Barbara G, Buurman W, Ockhuizen T, Schulzke J-D, Serino M, Tilg H, Watson A, Wells JM (2014) Intestinal permeability—a new target for disease prevention and therapy. BMC Gastroenterol 14:189. https://doi.org/10.1186/s12876-014-0189-7

Byanyima JI, Li X, Vesslee SA, Kranzler HR, Shi Z, Wiers CE (2023) Metabolic profiles associated with opioid use and opioid use disorder: a narrative review of the literature. Curr Addict Rep 10(3):581. https://doi.org/10.1007/s40429-023-00493-4

Cani PD, Delzenne NM (2009) The role of the gut microbiota in energy metabolism and metabolic disease. Curr Pharm Des 15(13):1546–1558. https://doi.org/10.2174/138161209788168164

Ciubotariu D, Ghiciuc CM, Lupuşoru CE (2015) Zinc involvement in opioid addiction and analgesia—should zinc supplementation be recommended for opioid-treated persons? Subst Abuse Treat Prev Policy 10:29. https://doi.org/10.1186/s13011-015-0025-2

Coluzzi F, LeQuang JAK, Sciacchitano S, Scerpa MS, Rocco M, Pergolizzi J (2023) A closer look at opioid-induced adrenal insufficiency: a narrative review. Int J Mol Sci 24(5):Article 5. https://doi.org/10.3390/ijms24054575

Cowan JA, Devine CM (2012) Process evaluation of an environmental and educational nutrition intervention in residential drug-treatment facilities. Public Health Nutr 15(7):1159–1167. https://doi.org/10.1017/S1368980012000572

Cryan JF, Dinan TG (2012) Mind-altering microorganisms: the impact of the gut microbiota on brain and behaviour. Nat Rev Neurosci 13(10):701–712. https://doi.org/10.1038/nrn3346

Cunningham PM (2016) The use of sobriety nutritional therapy in the treatment of opioid addiction. J Addict Res Ther 7(3):Article 3. https://doi.org/10.4172/2155-6105.1000282

D'Aunno T, Pollack HA, Frimpong JA, Wutchiett D (2014) Evidence-based treatment for opioid disorders: a 23-year national study of methadone dose levels. J Subst Abus Treat 47(4):245–250. https://doi.org/10.1016/j.jsat.2014.06.001

de Vos WM, Tilg H, Van Hul M, Cani PD (2022) Gut microbiome and health: mechanistic insights. Gut 71(5):1020–1032. https://doi.org/10.1136/gutjnl-2021-326789

Degenhardt L, Grebely J, Stone J, Hickman M, Vickerman P, Marshall BDL, Bruneau J, Altice FL, Henderson G, Rahimi-Movaghar A, Larney S (2019) Global patterns of opioid use and dependence: harms to populations, interventions, and future action. Lancet 394(10208):1560–1579. https://doi.org/10.1016/S0140-6736(19)32229-9

Dole VP, Nyswander ME, Kreek MJ (1966) Narcotic blockade. Arch Intern Med 118(4):304–309

Donroe JH, Bhatraju EP, Tsui JI, Edelman EJ (2020) Identification and management of opioid use disorder in primary care: an update. Curr Psychiatry Rep 22(5):23. https://doi.org/10.1007/s11920-020-01149-0

Drug harms in the UK: a multicriteria decision analysis—PubMed (n.d.) Retrieved October 18, 2024, from https://pubmed.ncbi.nlm.nih.gov/21036393/

Dufort A, Samaan Z (2021) Problematic opioid use among older adults: epidemiology, adverse outcomes and treatment considerations. Drugs Aging 38(12):1043–1053. https://doi.org/10.1007/s40266-021-00893-z

Elnimr T, Hashem A, Assar R (1996) Heroin dependence effects on some major and trace elements. Biol Trace Elem Res 54(2):153–162. https://doi.org/10.1007/BF02786262

Fadnes LT, Furulund E, Druckrey-Fiskaaen KT, Madebo T, Vold JH, Olsvold M, Pierron MC, Carlsen S-EL, Blomhoff R, Lid TG, ATLAS4LAR Study Group (2022) Effect of fruit smoothie supplementation on psychological distress among people with substance use disorders receiving opioid agonist therapy: protocol for a randomised controlled trial (FruktBAR). BMC Nutr 8(1):97. https://doi.org/10.1186/s40795-022-00582-z

Finnegan J, Gray D (1990) Recovery from addiction: a comprehensive understanding of substance abuse with nutritional therapies for recovering addicts and co-dependents. Celestial Arts

Fishbein: The effects of diet on behavior: Implications... - Google Scholar (n.d.) Retrieved October 18, 2024, from https://scholar.google.com/scholar_lookup?journal=Research%20in%20

Corrections&title=The%20Effects%20of%20Diet%20on%20Behavior:%20Implications%20for%20Criminology%20and%20Corrections&author=D%20Fishbein&author=S%20Pease&volume=1&issue=2&publication_year=1988&

Fountas A, Uum SV, Karavitaki N (2020) Opioid-induced endocrinopathies. Lancet Diabetes Endocrinol 8(1):68–80. https://doi.org/10.1016/S2213-8587(19)30254-2

Gadelha MR, Karavitaki N, Fudin J, Bettinger JJ, Raff H, Ben-Shlomo A (2022) Opioids and pituitary function: expert opinion. Pituitary 25(1):52–63. https://doi.org/10.1007/s11102-021-01202-y

Ghaderi A, Rasouli-Azad M, Farhadi MH, Mirhosseini N, Motmaen M, Pishyareh E, Omidi A, Asemi Z (2020) Exploring the effects of vitamin D supplementation on cognitive functions and mental health status in subjects under methadone maintenance treatment. J Addict Med 14(1):18. https://doi.org/10.1097/ADM.0000000000000550

Gómez-Sirvent JL, Santolaria-Fernández FJ, González-Reimers CE, Batista-López JN, Jorge-Hernández JA, Rodríguez-Moreno F, Martínez-Riera A, Hernández-García MT (1993) Nutritional assessment of drug addicts. Relation with HIV infection in early stages. Clin Nutr 12(2):75–80. https://doi.org/10.1016/0261-5614(93)90055-9

Hidalgo J, Giralt M, Garvey JS, Armario A (1991) Effect of morphine administration on rat liver metallothionein and zinc metabolism. J Pharmacol Exp Ther 259(1):274–278

Himmelgreen DA, Pérez-Escamilla R, Segura-Millán S, Romero-Daza N, Tanasescu M, Singer M (1998) A comparison of the nutritional status and food security of drug-using and non-drug-using Hispanic women in Hartford, Connecticut. Am J Phys Anthropol 107(3):351–361. https://doi.org/10.1002/(SICI)1096-8644(199811)107:3<351::AID-AJPA10>3.0.CO;2-7

Holzer P (2009) Opioid receptors in the gastrointestinal tract. Regul Pept 155(1–3):11–17. https://doi.org/10.1016/j.regpep.2009.03.012

Jain A, Dhingra A, Sivakumar A (2023) Opioids and Covid-19: perfect storm, or perfect opportunity? The Columbia University J Glob Health 13(1):Article 1. https://doi.org/10.52214/cujgh.v13i1.10687

Kakko J, Alho H, Baldacchino A, Molina R, Nava FA, Shaya G (2019) Craving in opioid use disorder: from neurobiology to clinical practice. Front Psych 10. https://doi.org/10.3389/fpsyt.2019.00592

Kolarzyk E, Pach D, Wojtowicz B, Szpanowska-Wohn A, Szurkowska M (2005) Nutritional status of the opiate dependent persons after 4 years of methadone maintenance treatment. Przegl Lek 62(6):373–377

Kolarzyk E, Szpanowska-Wohn A, Chrostek-Maj J (n.d.) Zinc, copper and magnesium intake in a daily diet in opiate-addicted persons before and after 5 years of methadone treatment

Lagzi N, Bateni A, Goli R, Talebiazar N (2023) The effect of multivitamins on anxiety and depression in patients undergoing methadone maintenance treatment: a double-blind randomized controlled trial. Int J Psychiatry Med 58(6):576–590. https://doi.org/10.1177/00912174231179320

Larson AA, Kitto KF (1997) Manipulations of zinc in the spinal cord, by intrathecal injection of zinc chloride, disodium-calcium-EDTA, or dipicolinic acid, alter nociceptive activity in mice. J Pharmacol Exp Ther 282(3):1319–1325

Le TA, Le MQT, Dang AD, Dang AK, Nguyen CT, Pham HQ, Vu GT, Hoang CL, Tran TT, Vuong Q-H, Tran TH, Tran BX, Latkin CA, Ho CSH, Ho RCM (2019) Multi-level predictors of psychological problems among methadone maintenance treatment patients in difference types of settings in Vietnam. Subst Abuse Treat Prev Policy 14(1):39. https://doi.org/10.1186/s13011-019-0223-4

Madras BK, Ahmad NJ, Wen J, Sharfstein JS (2020) Improving access to evidence-based medical treatment for opioid use disorder: strategies to address key barriers within the treatment system. NAM Perspect 2020. https://doi.org/10.31478/202004b

Mahboub N, Rizk R, Karavetian M, de Vries N (2021) Nutritional status and eating habits of people who use drugs and/or are undergoing treatment for recovery: a narrative review. Nutr Rev 79(6):627–635. https://doi.org/10.1093/nutrit/nuaa095

Mangili A, Murman DH, Zampini AM, Wanke CA (2006) Nutrition and HIV infection: review of weight loss and wasting in the era of highly active antiretroviral therapy from the nutrition for healthy living cohort. Clin Infect Dis 42(6):836–842. https://doi.org/10.1086/500398

Mattick RP, Breen C, Kimber J, Davoli M (2009) Methadone maintenance therapy versus no opioid replacement therapy for opioid dependence. Cochrane Database Syst Rev 2009(3):CD002209. https://doi.org/10.1002/14651858.CD002209.pub2

Mattick RP, Breen C, Kimber J, Davoli M (2014) Buprenorphine maintenance versus placebo or methadone maintenance for opioid dependence. Cochrane Database Syst Rev 2014(2):CD002207. https://doi.org/10.1002/14651858.CD002207.pub4

Mayer EA, Tillisch K (2011) The brain-gut axis in abdominal pain syndromes. Annu Rev Med 62:381–396. https://doi.org/10.1146/annurev-med-012309-103958

McClain CJ, Barve SS, Barve A, Marsano L (2011) Alcoholic liver disease and malnutrition. Alcohol Clin Exp Res 35(5):815–820. https://doi.org/10.1111/j.1530-0277.2010.01405.x

Medici V, Halsted CH (2013) Folate, alcohol, and liver disease. Mol Nutr Food Res 57(4):596–606. https://doi.org/10.1002/mnfr.201200077

Microbes and Mental Health: The Link Between Biome and Brain (n.d.) Retrieved October 18, 2024, from https://www.medscape.com/viewarticle/859484

Migdanis A, Migdanis I, Papadopoulou SK, Hadjivasiliou L, Trifonova N, Villioti M, Giaginis C, Kosti RI, Androutsos O (2023) Assessment of dietary intake and nutritional status of former opioid users undergoing detoxification process. Cureus 15(12):e50068. https://doi.org/10.7759/cureus.50068

Miller RP (n.d.) Nutrition in addiction recovery. Barre, MA: Many Hands Sustainability Center, 2010. https://mhof.net/wp-content/uploads/2020/12/Nutrition_in_Addiction_Recovery.pdf

Molavi N, Rasouli-Azad M, Mirzaei H, Matini AH, Banafshe HR, Valiollahzadeh M, Hassanzadeh M, Saghazade AR, Abbaszadeh-Mashkani S, Mamsharifi P, Ghaderi A (2022) The effects of probiotic supplementation on opioid-related disorder in patients under methadone maintenance treatment programs. Int J Clin Pract 2022:1206914. https://doi.org/10.1155/2022/1206914

Nabipour S, Ayu Said M, Hussain Habil M (2014) Burden and nutritional deficiencies in opiate addiction- systematic review article. Iran J Public Health 43(8):1022–1032

Nazrul Islam SK, Jahangir Hossain K, Ahmed A, Ahsan M (2002) Nutritional status of drug addicts undergoing detoxification: prevalence of malnutrition and influence of illicit drugs and lifestyle. Br J Nutr 88(5):507–513. https://doi.org/10.1079/BJN2002702

Negi PS, Mathur R, Sarkar S (2023) Health-promoting lifestyle factors among opioid-dependent patients: a scoping review. J SAARC Psychiatr Fed 1(2):72. https://doi.org/10.4103/jspf.jspf_1_24

Nutrition Therapy for Eating Disorders, Substance Use Disorders, and Addictions | Request PDF (n.d.) Retrieved October 18, 2024, from https://www.researchgate.net/publication/287305806_Nutrition_Therapy_for_Eating_Disorders_Substance_Use_Disorders_and_Addictions

Nutritional effects of marijuana, heroin, cocaine, and nicotine—PubMed (n.d.) Retrieved October 18, 2024, from https://pubmed.ncbi.nlm.nih.gov/2204648/

Opioid agonist treatment for pharmaceutical opioid dependent people—Nielsen, S - 2016 | Cochrane Library (n.d.) Retrieved October 18, 2024, from https://www.cochranelibrary.com/cdsr/doi/10.1002/14651858.CD011117.pub2/full

Organization, W. H (2014) Community management of opioid overdose. World Health Organization. https://iris.who.int/handle/10665/137462

Paul AK, Smith CM, Rahmatullah M, Nissapatorn V, Wilairatana P, Spetea M, Gueven N, Dietis N (2021a) Opioid analgesia and opioid-induced adverse effects: a review. Pharmaceuticals (Basel) 14(11):Article 11. https://doi.org/10.3390/ph14111091

Paul AK, Smith CM, Rahmatullah M, Nissapatorn V, Wilairatana P, Spetea M, Gueven N, Dietis N (2021b) Opioid analgesia and opioid-induced adverse effects: a review. Pharmaceuticals (Basel) 14(11):1091. https://doi.org/10.3390/ph14111091

Perez V, Bourgeois JA (2022) Substance use disorders and the kidney. In: Hategan A, Bourgeois JA, Gangji AS, Woo TKW (eds) Psychonephrology: a guide to principles and practice. Springer International Publishing, pp 179–201. https://doi.org/10.1007/978-3-030-84740-1_11

Potatoes Not Prozac: A Natural Seven-step Dietary Plan to Control Your ... - Kathleen DesMaisons—Google Books (n.d.) Retrieved October 18, 2024, from https://books.google.com.eg/books?hl=en&lr=&id=i5Og1muO14sC&oi=fnd&pg=PA11&ots=EjU6MZWrJs&sig=z-_9SZLd38XN1FcLhoE1jSRFeZY&redir_esc=y#v=onepage&q&f=false

Potatoes not Prozac: Simple solutions for sugar sensitivity / | Fayetteville Public Library (n.d.) Retrieved October 18, 2024, from https://catalog.faylib.org/Record/80549?searchId=1425966&recordIndex=1&page=&referred=resultIndex

Probiotics reduce psychological stress in patients before laryngeal cancer surgery—PubMed (n.d.) Retrieved October 18, 2024, from https://pubmed.ncbi.nlm.nih.gov/24571169/

Rhee SH, Pothoulakis C, Mayer EA (2009) Principles and clinical implications of the brain-gut-enteric microbiota axis. Nat Rev Gastroenterol Hepatol 6(5):306–314. https://doi.org/10.1038/nrgastro.2009.35

Rose-Jacobs R, Trevino-Talbot M, Vibbert M, Lloyd-Travaglini C, Cabral HJ (2019) Pregnant women in treatment for opioid use disorder: material hardships and psychosocial factors. Addict Behav 98:106030. https://doi.org/10.1016/j.addbeh.2019.106030

Ruiz MK (2021) Nutritional concerns in substance use disorders. University of Washington, Washington, DC. https://foodsystems.uw.edu/wp-content/uploads/2021/09/Ruiz_MS-Capstone-Report_Final_2021.pdf

Ruiz Martínez M, Gil Extremera B, Maldonado Martín A, Cantero-Hinojosa J, Moreno-Abadía V (1990) Trace elements in drug addicts. Klin Wochenschr 68(10):507–511. https://doi.org/10.1007/BF01648242

Sadlik J, Pach J, Winnik L, Piekoszewski W (2000) [Concentration of zinc, copper and magnesium in the serum of drug addicts]. Przegl Lek 57(10):563–564

Saeland M, Haugen M, Eriksen F-L, Smehaugen A, Wandel M, Böhmer T, Oshaug A (2009) Living as a drug addict in Oslo, Norway—a study focusing on nutrition and health. Public Health Nutr 12(5):630–636. https://doi.org/10.1017/S1368980008002553

Sason A, Adelson M, Herzman-Harari S, Peles E (2018) Knowledge about nutrition, eating habits and weight reduction intervention among methadone maintenance treatment patients. J Subst Abus Treat 86:52–59. https://doi.org/10.1016/j.jsat.2017.12.008

Savoie-Roskos M, Yaugher A, Condie A, Murza G, Voss M, Atisme K (2020) Diet, nutrition, and substance use disorder. All Current Publications, pp 1–6

Seven weeks to sobriety: The proven program to fight alcoholism through nutrition (n.d.)

Steenbergen L, Sellaro R, van Hemert S, Bosch JA, Colzato LS (2015) A randomized controlled trial to test the effect of multispecies probiotics on cognitive reactivity to sad mood. Brain Behav Immun 48:258–264. https://doi.org/10.1016/j.bbi.2015.04.003

Suk KT, Kim MY, Baik SK (2014) Alcoholic liver disease: treatment. World J Gastroenterol 20(36):12934–12944. https://doi.org/10.3748/wjg.v20.i36.12934

Sun Y, Bao Y, Kosten T, Strang J, Shi J, Lu L (2020) Editorial: challenges to opioid use disorders during COVID-19. Am J Addict 29(3):174. https://doi.org/10.1111/ajad.13031

Tao HG, Fox HM (1976) Measurements of urinary pantothenic acid excretions of alcoholic patients. J Nutr Sci Vitaminol 22(4):333–337. https://doi.org/10.3177/jnsv.22.333

The Eating for Recovery: The Essential Nutrition Plan to Reverse the Physical Damage of Alcoholism by Molly Siple, Paperback | Barnes & Noble® (n.d.) Retrieved October 18, 2024, from https://www.barnesandnoble.com/w/the-eating-for-recovery-molly-siple/1111374312

Varela P, Marcos A, Ripoll S, Santacruz I, Requejo AM (1997) Effects of human immunodeficiency virus infection and detoxification time on anthropometric measurements and dietary intake of male drug addicts. Am J Clin Nutr 66(2):509S–514S. https://doi.org/10.1093/ajcn/66.2.509S

Waddington F, Naunton M, Thomas J, Kyle G, Wheatley B, Oguoma V (2022) Examination of the nutritional intake of patients undergoing opioid replacement therapy: a systematic review. Nutr Diet 80(1):55. https://doi.org/10.1111/1747-0080.12784

Williams RJ (1956) Biochemical individuality; the basis for the genetotrophic concept. Wiley, p xiii, 214

Wiss DA (2019) A biopsychosocial overview of the opioid crisis: considering nutrition and gastrointestinal health. Front Public Health 7. https://doi.org/10.3389/fpubh.2019.00193

Wiss DA, Schellenberger M, Prelip ML (2019) Rapid assessment of nutrition services in Los Angeles substance use disorder treatment centers. J Community Health 44(1):88–94. https://doi.org/10.1007/s10900-018-0557-2

Young AL (2023) Understanding the role of nutrition in attitudes, behaviors, and choices of individuals with substance use disorders (SUD) in recovery to promote remission [PhD Thesis, Saybrook University]. https://search.proquest.com/openview/05b77655d38d5a48a632ae242d55f0a3/1?pq-origsite=gscholar&cbl=18750&diss=y

Zhang M, Zheng H, Wang J (2020) Considerations when COVID-19 pandemic collides opioid crisis: what we should know. J Anes Perio Manag 4(1):100009. Henry Publishing Groups

Zhou JF, Yan XF, Ruan ZR, Peng FY, Cai D, Yuan H, Sun L, Ding DY, Xu SS (2000) Heroin abuse and nitric oxide, oxidation, peroxidation, lipoperoxidation. Biomed Environ Sci 13(2):131–139

The Role of Diet in Managing Withdrawal Symptoms

21

Nada Fath, Hasna Tirichen, Hasnaa Yaigoub,
and Ismail Louragli

Abstract

Withdrawal syndrome is characterized by a range of physiological and psychological symptoms, and it poses a significant challenge during recovery from substance use disorders. This chapter delves into the neurobiological mechanisms underlying withdrawal symptoms and examines the potential role of nutrition in their management. Specific nutrients, such as omega-3 fatty acids, amino acids, and micronutrients, are highlighted for their impact on neurotransmitter regulation, inflammation, and stress response pathways, all of which are implicated in the withdrawal process. Practical dietary strategies and evidence-based recommendations are presented to support individuals in mitigating the severity of withdrawal and enhancing recovery outcomes. By integrating nutritional approaches into withdrawal management, this chapter underscores the potential of diet as a complementary tool in the treatment plans, providing insights into nutrition therapy during detoxification as a pathway toward improved health and recovery success.

N. Fath (✉)
School of Basic Medical Sciences, Fujian Medical University,
Fuzhou, Fujian Province, China

H. Tirichen · H. Yaigoub
Institutes of Biomedical Sciences, Shanxi University, Taiyuan, China

I. Louragli
Higher Institute of Nursing and Health Techniques, Ministry of Health and Social Protection,
Dakhla, Morocco

W. Mohamed, A. Özge (eds.), *Feeding the Mind: The Connection Between Diet, Drugs, and Mental Health Volume 1*, Nutritional Neurosciences,
https://doi.org/10.1007/978-981-95-1414-4_21

Keywords

Withdrawal Syndrome · Substance Use Disorders · Neurobiological Mechanisms · Nutritional Therapy · Omega-3 Fatty Acids · Detoxification · Recovery

21.1 Introduction

Substance withdrawal or substance withdrawal syndrome refers to a group of physical and psychological symptoms that occur following an abrupt or gradual cessation of an addictive substance. The severity of these symptoms varies based on the substance of dependence, as well as the frequency and duration of use, and other factors, including genetic predisposition, gender, and health status. Common substances that might cause withdrawal include alcohol, opioids, cannabis, and tobacco (American Psychiatric Association 2013). Discontinuing the use of addictive substances may result in multiple symptoms, some of which are common, including anxiety, irritability, insomnia, headaches, and nausea (Lerner and Klein 2019).

Managing withdrawal symptoms is crucial to achieve complete recovery and reduce the risk of relapse (Melemis 2015). Although the withdrawal management strategies are various and depend on the type of substance of dependence and level of addiction in users, all management strategies follow the same steps. These steps include assessing the risk of developing severe complications during withdrawal, medical supervision, gradual decrease in substance dosage, use of medication to alleviate specific symptoms, and psychological therapy (Geneva: World Health Organization 2009).

Diet has emerged as a promising nonpharmacological strategy alongside classic withdrawal management approaches (Wiers et al. 2021). Proper dietary approaches can effectively restore the nutritional balance depleted during the withdrawal process, improve sleep quality, and support immune functions (Wiss 2019). Introducing the right dietary elements helps stabilize the blood sugar level which reduces addictive substance cravings (Wiss and Waterhous 2014).

This chapter aims to elucidate the intricate interplay between diet and withdrawal symptoms, which can open a promising avenue for managing the physiological and psychological challenges associated with withdrawal syndrome. Herein, we explored the neurobiological mechanisms of withdrawal, the specific effects of various nutrients on the modulation of various symptoms, and evidence-based nutritional strategies to mitigate their intensity. By integrating dietary interventions into broader treatment frameworks, individuals undergoing withdrawal may find enhanced support for their detoxification and recovery journey, fostering resilience and improving overall well-being.

21.2 Understanding Withdrawal Symptoms

Withdrawal symptoms can be divided into physiological and psychological symptoms, each of which affects users differently. Physiological symptoms include changes in the nervous system, hormonal regulation, and other biological processes, while psychological symptoms arise primarily from altered neurotransmitter levels, affecting mood and cognition. The severity of withdrawal symptoms can vary widely, influenced by factors such as the type of substance used (e.g., opioids, cannabis, alcohol, nicotine, caffeine, benzodiazepines, methamphetamines, cocaine), the frequency and duration of use, and individual characteristics like genetic predisposition, gender, and overall health. On the one hand, withdrawal symptoms are manifested by a range of physiological symptoms, including insomnia, headaches, dizziness, muscle pain, increased sensitivity to pain, chills, muscle tension, tremors, sweating, fatigue, decreased alertness, extreme fatigue, disrupted sleep patterns, psychomotor slowing, intense cravings, and seizures (Weber et al. 2024), in addition to gastrointestinal discomfort and digestive issues such as nausea and diarrhea, as well as tingling, numbness, or muscle twitching. Some people report fluctuations in appetite, weight changes, and other sensory changes, such as hypersensitivity to light and sound (Edinoff et al. 2021). On the other hand, withdrawal can lead to various psychological symptoms including severe and persistent mood disturbances such as depression, anxiety, irritability, mood swings, emotional distress, impulsivity, and cravings (Cheeks et al. 2024; Smith et al. 2021). These mood disturbances are often accompanied by cognitive impairments, including memory and concentration problems, decreased cognitive performance, and attention deficits. Evidence suggests that these mood symptoms may persist, making psychological support essential to prevent relapse and address mental health issues associated with withdrawal (American Psychiatric Association 2013; Schwartz et al. 2022). In extreme cases, withdrawal can lead to psychosis and delirium, panic attacks, paranoid thoughts, hallucinations, and social phobia (Edinoff et al. 2021). Anorexia and sleep disturbances are prevalent upon cessation of drug use, drowsiness, and difficulty concentrating, symptoms that can potentially exacerbate stress and anxiety levels, and anhedonia (Connor et al. 2022).

21.3 Neurobiological Mechanisms of Withdrawal

Addiction can be conceptualized as a repetitive, three-phase cycle as described by Koob and Volkow (2016): binge/intoxication, withdrawal/negative affect, and preoccupation/anticipation (craving). Over time, this cycle intensifies and is driven by neuroplastic changes in the brain's reward, stress, and executive function systems. The aforementioned symptoms of withdrawal occur as a result of the body's adaptation to the absence of the substance to which it has become dependent. During this process, the brain's neurochemistry undergoes significant changes, and it is associated with the imbalance of specific neurotransmitter systems. One of the primary neurobiological mechanisms underlying withdrawal is the dysregulation of these neurotransmitter systems (Dahchour and Ward 2024; Sharma et al. 2023). The withdrawal/negative affect stage of addiction is characterized by significant disruptions in the brain's reward and stress systems, contributing to dysphoric emotional states and compulsive substance-seeking behavior. Withdrawal from chronic drug use leads to a decreased synaptic availability of dopamine (DA) and serotonin, leading to a disruption of regulatory processes in the prefrontal cortex. This double deficiency contributes to withdrawal symptoms, drug craving, and ultimately relapse (Dresp-Langley and Hutt 2022). During this stage, the function of the DA-based reward system declines, while brain stress systems, particularly those involving corticotropin-releasing factor (CRF) and dynorphin, are activated in the extended amygdala (Monroe and Radke 2023). Substances of abuse often enhance or inhibit the release of key neurotransmitters, including DA, serotonin, norepinephrine, and gamma-aminobutyric acid (GABA). For instance, stimulants like cocaine increase DA levels, reinforcing behaviors associated with drug use (Volkow and Blanco 2023). Upon cessation, there is a sudden deficit of these neurotransmitters, leading to symptoms such as anxiety, depression, and irritability (Koob and Volkow 2016). Diet can play a significant role in modulating these neurotransmitter levels. Nutrients such as omega-3 fatty acids, amino acids (like tryptophan and tyrosine), and B vitamins are essential for the synthesis and regulation of neurotransmitters (Stachowicz 2023). Therefore, a diet lacking in these nutrients may exacerbate withdrawal symptoms by further impairing neurotransmitter balance. In addition, withdrawal involves complex changes in brain structures such as the ventral tegmental area (VTA) and nucleus accumbens (NAc), which play key roles in the reinforcing effects of opioids. During withdrawal, impaired DA signaling leads to a reduction in reward sensitivity. This is partly due to decreased firing of DA neurons and a compensatory increase in activity within the cAMP signaling pathway, which affects GABAergic neurons in the VTA, further inhibiting DA neuron activity. Other factors contributing to withdrawal include changes in cholinergic interneurons in the striatum, which influence DA release, and plasticity in glutamatergic signaling in the NAc. Increased glutamate signaling, particularly via AMPA receptors, also plays a significant role in withdrawal-induced aversion, suggesting that altered excitatory signaling contributes to the negative emotional state associated with withdrawal (Monroe and Radke 2023). Decreased stimulation of these pathways (He et al. 2022) can also lead to anhedonia, which is defined as the reduced ability to experience pleasure, and this contributes to cravings and emotional distress (He et al. 2022). However, dietary factors can influence the health and functioning of these neural circuits. Thus, a diet rich in antioxidants may help reduce oxidative stress in the brain, promoting neuroprotection and potentially alleviating these withdrawal symptoms (Amiry et al. 2023). Another critical player in withdrawal is the hypothalamic-pituitary-adrenal (HPA) axis. Withdrawal often leads to increased stress hormone levels, particularly cortisol, which can exacerbate anxiety and other symptoms (Sinha 2024). Furthermore, recent research emphasizes the role of inflammation and the gut-brain axis in withdrawal syndrome (Barkus et al. 2024; Vatsalya et al. 2023). Substance abuse can lead to increased systemic inflammation, which has been linked to mood disturbances and cognitive deficits. An anti-inflammatory diet rich in fruits, vegetables, whole grains, and healthy fats may help reduce inflammation, which will alleviate withdrawal symptoms. The gut microbiome significantly impacts brain health (Sasso et al. 2023), and a diet that supports a diverse and healthy gut microbiome may improve overall mental health and resilience to withdrawal symptoms. Thus, fermented foods, probiotics, and prebiotics can enhance gut health, thereby influencing mood and behavior. To sum up, both physiological and psychological symptoms of withdrawal are associated with alterations in brain regions, mainly the noradrenergic system in the locus coeruleus and the dopaminergic system in the mesolimbic area, the serotoninergic system, cholinergic interneurons in the striatum, the GABAergic and glutamatergic systems, in addition to dysregulation of adenosine in the case of decreased cognitive performance and alertness due to caffeine withdrawal, and the brain's endocannabinoid system in case of cannabis withdrawal.

21.4 Role of Specific Nutrients in Managing Withdrawal Symptoms

Effective management of withdrawal symptoms requires addressing the underlying nutritional deficiencies caused by prolonged substance use and addiction. These deficiencies impair physical and psychological health, exacerbate withdrawal symptoms, and hinder recovery (Karaagac and Özçelik 2024; White and Sirohi 2024). Prolonged substance use disrupts nutrient absorption and metabolism, leading to deficiencies that alter the body's ability to repair itself. For example, alcohol dependence is linked to decreased levels of B vitamins, magnesium, calcium, and antioxidants, which are important micronutrients for neurological processes, cognitive function, and muscular health (Baj et al. 2020; Kamiński et al. 2024). Moreover, chronic alcohol use damages the gastrointestinal system, leading to impaired nutrient absorption and increased risk of diseases, such as alcoholic cardiomyopathy and osteoporosis (Butts et al. 2023; Wiss 2019). Similarly, methamphetamine use alters essential antioxidants like vitamin C and glutathione, exposing the brain to oxidative stress and increasing the severity of the neuropsychiatric symptoms such as anxiety and psychosis (Altufailia et al. 2023). Nicotine addiction further contributes to nutritional deficits, particularly reduced vitamin C levels, which weaken immune function and cellular repair mechanisms (Astori et al. 2022). Moreover, deficiencies in B vitamins and magnesium exacerbate fatigue, mood instability, and cognitive challenges, complicating recovery efforts (Wiss 2019).

Amino acids such as tyrosine and tryptophan are crucial for neurotransmitter production, regulating mood, cognition, and emotional stability. Substance use disrupts the metabolism of these amino acids, leading to imbalances in DA, serotonin, and norepinephrine levels (Mahboub et al. 2021). Tryptophan is essential for serotonin synthesis; tryptophan supplementation has demonstrated benefits in reducing alcohol withdrawal symptoms, stabilizing mood, improving sleep, and enhancing cognitive functions such as visual memory (Relojo-Howell et al. 2023). Foods rich in tryptophan, such as bananas, milk, sunflower seeds, and turkey, support serotonin production during recovery. In addition, tyrosine and phenylalanine, precursors to DA and norepinephrine, are vital for mental alertness and energy, and supplementation has been shown to improve mood, motivation, and emotional stability while reducing cravings (Relojo-Howell et al. 2023). Besides amino acids and neurotransmitter balance, micronutrients have a crucial role in managing withdrawal symptoms. Micronutrients such as vitamins C, E, and B-complex, along with magnesium, zinc, and selenium, are essential for cellular repair, immune defense, and antioxidant protection (Baj et al. 2020). For instance, vitamin C enhances immune function and reduces oxidative stress, particularly in individuals recovering from stimulant addiction, B-complex vitamins are critical for energy metabolism and neurological health, with supplementation shown to alleviate fatigue, depression, and cognitive difficulties (Ali et al. 2022; Wiss 2019), and finally magnesium, supporting muscle relaxation, nerve function, and mood stabilization, reduces anxiety and irritability during withdrawal (Baj et al. 2020). Research indicates that it is possible to reduce withdrawal symptoms with amino acid supplementation, which will give the individuals in detoxification a generally improved feeling of well-being. Taken together, these findings show that integrating amino acid therapy with multivitamin and mineral supplementation during withdrawal can restore neurotransmitter balance, repair tissue damage, and alleviate withdrawal symptoms like cravings and mood disturbances. Furthermore, the gut-brain axis plays a critical role in recovery. Addictive substances damage the gastrointestinal system, impairing nutrient absorption and causing symptoms like diarrhea, constipation, and poor appetite (Butts et al. 2023). Restoring gut health through a nutrient-dense diet helps rebuild tissues, enhance digestive function, and nourish the brain, reducing withdrawal symptoms and supporting cognitive recovery. High-protein, high-carbohydrate, and fiber-rich foods are particularly beneficial during early recovery, as they stabilize blood sugar levels and promote satiety. Research supports the value of integrating nutritional interventions aimed at replenishing these essential nutrients to alleviate withdrawal symptoms, stabilize mood, and improve overall cognitive function. Thus, a nutrient-rich diet and targeted supplementation can significantly support the recovery process, providing a foundation for increased resilience and long-term recovery outcomes (Wiss and Waterhous 2014).

21.5 Dietary Strategies for Managing Withdrawal Symptoms

Individuals with substance use disorders are particularly vulnerable to poor nutrition and malnutrition. This susceptibility is often due to factors such as reduced appetite, impaired nutrient absorption, and insufficient dietary intake (Neale et al. 2012; Preedy and Watson 1996). Additionally, preexisting nutrient deficiencies may further increase the likelihood of developing SUDs (Schroeder and Higgins 2017). A study conducted with 140 hospitalized individuals who used heroin and cocaine found that 18% were classified as severely malnourished, based on serum albumin levels below 3.5 g/dL and/or serum transferrin levels below 200 mg/dL. Severe malnutrition is more prevalent among females than males, and the most severe cases were observed in individuals with the highest levels of drug use (Santolaria-Fernández 1995). The body can benefit from various foods and nutrients throughout detox and recovery; thus, nutrition can serve as a critical tool to manage withdrawal symptoms by stabilizing blood sugar levels, promoting neurotransmitter release, and reducing oxidative stress. The following are some effective strategies:

21.5.1 Hydration: The Cornerstone of Recovery

Hydration is considered a crucial component in the management of alcohol withdrawal symptoms, as it plays a crucial role in alcohol withdrawal syndrome by supporting general physical health and helping alleviate particular withdrawal effects like headaches, nausea, and general fatigue (Sachdeva 2015). Because alcohol consumption usually causes the body to lose fluids and electrolytes, maintaining water intake during withdrawal can help balance electrolytes and avoid dehydration, which can exacerbate symptoms (Baj et al. 2020). In addition to water, electrolyte-rich beverages such as coconut water, oral rehydration treatments, or sugar-free sports drinks are also advantageous (Kalman et al. 2012). Additionally, even mild dehydration can have a noticeable negative impact on mood and cognitive performance, according to a study by Ganio et al. (2011). This highlights the importance of maintaining proper hydration for optimal mental functioning and emotional well-being, particularly in situations where dehydration risks are high.

21.5.2 Complex Carbohydrates: Stabilizing Energy and Mood

Cocaine dependence disrupts carbohydrate metabolism and alters food preferences, as individuals often reduce their intake of carbohydrate-rich foods due to appetite suppression and changes in the reward pathway (Ersche et al. 2013). However, they may develop an increased preference for sucrose and high-glycemic-index (GI) carbohydrates during withdrawal to compensate for lower DA stimulation (Janowsky et al. 2003). This shift in preference can lead to unstable blood sugar levels, exacerbating irritability, anxiety, and fatigue. Incorporating complex carbohydrates, such as whole grains and legumes, can help stabilize glucose release, preventing hypoglycemia and providing sustained energy. Low-GI diets are particularly beneficial, as they regulate insulin release and reduce rapid sugar crashes, thereby improving mood and alleviating cravings, as noted by Ludwig (2002). Addressing these dietary needs is crucial for managing nutritional imbalances and supporting recovery in substance-dependent individuals (Lennerz and Lennerz 2018).

21.5.3 Protein and Neurotransmitter Production

Proteins are very important for the production of neurotransmitters such as DA, serotonin, and GABA (Gasmi et al. 2022), which are often depleted in individuals recovering from substance use disorders. Dopaminergic pathways are disrupted in addiction, and protein-derived amino acids like tyrosine and tryptophan help restore neurotransmitter balance; therefore, foods rich in proteins such as lean meats, eggs, dairy products, tofu, beans, and nuts are essential to manage withdrawal symptoms (Gasmi et al. 2022).

21.5.4 Healthy Fats: Supporting Brain Health

The brain exhibits noticeable neuroinflammation and oxidative stress during withdrawal (Berríos-Cárcamo et al. 2020). Omega-3 fatty acids, particularly eicosapentaenoic acid (EPA) and docosahexaenoic acid (DHA), are important nutrients that offer a variety of neurobiological functions, through anti-inflammation, antioxidation, neuroplasticity, and modulation of neurotransmitters (Lee et al. 2020). Due to its regulatory effects on immunomodulation, anti-inflammation, signal transduction, neurotransmission, and neuroprotection, a meta-analysis conducted by Su et al. (2015) demonstrated that omega-3 supplements had preventive advantages on mood and anxiety disorders.

21.5.5 Fruits and Vegetables Rich in Antioxidant

One of the primary pathophysiological mechanisms that SUDs cause in several brain regions is oxidative stress, which is caused by the overproduction of reactive oxygen species (ROS) (Lirio et al. 2023). Fruits and vegetables rich in antioxidants like vitamins A, C, and E help in neutralizing ROS and protecting cells from damage. Antioxidants, including vitamins A, C, and E, found in fruits and vegetables, such as blueberries, spinach, kale, and carrots, aid in neutralizing ROS and protect cells from damage (Grabska-Kobyłecka et al. 2023). Research has demonstrated the beneficial effects of plant polyphenols on cognitive function and oxidative damage reduction, particularly during withdrawal recovery (Grabska-Kobyłecka et al. 2023).

21.5.6 Magnesium-Rich Foods: Easing Anxiety and Muscle Cramps

Magnesium plays a pivotal role in various physiological aspects of the nervous system, including nerve transmission and neuromuscular conduction (Kirkland et al. 2018). Individuals with substance use disorders are particularly vulnerable to magnesium deficiency due to malnutrition, which can worsen withdrawal symptoms such as anxiety, insomnia, and muscle cramps. It was found that magnesium supplementation improved mild depression, anxiety symptoms, and mood (Tarleton et al. 2017). Dietary sources include green leafy vegetables, such as spinach, legumes, nuts, seeds, and whole grains.

21.6 Conclusion

The neurobiological mechanisms of withdrawal are complex, involving neurotransmitter dysregulation, alterations in neural circuitry, hormonal responses, and inflammatory processes. Diet can significantly influence these mechanisms, providing a potential avenue for mitigating these withdrawal symptoms. Understanding this interplay offers valuable insights for developing dietary strategies to support individuals undergoing withdrawal, ultimately aiding in recovery and improving quality of life. By replenishing depleted nutrients, restoring neurotransmitter balance, and repairing gastrointestinal health, targeted dietary strategies provide a foundation for resilience. Thus, given the evidence reviewed herein, future research should focus on long-term dietary interventions and their impact on relapse prevention.

References

Ali MA, Hafez HA, Kamel MA, Ghamry HI, Shukry M, Farag MA (2022) Dietary vitamin B complex: orchestration in human nutrition throughout life with sex differences. Nutrients 14(19):3940. https://doi.org/10.3390/nu14193940

Altufailia MF, Al-Hakeima HK, Twaij BAA-R (2023) Review on the oxidative stress in methamphetamine addicts. J Popul Ther Clin Pharmacol 30(7). https://doi.org/10.47750/jptcp.2023.30.07.046

American Psychiatric Association (2013) Diagnostic and statistical manual of mental disorders. American Psychiatric Association. https://doi.org/10.1176/appi.books.9780890425596

Amiry GY, Haidary M, Azhdari-Zarmehri H, Beheshti F, Ahmadi-Soleimani SM (2023) Omega-3 fatty acids prevent nicotine withdrawal-induced exacerbation of anxiety and depression by affecting oxidative stress balance, inflammatory response, BDNF and serotonin metabolism in rats. Eur J Pharmacol 947:175634. https://doi.org/10.1016/j.ejphar.2023.175634

Astori E, Garavaglia ML, Colombo G, Landoni L, Portinaro NM, Milzani A, Dalle-Donne I (2022) Antioxidants in smokers. Nutr Res Rev 35(1):70–97. https://doi.org/10.1017/S0954422421000093

Baj J, Flieger W, Teresiński G, Buszewicz G, Sitarz R, Forma A, Karakuła K, Maciejewski R (2020) Magnesium, calcium, potassium, sodium, phosphorus, selenium, zinc, and chromium levels in alcohol use disorder: a review. J Clin Med 9(6):1901. https://doi.org/10.3390/jcm9061901

Barkus A, Baltrūnienė V, Baušienė J, Baltrūnas T, Barkienė L, Kazlauskaitė P, Baušys A (2024) The gut-brain axis in opioid use disorder: exploring the bidirectional influence of opioids and the gut microbiome—a comprehensive review. Life (Basel) 14(10):1227. https://doi.org/10.3390/life14101227

Berríos-Cárcamo P, Quezada M, Quintanilla ME, Morales P, Ezquer M, Herrera-Marschitz M, Israel Y, Ezquer F (2020) Oxidative stress and neuroinflammation as a pivot in drug abuse. A focus on the therapeutic potential of antioxidant and anti-inflammatory agents and biomolecules. Antioxidants (Basel) 9(9):830. https://doi.org/10.3390/antiox9090830

Butts M, Sundaram VL, Murughiyan U, Borthakur A, Singh S (2023) The influence of alcohol consumption on intestinal nutrient absorption: a comprehensive review. Nutrients 15(7):1571. https://doi.org/10.3390/nu15071571

Cheeks SN, Buzzi B, Valdez A, Mogul AS, Damaj MI, Fowler CD (2024) Cannabidiol as a potential cessation therapeutic: effects on intravenous nicotine self-administration and withdrawal symptoms in mice. Neuropharmacology 246:109833. https://doi.org/10.1016/j.neuropharm.2023.109833

Connor JP, Stjepanović D, Budney AJ, Le Foll B, Hall WD (2022) Clinical management of cannabis withdrawal. Addiction 117(7):2075–2095. https://doi.org/10.1111/add.15743

Dahchour A, Ward RJ (2024) Changes in serotonin neurotransmission as assayed by microdialysis after acute, intermittent or chronic ethanol administration and withdrawal. Fundam Clin Pharmacol 38(1):42–59. https://doi.org/10.1111/fcp.12949

Dresp-Langley B, Hutt A (2022) Digital addiction and sleep. Int J Environ Res Public Health 19(11):6910. https://doi.org/10.3390/ijerph19116910

Edinoff AN, Nix CA, Hollier J, Sagrera CE, Delacroix BM, Abubakar T, Cornett EM, Kaye AM, Kaye AD (2021) Benzodiazepines: uses, dangers, and clinical considerations. Neurol Int 13(4):594–607. https://doi.org/10.3390/neurolint13040059

Ersche KD, Stochl J, Woodward JM, Fletcher PC (2013) The skinny on cocaine: insights into eating behavior and body weight in cocaine-dependent men. Appetite 71:75–80. https://doi.org/10.1016/j.appet.2013.07.011

Ganio MS, Armstrong LE, Casa DJ, McDermott BP, Lee EC, Yamamoto LM, Marzano S, Lopez RM, Jimenez L, Le Bellego L, Chevillotte E, Lieberman HR (2011) Mild dehydration impairs cognitive performance and mood of men. Br J Nutr 106(10):1535–1543. https://doi.org/10.1017/S0007114511002005

Gasmi A, Nasreen A, Menzel A, Gasmi Benahmed A, Pivina L, Noor S, Peana M, Chirumbolo S, Bjørklund G (2022) Neurotransmitters regulation and food intake: the role of dietary sources in neurotransmission. Molecules 28(1):210. https://doi.org/10.3390/molecules28010210

Geneva: World Health Organization. (2009). Clinical guidelines for withdrawal management and treatment of drug dependence in closed settings

Grabska-Kobyłecka I, Szpakowski P, Król A, Książek-Winiarek D, Kobyłecki A, Głąbiński A, Nowak D (2023) Polyphenols and their impact on the prevention of neurodegenerative diseases and development. Nutrients 15(15):3454. https://doi.org/10.3390/nu15153454

He H, Zhou S, Peng C, Ran W, Tong S, Hong L, Cai F, Jin W, Jiang Y, Li M, Wang X, Luo M, Wang W, Zhao K (2022) Effects of resilience on impulsivity, cognition and depression during protracted withdrawal among Chinese male methamphetamine users. BMC Psychiatry 22(1):414. https://doi.org/10.1186/s12888-022-04041-8

Janowsky DS, Pucilowski O, Buyinza M (2003) Preference for higher sucrose concentrations in cocaine abusing-dependent patients. J Psychiatr Res 37(1):35–41. https://doi.org/10.1016/S0022-3956(02)00063-8

Kalman DS, Feldman S, Krieger DR, Bloomer RJ (2012) Comparison of coconut water and a carbohydrate-electrolyte sport drink on measures of hydration and physical performance in exercise-trained men. J Int Soc Sports Nutr 9(1). https://doi.org/10.1186/1550-2783-9-1

Kamiński P, Lorek M, Baszyński J, Tadrowski T, Gorzelańczyk EJ, Feit J, Tkaczenko H, Owoc J, Woźniak A, Kurhaluk N (2024) Role of antioxidants in the neurobiology of drug addiction: an update. Biomed Pharmacother 175:116604. https://doi.org/10.1016/j.biopha.2024.116604

Karaagac Y, Özçelik AÖ (2024) Nutritional status changes in the treatment of substance use disorder: relationship with substance craving. Nutr Res 123:120–129. https://doi.org/10.1016/j.nutres.2024.01.007

Kirkland AE, Sarlo GL, Holton KF (2018) The role of magnesium in neurological disorders. Nutrients 10(6):730. https://doi.org/10.3390/nu10060730

Koob GF, Volkow ND (2016) Neurobiology of addiction: a neurocircuitry analysis. Lancet Psychiatry 3(8):760–773. https://doi.org/10.1016/S2215-0366(16)00104-8

Lee J, Lee A, Kim J-H, Shin YM, Kim S-J, Cho WD, Lee SI (2020) Effect of omega-3 and Korean red ginseng on children with attention deficit hyperactivity disorder: an open-label pilot study. Clin Psychopharmacol Neurosci 18(1):75–80. https://doi.org/10.9758/cpn.2020.18.1.75

Lennerz B, Lennerz JK (2018) Food addiction, high-glycemic-index carbohydrates, and obesity. Clin Chem 64(1):64–71. https://doi.org/10.1373/clinchem.2017.273532

Lerner A, Klein M (2019) Dependence, withdrawal and rebound of CNS drugs: an update and regulatory considerations for new drugs development. Brain Commun 1(1). https://doi.org/10.1093/braincomms/fcz025

Lirio P, Santos M, Silva I, Rodrigues L (2023) Oxidative stress in substance use disorders: endogenous and exogenous mechanisms of repair. https://doi.org/10.5772/intechopen.108886

Ludwig DS (2002) The glycemic index. JAMA 287(18):2414. https://doi.org/10.1001/jama.287.18.2414

Mahboub N, Rizk R, Karavetian M, de Vries N (2021) Nutritional status and eating habits of people who use drugs and/or are undergoing treatment for recovery: a narrative review. Nutr Rev 79(6):627–635. https://doi.org/10.1093/nutrit/nuaa095

Melemis SM (2015) Relapse prevention and the five rules of recovery. Yale J Biol Med 88(3):325–332

Monroe SC, Radke AK (2023) Opioid withdrawal: role in addiction and neural mechanisms. Psychopharmacology 240(7):1417–1433. https://doi.org/10.1007/s00213-023-06370-2

Neale J, Nettleton S, Pickering L, Fischer J (2012) Eating patterns among heroin users: a qualitative study with implications for nutritional interventions. Addiction 107(3):635–641. https://doi.org/10.1111/j.1360-0443.2011.03660.x

Preedy VR, Watson RR (1996) Alcohol and the gastrointestinal tract. Alcohol Clin Exp Res 20(s8). https://doi.org/10.1111/j.1530-0277.1996.tb01744.x

Relojo-Howell D, Zehtabian K, Yrog-irog RC, Kazemizadeh A (2023) Exploring the therapeutic potential of amino acids in the treatment of psychoactive substance addiction: a comprehensive review. Univ Libr Med Health Sci 01(01):01–03. https://doi.org/10.70315/uloap.ulmhs.2023.0101001

Sachdeva A (2015) Alcohol withdrawal syndrome: benzodiazepines and beyond. J Clin Diagn Res. https://doi.org/10.7860/JCDR/2015/13407.6538

Santolaria-Fernández F (1995) Nutritional assessment of drug addicts. Drug Alcohol Depend 38(1):11–18. https://doi.org/10.1016/0376-8716(94)01088-3

Sasso JM, Ammar RM, Tenchov R, Lemmel S, Kelber O, Grieswelle M, Zhou QA (2023) Gut microbiome–brain alliance: a landscape view into mental and gastrointestinal health and disorders. ACS Chem Neurosci 14(10):1717–1763. https://doi.org/10.1021/acschemneuro.3c00127

Schroeder RD, Higgins GE (2017) You are what you eat: the impact of nutrition on alcohol and drug use. Subst Use Misuse 52(1):10–24. https://doi.org/10.1080/10826084.2016.1212603

Schwartz EKC, Wolkowicz NR, De Aquino JP, MacLean RR, Sofuoglu M (2022) Cocaine use disorder (CUD): current clinical perspectives. Subst Abus Rehabil 13:25–46. https://doi.org/10.2147/SAR.S337338

Sharma R, Tikka SK, Bhute AR, Dhamija P, Bastia BK (2023) Characterization of cannabis withdrawal symptoms and serum levels of neurotransmitters among cannabis-dependent smokers during sustained abstinence within a controlled residential environment. Eur Psychiatry 66(S1):S569–S569. https://doi.org/10.1192/j.eurpsy.2023.1195

Sinha R (2024) Stress and substance use disorders: risk, relapse, and treatment outcomes. J Clin Invest 134(16). https://doi.org/10.1172/JCI172883

Smith LC, Tieu L, Suhandynata RT, Boomhower B, Hoffman M, Sepulveda Y, Carrette LLG, Momper JD, Fitzgerald RL, Hanham K, Dowling J, Kallupi M, George O (2021) Cannabidiol reduces withdrawal symptoms in nicotine-dependent rats. Psychopharmacology 238(8):2201–2211. https://doi.org/10.1007/s00213-021-05845-4

Stachowicz K (2023) The role of polyunsaturated fatty acids in neuronal signaling in depression and cognitive processes. Arch Biochem Biophys 737:109555. https://doi.org/10.1016/j.abb.2023.109555

Su K-P, Matsuoka Y, Pae C-U (2015) Omega-3 polyunsaturated fatty acids in prevention of mood and anxiety disorders. Clin Psychopharmacol Neurosci 13(2):129–137. https://doi.org/10.9758/cpn.2015.13.2.129

Tarleton EK, Littenberg B, MacLean CD, Kennedy AG, Daley C (2017) Role of magnesium supplementation in the treatment of depression: a randomized clinical trial. PLoS One 12(6):e0180067. https://doi.org/10.1371/journal.pone.0180067

Vatsalya V, Verster JC, Sagaram M, Royer AJ, Hu H, Parthasarathy R, Schwandt ML, Kong M, Ramchandani VA, Feng W, Agrawal R, Zhang X, McClain CJ (2023) Novel paradigms for the gut–brain axis during alcohol withdrawal, withdrawal-associated depression, and craving in patients with alcohol use disorder. Front Psych 14. https://doi.org/10.3389/fpsyt.2023.1203362

Volkow ND, Blanco C (2023) Substance use disorders: a comprehensive update of classification, epidemiology, neurobiology, clinical aspects, treatment and prevention. World Psychiatry 22(2):203–229. https://doi.org/10.1002/wps.21073

Weber A, Trebach J, Brenner M, Thomas M, Bormann N (2024) Managing opioid withdrawal symptoms during the fentanyl crisis: a review. Subst Abus Rehabil 15:59–71. https://doi.org/10.2147/SAR.S433358

White B, Sirohi S (2024) A complex interplay between nutrition and alcohol use disorder: implications for breaking the vicious cycle. Curr Pharm Des 30(23):1822–1837. https://doi.org/10.2174/0113816128292367240510111746

Wiers CE, Vendruscolo LF, van der Veen J-W, Manza P, Shokri-Kojori E, Kroll DS, Feldman DE, McPherson KL, Biesecker CL, Zhang R, Herman K, Elvig SK, Vendruscolo JCM, Turner SA, Yang S, Schwandt M, Tomasi D, Cervenka MC, Fink-Jensen A, Benveniste H, Diazgranados N, Wang G-J, Koob GF, Volkow ND (2021) Ketogenic diet reduces alcohol withdrawal symptoms in humans and alcohol intake in rodents. Sci Adv 7(15). https://doi.org/10.1126/sciadv.abf6780

Wiss DA (2019) The role of nutrition in addiction recovery. In: The assessment and treatment of addiction. Elsevier, pp 21–42. https://doi.org/10.1016/B978-0-323-54856-4.00002-X

Wiss DA, Waterhous TS (2014) Nutrition therapy for eating disorders, substance use disorders, and addictions. In: Eating disorders, addictions and substance use disorders. Springer Berlin Heidelberg, Berlin, Heidelberg, pp 509–532. https://doi.org/10.1007/978-3-642-45378-6_23

Integrative Approaches to Treatment and Recovery of Substance Use Disorders (SUD)

22

Salah Abdelfatah Mostafa, Esraa Ayman Elemary, Rawan Ashraf Allam, and Nancy Ashraf Abd El-Fattah

Abstract

The complicated interaction among nutrition, medical treatment, and psychological health, especially in the case of SUDs, emphasizes the need for an integrated treatment plan for recovery. SUDs are commonly associated with a variety of mental health conditions, further compounding the challenges during treatment and recovery. This chapter discusses dietary contributions to better mental health by delineating relationships between nutrients, pharmacological treatments, and behavioral interventions in addressing mental health problems and SUDs. It underlines, especially, the neurobiological mechanisms through which some dietary components—in the shape of amino acids and growth regulators—meet brain functioning and mood modulation. This approach addresses an integrative viewpoint where pharmacological, nutritional, and behavioral interventions become essential for a comprehensive treatment of SUDs aimed at long-lasting promotion of recovery. Through the examination of the latest research and therapeutic practices, this work undertakes an attempt to emphasize the potential of diet and nutrition to act as adjuncts to traditional treatment modalities with the ultimate goal of improving patient outcomes in the management of SUDs, mental health disorders, and related comorbidities.

Keywords

Substance Use Disorders · Nutrition · Lifestyle · Cognitive Behavioral Therapy · Mental Health

---

S. A. Mostafa · E. A. Elemary (✉) · R. A. Allam · N. A. A. El-Fattah
Biotechnology Program, Faculty of Agriculture, Ain Shams University, Cairo, Egypt

W. Mohamed, A. Özge (eds.), *Feeding the Mind: The Connection Between Diet, Drugs, and Mental Health Volume 1*, Nutritional Neurosciences,
https://doi.org/10.1007/978-981-95-1414-4_22

22.1 Introduction

Substance use disorders (SUDs) are complex disorders that often co-occur with comorbid psychiatric disorders, commonly referred to as "dual" or "co-occurring" diagnoses. Epidemiological studies show a strong association of SUD with various mental health diagnoses, such as borderline and antisocial personality disorders, depression, anxiety, bipolar disorder, and psychotic disorders (Kessler et al. 2005). These dual diagnoses underline the complex interrelationship between the psychological, physiological, and behavioral dimensions in addiction that further complicates the pathways toward recovery. Physiologically, SUDs leave a deep imprint on the individual with chronic stress, having a neuroplastic impact. To be chronically exposed to adverse psychosocial stressors is found to alter brain structures dealing with decision-making, the regulation of emotions, and cognitive functions. Although these adaptations might have adaptive functions, they simultaneously increase vulnerability to poor mental health and addiction through epigenetic alterations (Chavez and Rigg 2020; Mörkl et al. 2020).

Nutrition is an often overlooked but critically important part of the recovery process. Substance misuse typically disrupts normal metabolic processes, leading to malnutrition and poor gut health. Opioids, for instance, can suppress appetite, whereas chronic alcohol use may damage the gastric mucosa, impairing nutrient absorption (Egerer et al. 2005; Neale et al. 2012). The relationship between poor nutrition and drug-seeking behavior becomes cyclical in nature, as malnourishment increases cravings while undermining recovery due to poor treatment outcomes (Jeynes and Gibson 2017). However, tailored nutrition has been found to effectively reduce symptoms of withdrawal, rebalance neurotransmitters, and support abstinence.

Despite its importance, nutritional therapy is still underutilized in most SUD treatment programs. Many treatment facilities do not have nutritionists or include dietary education as part of the recovery plans (Wiss et al. 2019). That presents an almost singular opportunity to integrate the holistic strategies—dietary interventions combined with medical, psychological, and behavioral therapies. Tapping into the embedded neuroprotective and anti-inflammatory faculties of a nutrient-dense diet, rehabilitation efforts may promote continued resilience against relapse while simultaneously improving general health (Du et al. 2016). In this chapter, integrative treatment approaches to SUD will be considered with a focus on transformative potential when merging medical treatments, psychological therapies, and nutritional interventions. To address addiction's biological, psychological, and social levels makes it possible to derive a comprehensive recovery framework.

22.2 Substance Craving

"Craving" is considered a crucial element in SUDs. Despite widespread agreement on its importance among researchers, clinicians, and patients, multiple definitions and theoretical models exist (Anton 1999; Potgieter et al. 1999; Rosenberg 2009;

Skinner and Aubin 2010; Tiffany et al. 2000). Skinner and Aubin (2010) reviewed 18 craving models developed across six decades, concluding that while each contributed to understanding and treating addictive behaviors, no single model fully explained craving. This chapter defines craving as an individual's experience of the urge or desire to use substances. It can take various forms, including intrusive thoughts (Kavanagh et al. 2006), impulsive urges or motivations (Cox and Klinger 2002), substance desires (Robinson and Berridge 1993), emotional states, physical sensations, and stress reactions (Sinha and Li 2007; Tiffany and Wray 2009; Paulus 2007). Craving is thought to be initiated by biological, affective, or cognitive factors. Biological theories of craving conceptualize addiction as a neurological disorder where drug wanting and use originate from neurobiological and physiological conditions (Robinson and Berridge 1993; Wise 1988).

The relationships between neurotransmitters such as dopamine, serotonin, and gamma-aminobutyric acid (GABA) and cravings and substance use have been extensively researched, emphasizing the neurobiological basis of addiction (Johnson et al. 1992; Wise 1988; Addolorato et al. 2005; Verheul et al. 1999).

22.3 Behavioral Therapies and Psychosocial Support

Behavioral therapies, which build on the central ideas of B.F. Skinner, more specifically, Behavior Analysis (BA) and Radical Behaviorism (RB), focus on understanding the individual as a unique entity with a distinct history and challenges. These therapies emphasize the modification of maladaptive behaviors through functional analysis, which identifies behavioral deficits and excesses contributing to psychological distress (Skinner 1953, 1945). The cornerstone of this approach lies in manipulating reinforcement contingencies to promote adaptive behaviors while reducing problematic ones (Ferster 1972). Behavioral interventions have shown considerable effectiveness in addressing SUDs, whether utilized as standalone treatments or in conjunction with pharmacological approaches.

Behavioral interventions have proven highly effective in addressing SUDs, whether used alone or alongside pharmacological treatments. The most recognized evidence-based approaches include motivational interviewing (MI), cognitive behavioral therapy (CBT), contingency management (CM), and twelve-step facilitation. CBT has been extensively validated for bulimia nervosa (BN) and binge eating disorder (BED), as outlined in DSM-5 (APA 2013). For BN, CBT yields remission rates of 30–50% for binge eating and purging behaviors, outperforming pharmacological interventions in achieving symptom abstinence (Wilson et al. 2007). Beyond reducing symptoms, CBT also enhances comorbid psychiatric conditions, notably depression, while improving self-esteem and social functioning (Wilson et al. 2010). Enhanced cognitive behavioral therapy (CBT-E) addresses the shortcomings of earlier CBT models and shows improved outcomes in complex cases (Fairburn et al. 2009).

22.3.1 Motivational Interviewing (MI) and Cognitive Behavioral Therapy (CBT)

MI is a goal-oriented, client-centered counseling style designed to elicit resolve to change. MI was primarily used in the environment of primary care clinics or emergency rooms with the objective of promoting engagement in treatment and enhancing treatment adherence by fostering self-efficacy and autonomy among the clients. Most research indicates that combining MI with CBT is exceptionally effective in its practice, which together forms a new therapeutic tool called Motivational Enhancement Therapy (MET). While CBT concentrates on recognizing and altering unhelpful thought patterns, it complements MI by addressing current issues and goal-oriented tactics. Studies indicate that combining MI and CBT enhances treatment outcomes in addiction recovery (Welsh and Hadland 2019).

22.3.2 Cognitive Behavioral Therapy (CBT) in Substance Use Treatment

Studies have shown that CBT is highly effective in treating BN and BED, outperforming pharmacological interventions in symptom remission (Wilson et al. 2007). In the realm of SUDs, CBT-inspired approaches with strong empirical support include cue exposure, CM, and cognitive coping skills training. Cue exposure involves the methodical presentation of substance-related stimuli to reduce conditioned cravings and enhance coping abilities (Monti et al. 2002; Rohsenow et al. 2001). However, this method may be hindered by limited generalization to real-world situations and potential relapse risks.

Aaron Beck's cognitive model enhances our understanding of SUDs by elucidating core beliefs, perceptions of helplessness, and low frustration tolerance that predispose individuals to substance misuse. These cognitive vulnerabilities often exist before SUDs develop and contribute to the high prevalence of psychiatric comorbidity in affected populations (Wright et al. 1993). Additionally, Beck's model incorporates addiction-related beliefs, comprising anticipatory, relief-oriented, and permissive cognitive patterns that trigger substance use (Knapp et al. 2001).

22.3.3 Family Therapy in Substance Use Treatment

In treating SUDs, family therapy is crucial due to addiction's significant effect on family relationships. This therapeutic approach considers the family as an interconnected unit, where changes in one member can lead to healthier interactions throughout. Family therapy sessions strive to enhance communication, develop coping mechanisms, and increase awareness of addiction's consequences. Studies have shown that family therapy models, especially when combined with CBT, are effective in decreasing substance use and enhancing overall family functioning (Tanner-Smith et al. 2013; Waldron and Turner 2008).

CBT incorporates principles of classical and operant conditioning to modify substance use behaviors in the treatment of alcohol use disorder (AUD) and other SUDs. Three key CBT-based techniques have been empirically validated: cue exposure, CM, and cognitive behavioral coping skills training (McHugh et al. 2010). While cue exposure has been shown to reduce cravings (Loeber et al. 2006), it has limitations, including the potential for poor generalization across different settings and the risk of relapse outside the therapeutic environment.

22.3.4 Psychoeducation and the Therapeutic Alliance

Psychoeducation is a fundamental component of CBT for SUD, enabling patients to identify and challenge maladaptive thoughts, beliefs, and behaviors. By recognizing the interdependent relationship among situations, thoughts, and emotions, individuals learn to develop healthier coping strategies (Araujo et al. 2013). Cognitive restructuring techniques aim to replace dysfunctional cognitions with healthier alternatives, fostering adaptive behaviors and improving psychological functioning (Gehlhaar and McMullin 2005). Furthermore, the therapeutic relationship, built through empathy and MI, is vital in establishing a strong therapeutic alliance that enhances treatment engagement and success, particularly among individuals with SUDs who may exhibit ambivalence toward treatment (Araujo et al. 2013; Lopes et al. 2021).

Behavioral and cognitive behavioral therapies offer robust evidence-based frameworks for addressing the complexities of SUD. By integrating functional behavioral analysis, cognitive restructuring, cue exposure, and coping skills training within a supportive therapeutic alliance, these interventions facilitate meaningful behavioral changes and improve psychological functioning in individuals with SUDs. Continued research into personalized treatment approaches, combined with continued evaluations of efficacy and generalizability, remains essential for advancing the field of SUD therapy.

22.4 Mind-Body Practices and Alternative Therapies

Originating from Eastern practices and brought to mainstream attention by Jon Kabat-Zinn, mindfulness stresses the importance of being aware and accepting of one's thoughts, emotions, and bodily sensations. In contrast to conventional CBT, which targets the identification and modification of maladaptive thoughts, mindfulness encourages embracing internal experiences and promotes acceptance over resistance. This method has shown potential in treating both Eating Disorders (EDs) and SUDs through various therapeutic approaches, including Mindfulness-Based Relapse Prevention (MBRP), Acceptance and Commitment Therapy (ACT), and Dialectical Behavior Therapy (DBT) (Keng et al. 2011).

22.4.1 Mindfulness-Based Interventions

Mindfulness-Based Stress Reduction (MBSR), created to boost awareness and manage emotional responses through meditation, forms the basis for many mindfulness-centered interventions (Grossman et al. 2004). While MBSR's effect on eating habits remains uncertain, research suggests it may help decrease binge eating and anxiety (Smith et al. 2006). For SUDs, variations like Mindfulness-Based Therapeutic Community (MBTC) have successfully reduced stress and emotional irritability (Marcus et al. 2009).

Initially developed for persistent depression, mindfulness-based cognitive therapy (MBCT) has expanded its application to EDs. By integrating mindfulness techniques with cognitive therapy concepts, MBCT has demonstrated effectiveness in reducing binge eating and enhancing body image perception (Baer et al. 2005; Alberts et al. 2012). These outcomes highlight its potential as a dual-purpose treatment for addressing disordered eating and the related emotional dysregulation.

22.4.2 Mindfulness-Based Relapse Prevention (MBRP)

Grounded in Buddhist practices, Marlatt (2002) hypothesized that mindfulness meditation may reduce cravings and addiction. This hypothesis spurred the development of MBRP—a structured, post-treatment program to reduce relapse risk following intensive substance abuse treatment. The MBRP curriculum includes eight 2-hour sessions that combine formal mindfulness exercises with cognitive behavioral skills. Sessions start by teaching clients to identify specific environmental triggers and related physical, emotional, and cognitive reactions. As clients progress, they learn to use informal mindfulness practices, such as urge surfing and mindful awareness of daily activities, to interrupt patterns of habitual responding.

Clients learn exercises developed specifically for coping with cravings in a manner that gradually exposes them to triggers while they practice non-reactive responses. Approaching cravings with curiosity and acceptance, clients develop alternative, skillful reactions that, over time, dampen the amplitude of the craving response. This increases their tolerance of discomfort and better prepares them to respond adaptively during high-risk relapses (Bowen et al. 2010; Witkiewitz et al. 2005).

22.4.3 Role of Mindfulness in Treating Substance Use Disorders

Mindfulness-based interventions, hence, have great potential to improve the treatment of EDs and SUD through MBSR, MBCT, and MBRP. Increased self-awareness, reinforcement of emotional control, and a decrease in the frequency of cravings by the resolution of core roots that trigger these problems are the emphases of these modalities. For instance, as it pertains to ED, mindfulness helps in working out strategies to handle reconnective feelings of one's body signals and decreases episodes of binge eating. Similarly, in SUD, mindfulness practices have been proven to reduce emotional reactivity and provide means for the management of urges, hence fostering recovery (Grossman et al. 2004; Bowen et al. 2010). These approaches develop self-awareness, strengthen emotional resilience, and reduce the intensity of triggers and cravings (Weber et al. 2017). More research is needed to establish the place of mindfulness in treatment modalities and, therefore, possibly expand the range of other mental and behavioral disorders.

22.5 Nutrition and Substance Use Disorders (SUDs)

Nutrition plays a crucial role in overall health, yet people who use drugs (PWUD) often lack adequate nutritional guidance, despite their typically compromised dietary habits that increase their risk of malnutrition (Jeynes and Gibson 2017). Multiple factors contribute to nutritional deficits in PWUD, including disorganized lifestyles affecting food choices and availability (Tang et al. 2011), impaired liver function limiting nutrient storage (Hossain et al. 2007), reduced appetite (Nazrul Islam et al. 2007), and metabolic changes (Tang et al. 2011). Consequently, research consistently indicates a trend of malnutrition among PWUD (Jeynes and Gibson 2017). For instance, individuals who use heroin and cocaine typically consume less energy and protein than non-users (Forrester et al. 2004; Saeland et al. 2009), and their intake of vitamins and minerals often falls short of recommended amounts (Saeland et al. 2009).

22.5.1 The Impact of SUD on Sleep and Physical Activity

Substance use disorders (SUDs) not only impact nutrition but also affect other health aspects, including sleep patterns and physical exercise. The use of drugs can interfere with the neural systems responsible for regulating sleep-wake cycles, leading to sleep problems and disrupted circadian rhythms (Conroy and Arnedt 2014). For instance, individuals who use cocaine often experience difficulties falling asleep and reduced overall sleep duration (Schierenbeck et al. 2008). However, the impact of SUDs on physical activity levels remains unclear based on current studies.

22.5.2 The Role of Nutrition in Mood Regulation and Recovery

Epidemiological research has consistently shown a high occurrence of mood disorders, such as depression and anxiety, among individuals who use drugs (PWUD) (Regier et al. 1990; Grant et al. 2004; Kessler et al. 1997). These mental health issues frequently hinder recovery efforts and often result in relapses (Jeynes and Gibson 2017; Tolliver and Anton 2015). Micronutrients play an essential role in the production and regulation of neurotransmitters like serotonin and dopamine, which are known to affect mood and mental well-being. Inadequate intake or deficiencies of nutrients such as copper, selenium, manganese, magnesium, folate, and B vitamins have been associated with depression and other mental health disorders that can impede the recovery process (Herbison et al. 2012; Młyniec et al. 2015; Pasco et al. 2012; Almeida et al. 2015).

22.5.3 Amino Acids and Mood Regulation

Amino acids are the precursors of major neurotransmitters, in particular, tryptophan, tyrosine, and phenylalanine as precursors for serotonin, dopamine, and norepinephrine, respectively. Serotonin, a neurotransmitter derived from tryptophan, is an important neurotransmitter that mediates mood, sleep, and appetite. Increasing intake of tryptophan has been associated with increasing the levels of serotonin and has been attempted to treat and/or improve mood and behavior. The efficacy of tryptophan supplementation in treating depression, however, remains inconsistent for the purpose of treating SUD (Capello and Markus 2014; Markus and De Raedt 2011; Ormstad et al. 2016). The amino acids tyrosine and phenylalanine are equally important because they form dopamine and catecholamines, respectively—neurotransmitters implicated in motivation, attention, and mood.

22.5.4 Micronutrient Deficiencies and Their Role in Mental Health

Research has also identified that deficiencies in micronutrient intake, such as zinc, magnesium, chromium, and B vitamins, may also contribute to the development of mental health disorders, including anxiety and depression (Młyniec et al. 2014; Muss et al. 2016). Low magnesium levels have been associated with increased anxiety (Eby and Eby 2006; Młyniec et al. 2014), while deficiencies in zinc and chromium have been associated with depression and mood disorders (Młyniec et al. 2014). Moreover, low levels of omega-3 fatty acids have also been associated with mood disorders and increased relapse rates among AUD and DUD patients (Barbadoro et al. 2013; Buydens-Branchey et al. 2009).

22.5.5 The Chaotic Lifestyle of PWUD and Its Impact on Recovery from SUD

The chaotic nature of the lifestyles led by PWUD is normally marked by lots of problems, including unstable housing, unemployment, economic troubles, and disturbed social networks (Smith and Flatley 2011; Green et al. 2007; Sutton et al. 2004; Corneil et al. 2006). Most of these challenges stick people to the substances and make it hard for them to recover from the use of the same substances. For example, there are those who try to escape from stressors associated with losing a job, unstable housing, and mental health issues using drugs (Degenhardt and Hall

2012; Lauder et al. 2009). The social environment is also another key determinant of the lifestyle of SUDs individuals. Where there is extensive drug use in an individual's social network, they are more likely to continue using substances (Becker 1967), hence further contributing to disorganization in lifestyle.

22.5.6 Nutritional Changes During Recovery

Although harm reduction is a highly effective strategy, studies show that patients receiving treatment for SUD usually experience an increase in body weight and body mass index (Montazerifar et al. 2012). Among individuals with higher levels of education and income, these same trends are more severe, which again suggests the positive influence of social environments on nutrition (Himmelgreen et al. 1998; Montazerifar et al. 2012; Kheradmand et al. 2010; Nazrul Islam et al. 2002). Detoxification itself as a process increases weight gain and higher food intake but varies across recovery stages. For instance, hyperphagia, or excess eating, has been highly reported during early recovery because people often substitute drugs for food at this stage of recovery. This may serve as a reaction to challenging the old pattern of nutrition disrupted by active drug consumption. At this stage of recovery, a tendency to consume high-calorie food with high sugar content can also be understood as compensation by the brain for missing dopamine production interrupted by abstinence from drugs (Orsini et al. 2014; Hodgkins et al. 2004; Warren et al. 2013).

Individual and environmental factors, which include reduced levels of physical activity, overeating induced by stress, and accessibility to affordable high-fat foods, further promote weight gain post-recovery. These can be addressed by adopting comprehensive strategies, which would include nutritional education, behavioral therapy, and an exercise regimen for better long-term outcomes (Emerson et al. 2009; Collingwood et al. 1991; Mahboub et al. 2021).

22.5.7 Social Networks and Recovery

The impact of social networks on substance use and recovery has been immense. Social epidemiology studies report that individuals who are part of social networks that include substance users are at a higher risk for SUD (Galea et al. 2004, 2009). The acceptance of substance use in their networks further perpetuates addiction (Fuchs et al. 2008; Steele et al. 2011; Measham and Shiner 2009). The concept of "recovery capital" is important to the understanding of lifestyle in recovery. It comprises not only internal but also external resources, for example, self-efficacy, motivational resources, social support, and living arrangement stability, that contribute to recovery (Cloud and Granfield 2008; Best and Laudet 2010).

22.5.8 Lifestyle Factors and Recovery-Oriented Interventions

Recovery-oriented interventions often aim to improve lifestyle factors in order to enhance overall well-being. Such interventions build on resilience and recovery capital by helping with things like housing, social support, and work (Laudet and White 2010; Groh et al. 2007; Tsai et al. 2010; Bush et al. 2009). One national U.S. study reported stepwise improvements in quality of life, with social support, housing, and job satisfaction increasing across successive stages of recovery. Stress is a major factor in reducing relapse, which declined with the recovery stage (Laudet et al. 2014; Laudet et al. 2004).

22.5.9 National and International Policies Supporting Recovery

A reflection of the importance of lifestyle factors in supporting recovery is hence to be found in national drug policies internationally. In the United States, for example, the national drug control strategy supports such initiatives as Recovery-Oriented Systems of Care, or ROSC, expanding recovery support through community-based programs (Davidson et al. 2007). In the UK, for example, Public Health England identifies three fundamental lifestyle factors in building recovery capital—jobs, homes, and friendship—that support people in recovery (Davies et al. 2015).

22.6 Co-occurring Substance Use Disorders (SUDs) and Mental Health Conditions (MHCs)

Co-occurring SUDs and mental health conditions (MHCs) pose significant challenges in primary care settings. Patients with SUDs often present with high clinical complexity due to the high prevalence of chronic comorbidities, such as diabetes, hypertension, and obesity. For instance, almost half of all visits to primary care physicians are related to chronic conditions like hypertension, for which about 33.5% of active patients are diagnosed with the disease (Ashman et al. 2017; Ornstein et al. 2013). Among patients with diabetes and hypertension, co-occurring SUDs—such as alcohol, opioid, cocaine, and cannabis use disorders—are reported in 1.1–8.8% of cases. However, these numbers are likely underestimates due to reliance on provider-documented diagnoses rather than comprehensive screenings (Winhusen et al. 2019).

Outcomes for patients with AUD and MHCs are generally poorer compared with those having only one of these disorders; this includes higher relapse rates, increased utilization of emergency services, and higher healthcare costs (Urbanoski et al. 2018; Suter et al. 2011). Despite the significant need for care, only 17% of people who need SUD treatment receive any form of treatment. Reasons identified include stigma, lack of finances, a shortage of providers, and high treatment costs (SAMHSA 2019). Among adults with serious mental illness, 49% report illicit drug use, while 16% of adults without mental illness do (SAMHSA 2019). Clearly, understanding the interrelatedness between the use of substances and mental health is crucial in providing good care for patients. Patients with SUD are more likely to attain sustained recovery if both conditions are treated in an integrated, multidisciplinary way. With the increasing integration of SUD care into the primary care setting, the priorities continue to be expanding access to comprehensive screening, increasing behavioral health supports, and developing a treatment plan tailored to individual needs.

22.6.1 Evolution of Treatment Approaches

The treatment landscape for individuals with co-occurring disorders has evolved over time. Traditionally, treatment plans did not focus on both disorders concurrently, and thus there was generally little improvement in either disorder (Rounsaville et al. 1987; Tómasson and Vaglum 1996). More contemporary approaches include sequential treatment of one disorder before the other, parallel treatment by different care teams, and integrated treatment where both disorders are treated at the same time by the same provider or treatment team. Integrated behavioral health in primary care holds some promise for improving outcomes for patients with co-occurring SUDs and chronic medical or MHCs, based on evidence-based screening and treatment strategies (Polydorou et al. 2008; LaBelle et al. 2016).

22.6.2 Anxiety Disorders and Substance Use

The relationship between anxiety disorders and substance use is complex and only partly understood. Findings about the association of anxiety disorders and substance use are mixed; for instance, some studies suggest that anxiety symptoms may also protect some adolescents from substance use, while in some, they seem to constitute a risk for future substance misuse (Hussong et al. 2017). There are strong associations, for example, between SAD and both AUDs and PTSD (Wolitzky-Taylor et al. 2012), and when identifying patients who sought substance-specific treatment from adolescence into young adulthood, any anxiety-related diagnosis, mainly social anxiety, has been associated significantly with an increased risk of opioid use (Welsh et al. 2017). These findings illustrate the complex relationships between anxiety and substance use, furthering the need to study the co-occurring conditions.

22.6.3 Bipolar Disorder and Co-occurring Substance Use

Studies have shown that as many as 48% of patients with bipolar disorders have a comorbid SUD (Stephens et al. 2014). Despite such a high prevalence rate of the comorbidity, very little empirical data exists concerning the underlying risk factors for substance misuse among such patients. For example, it remains unclear whether BD generally precedes substance use or whether substance use is a contributing factor to the onset of bipolar disorder. Adolescents with BD are particularly at risk; for example, studies have associated this condition with higher rates of suicide attempts, more frequent hospitalizations, and significant functional impairments (Lewinsohn et al. 1995). Such high rates of comorbidity between BD and SUD point to the necessity of incorporating multiple therapeutic strategies so that both conditions are treated simultaneously.

22.6.4 Integrated Treatment Strategies

Critical to improving recovery outcomes. Untreated mental health symptoms—whether they are at a level to meet full symptom criteria for a disorder—can negatively affect the course of SUD treatment. Traditionally, there has been a tendency to focus on achieving abstinence from substance use before attempting to address co-occurring MHCs. However, delaying treatment of mental health disorders until substance use is stabilized significantly disrupts effective recovery from SUDs. Current approaches to treatment stress the importance of addressing both disorders simultaneously because treating both concurrently is now widely regarded as the most effective strategy for long-term recovery. Treatment plans for patients with co-occurring AUDs and mental health disorders must be individualized. They usually include some combination of behavioral therapy, pharmacotherapy, and community-based recovery support. A key element of treatment planning is considering the severity of both the substance use and mental health disorders, along with patient preference and resources available for stabilization and recovery (Groshkova et al. 2013; Vilsaint et al. 2017).

22.6.4.1 Psychological Interventions

Approaching co-occurring SUDs and psychiatric conditions within a broader perspective will include both psychological and pharmacological interventions. Various psychotherapeutic approaches, such as CBT, DBT, MI, and family-based interventions, target different areas of addiction and mental health. Although CBT seeks to modify unhelpful ways of thinking, DBT places a lot of emphasis on dealing with emotions and distress. At the same time, MI increases the urge for change, while the family-based interventions target the familial risk factors, most especially among younger patients. Other techniques, such as CM and trauma-focused therapy, are promising in improving the outcome of treatment by encouraging positive behaviors and helping to process traumatic experiences. Results of research indicate that such psychosocial treatments confer modest to small benefits over the conditions of control, with therapies such as CBT, DBT, MI, and family-based treatments being particularly helpful in realizing a reduction in substance use and sustaining recovery (Krausz et al. 2015; Lee et al. 2015). This is because such approaches target not just the substance abuse but also the underlying psychological issues most often driving the addiction; hence, the more sustainable treatment outcomes.

22.6.4.2 Pharmacological Interventions

Pharmacological interventions are fundamental in the treatment of both SUD and other psychiatric disorders. Medication may be used to treat withdrawal, to reduce craving, or to stabilize mood or anxiety disorders. There are few studies addressing the use of medications with dually diagnosed adolescents, and treatment usually follows adult-based research findings. Care should be taken in the selection of medications, considering factors like potential diversion and abuse, the side effect profile, and the needs of the patient. Psychosocial intervention, together with medication, could be very effective. This has been documented by research when medications are incorporated into the treatment of opioid use disorder, in particular (Kampman and Jarvis 2015; Otasowie 2021). No single medication can resolve a complex co-occurring disorder, yet incorporating pharmacologic treatment with psychotherapeutic and psychosocial support may greatly improve treatment outcomes (Vitali et al. 2018; Danilewitz et al. 2021).

22.7 Medications for Substance Use and Relapse Prevention

Over the past 30 years, there has been an increasing recognition of medications' potential in treating SUDs. Recent developments include drugs for managing withdrawal symptoms, diminishing cravings, and preventing relapses, thus enabling longer periods of abstinence. Usually, these pharmacological interventions are combined with psychosocial treatments, and the role of each of them depends much on the type of addiction (Barber et al. 1999). The whole field of pharmacotherapy experienced considerable progress in the course of the last few decades, which expanded the options for healthcare providers and empowered them to offer more differentiated treatment solutions.

The major goals of SUD medications are to alleviate acute withdrawal symptoms, reduce cravings, and, most importantly, prevent relapse. These drugs help stabilize the patients during the critical early recovery stages by easing the physical and psychological effects of withdrawal. Some medications also target the neurobiological foundations of addiction, lessening the urge and compulsion to use substances. In long-term SUD management, pharmacological interventions are important for maintaining recovery by preventing a return to compulsive substance use (O'Brien 2005). The combination of pharmacological agents with psychosocial and behavioral therapies is taken into consideration as an approach to comprehensive treatment plans. Despite the significant advances made in the development of medication over the last 25 years, most health professionals are not aware of its effectiveness. Medications like acamprosate, which was approved by the Food and Drug Administration in 2004, and buprenorphine, introduced in 2002, have been proven effective in randomized controlled trials. But a number of misconceptions about these treatments persist, such as "you can't treat a drug problem with a drug," which impede more widespread acceptance and use.

22.7.1 Neurobiology of Addiction

Neurobiological mechanisms involved in behavior and decision-making make addiction a chronic relapsing disorder. Addiction affects the brain's reward system significantly. The understanding of neurobiological changes has important implications for the development of pharmacotherapy.

22.7.1.1 Alcohol Use Disorder (AUD)

AUD is clinically defined in DSM-5 as a pathological condition manifesting through 11 diagnostic criteria, ranging from impaired control over alcohol consumption to physical and psychological dependence (American Psychiatric Association 2013). Individuals often initiate alcohol use for its euphoric or recreational effects, which, over time, result in tolerance. Tolerance occurs due to desensitization of GABA receptors and/or activation of liver enzymes involved in alcohol metabolism. This desensitization leads to increased consumption to achieve the same effects, creating a cycle of addiction. Alcohol withdrawal is associated with a variety of symptoms, from mild manifestations like anxiety, shakiness, sweating, tachycardia, and fever to more severe symptoms like seizures, hallucinations, and delirium tremens (National Institute for Clinical Excellence [NICE] 2010). These withdrawal symptoms are mediated by neurochemical imbalances, particularly involving GABA and N-methyl-D-aspartic acid (NMDA) receptors. Severe alcohol dependence can lead to very severe withdrawal syndromes, including cardiovascular instability, anxiety, seizures, and, in extreme cases, death due to delirium tremens (Kosten and O'Conner 2003).

22.7.1.2 Nicotine Addiction

Tobacco use, particularly nicotine addiction, is highly comorbid with all SUDs and is among the leading causes of premature death and disability among patients recovering from other forms of addiction (Le Strat et al. 2010). Smoking cessation may often become an important goal within the process of recovery, because quitting smoking is likely to lead to a more sustained abstinence from alcohol and other substances. Indeed, the existing literature suggests that smoking cessation seems to improve the results obtained during recovery; accordingly, individuals in recovery with a tobacco use problem need to be informed of the potential benefits of smoking reduction or cessation (Bobo et al. 1998; Carmody et al. 2012; Holt et al. 2012). Pharmacotherapy has emerged as an essential factor in smoking cessation: use of combination therapy—that is, medication and behavioral counseling—was associated with approximately a 40% one-year abstinence rate (Batra 2011; George and O'Malley 2004).

22.7.2 Role of Neurotransmitters

Neurotransmitters play a crucial role in the development and maintenance of SUD. The primary neurotransmitters involved in SUD's impact on the brain include GABA, glutamate, serotonin, and dopamine.

22.7.2.1 GABA and Glutamate

Key Neurotransmitters in Brain Function: GABA and glutamate are crucial neurotransmitters with contrasting roles in regulating brain activity. GABA functions as the main inhibitory neurotransmitter, whereas glutamate is the primary excitatory neurotransmitter. The equilibrium between these two chemicals is vital for proper brain function, and alcohol intake significantly disturbs this balance, contributing to the onset of AUD. Alcohol consumption enhances GABA inhibition and reduces glutamate excitation, resulting in sedation, impaired motor coordination, and the relaxing effects typically associated with alcohol intoxication. This interaction is responsible for the immediate behavioral effects of alcohol, such as decreased anxiety and increased relaxation. However, long-term alcohol use leads to neuroadaptive changes, including increased glutamate transmission and decreased GABA activity. These adaptations are the brain's attempt to counteract alcohol's inhibitory effects, ultimately leading to tolerance and dependence (Grobin et al. 1998; Boyle et al. 1993). These neuroplastic alterations contribute to intensified alcohol cravings and more severe withdrawal symptoms, including anxiety, tremors, and seizures, when alcohol use is discontinued.

22.7.2.2 Serotonin

Serotonin or 5-HT is a neurotransmitter mainly produced in the raphe nuclei and is projected throughout the brain and has been implicated in the modulation of alcohol consumption and other alcohol addiction-related behaviors. Serotonergic signaling is involved in the control of mood, which, in turn, affects alcohol consumption and addiction (Sari et al. 2011).

Various studies reveal a relationship between changes in serotonin transmission and increased alcohol consumption with the development of AUD. More precisely, the lower the serotonergic activity, the higher the possibility of consumption and vulnerability to addiction will be (Goodman 2008; Holmes et al. 2013; Worst and Vrana 2005). Moreover, individuals with especially low levels of serotonin are those who also have impulsive and aggressive behaviors; these are very common among alcohol addicts (Mosner et al. 1997; Johnson 2000). Based on this, it would be that serotonergic dysregulation has effects on alcohol craving/consumption but also on mood disorders and behavioral problems common in AUD.

22.7.2.3 Dopamine

Dopamine acts in the brain's reward system as a neurotransmitter to reinforce pleasurable sensations that result from alcohol consumption. The main dopamine-producing neurons are found in the ventral tegmental area—VTA—of the mesencephalon. These neurons extend axons to several regions, including the nucleus accumbens, the prefrontal cortex, and the amygdala, all of which are regions in the brain that are implicated in reward and reinforcement. Alcohol consumption activates the dopamine A10 neurons in the VTA, and this activation results in the release of dopamine in the terminal regions, such as the NAc and PFC. This dopaminergic activity mediates rewarding aspects of alcohol and, therefore, reinforces the behavior and encourages repeated intake of alcohol (McBride et al. 1991; Tupala and Tiihonen 2004; Nestler et al. 1993). The release of dopamine is rewarding—it gives a sense of pleasure, therefore motivating further alcohol intake. Because of methodological problems, it is hard to determine dopamine directly in alcohol-intoxicated humans, but evidence indicates that endogenous opioids facilitate alcohol-induced dopamine release. These opioids activate the dopamine system and augment the rewarding aspects of alcohol (Gianoulakis et al. 1996; O'Brien 2005).

22.7.3 Medications for Withdrawal Syndromes, Recovery, and Relapse Prevention Medications

In the treatment of severe substance dependence, medications are important in the management of withdrawal symptoms. These medications decrease discomfort during withdrawal and thereby strengthen patients in the initial stages of treatment following cessation of substance use. This contributes to maintaining abstinence and increases the possibility of patients remaining in treatment, while it decreases the possibility of relapse to substance use, which often occurs in response to the distress produced by withdrawal. By alleviating these symptoms, pharmacological treatments prolong the recovery process and reduce relapse rates. Pharmacological treatments for SUDs primarily act on the receptors in the brain for neurotransmitters or neuromodulators that have been affected by the addiction. These pharmacological agents can be classified into three major classes: full agonists, partial agonists, and antagonists. Full agonists act directly on the receptor sites in the brain, thereby mimicking the effects of the substance being abused. These chemicals are sometimes used as substitutes for addictive drugs, helping to stabilize neurochemical activity for a long time while reducing cravings and thus supporting recovery. Partial agonists also interact with the same receptors but with lower strength. These drugs are usually used in the detox phase to help alleviate withdrawal symptoms and cravings. Antagonists, such as naltrexone, bind to the receptors without causing activation, thus blocking the actions caused by the agonist drugs. Antagonists help to curb cravings and decrease the potential for relapse by blocking the response of the abused substance on the receptor. A specific case in this pharmacological class is naltrexone, which was actually developed for the treatment of addiction to opioids. Antagonists are usually prescribed after the initial recovery phase to help maintain sobriety and prevent relapse (McCaul and Petry 2003; Douaihy et al. 2013).

22.7.3.1 Buprenorphine

Buprenorphine is a high-affinity partial agonist of the μ-opioid receptor. Unique pharmacological properties of the drug include a slow dissociation from the receptor, which confers longer acting clinical effects and effectiveness with less frequent dosing (Amass et al. 2000, 2001; Petry et al. 2001). For example, administration of buprenorphine to a person who has used heroin recently displaces heroin from the receptor site, inducing symptoms of withdrawal. Today, buprenorphine is typically administered in a preparation with naloxone in a 4:1 ratio to discourage the diversion of buprenorphine for injecting. Naloxone is an opioid antagonist that reverses opioid-like actions and induces withdrawal if administered by injection. However, poor gastrointestinal absorption following sublingual administration greatly limits the withdrawal-precipitating effects of naloxone (Welsh and Hadland 2019). The partial agonism of buprenorphine carries several therapeutic advantages, including lower overdose risk due to decreased respiratory depression. Compared to full opioid agonists, buprenorphine produces less subjective euphoria, has a lower potential to cause physical dependence, and reduces the intensity of the withdrawal syndrome (Walsh et al. 1994, 1995). Since the Drug Abuse Treatment Act of 2000, buprenorphine/naloxone (Subutex) was granted approval from the FDA in October 2002 for the treatment of opioid dependence and may be prescribed in outpatient office-based settings.

22.7.3.2 Nicotine Replacement Therapy (NRT)

Nicotine replacement therapies are medications that provide nicotine from sources other than tobacco to relieve withdrawal symptoms and enhance quitting rates. NRT is considered because few people can stop using nicotine abruptly and remain abstinent. As an FDA-approved agonist-based harm reduction medication, NRT is approved for adults, ages 18 and over, for the indication of quitting smoking. These medications can be prescribed for persons under age 18 but cannot be sold over-the-counter to minors. NRT is available as transdermal nicotine patches, nicotine gum, nicotine lozenges, nicotine vapor inhalers, and nicotine nasal sprays. Quit rates have been shown to be increased when behavioral therapies are combined with NRT. These are generally conducted in conjunction with setting a quit date and can include counseling as well as recommendations for quit support strategies (Lancaster et al. 2000; Nides 2008). Most people start using NRT on their quit date, and quitting on the quit date is a strong predictor of long-term quitting success (George and O'Malley 2004; Wu et al. 2006). While NRT is usually used as a short-term medication, longer use has been found to be beneficial for those with more difficult nicotine addictions.

22.7.3.3 Disulfiram (Antabuse)

For over five decades, disulfiram has been utilized as the first medication approved by the FDA for treating alcohol dependence. Its mechanism of action involves blocking aldehyde dehydrogenase, an enzyme that converts acetaldehyde to acetate during the metabolism of alcohol. This blockage causes acetaldehyde to build up, triggering the disulfiram-ethanol reaction (DER) when alcohol is ingested. The DER produces uncomfortable and potentially dangerous symptoms, including perspiration, nausea, emesis, flushing of the face, increased heart rate, rapid breathing, dyspnea, and low blood pressure. In severe cases, the reaction can lead to irregular heartbeats, heart attacks, convulsions, and even death. This adverse response acts as a negative reinforcement, aiming to eliminate addictive behavior through behavioral counterconditioning. Daily intake of disulfiram conditions individuals to avoid alcohol consumption due to the anticipated negative consequences. Disulfiram treatment is to help achieve initial sobriety, thereby enabling participation in psychosocial interventions (Banys 1988; Fuller and Gordis 2004).

22.7.3.4 Naltrexone

Initially developed for the treatment of opioid addiction, naltrexone has demonstrated efficacy as a pharmacological intervention for reducing alcohol cravings. Animal studies elucidated that alcohol stimulates the endogenous opioid system, leading to its application in alcohol treatment (Altshuler et al. 1980). The efficacy of naltrexone was substantiated through a series of open trials in the 1980s and a subsequent randomized, controlled study at the Philadelphia Veterans Administration Medical Center (Volpicelli et al. 1992; O'Malley et al. 1992). These investigations demonstrated that naltrexone decreased alcohol cravings and consumption, resulting in FDA approval for treating alcohol dependence. Naltrexone functions by inhibiting opioid receptors that regulate dopamine release in the brain's reward system, particularly in the nucleus accumbens. While alcohol consumption typically elevates extracellular dopamine levels in this area, naltrexone administration prevents this dopamine increase (Gonzales and Weiss 1998). Additionally, mice lacking opioid receptors due to mu receptor knockout do not self-administer alcohol, further supporting the opioid system's involvement in alcohol reward (Roberts et al. 2000).

22.7.3.5 Acamprosate

Acamprosate is an NMDA receptor antagonist; it has been approved by the FDA for the treatment of AUDs for individuals aged 18 and above (Kranzler and Van Kirk 2001). It is considered to restore the balance between excitatory and inhibitory neurotransmission by affecting both GABA and glutamatergic systems. It is believed that this diminishes "protracted" withdrawal symptoms and craving (Witkiewitz et al. 2012). Overall, research findings indicate that acamprosate improves alcohol-related outcomes, increases abstinence rates, and reduces relapse and heavy drinking. In the context of AUD, acamprosate exhibits comparable efficacy with oral naltrexone (Kranzler and Van Kirk 2001).

22.8 Conclusion

The connection between diet, drugs, and mental health is an exciting frontier in the treatment of SUD and MHCs. More and more, nutrition plays a modulatory role in neurobiological pathways, enhancing neurotransmitter function and promoting recovery from SUDs. The integration of dietetic management with conventional pharmacological and psychosocial treatment holds tremendous promise to enhance recovery rates. Growth regulators, hormones, and amino acids can boost the body's natural defenses, raising resistance against mental illness and addiction alike. Tailored nutritional programs, on top of everything else, were designed and implemented to take care of specific needs in a comorbidity's treatment. As research continues to progress, it is envisaged that the amalgamation of dietary interventions with pharmacological treatments and behavioral therapies will become a cornerstone of treatment plans, offering a holistic approach that incorporates both the psychological and physiological aspects of addiction and mental health. More research is needed to improve treatment methods and increase long-term outcomes for those suffering from these disorders by investigating the complex interplay between diet, medication, and mental health.

References

Addolorato G, Leggio L, Abenavoli L, Gasbarrini G, Alcoholism Treatment Study Group (2005) Neurobiochemical and clinical aspects of craving in alcohol addiction: a review. Addict Behav 30(6):1209–1224. https://doi.org/10.1016/j.addbeh.2004.12.011

Alberts HJ, Thewissen R, Raes L (2012) Dealing with problematic eating behaviour. The effects of a mindfulness-based intervention on eating behaviour, food cravings, dichotomous thinking and body image concern. Appetite 58(3):847–851. https://doi.org/10.1016/j.appet.2012.01.009

Almeida OP, Ford AH, Flicker L (2015) Systematic review and meta-analysis of randomized placebo-controlled trials of folate and vitamin B12 for depression. Int Psychogeriatr 27(5):727–737. https://doi.org/10.1017/S1041610215000046

Altshuler HL, Phillips PE, Feinhandler DA (1980) Alteration of ethanol self-administration by naltrexone. Life Sci 26(9):679–688. https://doi.org/10.1016/0024-3205(80)90257-x

Amass L, Kamien JB, Mikulich SK (2000) Efficacy of daily and alternate-day dosing regimens with the combination buprenorphine-naloxone tablet. Drug Alcohol Depend 58(1–2, 152):143. https://doi.org/10.1016/s0376-8716(99)00074-5

Amass L, Kamien JB, Mikulich SK (2001) Thrice-weekly supervised dosing with the combination buprenorphine-naloxone tablet is preferred to daily supervised dosing by opioid-dependent humans. Drug Alcohol Depend 61(2):173–181. https://doi.org/10.1016/s0376-8716(00)00141-1

American Psychiatric Association (2013) Diagnostic and statistical manual of mental disorders (DSM-5®). American Psychiatric Pub, Washington, DC. https://psycnet.apa.org/record/2013-14907-000

Anton RF (1999) What is craving? Models and implications for treatment. Alcohol Res Health 23(3):165–173

Araujo RB, Lucena-Santos P, Castro MG (2013) Terapia cognitivo-comportamental de Beck. In: Araujo RB (ed) Guia de terapias cognitivo-comportamentais para os transtornos do exagero: tratando pacientes da vida real. Sinopsys, Novo Hamburgo, pp 87–108

Ashman JJ, Rui P, Schappert SM, Strashny A (2017) Characteristics of visits to primary care physicians by adults diagnosed with hypertension. National Health Statistics Reports, 106. Retrieved from https://www.cdc.gov/nchs/data/nhsr/nhsr106.pdf

Baer RA, Fischer S, Huss DB (2005) Mindfulness and acceptance in the treatment of disordered eating. J Ration-Emot Cogn-Behav Ther 23:281–300. https://doi.org/10.1007/s10942-005-0015-9

Banys P (1988) The clinical use of disulfiram (Antabuse): a review. J Psychoactive Drugs 20(3):243–261. https://doi.org/10.1080/02791072.1988.10472495

Barbadoro P, Annino I, Ponzio E, Romanelli RM, D'Errico MM, Prospero E, Minelli A (2013) Fish oil supplementation reduces cortisol basal levels and perceived stress: a randomized, placebo-controlled trial in abstinent alcoholics. Mol Nutr Food Res 57(6):1110–1114. https://doi.org/10.1002/mnfr.201200676

Barber WS, O'Brien CP, McCrady BS, Epstein EE (1999) Addictions: a comprehensive guidebook. Oxford University Press. 1

Batra A (2011) Treatment of tobacco dependence. Dtsch Arztebl Int 108(33):555–564. https://doi.org/10.3238/arztebl.2011.0555

Becker HS (1967) History, culture, and subjective experience: an exploration of the social bases of drug-induced experiences. J Health Soc Behav 8(3):163–176

Best D, Laudet A (2010) The potential of recovery capital. RSA, London

Bobo JK, McIlvain HE, Lando HA, Walker RD, Leed-Kelly A (1998) Effect of smoking cessation counseling on recovery from alcoholism: findings from a randomized community intervention trial. Addiction 93(6):877–887. https://doi.org/10.1046/j.1360-0443.1998.9368779.x

Bowen S, Chawla N, Marlatt GA (2010) Mindfulness-based relapse prevention for the treatment of substance use disorders: a clinician's guide. Guilford Press

Boyle AE, Segal R, Smith BR, Amit Z (1993) Bidirectional effects of GABAergic agonists and antagonists on maintenance of voluntary ethanol intake in rats. Pharmacol Biochem Behav 46(1):179–182. https://doi.org/10.1016/0091-3057(93)90338-t

Bush PW, Drake RE, Xie H, McHugo GJ, Haslett WR (2009) The long-term impact of employment on mental health service use and costs for persons with severe mental illness. Psychiatr Serv 60(8):1024–1031. https://doi.org/10.1176/ps.2009.60.8.1024

Buydens-Branchey L, Branchey M, Hibbeln JR (2009) Low plasma levels of docosahexaenoic acid are associated with an increased relapse vulnerability in substance abusers. Am J Addict 18(1):73–80. https://doi.org/10.1080/10550490802544003

Capello AE, Markus CR (2014) Effect of sub chronic tryptophan supplementation on stress-induced cortisol and appetite in subjects differing in 5-HTTLPR genotype and trait neuroticism. Psychoneuroendocrinology 45:96–107. https://doi.org/10.1016/j.psyneuen.2014.03.005

Carmody TP, Delucchi K, Duncan CL, Banys P, Simon JA, Solkowitz SN, Huggins J, Lee SK, Hall SM (2012) Intensive intervention for alcohol-dependent smokers in early recovery: a randomized trial. Drug Alcohol Depend 122(3):186–194. https://doi.org/10.1016/j.drugalcdep.2011.09.026

Chavez MN, Rigg KK (2020) Nutritional implications of opioid use disorder: a guide for drug treatment providers. Psychol Addict Behav 34(6):699–707. https://doi.org/10.1037/adb0000575

Cloud W, Granfield R (2008) Conceptualizing recovery capital: expansion of a theoretical construct. Subst Use Misuse 43(12–13):1971–1986. https://doi.org/10.1080/10826080802289762

Collingwood TR, Reynolds R, Kohl HW, Smith W, Sloan S (1991) Physical fitness effects on substance abuse risk factors and use patterns. J Drug Educ 21(1):73–84. https://doi.org/10.2190/HV5J-4EYN-GPP7-Y3QG

Conroy DA, Arnedt JT (2014) Sleep and substance use disorders: an update. Curr Psychiatry Rep 16(10):487. https://doi.org/10.1007/s11920-014-0487-3

Corneil TA, Kuyper LM, Shoveller J, Hogg RS, Li K, Spittal PM, Schechter MT, Wood E (2006) Unstable housing, associated risk behaviour, and increased risk for HIV infection among injection drug users. Health Place 12(1):79–85. https://doi.org/10.1016/j.healthplace.2004.10.004

Cox WM, Klinger E (2002) Motivational structure. Relationships with substance use and processes of change. Addict Behav 27(6):925–940. https://doi.org/10.1016/s0306-4603(02)00290-3

Danilewitz M, Bahji A, Lamba W, Chopra N, George T (2021) Concurrent disorders management in psychiatric care: opportunities and challenges. Can J Addict 12(1):7–9. https://doi.org/10.1097/CXA.0000000000000122

Davidson L, Tondora J, O'Connell MJ, Kirk T Jr, Rockholz P, Evans AC (2007) Creating a recovery-oriented system of behavioral health care: moving from concept to reality. Psychiatr Rehabil J 31(1):23–31. https://doi.org/10.2975/31.1.2007.23.31

Davies G, Elison S, Ward J, Laudet A (2015) The role of lifestyle in perpetuating substance use disorder: the lifestyle balance model. Subst Abuse Treat Prev Policy 10:2. https://doi.org/10.1186/1747-597X-10-2

Degenhardt L, Hall W (2012) Extent of illicit drug use and dependence, and their contribution to the global burden of disease. Lancet 379(9810):55–70. https://doi.org/10.1016/S0140-6736(11)61138-0

Douaihy AB, Kelly TM, Sullivan C (2013) Medications for substance use disorders. Soc Work Public Health 28(3–4):264–278. https://doi.org/10.1080/19371918.2013.759031

Du J, Zhu M, Bao H, Li B, Dong Y, Xiao C, Zhang GY, Henter I, Rudorfer M, Vitiello B (2016) The role of nutrients in protecting mitochondrial function and neurotransmitter signaling: implications for the treatment of depression, PTSD, and suicidal behaviors. Crit Rev Food Sci Nutr 56(15):2560–2578. https://doi.org/10.1080/10408398.2013.876960

Eby GA, Eby KL (2006) Rapid recovery from major depression using magnesium treatment. Med Hypotheses 67(2):362–370. https://doi.org/10.1016/j.mehy.2006.01.047

Egerer E, Stikel F, Seitz HK (2005) Alcohol and the gastrointestinal tract. In: Preedy VR, Watson RR (eds) Comprehensive handbook of alcohol related pathology. Elsevier Ltd, Oxford, pp 557–570

Emerson MH, Glovsky E, Amaro H, Nieves R (2009) Unhealthy weight gain during treatment for alcohol and drug use in four residential programs for Latina and African American women. Subst Use Misuse 44(11):1553–1565. https://doi.org/10.1080/10826080802494750

Fairburn CG, Cooper Z, Doll HA, O'Connor ME, Bohn K, Hawker DM, Wales JA, Palmer RL (2009) Transdiagnostic cognitive-behavioral therapy for patients with eating disorders: a two-site trial with 60-week follow-up. Am J Psychiatry 166(3):311–319. https://doi.org/10.1176/appi.ajp.2008.08040608

Ferster CB (1972) An experimental analysis of clinical phenomena. Psychol Rec 22:1–16. https://doi.org/10.1007/BF03394059

Forrester JE, Tucker KL, Gorbach SL (2004) Dietary intake and body mass index in HIV-positive and HIV-negative drug abusers of Hispanic ethnicity. Public Health Nutr 7(7):863–870. https://doi.org/10.1079/phn2004617

Fuchs RA, Lasseter HC, Ramirez DR, Xie X (2008) Relapse to drug seeking following prolonged abstinence: the role of environmental stimuli. Drug Discov Today Dis Models 5(4):251–258. https://doi.org/10.1016/j.ddmod.2009.03.001

Fuller RK, Gordis E (2004) Does disulfiram have a role in alcoholism treatment today? Addiction 99(1):21–24. https://doi.org/10.1111/j.1360-0443.2004.00597.x

Galea S, Nandi A, Vlahov D (2004) The social epidemiology of substance use. Epidemiol Rev 26:36–52. https://doi.org/10.1093/epirev/mxh007

Galea S, Hall C, Kaplan GA (2009) Social epidemiology and complex system dynamic modelling as applied to health behaviour and drug use research. Int J Drug Policy 20(3):209–216. https://doi.org/10.1016/j.drugpo.2008.08.005

Gehlhaar P, McMullin RE (2005) Terapia de reestruturação cognitiva com pacientes adictos. In: McMullin RE (ed) Manual de técnicas em terapia cognitiva. Artmed, Porto Alegre, pp 243–260

George TP, O'Malley SS (2004) Current pharmacological treatments for nicotine dependence. Trends Pharmacol Sci 25(1):42–48. https://doi.org/10.1016/j.tips.2003.11.003

Gianoulakis C, Krishnan B, Thavundayil J (1996) Enhanced sensitivity of pituitary beta-endorphin to ethanol in subjects at high risk of alcoholism. Arch Gen Psychiatry 53(3):250–257. https://doi.org/10.1001/archpsyc.1996.01830030072011

Gonzales RA, Weiss F (1998) Suppression of ethanol-reinforced behavior by naltrexone is associated with attenuation of the ethanol-induced increase in dialysate dopamine levels in the nucleus accumbens. J Neurosci 18(24):10663–10671. https://doi.org/10.1523/JNEUROSCI.18-24-10663.1998

Goodman A (2008) Neurobiology of addiction. An integrative review. Biochem Pharmacol 75(1):266–322. https://doi.org/10.1016/j.bcp.2007.07.030

Grant BF, Stinson FS, Dawson DA, Chou SP, Dufour MC, Compton W, Pickering RP, Kaplan K (2004) Prevalence and co-occurrence of substance use disorders and independent mood and anxiety disorders: results from the National Epidemiologic Survey on Alcohol and Related Conditions. Arch Gen Psychiatry 61(8):807–816. https://doi.org/10.1001/archpsyc.61.8.807

Green BL, Rockhill A, Furrer C (2007) Does substance abuse treatment make a difference for child welfare case outcomes? A statewide longitudinal analysis. Child Youth Serv Rev 29(4):460–473. https://doi.org/10.1016/j.childyouth.2006.08.006

Grobin AC, Matthews DB, Devaud LL, Morrow AL (1998) The role of GABA(A) receptors in the acute and chronic effects of ethanol. Psychopharmacology 139(1-2):2–19. https://doi.org/10.1007/s002130050685

Groh DR, Jason LA, Davis MI, Olson BD, Ferrari JR (2007) Friends, family, and alcohol abuse: an examination of general and alcohol-specific social support. Am J Addict 16(1):49–55. https://doi.org/10.1080/10550490601080084

Groshkova T, Best D, White W (2013) The assessment of recovery capital: properties and psychometrics of a measure of addiction recovery strengths. Drug Alcohol Rev 32(2):187–194. https://doi.org/10.1111/j.1465-3362.2012.00489.x

Grossman P, Niemann L, Schmidt S, Walach H (2004) Mindfulness-based stress reduction and health benefits. A meta-analysis. J Psychosom Res 57(1):35–43. https://doi.org/10.1016/S0022-3999(03)00573-7

Herbison CE, Hickling S, Allen KL, O'Sullivan TA, Robinson M, Bremner AP, Huang RC, Beilin LJ, Mori TA, Oddy WH (2012) Low intake of B-vitamins is associated with poor adolescent mental health and behaviour. Prev Med 55(6):634–638. https://doi.org/10.1016/j.ypmed.2012.09.014

Himmelgreen DA, Pérez-Escamilla R, Segura-Millán S, Romero-Daza N, Tanasescu M, Singer M (1998) A comparison of the nutritional status and food security of drug-using and non-drug-using Hispanic women in Hartford, Connecticut. Am J Phys Anthropol 107(3):351–361. https://doi.org/10.1002/(SICI)1096-8644(199811)107:3<351::AID-AJPA10>3.0.CO;2-7

Hodgkins CC, Cahill KS, Seraphine AE, Frost-Pineda K, Gold MS (2004) Adolescent drug addiction treatment and weight gain. J Addict Dis 23(3):55–65. https://doi.org/10.1300/J069v23n03_05

Holmes A, Spanagel R, Krystal JH (2013) Glutamatergic targets for new alcohol medications. Psychopharmacology 229(3):539–554. https://doi.org/10.1007/s00213-013-3226-2

Holt LJ, Litt MD, Cooney NL (2012) Prospective analysis of early lapse to drinking and smoking among individuals in concurrent alcohol and tobacco treatment. Psychol Addict Behav 26(3):561–572. https://doi.org/10.1037/a0026039

Hossain KJ, Kamal MM, Ahsan M, Islam SKN (2007) Serum antioxidant micromineral (Cu, Zn, Fe) status of drug dependent subjects: influence of illicit drugs and lifestyle. Subst Abuse Treat Prev Policy 2(1):12. https://doi.org/10.1186/1747-597X-2-12

Hussong AM, Ennett ST, Cox MJ, Haroon M (2017) A systematic review of the unique prospective association of negative affect symptoms and adolescent substance use controlling for externalizing symptoms. Psychol Addict Behav 31(2):137–147. https://doi.org/10.1037/adb0000247

Jeynes KD, Gibson EL (2017) The importance of nutrition in aiding recovery from substance use disorders: a review. Drug Alcohol Depend 179:229–239. https://doi.org/10.1016/j.drugalcdep.2017.07.006

Johnson BA (2000) Serotonergic agents and alcoholism treatment: rebirth of the subtype concept–an hypothesis. Alcohol Clin Exp Res 24(10):1597–1601

Johnson SW, Seutin V, North RA (1992) Burst firing in dopamine neurons induced by N-methyl-D-aspartate: role of electrogenic sodium pump. Science (New York, NY) 258(5082):665–667. https://doi.org/10.1126/science.1329209

Kampman K, Jarvis M (2015) American Society of Addiction Medicine (ASAM) National Practice Guideline for the use of medications in the treatment of addiction involving opioid use. J Addict Med 9(5):358–367. https://doi.org/10.1097/ADM.0000000000000166

Kavanagh DJ, Sitharthan G, Young RM, Sitharthan T, Saunders JB, Shockley N, Giannopoulos V (2006) Addition of cue exposure to cognitive-behaviour therapy for alcohol misuse: a randomized trial with dysphoric drinkers. Addiction 101(8):1106–1116. https://doi.org/10.1111/j.1360-0443.2006.01488.x

Keng SL, Smoski MJ, Robins CJ (2011) Effects of mindfulness on psychological health: a review of empirical studies. Clin Psychol Rev 31(6):1041–1056. https://doi.org/10.1016/j.cpr.2011.04.006

Kessler RC, Crum RM, Warner LA, Nelson CB, Schulenberg J, Anthony JC (1997) Lifetime co-occurrence of DSM-III-R alcohol abuse and dependence with other psychiatric disorders in the

National Comorbidity Survey. Arch Gen Psychiatry 54(4):313–321. https://doi.org/10.1001/archpsyc.1997.01830160031005
Kessler RC, Chiu WT, Demler O, Merikangas KR, Walters EE (2005) Prevalence, severity, and comorbidity of 12-month DSM-IV disorders in the National Comorbidity Survey Replication. Arch Gen Psychiatry 62(6):617–627. https://doi.org/10.1001/archpsyc.62.6.617
Kheradmand A, Banazadeh N, Abedi H (2010) Physical effects of methadone maintenance treatment from the standpoint of clients. Addict Health 2(3-4):66–73
Knapp P, Luz E Jr, Baldisserotto GV (2001) Terapia cognitiva no tratamento da dependência química. In: Rangé B (ed) Psicoterapias cognitivo-comportamentais: um diálogo com a psiquiatria. Artmed, Porto Alegre, pp 332–350
Kosten TR, O'Connor PG (2003) Management of drug and alcohol withdrawal. N Engl J Med 348(18):1786–1795. https://doi.org/10.1056/NEJMra020617
Kranzler HR, Van Kirk J (2001) Efficacy of naltrexone and acamprosate for alcoholism treatment: a meta-analysis. Alcohol Clin Exp Res 25(9):1335–1341
Krausz M, Vogel M, Ploesser M (2015) Care for patients with addiction and concurrent disorders in Europe, The United States of America, and Canada: similarities and differences. In: Dom G, Moggi F (eds) Co-occurring addictive and psychiatric disorders. Springer, Berlin, Heidelberg. https://doi.org/10.1007/978-3-642-45375-5_4
LaBelle CT, Han SC, Bergeron A, Samet JH (2016) Office-based opioid treatment with buprenorphine (obot-b): statewide implementation of the Massachusetts collaborative care model in community health centers. J Subst Abus Treat 60:6–13. https://doi.org/10.1016/j.jsat.2015.06.010
Lancaster T, Stead L, Silagy C, Sowden A (2000) Effectiveness of interventions to help people stop smoking: findings from the Cochrane Library. BMJ 321(7257):355–358. https://doi.org/10.1136/bmj.321.7257.355
Lauder W, Roxburgh M, Harris J, Law J (2009) Developing self-neglect theory: analysis of related and atypical cases of people identified as self-neglecting. J Psychiatr Ment Health Nurs 16(5):447–454. https://doi.org/10.1111/j.1365-2850.2009.01397.x
Laudet AB, White W (2010) What are your priorities right now? Identifying service needs across recovery stages to inform service development. J Subst Abus Treat 38(1):51–59. https://doi.org/10.1016/j.jsat.2009.06.003
Laudet AB, Magura S, Vogel HS, Knight EL (2004) Perceived reasons for substance misuse among persons with a psychiatric disorder. Am J Orthopsychiatry 74(3):365–375. https://doi.org/10.1037/0002-9432.74.3.365
Laudet A, Harris K, Kimball T, Winters KC, Moberg DP (2014) Collegiate recovery communities programs: what do we know and what do we need to know? J Soc Work Pract Addict 14(1):84–100. https://doi.org/10.1080/1533256X.2014.872015
Le Strat Y, Ramoz N, Gorwood P (2010) In alcohol-dependent drinkers, what does the presence of nicotine dependence tell us about psychiatric and addictive disorders comorbidity? Alcohol Alcohol 45(2):167–172. https://doi.org/10.1093/alcalc/agp094
Lee NK, Cameron J, Jenner L (2015) A systematic review of interventions for co-occurring substance use and borderline personality disorders. Drug Alcohol Rev 34(6):663–672. https://doi.org/10.1111/dar.12267
Lewinsohn PM, Klein DN, Seeley JR (1995) Bipolar disorders in a community sample of older adolescents: prevalence, phenomenology, comorbidity, and course. J Am Acad Child Adolesc Psychiatry 34(4):454–463
Loeber S, Croissant B, Heinz A, Mann K, Flor H (2006) Cue exposure in the treatment of alcohol dependence: effects on drinking outcome, craving and self-efficacy. Br J Clin Psychol 45:515–529. https://doi.org/10.1348/014466505X82586
Lopes FM, da Luz WL, Bizarro L (2021) Cognitive-behavioral therapy in the treatment of substance use disorders. In: Andrade ALM, De Micheli D, Silva EA, Lopes FM, Pinheiro BO, Reichert RA (eds) Psychology of substance abuse. Springer, Cham. https://doi.org/10.1007/978-3-030-62106-3_10

Mahboub N, Rizk R, Karavetian M, de Vries N (2021) Nutritional status and eating habits of people who use drugs and/or are undergoing treatment for recovery: a narrative review. Nutr Rev 79(6):627–635. https://doi.org/10.1093/nutrit/nuaa095

Marcus MT, Schmitz J, Moeller G, Liehr P, Cron SG, Swank P et al (2009) Mindfulness-based stress reduction in therapeutic community treatment: a stage 1 trial. Am J Drug Alcohol Abuse 35(2):103–108. https://doi.org/10.1080/00952990902823079

Markus CR, De Raedt R (2011) Differential effects of 5-HTTLPR genotypes on inhibition of negative emotional information following acute stress exposure and tryptophan challenge. Neuropsychopharmacology 36(4):819–826. https://doi.org/10.1038/npp.2010.221

Marlatt GA (2002) Buddhist philosophy and the treatment of addictive behavior. Cogn Behav Pract 9(1):44–50. https://doi.org/10.1016/S1077-7229(02)80039-6

McBride WJ, Murphy JM, Gatto GJ, Levy AD, Lumeng L, Li TK (1991) Serotonin and dopamine systems regulating alcohol intake. Alcohol Alcohol Suppl 1:411–416

McCaul ME, Petry NM (2003) The role of psychosocial treatments in pharmacotherapy for alcoholism. Am J Addict 12(s1):s41–s52. https://doi.org/10.1111/j.1521-0391.2003.tb00495.x

McHugh RK, Hearon BA, Otto MW (2010) Cognitive behavioral therapy for substance use disorders. Psychiatr Clin North Am 33(3):511–525. https://doi.org/10.1016/j.psc.2010.04.012

Measham F, Shiner M (2009) The legacy of 'normalisation': the role of classical and contemporary criminological theory in understanding young people's drug use. Int J Drug Policy 20(6):502–508. https://doi.org/10.1016/j.drugpo.2009.02.001

Młyniec K, Davies CL, de Agüero Sánchez IG, Pytka K, Budziszewska B, Nowak G (2014) Essential elements in depression and anxiety. Part I. Pharmacol Rep 66(4):534–544. https://doi.org/10.1016/j.pharep.2014.03.001

Młyniec K, Gaweł M, Doboszewska U, Starowicz G, Pytka K, Davies CL, Budziszewska B (2015) Essential elements in depression and anxiety. Part II. Pharmacol Rep 67(2):187–194. https://doi.org/10.1016/j.pharep.2014.09.009

Montazerifar F, Karajibani M, Lashkaripour K (2012) Effect of methadone maintenance therapy on anthropometric indices in opioid dependent patients. Int J High Risk Behav Addict 1(3):100–103. https://doi.org/10.5812/ijhrba.4968

Monti PM, Kadden RM, Rohsenow DJ, Cooney NL, Abrams DB (2002) Treating alcohol dependence: a coping skills training guide, 2nd edn. Guilford Press

Mörkl S, Wagner-Skacel J, Lahousen T, Lackner S, Holasek SJ, Bengesser SA, Painold A, Holl AK, Reininghaus E (2020) The role of nutrition and the gut-brain axis in psychiatry: a review of the literature. Neuropsychobiology 79(1):80–88. https://doi.org/10.1159/000492834

Mosner A, Kuhlman G, Roehm C, Vogel WH (1997) Serotonergic receptors modify the voluntary intake of alcohol and morphine but not of cocaine and nicotine by rats. Pharmacology 54(4):186–192. https://doi.org/10.1159/000139486

Muss C, Mosgoeller W, Endler T (2016) Mood improving potential of a vitamin trace element composition: a randomized, double-blind, placebo-controlled clinical study with healthy volunteers. Neuroendocrinol Lett 37(1):18–28

National Institute for Clinical Excellence (NICE) (2010) Alcohol use disorders: diagnosis and clinical management of alcohol-related physical complications. NICE, Nice, France. https://www.nice.org.uk/guidance/cg100

Nazrul Islam SK, Jahangir Hossain K, Ahmed A, Ahsan M (2002) Nutritional status of drug addicts undergoing detoxification: prevalence of malnutrition and influence of illicit drugs and lifestyle. Br J Nutr 88(5):507–513. https://doi.org/10.1079/BJN2002702

Nazrul Islam SK, Hossain KJ, Ahmed A, Ahsan M (2007) Nutritional status of drug addicts undergoing detoxification: prevalence of malnutrition and influence of illicit drugs and lifestyle. Br J Nutr 88(5):507–513

Neale J, Nettleton S, Pickering L, Fischer J (2012) Eating patterns among heroin users: a qualitative study with implications for nutritional interventions. Addiction 107(3):635–641. https://doi.org/10.1111/j.1360-0443.2011.03660.x

Nestler EJ, Hope BT, Widnell KL (1993) Drug addiction: a model for the molecular basis of neural plasticity. Neuron 11(6):995–1006. https://doi.org/10.1016/0896-6273(93)90213-b

Nides M (2008) Update on pharmacologic options for smoking cessation treatment. Am J Med 121(4 Suppl 1):S20–S31. https://doi.org/10.1016/j.amjmed.2008.01.016
O'Brien CP (2005) Anticraving medications for relapse prevention: a possible new class of psychoactive medications. Am J Psychiatry 162(8):1423–1431. https://doi.org/10.1176/appi.ajp.162.8.1423
O'Malley SS, Jaffe AJ, Chang G, Schottenfeld RS, Meyer RE, Rounsaville B (1992) Naltrexone and coping skills therapy for alcohol dependence. A controlled study. Arch Gen Psychiatry 49(11):881–887. https://doi.org/10.1001/archpsyc.1992.01820110045007
Ormstad H, Dahl J, Verkerk R, Andreassen OA, Maes M (2016) Increased plasma levels of competing amino acids, rather than lowered plasma tryptophan levels, are associated with a non-response to treatment in major depression. Eur Neuropsychopharmacol 26(8):1286–1296. https://doi.org/10.1016/j.euroneuro.2016.05.005
Ornstein SM, Nietert PJ, Jenkins RG, Litvin CB (2013) The prevalence of chronic diseases and multimorbidity in primary care practice: a PPRNet report. J Am Board Fam Med 26(5):518–524. https://doi.org/10.3122/jabfm.2013.05.130012
Orsini CA, Ginton G, Shimp KG, Avena NM, Gold MS, Setlow B (2014) Food consumption and weight gain after cessation of chronic amphetamine administration. Appetite 78:76–80. https://doi.org/10.1016/j.appet.2014.03.013
Otasowie J (2021) Co-occurring mental disorder and substance use disorder in young people: aetiology, assessment and treatment. BJPsych Adv 27(4):272–281. https://doi.org/10.1192/bja.2020.64
Pasco JA, Jacka FN, Williams LJ, Evans-Cleverdon M, Brennan SL, Kotowicz MA, Nicholson GC, Ball MJ, Berk M (2012) Dietary selenium and major depression: a nested case-control study. Complement Ther Med 20(3):119–123. https://doi.org/10.1016/j.ctim.2011.12.008
Paulus MP (2007) Neural basis of reward and craving--a homeostatic point of view. Dialogues Clin Neurosci 9(4):379–387. https://doi.org/10.31887/DCNS.2007.9.4/mpaulus
Petry NM, Bickel WK, Badger GJ (2001) Examining the limits of the buprenorphine interdosing interval: daily, every-third-day and every-fifth-day dosing regimens. Addiction 96(6):823–834. https://doi.org/10.1046/j.1360-0443.2001.9668234.x
Polydorou S, Gunderson EW, Levin FR (2008) Training physicians to treat substance use disorders. Curr Psychiatry Rep 10(5):399–404. https://doi.org/10.1007/s11920-008-0064-8
Potgieter AS, Deckers F, Geerlings P (1999) Craving and relapse measurement in alcoholism. Alcohol Alcohol 34(2):254–260. https://doi.org/10.1093/alcalc/34.2.254
Regier DA, Farmer ME, Rae DS, Locke BZ, Keith SJ, Judd LL, Goodwin FK (1990) Comorbidity of mental disorders with alcohol and other drug abuse. Results from the Epidemiologic Catchment Area (ECA) Study. JAMA 264(19):2511–2518
Roberts AJ, McDonald JS, Heyser CJ, Kieffer BL, Matthes HW, Koob GF, Gold LH (2000) mu-Opioid receptor knockout mice do not self-administer alcohol. J Pharmacol Exp Ther 293(3):1002–1008
Robinson TE, Berridge KC (1993) The neural basis of drug craving: an incentive-sensitization theory of addiction. Brain Res Brain Res Rev 18(3):247–291. https://doi.org/10.1016/0165-0173(93)90013-p
Rohsenow DJ, Monti PM, Rubonis AV, Gulliver SB, Colby SM, Binkoff JA, Abrams DB (2001) Cue exposure with coping skills training and communication skills training for alcohol dependence: 6- and 12-month outcomes. Addiction 96(8):1161–1174. https://doi.org/10.1046/j.1360-0443.2001.96811619.x
Rosenberg H (2009) Clinical and laboratory assessment of the subjective experience of drug craving. Clin Psychol Rev 29(6):519–534. https://doi.org/10.1016/j.cpr.2009.06.002
Rounsaville BJ, Dolinsky ZS, Babor TF, Meyer RE (1987) Psychopathology as a predictor of treatment outcome in alcoholics. Arch Gen Psychiatry 44(6):505–513. https://doi.org/10.1001/archpsyc.1987.01800180015002
Saeland M, Haugen M, Eriksen FL, Smehaugen A, Wandel M, Bohmer T, Oshaug A (2009) Living as a drug addict in Oslo, Norway–a study focusing on nutrition and health. Public Health Nutr 12(5):630–636. https://doi.org/10.1017/s1368980008002553

Sari Y, Johnson VR, Weedman JM (2011) Role of the serotonergic system in alcohol dependence: from animal models to clinics. Prog Mol Biol Transl Sci 98:401–443. https://doi.org/10.1016/B978-0-12-385506-0.00010-7

Schierenbeck T, Riemann D, Berger M, Hornyak M (2008) Effect of illicit recreational drugs upon sleep: cocaine, ecstasy and marijuana. Sleep Med Rev 12(5):381–389. https://doi.org/10.1016/j.smrv.2007.12.004

Sinha R, Li CS (2007) Imaging stress- and cue-induced drug and alcohol craving: association with relapse and clinical implications. Drug Alcohol Rev 26(1):25–31. https://doi.org/10.1080/09595230601036960. PMID: 17364833

Skinner BF (1945) The operational analysis of psychological terms. Psychol Rev 52(5):270–277. https://doi.org/10.1037/h0062535

Skinner BF (1953) Science and human behavior. McMillan, New York

Skinner MD, Aubin HJ (2010) Craving's place in addiction theory: contributions of the major models. Neurosci Biobehav Rev 34(4):606–623. https://doi.org/10.1016/j.neubiorev.2009.11.024

Smith K, Flatley J (2011) Drug misuse declared: findings from the 2010/11 British crime survey. England and Wales. Home Office, 27

Smith BW, Shelley BM, Leahigh L, Vanleit B (2006) A preliminary study of the effects of a modified mindfulness intervention on binge eating. Complement Health Pract Rev. https://doi.org/10.1177/1533210106297217

Steele JL, Peralta RL, Elman C (2011) The co-ingestion of nonmedical prescription drugs and alcohol: a partial test of social learning theory. J Drug Issues 41(4):561–585

Stephens JR, Heffner JL, Adler CM, Blom TJ, Anthenelli RM, Fleck DE, Welge JA, Strakowski SM, DelBello MP (2014) Risk and protective factors associated with substance use disorders in adolescents with first-episode mania. J Am Acad Child Adolesc Psychiatry 53(7):771–779. https://doi.org/10.1016/j.jaac.2014.04.018

Substance Abuse and Mental Health Services Administration (2019) Key substance use and mental health indicators in the United States: results from the 2018 National Survey on Drug Use and Health (HHS Publication No. PEP19-5068, NSDUH Series H-54). Center for Behavioral Health Statistics and Quality, Substance Abuse and Mental Health Services Administration, Rockville. Retrieved from https://www.samhsa.gov/data/.

Suter M, Strik W, Moggi F (2011) Depressive symptoms as a predictor of alcohol relapse after residential treatment programs for alcohol use disorder. J Subst Abus Treat 41(3):225–232. https://doi.org/10.1016/j.jsat.2011.03.005

Sutton L, Cebulla A, Heaver C, Smith N (2004) Drug and alcohol use as barriers to employment: a review of the literature (CRSP 499s). Loughborough University

Tang AM, Bhatnagar T, Ramachandran R, Dong K, Skinner S, Kumar MS, Wanke CA (2011) Malnutrition in a population of HIV-positive and HIV-negative drug users 25 living in Chennai, South India. Drug Alcohol Depend 118(1):73–77. https://doi.org/10.1016/j.drugalcdep.2011.02.020

Tanner-Smith EE, Wilson SJ, Lipsey MW (2013) The comparative effectiveness of outpatient treatment for adolescent substance abuse: a meta-analysis. J Subst Abus Treat 44(2):145–158. https://doi.org/10.1016/j.jsat.2012.05.006

Tiffany ST, Wray J (2009) The continuing conundrum of craving. Addiction 104(10):1618–1619. https://doi.org/10.1111/j.1360-0443.2009.02588.x

Tiffany ST, Carter BL, Singleton EG (2000) Challenges in the manipulation, assessment and interpretation of craving relevant variables. Addiction 95 Suppl 2:S177–S187. https://doi.org/10.1080/09652140050111753

Tolliver BK, Anton RF (2015) Assessment and treatment of mood disorders in the context of substance abuse. Dialogues Clin Neurosci 17(2):181–190. https://doi.org/10.31887/DCNS.2015.17.2/btolliver

Tómasson K, Vaglum P (1996) Psychopathology and alcohol consumption among treatment-seeking alcoholics: a prospective study. Addiction 91(7):1019–1030. https://doi.org/10.1046/j.1360-0443.1996.91710198.x

Tsai J, Bond GR, Salyers MP, Godfrey JL, Davis KE (2010) Housing preferences and choices among adults with mental illness and substance use disorders: a qualitative study. Community Ment Health J 46(4):381–388. https://doi.org/10.1007/s10597-009-9268-6

Tupala E, Tiihonen J (2004) Dopamine and alcoholism: neurobiological basis of ethanol abuse. Prog Neuro-Psychopharmacol Biol Psychiatry 28(8):1221–1247. https://doi.org/10.1016/j.pnpbp.2004.06.022

Urbanoski K, Cheng J, Rehm J, Kurdyak P (2018) Frequent use of emergency departments for mental and substance use disorders. Emerg Med J 35(4):220–225. https://doi.org/10.1136/emermed-2015-205554

Verheul R, van den Brink W, Geerlings P (1999) A three-pathway psychobiological model of craving for alcohol. Alcohol Alcohol 34(2):197–222. https://doi.org/10.1093/alcalc/34.2.197

Vilsaint CL, Kelly JF, Bergman BG, Groshkova T, Best D, White W (2017) Development and validation of a Brief Assessment of Recovery Capital (BARC-10) for alcohol and drug use disorder. Drug Alcohol Depend 177:71–76. https://doi.org/10.1016/j.drugalcdep.2017.03.022

Vitali M, Mistretta M, Alessandrini G, Coriale G, Romeo M, Attilia F, Rotondo C, Sorbo F, Pisciotta F, Attilia ML, Ceccanti M, Interdisciplinary Study Group CRARL - SITAC - SIPaD - SITD - SIPDip (2018) Pharmacological treatment for dual diagnosis: a literature update and a proposal of intervention. Pharmacological treatment for dual diagnosis: a literature update and a proposal of intervention. Riv Psichiatr 53(3):160–169. https://doi.org/10.1708/2925.29419

Volpicelli JR, Alterman AI, Hayashida M, O'Brien CP (1992) Naltrexone in the treatment of alcohol dependence. Arch Gen Psychiatry 49(11):876–880. https://doi.org/10.1001/archpsyc.1992.01820110040006

Waldron HB, Turner CW (2008) Evidence-based psychosocial treatments for adolescent substance abuse. J Clin Child Adolesc Psychol 37(1):238–261. https://doi.org/10.1080/15374410701820133

Walsh SL, Preston KL, Stitzer ML, Cone EJ, Bigelow GE (1994) Clinical pharmacology of buprenorphine: ceiling effects at high doses. Clin Pharmacol Ther 55(5):569–580. https://doi.org/10.1038/clpt.1994.71

Walsh SL, Preston KL, Bigelow GE, Stitzer ML (1995) Acute administration of buprenorphine in humans: partial agonist and blockade effects. J Pharmacol Exp Ther 274(1):361–372

Warren CS, Lindsay AR, White EK, Claudat K, Velasquez SC (2013) Weight-related concerns related to drug use for women in substance abuse treatment: prevalence and relationships with eating pathology. J Subst Abus Treat 44(5):494–501. https://doi.org/10.1016/j.jsat.2012.08.222

Weber B, Sala L, Gex-Fabry M, Docteur A, Gorwood P, Cordera P, Bondolfi G, Jermann F, Aubry JM, Mirabel-Sarron C (2017) Self-reported long-term benefits of mindfulness-based cognitive therapy in patients with bipolar disorder. J Altern Complement Med 23(7):534–540. https://doi.org/10.1089/acm.2016.0427

Welsh JW, Hadland SE (eds) (2019) Treating adolescent substance use: a clinician's guide. Springer, Cham. https://doi.org/10.1007/978-3-030-01893-1

Welsh JW, Knight JR, Hou SS, Malowney M, Schram P, Sherritt L, Boyd JW (2017) Association between substance use diagnoses and psychiatric disorders in an adolescent and young adult clinic-based population. J Adolesc Health 60(6):648–652. https://doi.org/10.1016/j.jadohealth.2016.12.018

Wilson GT, Grilo CM, Vitousek KM (2007) Psychological treatment of eating disorders. Am Psychol 62(3):199–216. https://doi.org/10.1037/0003-066X.62.3.199

Wilson GT, Wilfley DE, Agras WS, Bryson SW (2010) Psychological treatments of binge eating disorder. Arch Gen Psychiatry 67(1):94–101. https://doi.org/10.1001/archgenpsychiatry.2009.170

Winhusen T, Theobald J, Kaelber DC, Lewis D (2019) Medical complications associated with substance use disorders in patients with type 2 diabetes and hypertension: electronic health record findings. Addiction 114(8):1462–1470. https://doi.org/10.1111/add.14607

Wise RA (1988) The neurobiology of craving: implications for the understanding and treatment of addiction. J Abnorm Psychol 97(2):118–132. https://doi.org/10.1037/0021-843x.97.2.118

Wiss DA, Schellenberger M, Prelip ML (2019) Rapid assessment of nutrition services in Los Angeles substance use disorder treatment centers. J Community Health 44(1):88–94. https://doi.org/10.1007/s10900-018-0557-2

Witkiewitz K, Marlatt GA, Walker D (2005) Mindfulness-based relapse prevention for alcohol and substance use disorders. J Cogn Psychother 19(3)

Witkiewitz K, Saville K, Hamreus K (2012) Acamprosate for treatment of alcohol dependence: mechanisms, efficacy, and clinical utility. Ther Clin Risk Manag 8:45–53. https://doi.org/10.2147/TCRM.S23184

Wolitzky-Taylor K, Bobova L, Zinbarg RE, Mineka S, Craske MG (2012) Longitudinal investigation of the impact of anxiety and mood disorders in adolescence on subsequent substance use disorder onset and vice versa. Addict Behav 37(8):982–985. https://doi.org/10.1016/j.addbeh.2012.03.026

Worst TJ, Vrana KE (2005) Alcohol and gene expression in the central nervous system. Alcohol Alcohol 40(1):63–75. https://doi.org/10.1093/alcalc/agh119

Wright FD, Beck AT, Newman CF, Liese BS (1993) Cognitive therapy of substance abuse: theoretical rationale. NIDA Res Monogr 137:123–146

Wu P, Wilson K, Dimoulas P et al (2006) Effectiveness of smoking cessation therapies: a systematic review and meta-analysis. BMC Public Health 6:300. https://doi.org/10.1186/1471-2458-6-300

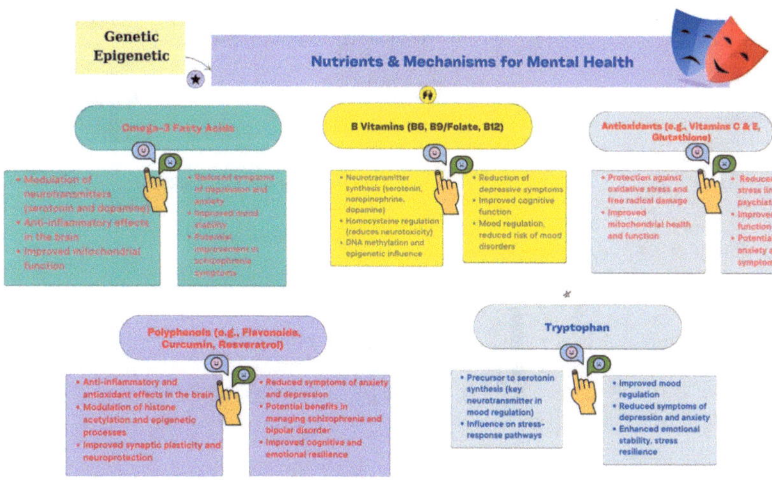

MIX
Papier aus verantwortungsvollen Quellen
Paper from responsible sources
FSC® C105338

If you have any concerns about our products,
you can contact us on
ProductSafety@springernature.com

In case Publisher is established outside the EU,
the EU authorized representative is:
Springer Nature Customer Service Center GmbH
Europaplatz 3, 69115 Heidelberg, Germany

Printed by Libri Plureos GmbH
in Hamburg, Germany